Sustainable Materials and Technology

Sustainable Materials and Technology (SMT) book series publishes research monographs (both edited and authored volumes) showcasing the latest developments in the field and comprehensively covering topics such as:

- Recycling of waste into useful material and their energy applications
- Catalytic action of Nano oxides for efficient carbon reforming process
- Sustainable technologies for plastic transformation
- Bifunctional nanoparticles for sustainable water splitting applications
- Sustainable dying and printing
- New materials from waste
- Sustainable Manure Management and Technology: Potentials, Uses and limitations
- Sustainable Mechanical Engineering Approach
- Sustainable biochemistry for the improvement of health
- Sustainable development of Mechanical recycling of automotive components
- Sustainable-waste recycling and conversion in useful materials for different applications
- Sustainable development of inexpensive Nano-photo catalysts
- Sustainable development of recycling of discarded lithium ion batteries
- Modern sustainable cement and concrete
- Sustainable adsorbent for hazardous removal
- Sustainable superior electromagnetic shielding materials
- Excellent sustainable nanostructured materials for energy storage device
- Sustainable development of heavy metal detoxification from water
- Carbon dioxide utilization for sustainable energy
- Sustainable development in green syntheses of materials
- Environment friendly and sustainable cloth for garments application
- Sustainable design and application of eco-materials
- Nanoparticles for sustainable environment applications
- Sustainable remediation of industrial contaminated water towards potential industrial applications
- Biomaterials for sustainable bioremediations

Malik Abdul Rub · Naved Azum · Anish Khan · Khalid A. Alzahrani
Editors

Industrial Applications of Individual and Mixed Surfactant Systems—Volume 2

Springer

Editors
Malik Abdul Rub
Center of Excellence for Advanced Materials Research
King Abdulaziz University
Jeddah, Saudi Arabia

Naved Azum
Center of Excellence for Advanced Materials Research
King Abdulaziz University
Jeddah, Saudi Arabia

Anish Khan
Center of Excellence for Advanced Materials Research
King Abdulaziz University
Jeddah, Saudi Arabia

Khalid A. Alzahrani
Department of Chemistry
King Abdulaziz University
Jeddah, Saudi Arabia

ISSN 2731-0426 ISSN 2731-0434 (electronic)
Sustainable Materials and Technology
ISBN 978-981-95-9194-7 ISBN 978-981-95-9195-4 (eBook)
https://doi.org/10.1007/978-981-95-9195-4

This Springer imprint is published by the registered company Springer Nature Singapore Pte Ltd.
The registered company address is: 152 Beach Road, #21-01/04 Gateway East, Singapore 189721, Singapore

Contents

Delivery of Drugs Using Micelles: A New Prototype in Therapeutic

Ikechukwu Ugbaga Nkole and Abhishek Srivastava

Abstract Drug deliveries have been revolutionized over time with the emergence of many carriers, such as liposomes (lipid bilayers), hydrogels (polymeric hydrophobic-active molecules), nanostructured molecules, biological polymers, and micelles. Each of the carriers has made an impact in the advancement of medicine and addresses the United Nations Sustainable Development Goal 3 (SDG-3), which focuses on human health and well-being. The challenges faced by these carriers, which include immune-system stimulation, material stability, carrier carrying capacity, potential toxicity, drug-carrier compatibility, biodistribution, and production shortfalls, have kept researchers on their toes for an alternate, friendly, and more efficient carrier. The setting of the micellar system in the front burner of the drug delivery scheme was after its first clinical trial (CT) in 2010 in Japan. Subsequently, many CT studies centring on the pharmacodynamics/kinetics of antitumour drug improvement with micelles have been underpinned. Paclitaxel-polyethylene glycol micelle encapsulation stands as a typical example. Others, such as Genexol-polymeric micelle, polyaspartate-polyethylene glycol, and polyglutamic acid-polyethylene glycol, have aimed to address the bioaccumulation and targeting of cell sites. Its polymeric form ushers in drug-loading capability, controllable drug release, and retention period in the field of modern medicine that stresses precision, less energy, and great impact. Many efforts have been made on the side of the researchers to harness, build, and reinforce this green and less toxic carrier as a new prototype of drug carrier by tuning the capabilities of both monomeric and polymeric micelles. The discussion around the micelle's structure, stability, cell interactions, delivery system, challenges, and prospective route is not left out.

I. U. Nkole (✉)
Basic Science Unit, Wigwe University, Isiokpo, Rivers State, Nigeria
e-mail: ikechukwu.ugbaga@wigweuniversity.edu.ng

A. Srivastava
Department of Chemistry, GLA University, Mathura, Uttar Pradesh, India

M. Abdul Rub et al. (eds.), *Industrial Applications of Individual and Mixed Surfactant Systems—Volume 2*, Sustainable Materials and Technology,
https://doi.org/10.1007/978-981-95-9195-4_1

Graphical Abstract

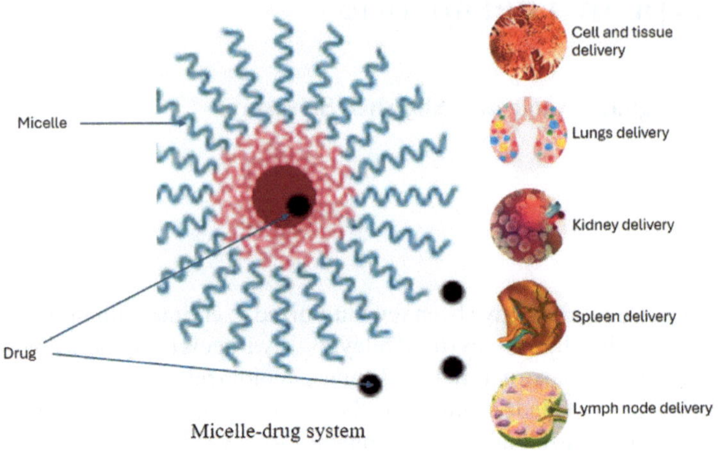

Keywords Micelles · Drug carriers · Therapeutics · Release mechanism · Pharmacokinetics

1 Introduction

Drugs exist in the form of chemical molecules and biological nutrients/dietaries that impact positively or negatively on a patient when administered. The administration can be through injection, inhalation, ingestion, and absorption [1, 2]. The need for drugs spans across medical treatment (disease extermination), performance enhancement (power booster), and recreation (relaxation and pleasure). Also, their delivery is anchored on materials with deliverable qualities such as biocompatibility, hydrophobicity, target-oriented, minimal toxicity, mechanical tunability, and biodegradability. The materials include hydrogels, liposomes, nanosized molecules, natural polymers, and micelles.

The hydrogel-based drug delivery scheme is characterized by a three-dimensional network of hydrophilic biocompatible systems that mimic a natural medium of live tissues, mechanical capability for drug encapsulation and release in an ordered manner, responsiveness to some biological stimuli that regulate haemostasis, and bio-adhesion to tissues that leads to biosensing, wound healing, and bio-tissue engineering [3–6]. They all strengthen it to act as a transdermal, gastroprotective, intestine and colon, cancer, protein, and peptide targeting drug delivery (Fig. 1).

Hydrogel-based drug carriers are synthetically and naturally sourced with a biodegradability view. Most of their shortfalls include loading difficulty, inadequate flexibility, dehydration inconsistency, weak oxygen permeability (limited regulation on porosity), uncontainable drug release, mechanical weakness (breaking under

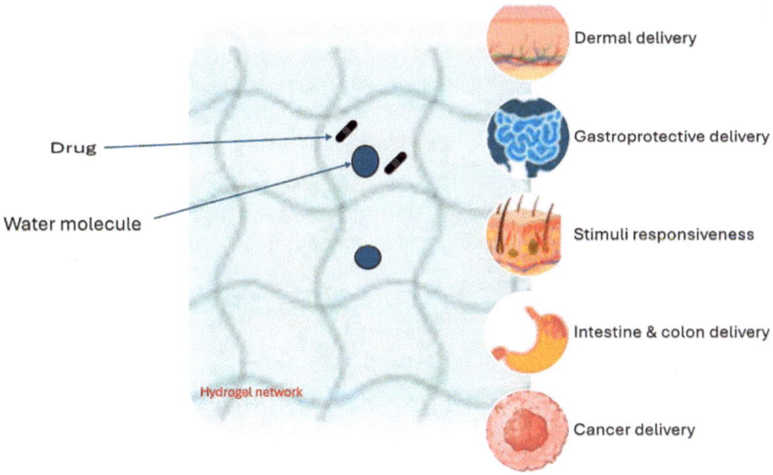

Fig. 1 Hydrogel-based drug delivery scheme

stress), low compatibility with body tissues, low stimuli response time, the low biodegradability of some hydrogels that can accumulate and lead to complications, and the high cost of scalability [7–12]. These materials include polyethylene glycol, polylactic acid, polyethylene oxide, poly(vinyl alcohol), gelatin methacrylate, poloxamer (pluronic), hyaluronic acid, chitosan, polyacrylic acid, hydroxyethyl cellulose, carbopol, hydroxylpropyl, dibenzylidene sorbitol, ascorbyl palmitate, diacylglycerols, phosphatidylserine, and agarose.

A liposome is a microscopic lipid bilayer sac that encapsulates and conveys molecules such as drugs. Its most common precursor is a hydrophilic phospholipid molecule with a twofold layer that permits molecular self-assembly [13–16]. Thus, the liposome-based drug delivery scheme (LBDDS) is featured with rich mechanical strength, controllable drug release, proper location targeting, good biodegradability, high drug impact, and minimal toxicity (Fig. 2). Its administrative pathways include oral, pulmonary, ophthalmic, parenteral, nasal, and transdermal ports. Research has shown that liposomes are capable of actively and/or passively delivering both hydrophobic and hydrophilic drugs to the desired location due to their mechanical strength, leading to a high therapeutic impact [17–20].

Liposome-based drug delivery schemes are not without shortfalls, and the most common one is liposome-liposome leakage and fusion, which reduces their reachability to targeted organs (liver, lungs, spleen, bone marrow, kidney, and lymph nodes). Others include susceptibility to oxidation and reduction, leading to reduced shelf life, and low scalability [21–23]. The liposome materials include dioleoylphosphatidylserine, 1,2-dioleoyl-3-trimethylammonium propane, distearoylphosphatidylethanolamine, dioctadecyldimethylammonium bromide, phosphatidylcholine, cholesterol, phosphatidylglycerol, dioleoylphosphatidic acid, and dioleoylphosphatidylethanolamine.

Fig. 2 Liposome-based drug delivery capabilities

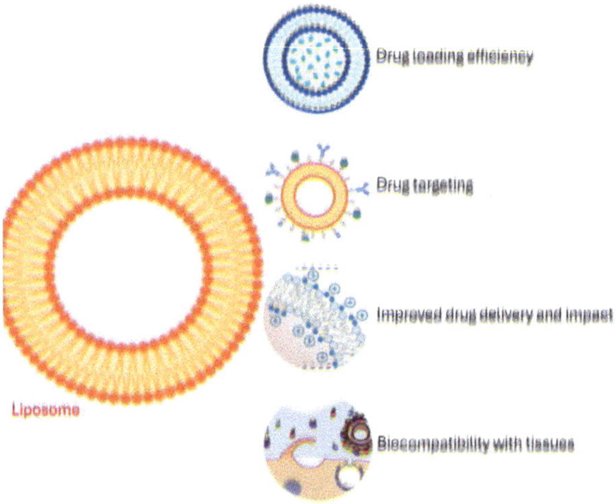

Nanoparticles are nano-scale dimensional particles sourced from both natural and synthetic materials (organic and inorganic-based). Owing to their small size nature, a large surface-area-to-volume-ratio endowment enhances the material efficiency in target delivery. Nanoparticle-based drug delivery scheme (NPBDDS) has made credible progress by positioning nanostructure molecules in drug delivery technologies (Fig. 3), with controllable release, targeted delivery, drug stability, improved pharmacokinetics, and minimal side effects. Drugs characterized by low metabolism and bioavailability leverage the NPBDDS bio-accessibility, convertibility, and stability. On the flip side, the challenges faced by NPBDDS include poor deep tissue penetration, shielding effect from mononuclear phagocytes in the spleen and liver, interruption by bloodstream protein, possibility of toxicity and insusceptible responses, large-scale reproducibility, probable out-of-target delivery, clumsiness, and unrich shelf life can negatively impact the operation [24–28].

Natural polymers (NP) are not left out in the drug delivery technologies as they possess biodegradability, bioavailability, biostability and compatibility. They are materials derived from animals and plants, which include starch, chitosan, gelatins, albumin, cellulose, dextran, pectin, alginate, collagen, fucoidan, elastin, guar-gum, dummer-gum, inulin, chondroitin sulphate, carrageenan, copal-gum, and alkyl glucosides [29–32]. The ease of their fabrication is relatively supportive, leading to the possibility of upscaling. The tunability of these natural polymeric-based drug carriers makes them preferable for enhancing site targeting, drug delivery and controllable release, and tissue absorptivity (Fig. 4).

On the other hand, a micelle is a product of surfactant aggregation within a specific concentration known as the critical micellar concentration (CMC). It is built up with a hydrophilic head (inorganic ion) and a hydrophobic tail (carbon chain) (Fig. 5). The surface-active surfactants make the micelle formation more water-loving (hydrophilic) and partly lipid-loving (lipophilic) by concentrating at interfaces within bodies of water and lipids. Thus, micellar technology is an emerging solution as

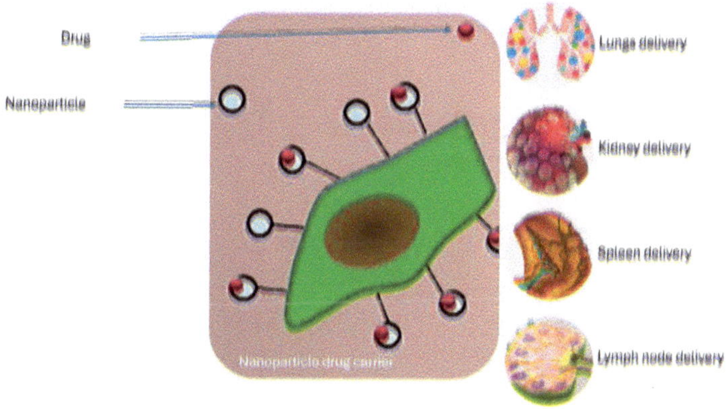

Fig. 3 Nanoparticle-based drug delivery scheme

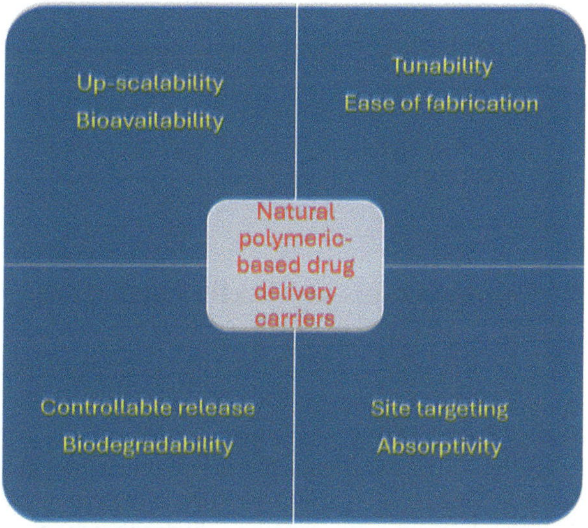

Fig. 4 Natural polymeric-based drug delivery capability

more diverse categories of surfactants, polymeric micelles, and protein-based micelle materials are being synthesized and modelled using artificial intelligence (AI) algorithms [33–36]. The materials include Tween-80 (polysorbate-80), poly(ethylene-glycol)-block-poly(lactic-co-glycolic acid), sodium dodecyl sulphate, poly(ethylene-glycol)-block-polyaspartate, pluronic-F127, poly(ethylene-glycol)-block-polylactic acid, polyamidoamine dendrimers, sodium taurocholate (bile salt), poly(ethylene-glycol)-block-polycaprolactone, and polyhistidine.

Micelle-based drug delivery system (MBDDS) employs polymeric-based and protein-based micelles with a concrete shape, loading capability, stability, and release index to match conventional delivery schemes discussed above. Some of the major strengths of MBDDS lie in its high kinetic stability in a lower CMC profile and

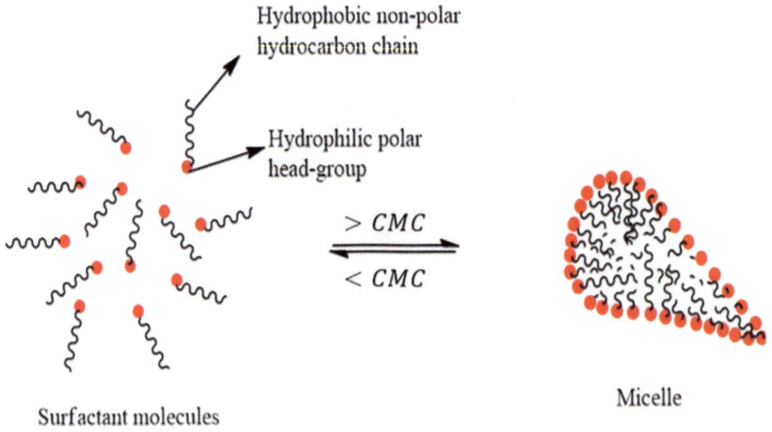

Fig. 5 Micelle formation in a bulk water medium

vehicle for hydrophobic drugs (low-soluble drugs) that promotes bioavailability [37–39]. Thus, the aim of the book chapter is to buttress the intricacies of micelles as the emerging drug delivery prototype and detail its fundamentals, method of preparation, drug-carrying capability, targeting strategies, therapeutic inputs, pharmacokinetics, toxicity, advances, challenges and imminent perspectives.

2 Fundamentals of Micelle Formation

The MBDDS uses a self-assembling amphiphilic precursor that has the capability of encapsulating poorly H_2O-soluble drugs, thereby improving the solubility and bioavailability of the drugs as the hydrophobic region hosts the drugs and the hydrophilic part constitutes the frame for the maintenance of the micelle structure [40–42]. Thus, amphiphilic precursors are compounds that contain both water-loving (hydrophilic) and water-repelling (hydrophobic) portions. This duality allows them to self-aggregate into various structures in aqueous milieus, depending on temperature and concentration. Surfactants, phospholipids, and block copolymers are the common amphiphilic precursors, and their arrangement can result in spherical aggregates (micelles) at CMC with the head (hydrophilic core) portraying outwardly and inwardly for the tail (hydrophobic extension), or bilayer vesicles (liposomes) with the hydrophilic core facing the exterior and interior aqueous milieus and hydrophobic extensions lying inside the bilayer, encapsulating both water-loving and water-repelling drugs, or bilayers (flat membranes) in the similitude of liposomes but without a closed sac within them (Fig. 6) [40–42].

Let's dive into normal and reverse micelles for a better understanding of the micelle structures. The aggregation of hydrophilic heads outwardly around a polar solvent (water) and the hydrophobic tails inwardly clustered in the core are denoted as normal

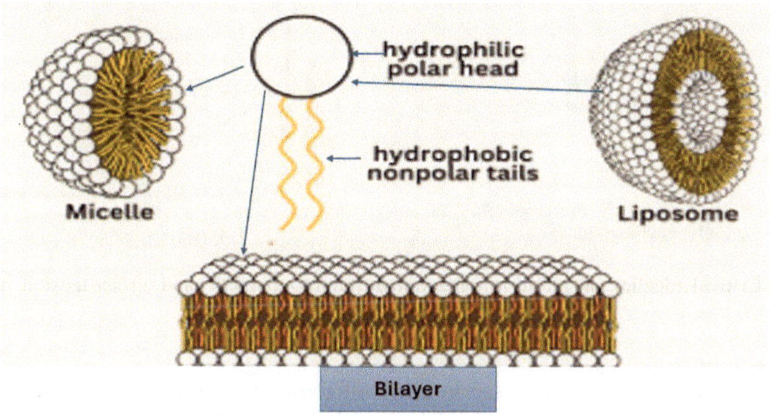

Fig. 6 Amphiphilic molecular arrangement

micelles. This orientation supports the delivery of sparingly soluble drugs to the targets. The reverse micelle is the opposite of the normal micelle, which possesses the hydrophilic head inwardly in the aqueous region and the hydrophobic tail outwardly around the non-polar solvent (Fig. 7). It aids the separation of proteins and lipids, as proteins solubilize in the inward region and their activity and configuration are maintained. It thereby encapsulates soluble proteins around the hydrophilic region, ensuring compatibility and stability within the protein membrane [43–45].

The possibility of micellar formation within and above a critical micelle concentration of the micelles' precursors is positive, and it is dependent on some factors such as ionic strength, temperature, and pH of the media. However, the concentration at which the amphiphilic precursor monomers aggregate in a solution is known as the critical micelle concentration. Its formation can be monitored through either conductivity or surface tension measurements. The surface tension decreases as the CMC approaches and eventually steadies, while conductivity increases as the CMC approaches and plateaus (Fig. 8) [46–48].

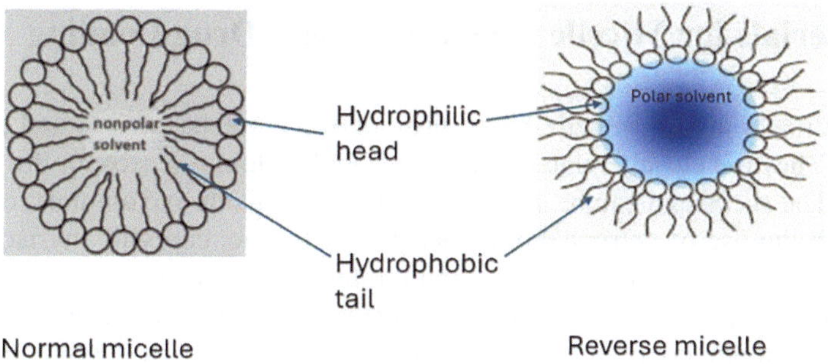

Fig. 7 Micelles structure

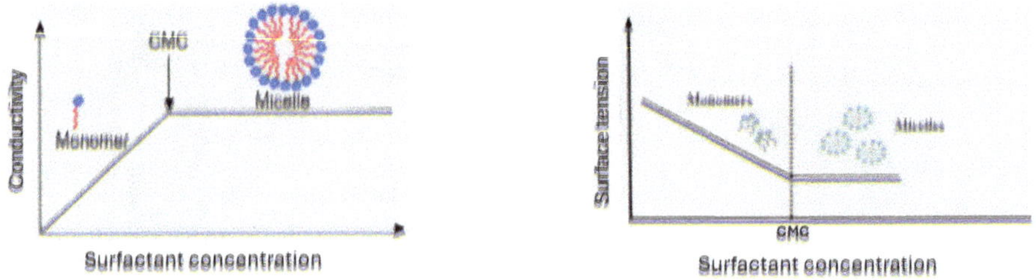

Fig. 8 Critical micelle concentration realization through conductivity and surface tension measurements

There is an observed decrease in CMC of ionic amphiphilic precursors with an increase in ionic strength of the media, as added ions initiate electrostatic repulsion within the hydrophilic head, making the aggregation faster and plateauing at high ionic strength. The effect of added ions on non-ionic amphiphilic precursor media is insignificant, although a little change is observed in the water phase structure. The increase in temperature decreases the CMC of ionic amphiphilic precursors due to the distortion of aqueous structures, which reduces hydrophilic head hydration and promotes micellization. The non-ionic amphiphilic precursors experience a rise in CMC, as it depends on H-bonding with water that weakens on temperature increase. In acidic media (low pH), the likelihood of anionic amphiphilic precursor protonation reduces charge repulsion, easing the micelle formation and lowering the CMC. In alkaline media (high pH), deprotonation of cationic amphiphilic precursors initiates charge repulsion, which affects the hydrophilic head charge and increases the CMC. The media pH of non-ionic amphiphilic precursors has an insignificant impact on CMC, and pH changes on zwitterionic amphiphilic precursors can lead to charge repulsion at the same charge point and lower the CMC, but increase the CMC at different charge points. The information derived from the CMC helps inform researchers/industrialists on the efficiency, stability, and self-aggregation behaviour of potential micellar precursors [49–52].

3 Materials for Micelle Preparation and Drug Loading

The main materials for micelle-based drug delivery are surfactants and polymeric micelles. They are closely related in interface and colloid sciences for drug delivery systems. The surfactant can be sourced naturally (in plants and animals) or synthetically with the use of petrochemical feedstocks. Lauric, caprylic, myristic, capric, stearic, palmitic, gum, linoleic, oleic, and linolenic acids can serve as precursors for surfactant production, whose monomers cluster to form micelles [29, 31]. The self-clustering of block copolymers from colloidal moieties forms polymeric micelles. Their preparation can be achieved using poly(ethyleneoxide)-poly(propyleneoxide), poly(butylmethacrylate), poly(isopropylacrylamide), phospholipids, or polystyrene.

They micellize beyond the critical micellization temperature (CMT) and CMC, and they are inert molecules below the CMC and CMT. They possess a semi-solid state hydrophobic core that enables them to accommodate drugs and a water-loving charged head for mobility and binding to the target in the system [53, 54].

The preparation of a synthetic surfactant follows the steps:

i. Surfactant precursor and reagent mixing: The surfactant precursors, such as amino acid esters, amino acids, or fatty acids, are mixed with alkaline (NaOH or KOH) and catalyst (orthophosphoric acid or its salts) solutions in an airtight system.
ii. Mixture dehydration: The mixture is heated within 373–573 K to aid the dehydration process for two to fifteen hours.
iii. Continuous reaction: An oxidizing agent or an alkylating reagent, such as diamines, is added while maintaining the system's condition.
iv. Adjustment of pH: The system's pH is maintained within 6–9 with ascorbic acid or sodium carbonate.
v. System cooling: The system is allowed to cool, filtered, and packaged to realize the product [55, 56].

The preparation of biosurfactant follows the steps:

a. Surfactant precursor hydration: The biosurfactant material, such as lauryl glucoside, is mixed with water while stirring.
b. Hydration support: The dissolved xanthan gum is added to the hydrating system while stirring continuously. The gum hydrates all available water molecules in the system, else there will be a partition of organic and inorganic layers.
c. Introduction of a preservative: The mixture is mixed with a preservative such as citric acid to lower the pH and thicken the mixture. It is important to reduce the pH to within 5–6 in order to lower the viscosity of the mixture [57, 58].

The preparation of polymeric micelle drug carriers can be made through:

I. Ultrasonication: The mixture of polymeric nanoparticles and drug solution is exposed to an ultrasonic source for 10–15 min, and the product is centrifuged for 3–5 min, which initiates a self-aggregation and loading of drug in the micelle core.
II. Emulsification: The polymeric precursor is dissolved in an H_2O-immiscible solvent, followed by the addition of a drug while stirring and the addition of a surfactant, which aids the oil-in-water emulsion formation. The mixture is therefore evaporated for 30–60 min in a vacuum to evacuate the organic solvent.
III. Dialysis: The polymeric precursor and drug are dissolved in a separate organic solvent (ethanol and acetone), mixed, and transferred to a dialysis bag containing water. The bag membrane permits the escape of the solvent while retaining the polymeric material and the drug, leading to self-aggregation of the polymeric material to micelles and harbouring the drug in the hydrophobic core.
IV. Solvent evaporation: The polymeric precursor and drug are dissolved in ethanol or acetone and placed on the evaporating dish, resulting in a solid product of

a polymer-drug complex, which is mixed with water and heated. Micelles are obtained, capturing the drug in the non-polar core region. It is useful for less water-soluble drugs.

V. Direct dissolution: The molar mass of the polymeric precursor and drug is solubilized in water, resulting in a concentration above the CMC and the realization of micelles loaded with drug. The process supports water-soluble drugs [59, 60].

The tunability of the polymer micelle materials is key to creating hybrid drug-micelle delivery systems, which foster a robust change in the chemical (interactiveness to stimuli) and physical properties (surface charge, size, stability, and shape) of the materials. Kazaryan and co-workers [61] report a tuned solvophobic-polymeric-micelle block material with a focus on improving the size and micellization of the micelle. The implementation of the desired property was guided by the length and solubility of the precursors. Singh and Aswal [62] reported the tunability of anionic micelles (sodium dodecyl sulphate) in a bid to improve their adsorptivity, distribution, and steric firmness by the conditional mixture of non-ionic and anionic surfactants, which is driven by electrostatic traits. Shalmani and co-workers [63] highlighted the tunability of polymer micelles for the delivery of corticosteroid (dexamethasone) and taxane (paclitaxel) in cancer treatment. Release rate and size of the micelle-drug carrier were majored in the tuning of the copolymeric materials (poly(ethylene-glycol)-b-(N-2-benzoyloxypropyl) methacrylamide), and the result underscores slow and fast drug release exhibited by bulky and small micelle polymers, which is attributed to molecular weight and hydrophobicity.

In general, loading of partially water-soluble drugs happens in the hydrophobic core of the micelles, and water-soluble drugs are accommodated in either the hydrophilic head or the hydrophobic core. The latter is dependent on the micellar structure. Thus, the precursor composition, size of micelles, and hydrophobicity of drugs determine the drug-loading capacity in the micelles. The drug-micelle binding is aided by the drug and polymer functional groups, medium pH, and enzymes [64].

3.1 Drug Release Mechanism

The micelle-driven drug release mechanism is a systematic route through which drugs are being uploaded into the micellized amphiphilic molecules' core and delivered to the target organs or tissues either via diffusion (the drugs leave the micelle core) or disassembly of the micelle initiated by system stimuli. The system stimuli are controlled by the temperature and alkalinity or acidity of the system. The drug release can also be motivated by a redox process, leading to bond breakage between the drug and the amphiphilic molecules and probable micelle degradation. The probability of bond breakage can be attributed to the type of bond existing between the drug and the micellized amphiphilic molecules and the factors (pH and temperature) moderating the system. Ionic or covalent bond drug-micelle compact shows a different way of release in a specific medium condition. Low pH environment influences the covalent

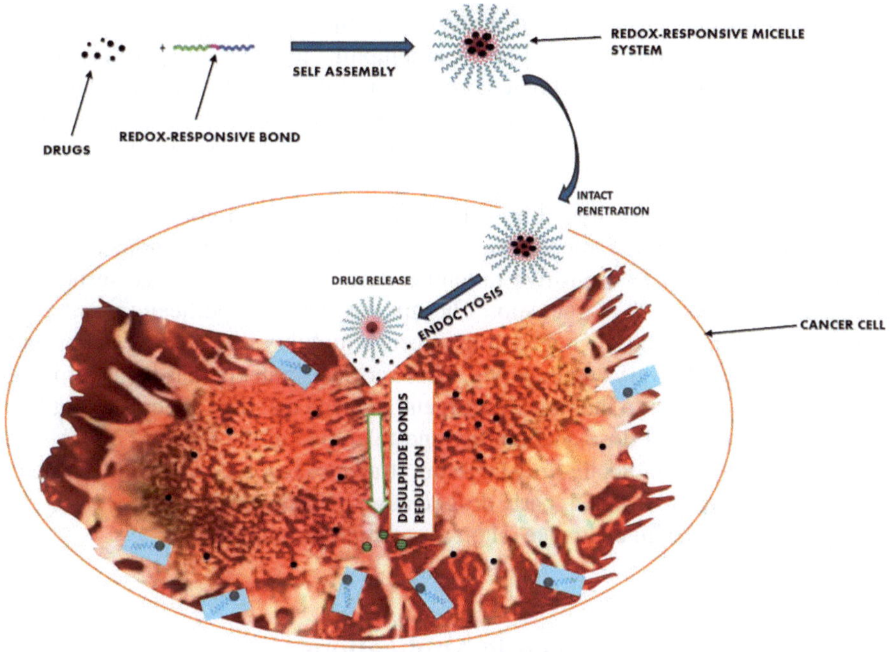

Fig. 9 Micelle-drug release through diffusion, disassembly of micelle, and redox processes

bond drug-micelle breakage that involves hydrophobic drugs such as camptothecin, tamoxifen, vitamin K3, triclosan, and porphyrin, and the opposite condition happens for ionic bond drug-micelle rupture that involves hydrophilic drugs (Fig. 9) [65–67]. Going forward, the targeting strategies for the drug delivery, which can follow active or passive protocols.

Micelle-drug host releasing protocol is prompted by the cell permeability/ distribution and redox activeness of the host and cell compositions. The limitations to this process include drug loading, stability, and release control, directing either a slow or fast drug release rate that determines the therapeutic impact of the drugs on the target site. It can be surmounted by structural and charge modification of the polymeric micelle materials [65].

4 Micelle-Drug Targeting Strategies

The micelle-drug target strategies involve the support structures within the micelle that aid in the seamless delivery of drugs. The uniqueness of the micelle's nanostructure offers drug delivery an improved transport means with increased solubility, permeability, and shielding from degradation. Thus, the critical targeting strategies centre on ligand (target cells)-micelle head region binding, stimuli-responsive micelle design, and micelle surface modification. Recall that the choice of the drug determines the carrying mode provided by the amphiphilic micelles, as sparingly soluble drugs are hosted in the hydrophobic core region of the micelle via a covalent bond and carried to the target cells, organs, or tissues. On the flip side, the soluble drugs are hosted at the hydrophilic head region via an ionic bond to the targets, improving bioavailability [68, 69].

However, the receptor ligand-micelle binding promotes a good engagement of drug release carried in the micelle via diffusion. The design of the molecular linker in the copolymers that is sensitive to temperature, pH, or ion change sustains the encapsulated drugs to the target spot. The micelle surface modification, especially with polyethylene glycol, is intended to improve mobility and binding opportunity for effective drug delivery and targeting. These critical strategies ensure an improved delivery concentration to the intended targets, reduced cell damage emanating from toxicity, and improved dose adaptability and tolerance in the system [70, 71]. The weighty limitations of micelle-drug targeting strategies revolve around the biological milieu instability, which spikes early drug release, inadequate tissue penetration, and slow phagocytic system clearance (short circulation time) [70].

5 Micelle-Drug Therapeutic Applications

The involvement of micelle-drug carriers in the treatment of health disorders and ailments such as neurodegenerative diseases, inflammation, cancer, and infectious diseases is apt, as it is poised with an improved target-site accumulation, minimal toxicity, increased solubilization of drugs, and versatile engagement of potent compounds for disease treatment [72].

The target-site accumulation of the micelle-drug carrier and its persistent circulation in a compromised system propel it to serve as a good alternative to cancer treatment in the reign of the conventional protocol. The invention of nanomedicine has supported and improved the conventional approach to cancer treatment. The spreading of cancer disease in a system is one of the major challenges of its treatment, and its stoppage is crucial in the survival of the patient. The accumulative and far-reaching micelle-drug carrier has curtailed the cancer and tumour cells by blocking their cell-tissue interface [73, 74]. Thus, the cancer cells' separation can be inhibited by the micellized-drug agent on the target, especially at the early stage of the detection (Fig. 10).

The correction of neurodegenerative disorder (NDD) is clinically under investigation with polymeric micelles carrying a therapeutic agent to the site of neuro-trigger (nerve cells in the spinal cord and brain). Some of the disorder is allergy-orientated and quick in action; hence, the use of drugs hosted in polymeric micelles is expected to deliver fast, efficient, stable, and defeat the blood–brain barrier (BBB) [75]. Most importantly, polymeric micelle-drug carriers are capable of scavenging reactive oxygen species, blocking neuro-inflammation and neuronal exhaustion, and impeding the protein aggregation that causes NDD, thereby improving cognitive

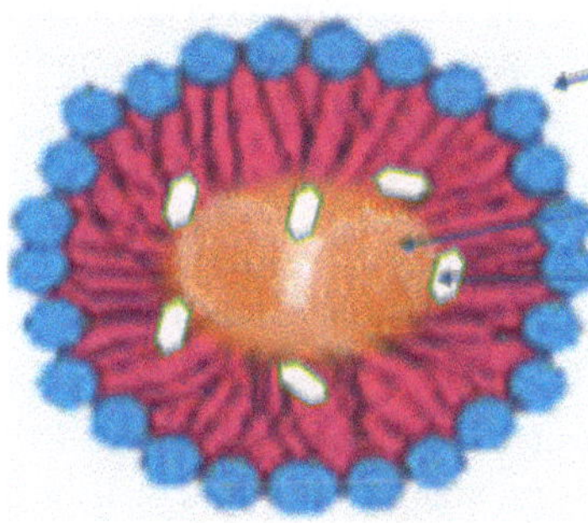

Fig. 10 Hydrophobic micelle-drug cancer therapeutic

and memory functions in Parkinson's, Alzheimer's, amyotrophic lateral sclerosis, and Huntington's disease patients. It is important to stress that the manifestation of NDD is driven by gene and environmental factors that tolerate inflammation, oxidative stress, redox and protein alteration. Nevertheless, neurological disorders are often sponsored by physiological barriers such as the BBB that impede the passage of harmful chemicals from the blood track and assist homeostasis. The micelle-drug system aims at the brain network with reticuloendothelial uptake, impacting an adequate-circulating effect [74].

Infectious diseases spread by parasites, bacteria and fungi are not left out in the application of the micelle-drug delivery system. The system is designed to transport sparingly/poorly soluble drugs, thus enhancing drug dissolution and site targeting. The micelle can either be a single polymeric host or a polymeric matrix host with the capacity to sustain supersaturation that permits bioavailability and decreases drug precipitation. Qian et al. [75] formulated polyzwitterion micelles from cis-5-Norbornene-exo-2,3-dicarboxylic anhydride, N α-Cbz-L-lysine, and tetrahydrofuran to address the biofilm resistance of drug-resistant bacteria strains. The responsiveness of the polyzwitterionic micelle improved the drug penetration with the aid of the hydrophilic and hydrophobic parts (amino acid-based and triclosan parts, respectively). Onat et al. [76] leveraged poly[3-dimethyl)methacryloxyethyl) ammonium propane sulfonate-b-2-(diisopropylamino)ethyl methacrylate] to achieve a polyzwitterionic micelle that is used to treat bacterial infections with multi- and mono-layer films, which have become resistant to many antibacterial drugs. Thus, the polymeric micelles have been able to provide the mobility needed for some antibacterial drugs to penetrate the biofilm of the bacteria. Similarly, Huang et al. [77] designed penetrative bacterial biofilm polymeric micelle-drug carriers (triclosan-cyclic trimethylene carbonate-caprolactone-poly(ethylene-glycol), with a strong attraction to the bacteria lipase that pushes for micelle opening, leading to a controllable drug release that targets both the bacteria and the biofilms. The highly H_2O-soluble, naturally charged non-toxic poly(ethylene-glycol) served as the hydrophilic region that diminishes non-directional interactions with protein and blood components. Thus, it responsively supports biocompatibility and less toxicity in the fight against drug-resistant bacteria. Alternatively, poly(N-isopropyl acrylamide) and poly(N-vinyl pyrrolidone) can be used in place of poly(ethylene-glycol). Some other micelle-drug systems are depicted in Table 1.

5.1 Micelle-Drug Pharmacokinetics

The description of a drug's job in the body and its exit can be referred to as pharmacokinetics. The process includes absorption, distribution, metabolism, and excretion of the drug's by-products. The micelle-drug stability and circulation time in the blood are enhanced because of reduced protein adsorption and quick clearance prompted by the hydrophilic core and, thus, the hydrophobic core hosting the drugs. There are two factors that ensure the improvement of the micelle in drug pharmacokinetics: structure, material composition, intramolecular interaction, and size. The structure of the micelle informs the intramolecular interaction happening within it, which could be electrostatic or crosslinking covalent interactions. The size is linked to either simple or complex micelle matrices, emanating from the nature of the hydrocarbon chain of the micelle. Thus, these factors contribute to micelle stability and invariably support drugs' longer circulation period, delivery, and targeting [98]. In comparison to liposomes, the micelle size aids the tumour vasculature penetration, giving it an advantage in the treatment of severe cancer cells and bacteria with biofilms [99]. Hou et al. [100] reported designed poly(ε-caprolactone)-methoxypolyethylene glycol polymeric micelles with a tuneable tail length ($n = 40$, 30, and 20) and the poly(ε-caprolactone) serving as the hydrophobic core that hosts the doxorubicin drug while following the dialysis method for its formulation. A seamless circulation and stability were recorded, with the micelle having 40 carbon atoms leading in performance, and the one with 10 carbon atoms being the least.

The seamless micelle-drug site accumulation and targeting lie on the hydrophilicity of the micelle to bind explicit target cell receptors, leading to an enhanced tissue permeability and retention. The contribution of hydrophilic head interaction with the receptors fosters a stealth layer effect that assists in prolonging micelle-drug system circulation within the host of the target cells. The process improves therapeutic index and reduces side effects, while receptor support endocytosis and eases intracellular delivery of drugs [101]. The challenges bedeviling micelle-drug pharmacokinetics include physiological environment complexity, drug-loading instability, and side interaction of micelles with some fluid proteins. These suggest that the drug delivery has direct side effects. Lu and co-workers [102] proffer strategies for micelle stability enhancement in drug delivery with the introduction of bonds crosslinking and non-crosslinking protocols on amphiphilic polymers to actualize block and copolymeric micelle precursors that are featured with moderately long carbon hydrophobic chains that do not lower the in vivo output of the hosted drugs. The protocols also lower CMC and stabilize micelles, while reducing the free energy via the sorting of hydrophobic layers from the unsuitable water milieu. Thus, the diluted micelle concentration can restrain crosslinking performance. Other

Table 1 Some micelle-drug therapeutic records

S. No.	Drug	Delivery host	Disease	References
1	Oridonin (Ori) and Phillyrin (Phi)	Polypeptide apolipoprotein E-modified micelle (APoE-Ori/Phi-Ms)	Alzheimer	[78]
2	Artemether	Soluplus-Artemether	Malaria	[79]
3	Ketoprofen	Soluplus-Ketoprofen	Inflammation and Pyretic BCS class II	[80]
4	Dioscorea bulbifera extract	Soluplus-Dioscorea bulbifera extract	Cancer	[81]
5	Luteolin	Soluplus-Luteolin	Cancer	[82]
6	Olaparib and Rapamycin	Soluplus-Olaparib/Rapamycin	Ovarian cancer	[83]
7	β-ionone	Soluplus-β-ionone	Cancer	[84]
8	Tamoxifen citrate	Soluplus-Chitosan-Tamoxifen citrate	Breast cancer	[85]
9	Caffeic acid	Pluronic P123-Caffeic acid; β-cyclodextrins caffeic acid	Inflammation and cancer; malaria	[86, 87]
10	Silibinin	Pluronic F68-Silibinin	Neuronal disease	[88]
11	Curcumin and Piperine	Soluplus-Curcumin/Piperine	Tumour and inflammation	[89]
12	Paclitaxel and Albendazole	Folic acid-Soluplus-Albendazole/Paclitaxel	Cancer	[90]
13	Aleo emodin	Glycyrrhizic acid-Soluplus-Aloe emodin	Hyperuricemia	[91]
14	Budesonide	Polycaprolactone-Budesonide	Bowell inflammation	[92]
15	Decoquinate	Soluplus-Decoquinate nanoparticles	Malaria	[93]
16	Emamectin benzoate	Sodium alginate-Emamectin benzoate	Parasite (*Caligus rogercresseyi*)	[94]

(continued)

Table 1 (continued)

S. No.	Drug	Delivery host	Disease	References
17	Lumefantrine	PoloxamerF68-piperine-Lumefantrine	Malaria	[95]
18	Tobramycin and Dexamethasone	Poloxamer407-Soluplus-Tobramycin/Dexamethasone	Rhinosinusitis	[96]
19	Naringin	Poloxamer188-Naringin/KollidonVA30-Naringin	Bacteria and Neuroblastoma	[97]
20	Agomelatine	Hydroxypropyl-β-cyclodextrine-poloxamer-Soluplus-Agomelatine	Depression	[98]

strategies include structural modification of polymeric micelle regions (hydrophobic and hydrophilic) ratios and increased crystallinity of hydrophobic regions.

6 Safety Considerations in Micelle-Drug Delivery System

The early release of drugs without reaching their targets is a concern, and the micelle-drug delivery system addresses it by safeguarding biocompatibility, averting hasty drug release, and evading immune complex (antibody-antigen) clearance. The kinetic and thermodynamic stability that is associated with micelles contributes to the delivery system of micelle-drug carriers to the targets. The kinetic stability is more pronounced in polymeric micelles than in monomeric micelles, and the impact resists micelles disassembling in a diluted solution below CMC. The thermodynamic stability, on the other hand, induces stable micellar assembly from unstable monomers with a low CMC. These factors minimize the toxicity of the micelles and improve enhanced permeability retention. However, the circulation of micelle-drug carriers in the bloodstream would be stable to prevent hasty disaggregation and untimely drug release [103, 104]. Moreover, the safety considerations spread across the screening of toxic fragments of micelle-drugs, the use of polyethylene glycol-based precursors, the choice of polymeric materials, ligand-receptor kinetic stability, the dominance of micelle-drug carriers at the disease spot, and the proper discharge of the drug carriers from the body. The outcome of these considerations will reduce adverse side effects, prevent hasty release of drugs, improve targeting efficiency, and reduce long-term organ or tissue accumulations [54].

7 Advances and Challenges in Micelle-Drug Delivery System

The micelle-drug delivery system is borne out of the quest for more efficient drug carriers with improved solubilization, precise targeting, delivery, and enhanced stability within the micelle structure [105]. More work is ongoing, both in vivo and in vitro, to improve micelle-drug delivery systems. Researchers are designing tuneable hydrophobic and hydrophilic cores that can address some of the challenges faced in the drug delivery process, which include untimely drug release, drug loading strength, and stability in the bloodstream. Advancing micelle-drug delivery technologies requires a mean feat in a reliable and manageable drug release, micelle-stimuli specificity for targeting, and upscaling in the manufacturing processes [106]. The summary of the challenges and the advances made thus far in the micelle-drug delivery system is presented in Table 2.

Table 2 Micelle-drug delivery system challenges and advances

S. No.	Challenges	Advances	References
1	Limited target permeability	Smart and high-affinity ligand design with mixed micelles PEGylation	[74]
2	Adequate drug-micelle encapsulation and stability	Tuneable covalent crosslinking polymeric micelle structure; difunctional crosslinker; UV-induced dimerization	[107]
3	High critical micelle concentration	Long polymeric micelle chain design with attachment of acrylics and benzyl groups	[71]
4	Minimal selectivity	Structural precision model for hydrophilic core enhancement with PEGylation	[108]
5	Sensitivity and toxicity	Dynamic alkaline-sensitive polymeric micelles; clean and degradable micelle precursors	[109]
6	Quick removal of drug-micelles from bloodstream by reticuloendothelial system	Design of a stable hydrophilic micellar core (e.g., polyethylene glycol) or multi/mixed hydrophilic micellar core; surface modification; lower nanosized micelles	[109]
7	Low targeting rate	Responsive polymeric micelle with specific stimuli-sensitive linker; dependable targeting ligand designs; intracellular responsive micelle-drug designs	[74]

8 Conclusion

The micelle-drug delivery system is an evolving drug delivery technology that has shown a significant value in modern medicine with its ability to enhance drug loading, encapsulation, targeting, delivery, tissue/cells permeability, circulation rate, compatibility, degradability, and minimal toxicity. The structural and size adjustment of the polymeric micelles has made the technology a potential therapeutic option for drug-resistant microbes with multi- or mono-layered biofilms, as well as enabling stimuli-responsive and target-specific drug release. The system is posed with some shortcomings, such as stability in biological fluids, up-scale production, clinical regulation, and immunogenicity. Integrating nanotechnology, material science, and pharmacology helps in overcoming the challenges. With the current innovations in drug delivery, micelle-based systems hold substantial potential to transform orthodox drug delivery approaches and contribute to the development of safer, more effective therapies in the future.

Abbreviations

AI	Artificial intelligence
BBB	Blood-brain barrier
CMC	Critical micelle concentration
CMT	Critical micelle temperature
CT	Clinical trial
LBDDS	Liposome-based drug delivery scheme
MBDDS	Micelle-based drug delivery system
NDD	Neurodegenerative disorder
NPBDDS	Nanoparticle-based drug delivery scheme
NP	Natural polymers
SDG-3	Sustainable development goal 3

References

1. Waheed SA, Mary I, Ogunwole CA (2022) A review on drug-diet interaction. Nat Sci 43:108–125
2. Stielow M, Witczyńska A, Kubryń N, Fijalkowski L, Nowaczyk J, Nowaczyk A (2023) The bioavailability of drug-the current state of knowledge. Molecules 28(24):8038. https://doi.org/10.3390/molecules28248038
3. Liu B, Chen K (2024) Advances in hydrogel-based drug delivery systems. Gels 10(4):262. https://doi.org/10.3390/gels10040262
4. Onaciu A, Munteanu RA, Moldovan AI, Moldovan CS, Berindan-Neagoe I (2019) Hydrogels based drug delivery synthesis, characterisation and administration. Pharmaceutics 11(9):432. https://doi.org/10.3390/pharmaceutics11090432
5. Lu P, Ruan D, Huang M, Tian M, Zhu K, Gan Z, Xiao Z (2024) Harnessing the potential of hydrogels for advanced therapeutic applications: current achievements and future directions. Signal Transduct Target Ther 9:166. https://doi.org/10.1038/s41392-024-01852-x
6. Thirumalaivasan N, Kanagaraj K, Nangan S, Pothu R, Rajendra SP, Karuppiah P, Boddula R (2025) Bioactive hydrogels (Bio-HyGs): emerging trends in drug delivery and wound healing applications. Polym Adv Technol 36(4):e70132. https://doi.org/10.1002/pat.70132
7. Munteanu C, Prifti E, Surd A, Mârza SM (2025) Tailoring therapy: hydrogels as tunable platforms for regenerative medicine and cancer intervention. Gels 11(9):679. https://doi.org/10.3390/gels11090679
8. Segneanu A, Bejenaru LE, Bejenaru C, Blendea A, Mogoşanu GD, Biţă A, Boia ER (2025) Advancements in hydrogels: a comprehensive review of natural and synthetic innovations for biomedical applications. Polymers 17(15):2026. https://doi.org/10.3390/polym17152026
9. Oyervides-Guajardo VG, Claudio-Rizo JA, Cabrera-Munguia DE, Cano-Salazar LF, León-Campos MI, Tamayo-Ordoñez MC (2025) Protein–polysaccharide hydrogels: a review on smart materials for the future of biotechnology. Polym Adv Technol 36(8):2025. https://doi.org/10.1002/pat.70295
10. Karthikeyan L, Kang HW (2025) Recent progress in multifunctional theragnostic hydrogels: the cornerstone of next-generation wound care technologies. Biomaterials Science 13(16):4358–4389. https://doi.org/10.1039/D5BM00781J
11. Fareed MM, Shityakov S (2025) Next-generation hydrogel design: computational advances in synthesis, characterization, and biomedical applications. Polymers 17:1373

12. Londhe PV, Londhe MV, Salunkhe AB, Laha SS, Thompson-Mefford O, Thorat ND, Khot VM (2025) Magnetic hydrogel (MagGel): an evolutionary pedestal for anticancer therapy. Coord Chem Rev 522:216228
13. Liu P, Chen G, Zhang J (2022) A review of liposomes as a drug delivery system: current status of approved products, regulatory environments, and future perspectives. Molecules 27(4):1372. https://doi.org/10.3390/molecules27041372
14. Pasenkiewicz M, Baczynski K, Markiewicz M, Murzyn K (2016) Computer modelling studies of the bilayer/water interface. Biochim Biophys Acta (BBA) Biomembr 1858(10):2305–2321. https://doi.org/10.1016/j.bbamem.2016.01.024
15. Nsairat H, Khater D, Odeh F, Al-Adaileh F, Al-Taher S, Jaber AM, Alshaer W, Al BA, Mubarak MS (2021) Lipid nanostructures for targeting brain cancer. Heliyon 2021:7. https://doi.org/10.1016/j.heliyon.2021.e07994
16. Nsairat H, Khater D, Sayed U, Odeh F, Al-Bawab A, Alshaer W (2022) Liposomes: structure, composition, types, and clinical applications. Heliyon 8(5):e09394. https://doi.org/10.1016/j.heliyon.2022.e09394
17. Nakhaei P, Margiana R, Bokov DO, Abdelbasset WK, Kouhbani MAJ, Varma RS, Marofi F, Jarahian M, Beheshtkhoo N (2021) Liposomes: structure, biomedical applications, and stability parameters with emphasis on cholesterol. Front Bioeng Biotechnol 9:705886. https://doi.org/10.3389/fbioe.2021.705886
18. Charumathy A, Ubaidulla U, Sinha P, Rathnam G (2022) Recent update on liposome-based drug delivery system. Int J Curr Pharm Res 14(3):22–27. https://doi.org/10.22159/ijcpr.2022v14i3.1991
19. Eugster R, Luciani P (2025) Liposomes: bridging the gap from lab to pharmaceuticals. Curr Opin Colloid Interface Sci 75:101875. https://doi.org/10.1016/j.cocis.2024.101875
20. Basak S, Das TK (2025) Liposome-based drug delivery systems: from laboratory research to industrial production—instruments and challenges. ChemEngineering 9(3):56. https://doi.org/10.3390/chemengineering9030056
21. Gatto MS, Johnson MP, Najahi-Missaoui W (2024) Targeted liposomal drug delivery: overview of the current applications and challenges. Life 14(6):672. https://doi.org/10.3390/life14060672
22. Basu B, Prajapati B, Dutta A, Paliwal H (2024) Medical application of liposomes. J Explor Res Pharmacol 9(1):13–22. https://doi.org/10.14218/JERP.2023.00002
23. Hamid MSS, Hatwar PR, Bakal RL, Kohale NB (2024) A comprehensive review on liposomes: as a novel drug delivery system. GSC Biol Pharm Sci 27(01):199–210. https://doi.org/10.30574/gscbps.2024.27.1.0121
24. Sultana A, Zare M, Thomas V, Kumar TSS, Ramakrishna S (2022) Nano-based drug delivery systems: conventional drug delivery routes, recent developments and future prospects. Med Drug Discov 15:100134. https://doi.org/10.1016/j.medidd.2022.100134
25. Yao Y, Zhou Y, Liu L, Xu Y, Chen Q, Wang Y, Wu S, Deng Y, Zhang J, Shao A (2020) Nanoparticle-based drug delivery in cancer therapy and its role in overcoming drug resistance. Front Mol Biosci 7:193. https://doi.org/10.3389/fmolb.2020.00193
26. Rahman MA, Jalouli M, Yadab MK, Al-Zharani M (2025) Progress in drug delivery systems based on nanoparticles for improved Gliobastoma therapy: addressing challenges and investigating opportunities. Cancer 17(4):701. https://doi.org/10.3390/cancers17040701
27. Kumar M, Jha A, Mishra B (2024) DNA-based nanostructured platforms as drug delivery systems. Chem Bio Eng 1(3):179–198. https://doi.org/10.1021/cbe.3c00023
28. Patra JK, Das G, Fraceto LF, Campos EVR, Rodriguez-Torres MDP, Acosta-Torres LS, Diaz-Torres LA, Grillo R, Swamy MK, Sharma S, Habtemariam S, Shin H (2018) Nano-based drug delivery systems: recent developments and future prospects. J Nanobiotechnol 16:71. https://doi.org/10.1186/s12951-018-0392-8
29. Jana S, Gandhi A, Sen KK, Basu SK (2010) Natural polymers and their application in drug delivery and biomedical fields. J PharmaSciTech 1(1):16–27
30. Satchanska G, Davidova S, Petrov PD (2024) Natural and synthetic polymer for biomedical and environmental applications. Polymers (Basel) 16(8):1159. https://doi.org/10.3390/polym16081159

31. Benabid FZ, Zouai F (2016) Natural polymers: cellulose, chitin, chitosan, gelatin, starch, carrageenan, xylan, and dextran. Alger J Nat Prod 4(3):348–357
32. Rajendran AT, Vadakkepushpakath AN (2024) Natural food components as biocompatible carriers: a novel approach to Glioblastoma drug delivery. Foods 13(17):2812. https://doi.org/10.3390/foods13172812
33. Pokhrel DR, Sah MK, Gautam B, Basak HK, Bhattarai A, Chatterjee A (2023) A recent overview of surfactant-drug interactions and their importance. RSC Adv 13(26):17685–17704. https://doi.org/10.1039/d3ra02883f
34. Kurangi B, Somnache S, Jangade N (2014) Micellar drug delivery system. Res J Pharm Dosage Forms Technol 6(4):267–281
35. Ghezzi M, Pescina S, Padula C, Santi P, Favero ED, Cantu L, Nicoli S (2021) Polymeric micelles in drug delivery: an insight of the techniques for their characterization and assessment in biorelevant conditions. J Control Release 332:312–336. https://doi.org/10.1016/j.jconrel.2021.02.031
36. Devi S, Saini V, Kumar M, Bhatt S, Gupta S, Deep A (2019) A novel approach of drug localization through development of polymeric micellar system containing Azelastine HCl for ocular delivery. Pharm Nanotechnol 7:314–327. https://doi.org/10.2174/2211738507666190726162000
37. Mustafai A, Zubair M, Hussain A, Ullah A (2023) Recent progress in proteins-based micelles as drug delivery carriers. Polymers (Basel) 15(4):836. https://doi.org/10.3390/polym15040836
38. Atanase LI (2021) Micellar drug delivery systems based on natural biopolymers. Polymers (Basel) 13(3):477. https://doi.org/10.3390/polym13030477
39. Isogiu IA, Ozsoy Y, Dincer S (2016) Advances in micelle-based drug delivery: cross-linked systems. Curr Top Med Chem 17(13):1468–1489. https://doi.org/10.2174/1568026616666161222110600
40. Middleton JM, Siefert RL, James MH, Schrand AM, Kolel-Veetil MK (2021) Micelle formation, structures, and metrology of functional metal nanoparticle compositions. AIMS Mater Sci 8(4):560–586. https://doi.org/10.3934/matersci.2021035
41. Hanafy NAN, El-Kemary M, Leporatti S (2018) Micelles structure development as a strategy to improve smart cancer therapy. Cancer 10(7):238. https://doi.org/10.3390/cancers10070238
42. Kim S, Cho Y, Kim JH, Song S, Lim J, Choi SH, Char K (2020) Structural analysis of bottlebrush block copolymer micelles using small-angle X-ray scattering. ACS Macro Lett 9:1261–1266. https://doi.org/10.1021/acsmacrolett.0c00442
43. Li J, Li Z, Zhou T, Zhang J, Xia H, Li H, He J, He S, Wang L (2015) Positively charged micelles based on a triblock copolymer demonstrate enhanced corneal penetration. Int J Nanomed 10:6027
44. Naharros-Molinero A, Caballo-Gonzalez MA, Mata FJD, Garcia-Gallego S (2022) Direct and reverse pluronic micelles: design and characterization of promising drug delivery nanosystems. Pharmaceuticals 14(12):2628. https://doi.org/10.3390/pharmaceutics14122628
45. Thapa RK, Cazzador F, Grønlien KG, Tønnesen HH (2020) Effect of curcumin and cosolvents on the micellization of pluronic F127 in aqueous solution. Colloids Surf B 195:111250
46. Xue X, Hu Y, Deng Y, Su J (2021) Recent advances in design of functional biocompatible hydrogels for bone tissue engineering. Adv Funct Mater 31:2009432
47. Tong X, Pan W, Su T, Zhang M, Dong W, Qi X (2020) Recent advances in natural polymer-based drug delivery systems. React Funct Polym 148:104501
48. Akhtar A, Andleeb A, Waris TS, Bazzar M, Moradi AR, Awan NR, Yar M (2021) Neurodegenerative diseases and effective drug delivery: a review of challenges and novel therapeutics. J Control Release 330:1152–1167
49. Ramireddy RR, Prasad P, Finne A, Thayumanavan S (2015) Zwitterionic amphiphilic homopolymer assemblies. Polym Chem 6:6083–6087. https://doi.org/10.1039/C5PY00879D
50. Wang D, Xu J, Tan J, Yang J, Zhou S (2021) In situ generation of amphiphilic coatings based on a self-catalytic zwitterionic precursor and their antifouling performance. Chem Eng J 422:130115. https://doi.org/10.1016/j.cej.2021.130115

51. Ghosh R, Wong W, Reimeers T, Rodzanowski A, Ruiz JC, Coughlin EB (2023) Amphiphilic-zwitterionic block polymers. Polym Chem 15:622–630. https://doi.org/10.1039/D3PY01179H
52. Jana P, Shyam M, Singh S, Jayaprakash V, Dev A (2021) Biodegradable polymers in drug delivery and oral vaccination. Eur Polym J 142:110155. https://doi.org/10.1016/j.eurpolymj.2020.110155
53. Li J, He W, Sun X (2007) Preparation of poly(styrene-b-N-isopropylacrylamide) micelles surface-linked with gold nanoparticles and thermos-responsive ultraviolet-visible absorbance. J Polym Sci Part A Polym Chem 45(22):5156–5163. https://doi.org/10.1002/pola.22260
54. Mourya VK, Inamdar N, Nawale RB, Kulthe SS (2011) Polymeric micelles: general considerations and their applications. Indian J Pharm Educ Res 45(2):128–138
55. Farias CBB, Almeida FCA, Silva IA, Souza TC, Meira HM, Silva RCFS, Luna JM, Santos VA, Converti A, Banat IM, Sarubbo LA (2021) Production of green surfactants: market prospects. Electron J Biotechnol 51:28–39. https://doi.org/10.1016/j.ejbt.2021.02.002
56. Jesus CF, Alves AAS, Fiuza SM, Murtinho D, Antunes FE (2021) Mini-review: synthetic methods for the production of cationic sugar-based surfactants. J Mol Liq 342:117389. https://doi.org/10.1016/j.molliq.2021.117389
57. Fakruddin M (2012) Biosurfactant: production and application. J Pet Environ Biotechnol 3(4):124. https://doi.org/10.4172/2157-7463.1000124
58. Ambaye TG, Vaccari M, Prasad S, Rtimi S (2021) Preparation, characterization and application of biosurfactant in various industries: a critical review on progress, challenges and perspectives. Environ Technol Innov 24:102090. https://doi.org/10.1016/j.eti.2021.102090
59. Salimi A, Zadeh BSM, Kazemi M (2019) Preparation and optimization of polymeric micelles as an oral drug delivery system for deferoxamine mesylate: *in vitro* and *ex vivo* studies. Res Pharm Sci 14(4):293–307. https://doi.org/10.4103/1735-5362.263554
60. Jassim ZE, Rajab NA (2025) An insight into polymeric micelles preparation methods and applications as drug delivery approach: a review. Int J Pharm Investig 15(3):849–857. https://doi.org/10.5530/ijpi.20250305
61. Kazaryan PS, Chertovich AV, Garvrilov AA (2024) Tuning polymer micelle size and dynamics with solvophoblic block structure. J Colloid Interface Sci 680:496–505. https://doi.org/10.1016/j.jcis.2024.10.186
62. Singh H, Aswal VK (2021) Tuning of micelle adsorption on nanoparticles by combination of surfactants. J Appl Phys 129:234703. https://doi.org/10.1063/5.0054181
63. Shalmani AA, Wang A, Ahmed Z, Sheybanifard M, Mihyar R, Buhl EM, Pohl M, Hennink WE, Kiessling F, Metselaar JM, Shi Y, Lammers T, Pena Q (2024) Tunable polymeric micelles for taxane and corticosteroid co-delivery. Drug Deliv Transl Res 14:2642–2654. https://doi.org/10.1007/s13346-023-01465-x
64. Vinchurkar RH, Kuchekar AB (2021) Polymeric micelles: a novel approach towards nano-drug delivery system. Biosci Biotechnol Res Asia 18(4). https://doi.org/10.13005/bbra/2947
65. Gaucher G, Dufresne MH, Sant VP, Kang N, Maysinger D, Leroux JC (2005) Block copolymer micelles: preparation, characterization and application in drug delivery. J Control Release 109:169–188
66. Kotta S, Aldawsari HM, Badr-Eldin SM, Nair AB, Kamal YT (2022) Progress in polymeric micelles for drug delivery applications. Pharmaceutics 14(8):1636. https://doi.org/10.3390/pharmaceutics14081636
67. Son G-H, Lee B-J, Cho C-W (2017) Mechanisms of drug release from advanced drug formulations such as polymeric-based drug-delivery systems and lipids nanoparticles. J Pharm Investig 47:287–296. https://doi.org/10.1007/s40005-017-0320-1
68. Saraf P, Li X, Wrischnik L, Jasti B (2015) *In vitro* and *in vivo* efficacy of self-assembling RGD peptide amphiphiles for targeted delivery of paclitaxel. Pharm Res 32:3087–3101
69. Shi Y, Lammers T, Storm G, Hennink WE (2016) Physico-chemical strategies to enhance stability and drug retention of polymeric micelles for tumor-targeted drug delivery. Macromol Biosci 17(1):1600160. https://doi.org/10.1002/mabi.201600160

70. Lu Y, Zhang E, Yang J, Cao Z (2018) Strategies to improve micelles stability for drug delivery. Nano Res 11:4985–4998. https://doi.org/10.1007/s12274-018-2152-3
71. Wang Y, Yang T, Wang X, Wang J, Zhang X, Zhang Q (2010) Targeted polymeric micelle system for delivery of combretastatin A4 to tumor vasculature *in vitro*. Pharm Res 27:1861–1868
72. Wu Y (2022) Application of polymer micelles in medical field. Highl Sci Eng Technol 26:328–334. https://doi.org/10.54097/hset.v26i.3993
73. Zheng Y, Oz Y, Gu Y, Ahamad N, Shariati K, Chevalier J, Kapur D, Annabi N (2024) Rational design of polymeric micelles for targeted therapeutic delivery. NanoToday 55:102147. https://doi.org/10.1016/j.nantod.2024.102147
74. Kaur P, Rajput A, Singh D, Singh G, Mehra A, Kaur S, Bedi N, Arora S (2024) The sojourn of polymeric micelles for effective brain drug delivery system. Curr Drug Deliv 21(1):65–79. https://doi.org/10.2174/1567201820666230413082352
75. Qian Y, Hu X, Wang J, Li Y, Liu Y, Xie L (2023) Polyzwitterionic micelles with antimicrobial-conjugation for eradication of drug-resistant bacterial biofilms. Colloids Surf B Biointerfaces 231:113542. https://doi.org/10.1016/j.colsurfb.2023.113542
76. Onat B, Bütün V, Banerjee S, Erel-Goktepe I (2016) Bacterial antiadhesive and pH-induced antibacterial agent releasing ultra-thin films of zwitterionic copolymer micelles. Acta Biomater 40:293–309. https://doi.org/10.1016/j.actbio.2016.04.033
77. Huang YQ, Li YF, Liu Y, Shi LQ (2024) Triclosan-conjugated, lipase-responsive polymeric micelles for eradication of Staphylococcal biofilms. Chin J Polym Sci 42:718–728. https://doi.org/10.1007/s10118-024-3094-8
78. Yu Y, He S, Kong L, Shi N, Liu Y, Zang J, Guo R, Zhang L, Li X, Li X (2023) Brain-targeted multifunctional micelles delivering Oridonin and Phillyrin for synergistic therapy of Alzheimer's disease. J Drug Deliv Sci Technol 87:104794. https://doi.org/10.1016/j.jddst.2023.104794
79. Kanojiya PS, Charde Y, Wadertwar RN (2022) Solid dispersion of Artemether in fast disintegrating tablet to enhance dissolution rate and oral bioavailability. Indian J Pharm Educ Res 56:153–165. https://doi.org/10.5530/ijper.56.1.18
80. Shamim R, Shafique S, Hussain K, Abbas N, Bukhari NI (2024) Surfactant-assisted wet granulation-based matrix tablets without exceptional additives: prolonging systemic exposure of model BCS class II Ketoprofen. AAPS PharmSciTech 25:241. https://doi.org/10.1208/s12249-024-02966-9
81. Trivedi S, Thool S, Wadher K, Bhalekar M, Bire P (2021) Self-assembling Dioscorea bulbifera loaded mixed micelles: formulation optimization, in-vitro cytotoxicity and in-vivo pharmacokinetics. J Drug Deliv Sci Technol 65:102722. https://doi.org/10.1016/j.jddst.2021.102722
82. Koromili M, Kapourani A, Barmpalexis P (2023) Preparation and evaluation of amorphous solid dispersions for enhancing Luteolin's solubility in simulated saliva. Polymers 15:169. https://doi.org/10.3390/polym15010169
83. Shin YB, Choi JY, Yoon MS, Yoo MK, Shin DH, Lee JW (2024) Evaluation of anticancer efficacy of D-α-Tocopheryl polyethylene-glycol succinate and Soluplus® mixed micelles loaded with Olaparib and Rapamycin against ovarian cancer. Int J Nanomed 19:7871–7893. https://doi.org/10.2147/IJN.S468935
84. Kendre P, Gite M, Jain S (2024) Solubility enhancement of β-ionone with lipidic, amphiphilic, and inclusion complex: extensive polymeric biomaterials to develop formulations for poorly soluble drugs. Polym Bull 81:14479–14498. https://doi.org/10.1007/s00289-024-05382-y
85. Twal S, Jaber N, Al-Remawi M, Hamad I, Al-Akayleh F, Alshaer W (2024) Dual stimuli-responsive polymeric nanoparticles combining soluplus and chitosan for enhanced breast cancer targeting. RSC Adv 14:3070. https://doi.org/10.1039/D3RA08074A
86. Mazzotta E, Chieffallo M, Muzzalupo R, Spingola M, Caputo P, Romeo M, Loele G (2024) Formulation of polymeric micelles to increase the solubility and photostability of caffeic acid. Molecules 29(14):3329. https://doi.org/10.3390/molecules29143329

87. Pinho E, Soares G, Henriques M (2015) Evaluation of antibacterial activity of caffeic acid encapsulated by β-cyclodextrins. J Microencapsul 32:804–810
88. Shankar G, Agrawal YK (2015) Formulation of and evaluation of polymeric micelles for a poorly absorbed drug. Res J Pharm Biol Chem Sci 6(3):1314–1321
89. Althobaiti AA, Ashour EA, Almutairi M, Almotairy A, Al YM, Repka MA (2022) Formulation development of curcumin-piperine solid dispersion via hot-melt extrusion. J Drug Deliv Sci Technol 76:103753. https://doi.org/10.1016/j.jddst.2022.103753
90. Gaikwad NM, Chaudhari PD, Shaikh KS, Chaudhari SY, Pathare SS, Saikh AA, Aljarba NH, Kumer A, Dhara B (2024) Dual drug-loaded polymeric mixed micelles for ovarian cancer: approach to enhanced therapeutic efficacy of albendazole and paclitaxel. J Cell Mol Med 28:e18389. https://doi.org/10.1111/jcmm.18389
91. Shi F, Chen L, Wang YP, Liu J, Adu-Frimpong M, Ji H, Toreniyazov E, Wang QL, Yu JN, Xu XM (2021) Enhancement of oral bioavailability and anti-hyperuricemic activity of aloe emodin via novel Soluplus®-glycyrrhizic acid mixed micelle system. Drug Deliv Transl Res 12:603–614. https://doi.org/10.1007/s13346-021-00969-8
92. Dirany Z, González-Benito J, Ginatta P, Nguewa P, González-Gaitano G (2025) Solution blow spun polymeric nanofibres embedding cyclodextrin complexes of miltefosine: an approach to the production of sprayable dressings for the treatment of cutaneous leishmaniasis. Carbohydr Polym 353:123173. https://doi.org/10.1016/j.carbpol.2024.123173
93. Wang HX, Fan YZ, Qin L, Cheng ZP, Chen XQ, Zeng SM, Liang SH, Tong Y, Tao Z, Liu YC, Ning X, Chen P, Zhao S, Chen X (2022) Preparation of Decoquinate solid dispersion by hot-melt extrusion as an oral dosage form targeting liver-stage plasmodium infection. Antimicrob Agents Chemother 66:e02218-e2221. https://doi.org/10.1128/aac.02218-21
94. Molina V, von Plessing C, Romero A, Benavides S, Troncoso JM, Pérez-Correa JR, Franco W (2022) Determination of the dissolution/permeation and apparent solubility for microencapsulated emamectin benzoate using *in vitro* and *ex vivo* salmo salar intestine membranes. Pharmaceuticals 15:652. https://doi.org/10.3390/ph15060652
95. Takale NR, Aji A, Jane K, Deshmukh PR, Pendharkar VV, Khade RR, Ghule BV, Inamdar NN, Kotagale NR (2022) Lumefantrine solid dispersions with piperine for the enhancement of solubility, bioavailability and anti-parasite activity. Int J Pharm 628:122354. https://doi.org/10.1016/j.ijpharm.2022.122354
96. Sipos B, Földes F, Budai-Szucs M, Katona G, Csóka I (2024) Comparative study of TPGS and soluplus polymeric micelles embedded in poloxamer 407 *in situ* gels for intranasal administration. Gels 10:521. https://doi.org/10.3390/gels10080521
97. Alanazi SA, Imam SS, Alshehri S, Alanazi MM, Yassin AEB, Mahdi WA, Alzhrani RF, Al-Agamy MH, Ghoneim MM (2024) Optimized freeze-dried solid-dispersions boost naringin solubility, antibacterial and in-vitro anticancer activity. J Drug Deliv Sci Technol 102:106335. https://doi.org/10.1016/j.jddst.2024.106335
98. Fathi AM, Eissa RG, Balata GF, Ghazy FES, Eissa NG (2023) Intranasal thermosensitive hydrogel of agomelatine solid dispersion for better management of depression. J Drug Deliv Sci Technol 88:104974. https://doi.org/10.1016/j.jddst.2023.104974
99. Zhang Y, Huang Y, Li S (2014) Polymeric micelles: nanocarriers for cancer-targeted drug delivery. AAPS PharmSciTech 15(4):862–871. https://doi.org/10.1208/s12249-014-z
100. Hou Z, Zhou W, Guo X, Zhong R, Wang A, Li J, Cen Y, You C, Tan H, Tian M (2022) Poly(ε-caprolactone)-methoxypolyethylene glycol (PCL-MPEG)-based micelles for drug-delivery: the effect of PCL chain length on blood components, phagocytosis, and biodistribution. Int J Nanomed 17:1613–1632. https://doi.org/10.2147/IJN.S349516
101. Prencipe F, Diaferia C, Rossi F, Ronga L, Tesauro D (2021) Forward precision medicine: micelles for active targeting delivery peptides. Molecules 26(13):4049. https://doi.org/10.3390/molecules26134049
102. Lu Y, Zhang E, Yang J, Cao Z (2019) Strategies to improve micelle stability for drug delivery. Nano Res 11(10):4985–4998. https://doi.org/10.1007/s12274-018-2152-3
103. Yokoyama M (2010) Polymeric micelles as a new drug carrier system and their required considerations for clinical trials. Expert Opin Drug Deliv 7(2):145–158. https://doi.org/10.1517/17425240903436479

104. Li Q, Qian X, Li HY, Lai KL, Gao Q, Lee WYT (2021) Safety assessment of polymeric micelles as an ophthalmic drug delivery system for intravitreal administration of Dasatinib. Int J Pharm 596:120226. https://doi.org/10.1016/j.ijpharm.2021.120226
105. Gavit LP, Yadav GS, Jain AS, More HR, Pukale PV, Patil PS (2024) Polymeric micelles for targeted drug delivery: recent advances and challenges. Indian J Drug Deliv 16(4):199–211
106. Owen SC, Chan DPY, Shoichet MS (2012) Polymeric micelle stability. Nanotoday 7(1):53–65. https://doi.org/10.1016/j.nantod.2012.01.002
107. Negut I, Bita B (2023) Polymeric micellar systems—a special emphasis on "smart" drug delivery. Pharmaceutics 15(3):976. https://doi.org/10.3390/pharmaceutics15030976
108. Cavalcante CH, Fernandes RS, Silva JD, Oda CMR, Leite EA, Cassali GD, Charlie-Silva I, Fernandes BHV, Ferreira LAM, De Barros ALB (2021) Doxorubicin-loaded pH-sensitive micelles: a promising alternative to enhance antitumor activity and reduce toxicity. Biomed Pharmacother 134:111076. https://doi.org/10.1016/j.biopha.2020.111076
109. Tang Y, Wang X, Li J, Nie Y, Liao G, Yu Y, Li C (2019) Overcoming the reticuloendothelial system barrier to drug delivery with a "don't-eat-us" strategy. ACS Nano 13(11):13015–13026. https://doi.org/10.1021/acsnano.9b05679

[illegible]

Fundamental Study of Amphiphilic Ionic Liquids as New Surfactants and Their Applications

Brinda and Lavanya Mulky

Abstract Amphiphilic ionic liquids (AImILs) have gained interest for functioning as both surfactants and functional materials. Their hydrophilic-lipophilic balance, structural tunability, and surface activity make them ideal for applications requiring controlled surface interactions. Among various applications, their role as corrosion inhibitors is particularly noteworthy. Their ability to form micelles and interfacial adsorption capabilities play a crucial role in corrosion control. By effectively adsorbing onto metals, AImILs create strong films that impede the electrochemical reactions responsible for corrosion. The environmentally benign nature and high efficacy of AImILs position them as promising candidates for sustainable corrosion control in industrial settings. This book chapter highlights the molecular structure, physicochemical properties, and mechanistic aspects of corrosion inhibition by amphiphilic ionic liquids, emphasizing their role in multiple sectors.

Keywords Amphiphilic ionic liquids · Surface-active ionic liquids · Corrosion · Industry · Corrosion inhibition

1 Introduction

Petrochemical-derived surfactants are among the most prevalent compounds in everyday life. These molecules are recognized for their ability to reduce surface and interfacial tension between two liquids. Their amphiphilic molecular structure makes surfactants indispensable in industries such as detergents, cosmetics, food, and pharmaceuticals [1]. Their amphiphilic structure enables functionalities such as wetting, suspension, foaming, emulsification, and lubrication [2]. Owing to their

Brinda · L. Mulky (✉)
Manipal Institute of Technology, Manipal Academy of Higher Education, Manipal 576104, Karnataka, India
e-mail: lavanya.m@manipal.edu

M. Abdul Rub et al. (eds.), *Industrial Applications of Individual and Mixed Surfactant Systems—Volume 2*, Sustainable Materials and Technology,
https://doi.org/10.1007/978-981-95-9195-4_2

versatility, they are essential in various industrial sectors, like petroleum, pharmaceuticals, detergents, firefighting foams, inks, paints, printing, and medical technologies [3]. Evolving from a basic Babylonian soap made of ash and oil, surfactants have developed a wide array of contemporary cleaning products [4]. However, their efficiency is frequently compromised by environmental factors. The presence of cations in aqueous environments often has a significant impact on their efficiency [1]. Over the past decades, researchers have made significant efforts towards designing amphiphilic molecules that unite hydrophilic and hydrophobic parts within one molecule. Reducing interfacial tension (IFT) has persuaded researchers in the development of environmentally friendly and stable surfactants. In this regard, amphiphilic ionic liquids have garnered significant interest due to their excellent amphiphilic characteristics and advantages such as recyclability and strong performance in extreme conditions. Surface-active ionic liquids or amphiphilic ionic liquids represent a new class of amphiphiles characterized by long alkyl chains. They have been designed to exhibit strong adsorption and significant aggregation in aqueous solutions [5, 6]. Amphiphilic ionic liquids have emerged as a promising green replacement for surfactants, particularly in petroleum applications such as demulsifiers and corrosion inhibitors [7]. Properties such as surface wetting and adsorption enable these molecules to form micelles. This process helps to reduce interfacial tension, making them highly effective in electrochemical applications like metal corrosion inhibition [8].

2 Properties and Structure of Amphiphilic Ionic Liquids

Amphiphilic Ionic Liquids (AImILs) are a class of ionic liquids characterized by a molecular structure with a hydrophilic head group and a hydrophobic tail, often composed of long alkyl or lipophilic chains [9, 10]. Figure 1 represents the general structure of amphiphilic ionic liquids.

Therefore, due to their molecular configuration and unique properties, they are reported to show corrosion-resistant behavior on copper, mild steel, and aluminum [11]. The cationic head group of amphiphilic ionic liquid surfactants, such as imidazolium, pyrrolidinium, pyridinium, or morpholinium, plays a crucial role in surface interactions. It promotes the adsorption of surfactants onto metallic surfaces and leads to the formation of a protective inhibiting film. For example, imidazole-based surface-active ionic liquids (SAILs) contain ionizable electrons on the imidazole ring

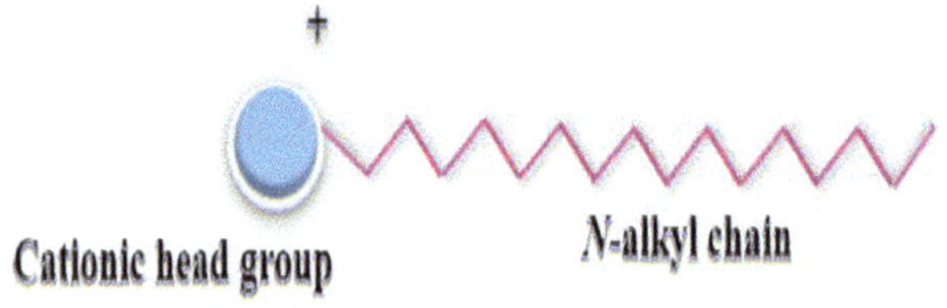

Fig. 1 The general structure of amphiphilic ionic liquids

and exhibit hydrogen bonding interactions between cations. As a result, they show notable surface activity and thermal stability, which is beneficial for corrosion inhibition [12]. Figure 2 represents a representative chemical structure of amphiphilic ionic liquid surfactants.

The longer alkyl chain length enhances hydrophobic interactions and promotes the development of a compact film on a metallic surface. However, the effect of hydrophobic chain lengths on anticorrosion has certain limitations. Research has demonstrated that the corrosion inhibition efficiency improves with an increase in alkyl chain length from C_1 to C_{10}/C_{12}, but it decreases beyond C_{12} [13]. The properties of amphiphilic ionic liquids, which are emerging as new surfactants, are illustrated in Fig. 3. The use of SAILs aligns with the principles of green chemistry and environmental sustainability.

Structurally, SAILs are classified into three types: cationic (such as imidazolium or pyridinium cations), anionic (including sulphate, sulfosuccinate, or sulfonate anions), and catanionic SAILs, which contain both surface-active anions and cations. Anionic SAILs are prepared from the metathesis of anionic surfactants (such as sodium dodecyl sulphate, etc.) and ordinary imidazolium ionic liquids (ILs), such

Fig. 2 The $[C_{10}MiM][I]$ illustrates a representative chemical structure of amphiphilic ionic liquids

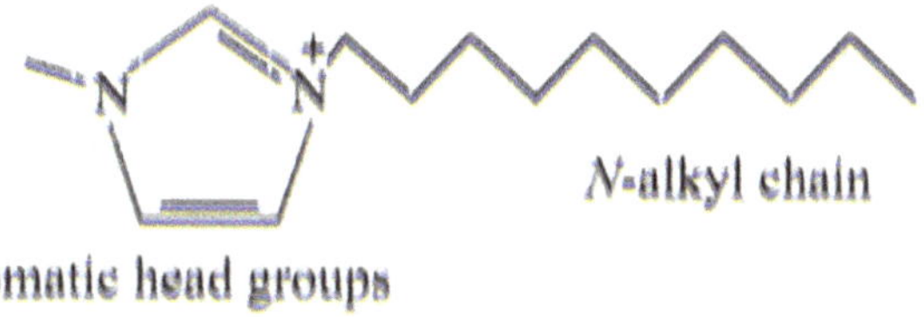

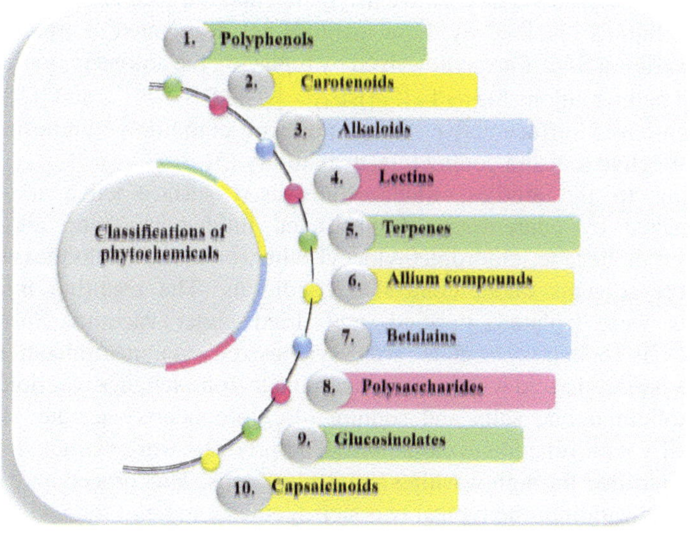

Fig. 3 Various properties of amphiphilic ionic liquids

as 1-ethyl-3-methylimidazolium chloride or 1-butyl-3-methylimidazolium chloride, which have been applied in many fields [14]. Besides common cationic SAILs with positively charged groups, other forms stand out due to unique properties. Gemini type of $SAIL_S$ has the benefits of both Gemini surfactants and SAILs. A distinct subclass of cationic surface-active ionic liquids is di-cationic ionic liquids, characterized by two charged head groups and two hydrophobic hydrocarbon chains. These compounds exhibit enhanced surface activity compared to their monocationic counterparts, which have similar polar and non-polar moieties. Structural modification of the head group can facilitate the formation of zwitterionic surfactants by repositioning the counterion to the side chain, resulting in amphiphiles with neutral net charge and enhanced surface properties [15, 16].

3 Synthesis of Amphiphilic Ionic Liquids

The first comprehensive study focused on designing green surface-active ionic liquids (SAILs) through a 'benign-by-design' strategy. The synthesis of amphiphilic ionic liquids is performed by two steps involving SN_2 quaternization of amines or phosphines with alkyl halides or alkyl sulphates, followed by an anion exchange step if required. The process of quaternization is initiated by the string and mixing amine or phosphine with alkyl halides. Improved reaction performance is achieved with shorter alkyl chains and bulkier ions. The interfacial activity of alkyl halides decreases in the order: 1-iodoalkanes > 1-bromoalkanes > 1-chloroalkanes [17]. The reaction is often performed under reflux conditions in a suitable solvent such as acetonitrile for 2–3 days. In contrast, solvent-free synthesis of amphiphilic ionic liquids can be achieved via microwave irradiation and ultrasound methods. Surface-active ionic liquids obtained by these methods are designated as first-generation. Second-generation SAILs are synthesized by replacing the anions in first-generation SAILs with bulkier anions like $[PF_6]^-$, $[BF_4]^-$, $\left[(CF_3SO_2)_2N\right]^-$, and $[C_6H_5CO_2]^-$. Imidazolium-based surface-active ionic liquids are commonly synthesized due to their cost-effectiveness and commercial availability [5].

According to the studies, imidazolium-based surface-active ionic liquids are synthesized by quaternizing 1-substituted imidazoles using alkyl halides such as 1-bromodecane, chloromethyldecyl ether, or chloromethylhexyl sulphide with suitable solvents under controlled conditions. The resulting imidazolium halide salts were isolated, purified, and dried under vacuum. Subsequently, sodium (4-chlorophenoxy)acetate was synthesized via neutralization of (4-chlorophenoxy)acetic acid with sodium bicarbonate. Ion exchange reactions between the imidazolium halide salts and sodium (4-chlorophenoxy)acetate yielded the desired SAILs with functionalized anions. The products were extracted with chloroform and purified through washing and drying steps. This procedure exemplifies a practical realization of the typical two-step synthesis strategy for SAILs [18].

4 Comparative Assessment of Amphiphilic Ionic Liquids with Conventional Surfactants

Like conventional surfactants, ionic liquids can be used as amphiphilic ionic liquids due to their amphiphilic nature. Limitations such as solubility, stability of aggregates, and biodegradability hinder the performance of conventional surfactants [19]. Advantages of Surface-Active Ionic Liquids (SAILs) over conventional surfactants

- The capability to create customized SAILS with specific physicochemical properties.
- An eco-friendly, sustainable alternative to toxic and non-degradable conventional surfactants.
- Neutralization enables quick and easy synthesis.
- Exhibits broad miscibility and solubility characteristics.
- It is user-friendly because of its low melting point.
- Generally, exhibit enhanced physicochemical properties.
- Their strong cohesion facilitates the formation of stable micelles without cosurfactants.

5 Industrial Applications of Amphiphilic Ionic Liquids as Corrosion Inhibitors

The degradation of metals via chemical and electrochemical processes in environmental conditions is referred to as corrosion. In 1978, the U.S. reported corrosion-related losses of $70 billion, representing nearly 5% of its GNP. Various countries, including Australia, the United Kingdom, Japan, Germany, Kuwait, Finland, India, China, and Sweden, have carried out corrosion-related expenses. The studies reveal a significant economic impact of corrosion on industrial nations [20]. Metal degradation is responsible for 3–4% of Gross Domestic Product (GDP) loss in industrialized countries [21]. The global cost of corrosion estimated by NACE is around 2.5 trillion USD, equivalent to 3.4% of the GDP. Effective prevention strategies could reduce this cost by 15–35% [22]. Corrosion affects vital sectors, including aviation, maritime, and municipal water distribution systems [23]. For example, since the 1970s, corrosion has caused multiple failures in nuclear power plants, resulting in significant financial losses. In recent decades, degradation of polymers and composites in seawater has been recognized as a form of corrosion. However, in marine environments, aqueous corrosion of metals is an electrochemical process resulting from interactions between the metal surface and the components of saline water [24]. In marine environments, the petrochemical sector is one of the most vulnerable sectors to corrosion. The presence of water, chloride ions, oxygen, and bacteria initiates degradation of the metallic surface. Metal degradation is a critical issue in offshore wind power plants, as 89% of their components are metallic [25]. In the

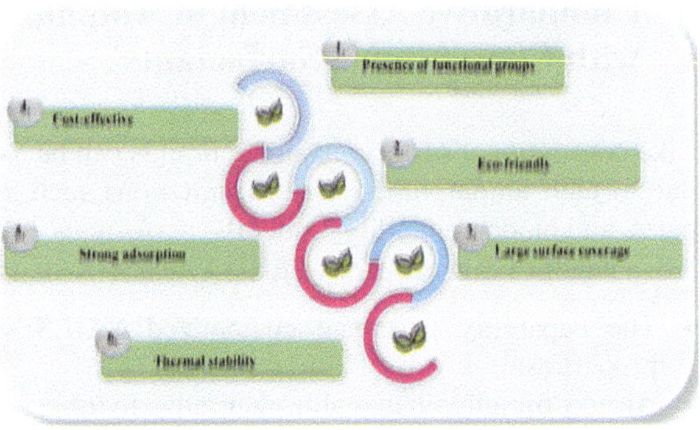

Fig. 4 General requirements for selecting environmentally sustainable inhibitors

petroleum industry, acidic solutions are extensively employed for cleaning and acidifying oil wells. Among various acids, hydrochloric acid (HCl) is widely used for pickling metals and alloys. In such corrosive environments, managing and inhibiting metallic corrosion presents a complex challenge. Therefore, injecting an appropriate corrosion inhibitor is an effective method to protect these oil well tubular materials [26]. General requirements for selecting environmentally sustainable inhibitors are presented in Fig. 4.

Researchers have discovered numerous novel Gemini ionic liquid surfactants that exhibit effective corrosion inhibition properties for metals in acidic environments. The strong surface adhesion capacity of these SAILs significantly prevented mild steel degradation in the HCl solutions [27]. Researchers have explored the corrosion-mitigating properties of the imidazolium-based surface-active ionic liquid (SAIL), 1-dodecyl-3-methylimidazolium bromide ($C_{12}MImBr$), on mild steel (MS) in an acidic environment (1 M HCl). Electrochemical techniques and weight loss measurement data demonstrated that, from pre- to post-critical micelle concentration (CMC) levels, $C_{12}MImBr$ adsorbed onto the steel surface. Their absorption formed a protective film and increased the inhibition efficiency to approximately 96%. Scanning electron microscopy (SEM) revealed smoother steel surfaces, while thermodynamic analysis indicated spontaneous Langmuir-type adsorption. Additionally, computational simulations had supported the experimental findings [28]. The anticorrosive performance of carbon steel (CS) in acidic conditions was achieved by increasing the concentration of 1-dodecyl-3-(hydroxymethyl) pyridin-1-ium bromide (DHPB) and temperature. Reduced admittance values, along with Atomic force microscopy (AFM) and scanning electron microscopy analyses, verified the adsorption of DHPB on CS. The adsorption followed a Langmuir isotherm. At lower temperatures, it involved both physisorption and chemisorption mechanisms, but as the temperature rose, the adsorption predominantly shifted to chemisorption [29]. Biocorrosion in oil fields and production is associated with various microbial species [30]. Microbial colonization of industrial metals occurs via biofilm formation, consisting of bacteria, EPS, and water. Biocorrosion and biofouling can lead to microbial contamination, energy loss, reduced efficiency, and structural damage [31]. To mitigate biocorrosion, the biofilm growth must be limited by utilizing suitable inhibitors that align with the operational needs, such as antimicrobial activity, compatibility, cost, and ecological impact [32]. SAILs, due to their structural resemblance to conventional amphiphilic quaternary ammonium compounds (QACs), exhibit significant antibacterial properties. The bioactive performance of these agents is mainly linked to their lipophilic or hydrophobic nature, with the cation primarily responsible for cell wall interactions [33]. In recent years, Imidazolium-based Surface-Active Ionic Liquids (Im-SAILs) have emerged as an effective antimicrobial agent against diverse microbial species. A systematic disk diffusion test against various Gram-positive and Gram-negative strains revealed that 1-dodecyl-3-methylimidazolium tetrafluoroborate exhibited significantly larger inhibition zones compared to its shorter-chain analogs [34]. Therefore, this research provides valuable insights into less toxic and efficient corrosion inhibitors, which can be broadly applied in industries where corrosion poses a significant challenge.

6 Mechanistic Insights into Corrosion Inhibition by Amphiphilic Ionic Liquids

Surfactants formulated from ionic liquids have garnered significant attention as innovative and eco-friendly corrosion inhibitors. Their adsorption on metal surfaces occurs through chemisorption, which obeys the Langmuir adsorption isotherm model [35]. The majority of corrosion inhibitors are amphiphilic in nature, characterized by a polar head group and a non-polar tail structure. Adsorption on a metallic surface is primarily carried out by a polar head group, while a non-polar hydrocarbon tail ensures hydrophobic protection. Heteroatoms (N, O, and S) in the polar groups of corrosion inhibitors contribute π electrons that enable bonding interactions with the d-orbitals of metals [36]. The cationic portion of surface-active ionic liquids usually exhibits an amphiphilic characteristic, similar to that of cationic surfactants. Similar to conventional surfactants, the adsorption of SAILs at the interface leads to a decrease in interfacial tension (IFT). SAILs form micelles similar to traditional surfactants, but their bulky heteroatom head groups enhance hydrophobicity and facilitate accumulation [5, 37]. The adsorption of amphiphilic ionic liquids can significantly alter the corrosion-resisting property of a metal [27]. Amphiphilic ionic liquids inhibit corrosion by enhancing carbon steel's resistance in hydrochloric acid (HCl) solution. Tetra-cationic imidazolium ionic liquids, 2-(4-dodecyldimethylamino) phenyl-1,3-bis (3-dodecyldimethylammnonio) propyl bromide-1-H-imidazol-3-ium acetate (DDAP-ImA) and 2-(4-hexyldimethylamino) phenyl-1,3-bis(3-hexeyldimethylammnonio) propyl bromide-1 H-imidazol-3-ium acetate (HDAP-ImA) was employed as a corrosion inhibitor.

The corrosion inhibitors consist of two types of anions: bromide and acetate. They also comprise two classes of compounds, imidazolium and quaternary ammonium, each featuring three different long alkyl chain lengths. These characteristics contribute to enhanced corrosion inhibition efficiencies even at lower dosages.

The anodic dissolution of iron in an acidic medium is described by the following Eqs. (1)–(5).

$$Fe + nH_2O \leftrightarrow Fe(H_2O)n\,(ads) \quad (1)$$

$$Fe + (H_2O)n\,(ads) + 2Cl^- \leftrightarrow Fe\left[(H_2O)n\,2Cl^-\right] ads \quad (2)$$

$$Fe\left[(H_2O)n\,2Cl^-\right] ads \leftrightarrow Fe[(H_2O)n\,2Cl]\,ads + 2e^- \quad (3)$$

$$Fe[(H_2O)n\,2Cl]\,ads \leftrightarrow Fe[(H_2O)n\,2Cl]^+ + 2e^- \quad (4)$$

$$Fe[(H_2O)n\,2Cl]^+ \leftrightarrow Fe^{2+} + 2Cl^- + H^+ + OH^- \quad (5)$$

The cathodic hydrogen evolution proceeds through the following Eqs. (6)–(8).

$$Fe + H^+ \leftrightarrow \left(FeH^+\right) ads \quad (6)$$

$$\left(FeH^+\right) ads + e^- \rightarrow (FeH)ads \quad (7)$$

$$(FeH)ads + H^+ + e^- \rightarrow Fe + H_2 \quad (8)$$

CH_3COO^-, Br^- and Cl^- ions adsorb and desorb at anodic sites on positively charged metal surfaces through coulombic attraction. In acidic environments, a high concentration of chloride ions leads to the dissolution of Fe to Fe^{2+}. The complex $Fe\left[(H_2O)n\,2Cl^-\right]$ ads interacts with protonated amphiphilic ionic liquids, which are adsorbed through electrostatic interactions with negatively charged metallic surfaces. Protonated AImILs then formed complexes with ($FeCl^-$) ads, creating a monolayer that inhibited corrosion, as shown in the following Eqs. (9)–(10).

$$Fe\left[(H_2O)n\,2Cl^-\right] ads + AImIL^+ \rightarrow Fe\left[(H_2O)n\,2Cl^- AImIL\right]^- ads \quad (9)$$

$$Fe\left[(H_2O)n\,2Cl^- AImIL\right]^- ads + 2ALImIL^+ + 2Cl^- \rightarrow Fe\left[(H_2O)n\,2Cl^- AImIL\right]^- ads + [2AImIL]^+ 2Cl^- \quad (10)$$

The adsorbed layers of $Fe\left[(H_2O)n\,2Cl^- AImIL\right]^- ads + [2AImIL]^+ 2Cl^-$ inhibited Fe Oxidation via electrostatic and Van der Waals forces. DDAP-ImA showed better anodic protection. Protonated amphiphilic ionic liquids adsorbed at the cathodic site reduced the hydrogen evolution reaction. Larger DDAP-ImA cations blocked the adsorption of water and hydronium cations on the electrode surface. With greater steric effects, DDAP-ImA contributed to superior inhibition efficiency than HDAP-ImA [38].

7 Operational Parameters Influencing the Efficiency of Amphiphilic Ionic Liquids in Corrosion Inhibition

The effectiveness of organic corrosion inhibitors is strongly influenced by their adsorption characteristics. The choice of suitable inhibitors largely depends on the specific operating conditions [39]. Nevertheless, the efficiency of organic inhibitors can be impacted by variables such as pH, temperature, and inhibitor concentration [40]. Variations in operating parameters can modify the kinetic energy of inhibitory molecules, thereby affecting their bonding efficiency with the metallic surface. Therefore, a thorough understanding of the relationship between operating conditions and the adsorption behavior of surface-active ionic liquids is crucial for optimizing their corrosion inhibition performance.

Temperature

An increase in inhibition efficiency with elevated temperature has been reported in numerous studies. This trend indicates that the interaction between the metal surface and the inhibitor molecules likely involves chemical bonding [41]. The efficiency of surface-active ionic liquids is strongly influenced by external conditions. Among those, temperature plays a crucial role in reducing their interfacial tension. The impact of IFT on reducing the ability of surface-active ionic liquids has been extensively studied by researchers. Temperature variations can lead to substantial changes in a system's characteristics by altering its entropy and molecular interactions. As the temperature rises, the IFT tends to decrease linearly. Because SAILs diffuse and adsorb more effectively under elevated thermal conditions [42]. Similarly, several researchers have reported that increasing temperature enhances IFT reduction in systems containing various SAILs with head groups including heterocyclic, hydroxyl ammonium, proline, and aromatic types. In contrast, aromatic heterocyclic head groups like imidazolium and pyridinium with alkyl chains longer ($n > 12$) exhibit increased IFT with temperature. Temperature has a critical effect on critical micelle concentration (CMC). Generally, elevated temperatures lead to micelle formation at lower concentrations [37, 43]. The minimized interfacial tension and micelle formation in aggressive media contribute to improved surface wetting and adsorption. These enhanced properties have a positive effect on exposed surfaces. As a result, they play a significant role in inhibiting metal degradation [44]. In contrast, the increased temperature is unfavorable for corrosion inhibition due to intense molecular motion. This results in weaker adsorption of the functional group on the metallic surface [45].

Electrolytes

Changes in the electrolyte layer's thickness impact multiple phenomena, such as oxygen diffusion, corrosion product accumulation, and metal ion hydration. Additionally, electrolytes can significantly affect the interfacial tension (IFT) and critical micelle concentration (CMC) of systems by modifying the adsorption behavior of surfactants at the interface. Electrolytes play a vital role in enhancing the adsorption of surface-active ionic liquids at the interface and their aggregation in the aqueous phase. Thus, this leads to a greater reduction in interfacial tension and critical micelle concentration [46]. This is attributed to the well-known "salting out" [47]. The cations and anions of surface-active ionic liquids and electrolytes capture water molecules. The high surface charge density of electrolyte cations leads to dehydrated SAIL cations. Smaller and harder cations exhibit a stronger effect due to the robust hydrogen bonding of water molecules around them. For example, sodium halides have a greater impact on minimizing IFT than potassium halides. Additionally, the presence of electrolyte anions reduces the repulsion between cationic head groups of surface-active ionic liquids. This results in denser interfacial packing and enhanced IFT reduction. Consequently, the adsorption parameters, involving adsorption tendency and saturated interface excess, significantly increase in the presence of electrolyte [48, 49].

pH

The functionality of surface-active ionic liquids is dependent on pH. Surface-active ionic liquids are known for their pronounced self-aggregation behavior. This enables them to form a wide variety of assemblies. Their aggregation behavior is strongly influenced by pH and their molecular structure [12]. Most of the studies have reported a greater reduction in IFT with increasing pH. The adsorption tendency and its effectiveness are influenced by rising pH. The maximum interfacial excess occurs due to the simultaneous adsorption of hydroxyl ions. The presence of OH^- ions reduces the electrostatic repulsion between SAIL molecules, thereby enhancing their adsorption at the interface. Higher pH levels facilitate micelle formation at lower concentrations. The phenomenon occurs because hydroxyl ions neutralize the positive charges of SAILs at the micelle interface [50]. Nano Gemini surface-active ionic liquids are more stable across a wide pH range. They exhibit increased activity at alkaline pH levels. The improved performance is primarily attributed to the decreased electrostatic repulsion among adsorbed molecules. Furthermore, the inherent interfacial activity of natural surfactants significantly contributes to the reduction in interfacial tension (IFT) [46]. The corrosion-inhibiting efficiency of amphiphilic ionic liquids increases with the rise in pH [51].

8 Future Perspectives and Challenges

The present insight into SAILs equips us to further explore and refine their use in multiple industrial applications. The application of SAILs as corrosion inhibitors in industrial metallic systems presents a promising yet underdeveloped frontier. While research on SAILs is rapidly gaining momentum, their large-scale deployment in corrosion control remains in its infancy, necessitating deeper exploration into their adsorption mechanisms, inhibition efficiency, and surface interaction with different metals under industrially relevant conditions. SAILs offer distinct advantages due to their amphiphilic nature, enabling them to form stable, protective films on metal surfaces, especially in aggressive environments such as those encountered in the oil and gas, chemical processing, and marine sectors.

Future research must focus on designing task-specific SAILs that exhibit strong surface affinity, low desorption under flow conditions, and compatibility with high-salinity or high-temperature systems. Structural modifications of branched alkyl chains, the design of multi-cationic head groups, or the incorporation of redox-responsive moieties can greatly enhance the corrosion inhibition efficiency of SAILs across diverse industrial environments. However, challenges persist in terms of their environmental impact, with particular attention needed towards developing fluorine-free, biodegradable SAILs that do not accumulate or introduce secondary toxicity. Moreover, the synergistic potential of SAILs with traditional inhibitors or polymers offers new pathways for enhancing inhibition efficiency and reducing costs. Evaluating their adsorption behavior on metal pipelines and equipment to minimize retention losses and ensure long-term protection. As SAILs move closer to industrial applicability, considerations of production cost, material availability, and regeneration or recyclability of spent SAILs will become critical for commercialization. Thus, pilot-scale testing, economic feasibility studies, and comparison with conventional corrosion technologies are essential steps moving forward. With advancements in synthesis, structural tuning, and sustainability, SAILs have the potential to revolutionize corrosion protection in high-performance and emerging industrial applications, offering a smart, multifunctional, and environmentally conscious solution for long-term metal preservation.

References

1. Du L, Wang S, Zhang T, Xu C (2025) Molecular dynamics simulation of the effect of calcium ions on the foamability of anionic surfactants. Colloids Surf A Physicochem Eng Asp 707. https://doi.org/10.1016/j.colsurfa.2024.135819
2. Romero Vega G, Gallo Stampino P (2025) Bio-based surfactants and biosurfactants: an overview and main characteristics. Molecules 30. https://doi.org/10.3390/molecules30040863
3. Czajka A, Hazell G, Eastoe J (2015) Surfactants at the design limit. Langmuir 31:8205–8217. https://doi.org/10.1021/acs.langmuir.5b00336
4. Rebello S, Asok AK, Mundayoor S, Jisha MS (2014) Surfactants: toxicity, remediation and green surfactants. Environ Chem Lett 12:275–287. https://doi.org/10.1007/s10311-014-0466-2

5. Kharazi M, Saien J, Asadabadi S (2022) Review on amphiphilic ionic liquids as new surfactants: from fundamentals to applications. Top Curr Chem 380. https://doi.org/10.1007/s41061-021-00362-6
6. Viradiya RA, Aswal VK, Patel N, Patel VK, Solanki JD, Soni SS, Vashi AS, Panjabi SH (2025) Surface-active ionic liquids with morpholinium cations and aromatic counterions: experimental and DFT approach to study their profound effect on micelle formation. Colloid Polym Sci. https://doi.org/10.1007/s00396-024-05362-8
7. Atta AM, Al-Lohedan HA, Ezzat AO, Tawfik AM (n.d.) Preparation and surface properties of amphiphilic green magnetite nanoparticles capped with branched ionic liquid
8. Yousefi A, Javadian S, Dalir N, Kakemam J, Akbari J (2015) Imidazolium-based ionic liquids as modulators of corrosion inhibition of SDS on mild steel in hydrochloric acid solutions: experimental and theoretical studies. RSC Adv 5:11697–11713. https://doi.org/10.1039/c4ra10995c
9. Kumar S, Scheidt HA, Kaur N, Kang TS, Gahlay GK, Huster D, Mithu VS (2019) Effect of the alkyl chain length of amphiphilic ionic liquids on the structure and dynamics of model lipid membranes. Langmuir 35:12215–12223. https://doi.org/10.1021/acs.langmuir.9b02128
10. Roy R, York E, Pacchini E, Rawling T (2024) Effects of cationic head group structure on cytotoxicity and mitochondrial actions of amphiphilic ionic liquids. Food Chem Toxicol 183. https://doi.org/10.1016/j.fct.2023.114202
11. Taghavikish M, Dutta NK, Choudhury NR (2017) Emerging corrosion inhibitors for interfacial coating. Coatings 7. https://doi.org/10.3390/coatings7120217
12. Cao H, Hu Y, Xu W, Wang Y, Guo X (2020) Recent progress in the assembly behavior of imidazolium-based ionic liquid surfactants. J Mol Liq 319. https://doi.org/10.1016/j.molliq.2020.114354
13. Xiong Y, Cao M (2024) Application of surfactants in corrosion inhibition of metals. Curr Opin Colloid Interface Sci 73. https://doi.org/10.1016/j.cocis.2024.101830
14. Hu K, Zhang H, Kong M, Qin M, Ouyang M, Jiang Q, Wang G, Zhuang L (2021) Effect of alkyl chain length of imidazolium cations on foam properties of anionic surface active ionic liquids: experimental and DFT studies. J Mol Liq 340. https://doi.org/10.1016/j.molliq.2021.117197
15. Buettner CS, Cognigni A, Schröder C, Bica-Schröder K (2022) Surface-active ionic liquids: a review. J Mol Liq 347. https://doi.org/10.1016/j.molliq.2021.118160
16. Zhou H, Liang Y, Huang P, Liang T, Wu H, Lian P, Leng X, Jia C, Zhu Y, Jia H (2018) Systematic investigation of ionic liquid-type Gemini surfactants and their abnormal salt effects on the interfacial tension of a water/model oil system. J Mol Liq 249:33–39. https://doi.org/10.1016/j.molliq.2017.11.004
17. El Seoud OA, Keppeler N, Malek NI, Galgano PD (2021) Ionic liquid-based surfactants: recent advances in their syntheses, solution properties, and applications. Polymers (Basel) 13. https://doi.org/10.3390/polym13071100
18. Wojcieszak M, Syguda A, Zięba S, Mizera A, Łapiński A, Materna K (2023) Effect of surface-active ionic liquids structure on their synthesis, physicochemical properties, and potential use as crop protection agents. J Mol Liq 383. https://doi.org/10.1016/j.molliq.2023.122050
19. Kaur G, Kumar H, Singla M (2022) Diverse applications of ionic liquids: a comprehensive review. J Mol Liq 351. https://doi.org/10.1016/j.molliq.2022.118556
20. Thompson NG, Yunovich M, Dunmire D (n.d.) Cost of corrosion and corrosion maintenance strategies
21. Fayomi OSI, Akande IG, Odigie S (2019) Economic impact of corrosion in oil sectors and prevention: an overview. J Phys Conf Ser. Institute of Physics Publishing. https://doi.org/10.1088/1742-6596/1378/2/022037
22. Kumar P, Holmberg K, Soni I, Islam N, Kumar M, Shandilya P, Sillanpää M, Chauhan V (2024) Advancements in ionic liquid-based corrosion inhibitors for sustainable protection strategies: from experimental to computational insights. Adv Colloid Interface Sci 333. https://doi.org/10.1016/j.cis.2024.103303

23. Liu P, Zhang H, Fan Y, Xu D (2023) Microbially influenced corrosion of steel in marine environments: a review from mechanisms to prevention. Microorganisms 11. https://doi.org/10.3390/microorganisms11092299
24. Valdez B, Ramirez J, Eliezer A, Schorr M, Ramos R, Salinas R (2016) Corrosion assessment of infrastructure assets in coastal seas. J Mar Eng Technol 15:124–134. https://doi.org/10.1080/20464177.2016.1247635
25. Dalmora GPV, Borges Filho EP, Maraschin Conterato AA, Roso WS, Pereira CE, Dettmer A (2025) Methods of corrosion prevention for steel in marine environments: a review. Results Surf Interfaces 18. https://doi.org/10.1016/j.rsurfi.2025.100430
26. Migahed MA, Nassar IF (2008) Corrosion inhibition of tubing steel during acidization of oil and gas wells. Electrochim Acta 53:2877–2882. https://doi.org/10.1016/j.electacta.2007.10.070
27. Tawfik SM (2016) Ionic liquids based Gemini cationic surfactants as corrosion inhibitors for carbon steel in hydrochloric acid solution. J Mol Liq 216:624–635. https://doi.org/10.1016/j.molliq.2016.01.066
28. Patel D, Makwana K, Shirdhonkar MB, Kuperkar KC (2020) Electrochemical response and computational approach on surface-active ionic liquid (SAIL) in metal corrosion inhibition. Emerg Mater 3:161–168. https://doi.org/10.1007/s42247-020-00099-w
29. Shalabi K, Helmy AM, El-Askalany AH, Shahba MM (2019) New pyridinium bromide mono-cationic surfactant as corrosion inhibitor for carbon steel during chemical cleaning: experimental and theoretical studies. J Mol Liq 293. https://doi.org/10.1016/j.molliq.2019.111480
30. Avelino-Jiménez IA, Hernández-Maya L, Larios-Serrato V, Quej-Ake L, Castelán-Sánchez H, Herrera-Díaz J, Garibay-Febles V, Rivera-Olvera JN, Zavala-Olivares G, Zapata-Peñasco I (2023) Biofouling and biocorrosion by microbiota from a marine oil pipeline: a metagenomic and proteomic approach. J Environ Chem Eng 11. https://doi.org/10.1016/j.jece.2023.109413
31. Videla A (2002) Prevention and control of biocorrosion. www.elsevier.com/locate/ibiod
32. Parthipan P, Elumalai P, Narenkumar J, Machuca LL, Murugan K, Karthikeyan OP, Rajasekar A (2018) *Allium sativum* (garlic extract) as a green corrosion inhibitor with biocidal properties for the control of MIC in carbon steel and stainless steel in oilfield environments. Int Biodeterior Biodegrad 132:66–73. https://doi.org/10.1016/j.ibiod.2018.05.005
33. Lo Schiavo S, De Leo F, Urzì C (2020) Present and future perspectives for biocides and antifouling products for stone-built cultural heritage: ionic liquids as a challenging alternative. Appl Sci (Switzerland) 10. https://doi.org/10.3390/APP10186568
34. Sivapragasam M, Shue Yee M, Bawa-Allah K (2023) Elucidation of the toxicity of imidazolium-based surface-active ionic liquids (Im-SAILs) against gram-positive and gram-negative bacteria. www.iscientific.org/Journal.html
35. Verma C, Ebenso EE, Quraishi MA (2017) Ionic liquids as green and sustainable corrosion inhibitors for metals and alloys: an overview. J Mol Liq 233:403–414. https://doi.org/10.1016/j.molliq.2017.02.111
36. Hernández-Bravo R, Miranda AD, Parra JG, Alvarado-Orozco JM, Domínguez-Esquivel JM, Mujica V (2022) Experimental and theoretical study on the effectiveness of ionic liquids as corrosion inhibitors. Comput Theor Chem 1210. https://doi.org/10.1016/j.comptc.2022.113640
37. Łuczak J, Hupka J, Thöming J, Jungnickel C (2008) Self-organization of imidazolium ionic liquids in aqueous solution. Colloids Surf A Physicochem Eng Asp 329:125–133. https://doi.org/10.1016/j.colsurfa.2008.07.012
38. Ghiaty EA, Shafek SH, Atta AM (2024) Morphological performances and corrosion inhibition mechanism of new amphiphilic tetra-cationic imidazolium ionic liquids for carbon steel in 1 M HCl medium. J Mol Struct 1300. https://doi.org/10.1016/j.molstruc.2023.137252
39. Puthalath R, Surendranathan AO, Rajeev P, Murthy CSN (2012) Corrosion mitigation of the oil well steels using organic inhibitors—a review. J Mater Environ Sci 3:856–869. https://www.researchgate.net/publication/288116318
40. Aydinsoy EA, Aghamaliyev ZZ, Aghamaliyeva DB, Abbasov VB (2024) A systematic review of corrosion inhibitors in marine environments: insights from the last 5 years. Process Petrochem Oil Refin 793. https://doi.org/10.62972/1726-4685.2024.3.793

41. Verma C, Ebenso EE, Quraishi MA, Hussain CM (2021) Recent developments in sustainable corrosion inhibitors: design, performance and industrial scale applications. Mater Adv 2:3806–3850. https://doi.org/10.1039/d0ma00681e
42. Nazar M, Ahmad A, Hussain SMS, Sulaimon AA, Moniruzzaman M (2025) Green solution for oil spills: a review on the role of surface-active ionic liquids. Adv Colloid Interface Sci 336. https://doi.org/10.1016/j.cis.2024.103362
43. Zhou H, Zhu Y, Peng T, Song Y, An J, Leng X, Yi Z, Sun Y, Jia H (2016) Systematic study of the effects of novel halogen-free anionic surface active ionic liquid on interfacial tension of water/model oil system. J Mol Liq 223:516–520. https://doi.org/10.1016/j.molliq.2016.08.080
44. Likhanova NV, Domínguez-Aguilar MA, Olivares-Xometl O, Nava-Entzana N, Arce E, Dorantes H (2010) The effect of ionic liquids with imidazolium and pyridinium cations on the corrosion inhibition of mild steel in acidic environment. Corros Sci 52:2088–2097. https://doi.org/10.1016/j.corsci.2010.02.030
45. Su W, Tang B, Fu F, Huang S, Zhao S, Bin L, Ding J, Chen C (2014) A new insight into resource recovery of excess sewage sludge: feasibility of extracting mixed amino acids as an environment-friendly corrosion inhibitor for industrial pickling. J Hazard Mater 279:38–45. https://doi.org/10.1016/j.jhazmat.2014.06.053
46. Kharazi M, Saien J (2022) Mechanism responsible altering in interfacial tension and emulsification of the crude oil-water system with nano Gemini surface active ionic liquids, salts and pH. J Pet Sci Eng 219. https://doi.org/10.1016/j.petrol.2022.111090
47. Yano YF, Uruga T, Tanida H, Terada Y, Yamada H (2011) Protein salting out observed at an air-water interface. J Phys Chem Lett 2:995–999. https://doi.org/10.1021/jz200111q
48. Howard SC, Atkin R, Craig VSJ (2009) Effect of electrolyte species on the adsorption of a cationic surfactant to silica: the common intersection point. Colloids Surf A Physicochem Eng Asp 347:109–113. https://doi.org/10.1016/j.colsurfa.2009.01.014
49. Zhao Y, Yue X, Wang X, Huang D, Chen X (2012) Micelle formation by N-alkyl-N-methylpiperidinium bromide ionic liquids in aqueous solution. Colloids Surf A Physicochem Eng Asp 412:90–95. https://doi.org/10.1016/j.colsurfa.2012.07.021
50. Asadabadi S, Saien J (2016) Effects of pH and salinity on adsorption of different imidazolium ionic liquids at the interface of oil-water. Colloids Surf A Physicochem Eng Asp 489:36–45. https://doi.org/10.1016/j.colsurfa.2015.10.021
51. Otmacic Curkovic H, Stupnisek-Lisac E, Takenouti H (2010) The influence of pH value on the efficiency of imidazole based corrosion inhibitors of copper. Corros Sci 52:398–405. https://doi.org/10.1016/j.corsci.2009.09.026

Application of Biosurfactants in the Preparation of Sustainable Agrochemicals

Urjita Sheth, Jagruti Barot, Chirag B. Mistry, Paresh Parekh, Vijay Patel, and Mehul Khimani

Abstract Human population growth requires higher agricultural productivity, a challenge that all nations face. Agrochemicals, such as fertilizers, pesticides, fungicides, and plant growth hormones, are used in significant quantities in agriculture to boost productivity and eliminate major threats, including insects, pests, pathogenic organisms, and weeds, as well as other diseases. The lower water solubility and higher hydrophobicity of these substances enable them to persist in soil and subsequently diffuse into water bodies. Overuse of these agrochemicals and their persistence in the environment harm the food chain, ecosystem, and consumers. Therefore, the use of green compounds, meaning eco-friendly or bio-based agrochemicals, which are integral to sustainable agriculture practices, is clearly essential in the current era. Thanks to low toxicity, exceptional environmental stability, and eco-friendliness, biosurfactants and the microorganisms that produce them are gaining popularity nowadays. They augment the efficacy of pesticides and fertilizers while mitigating ecological impacts by enhancing nutrient bioavailability in soil and promoting the

---

U. Sheth (✉)
Department of Biotechnology, Faculty of Science, Vanita Vishram Women's University, Surat, Gujarat, India
e-mail: urjita.sheth.68@gmail.com

J. Barot
Department of Bioscience, Veer Narmad South Gujarat University, Surat, Gujarat, India

C. B. Mistry
Department of Chemistry, Sir. P. T. Sarvajanik College of Science, Surat, India

P. Parekh
Department of Chemistry, Veer Narmad South Gujarat University, Surat, Gujarat, India

V. Patel
Department of Chemistry, Navyug Science College, Surat, Gujarat, India

M. Khimani (✉)
Chemical Division, P T H Consultancy Chemical LLP, Surat, Gujarat, India
e-mail: khimani_mehul@yahoo.com

M. Abdul Rub et al. (eds.), *Industrial Applications of Individual and Mixed Surfactant Systems—Volume 2*, Sustainable Materials and Technology,
https://doi.org/10.1007/978-981-95-9195-4_3

desorption and remediation of pesticides, attributable to their diverse functional properties, including wetting, spreading, and penetration capabilities. According to literature, some biosurfactants, including surfactin, rhamnolipids (RL), mannosylerythritol lipids (MELs), and sophorolipids (SL), exhibit broad-spectrum antibacterial activity against a variety of phytopathogens. Thus, the biosurfactants represent a sustainable alternative to conventional agrochemicals, which brings the global efforts in line to reduce environmental pollution and encourage sustainable/green agriculture. However, constant research/exploration, technological innovations, and regulatory support are crucial for translating their laboratory potential into large-scale agricultural applications. This chapter covers the organisms that produce biosurfactants; characteristics of biosurfactants; formulations of biosurfactant-based agrochemicals; their advantageous effects on the environment, humans, and cultivated crops—including plant growth promotion, challenges to phytopathogens, efficacy in solubilizing agrochemicals to mitigate toxicity, and minimal dosage with lesser side effects to ecosystems—and their applications within the agricultural sector. Also, a few statistics related to biosurfactant-based agrochemicals shall be discussed.

Keywords Agrochemical · Amphiphilic · Bioremediation · Biosurfactants · Metagenomics · Pesticide · Sustainable agriculture

1 Introduction to Agrochemicals

1.1 Agrochemicals: Their Need, Usages and Benefits

The modern agricultural era faces the challenge of meeting the increasing global demand for food while minimizing its negative environmental impacts, including soil erosion, pollution, and food contamination, to achieve sustainable development objectives [1, 2].

Agrochemicals are synthetic substances that farmers use to control weeds, pests, and diseases, as well as to enrich soil and crops with essential nutrients. They include a wide range of products:

- **Fertilizers** that supply essential nutrients to the soil, including nitrogen, phosphorus, and potassium
- **Pesticides, herbicides, insecticides, and fungicides**, which control pests, diseases, and weeds
- **Plant growth regulators**, which control the rate of plant development (e.g., for flowering or ripening)
- **Other agents** like **algaecides, industrial chemicals, soil supplements** (like lime), **acidifiers** designed to alter the pH, soil conditioners, etc. Modern agriculture relies on them, and their use has significantly increased crop yields [3–6].

1.2 *Synthetic Surfactants Versus Biosurfactants*

The world's population is growing at a faster rate than it can be managed, and this is harming agricultural productivity because of climate change. This includes things like higher average temperatures, less consistent rainfall, and other biotic and abiotic stress factors. One of the main things that worries agricultural scientists, researchers, and the scientific community is the possibility of a new pest, disease, or pathogen outbreak in plants. Chemical surfactants are those that are synthesized chemically, and Biosurfactants are the molecules that are produced from biological sources. Table 1 reflects a comparison between Synthetic Surfactants and Biosurfactants.

1.2.1 Risks Associated with Chemical Agrochemicals

Chemical pesticides and agrochemicals, with chemical surfactants as the main ingredient, are relied upon by many communities in both developed and developing nations to control pathogens and manage plant diseases [18]. Synthetic surfactants are chemically derived and comprise a hydrophobic chain of alcohols, paraffin, olefins, alkyl phenols, and alkyl benzenes. Their long-term use, however, poses the risk of agrochemical contamination as well as soil and environmental contamination [1]. Surfactants, which are amphiphilic molecules, are essential parts of many herbicides, wetting, dispersing, and emulsifying agents. They diminish interfacial tension and improve performance [18, 19].

Worries about the toxicity, environmental persistence, and non-biodegradability of synthetic surfactants have increased due to their frequent use in agrochemicals, which has led to significant soil and environmental pollution. These compounds can influence crop quality and soil health, while also contributing to the development of pathogens that are resistant to pesticides. These compounds make it easier for organic hydrocarbons and heavy metals like cadmium, lead, copper, nickel, and arsenic to

Table 1 Comparison between synthetic surfactants and biosurfactants

Parameter	Chemical surfactants	Biosurfactants	References
Source	Petroleum-based chemicals	Microbial, plant, or animal origin	[7, 8]
Biodegradability	Low to moderate	High	[9, 10]
Toxicity	Often high, aquatic, and soil toxicity	Low to negligible	[11, 12]
Cost	Low production cost	Higher (can be reduced using agro-industrial wastes)	[13, 14]
Environmental impact	Persistent pollutants, bioaccumulation	Eco-friendly, sustainable	[8, 15]
Functional range	Limited in extreme pH/salinity	Effective in extreme pH, salinity, and temperature	[16, 17]

accumulate in agricultural soils. When this pollution enters food chains, it lowers nutritional value, damages soil structure and native microbial communities, inhibits plant growth, and increases the risk of cancer, liver and kidney damage, neurological impairments, and birth defects, among other health issues [20]. Additionally, this type of pollution pushes environmental systems to their limits and contributes to the development of infectious agents that are resistant to chemicals. There is an immediate need for natural alternatives in agriculture that are effective, affordable, and environmentally benign due to these challenges [21].

1.2.2 Biosurfactants: Promising Biomolecules for Agriculture

Biosurfactants, which are biomolecules with a low molecular weight, amphiphilic nature, and surface-active properties, are attracting interest as a sustainable alternative to conventional detergents. They have the potential to increase crop yields and agricultural productivity while also reducing harmful environmental contaminants made by animals, plants, and microbes. They lower the interfacial tension and surface tension at the oil–water and air–water interfaces, respectively [13, 22]. Biosurfactants can be classified into several groups according to their physicochemical characteristics; these groups include glycolipids, lipopeptides, neutral lipids, phospholipids, fatty acids, and polymeric [23–25]. When compared to synthetic surfactants, biosurfactants are economically exploited due to their low or non-toxic nature, biodegradability, high specificity, effectiveness under extreme conditions, and ability to be reused through regeneration [26–28]. Superior to their chemically produced counterparts, these environmentally friendly surfactants offer a host of benefits. These biosurfactants are fantastically useful in many areas, including personal care products, agriculture, the food and household sectors, oil recovery, and surface activity and detergency compared to traditional surfactants [22, 29–31]. They provide exceptional advantages [32–37]:

- Biodegradable and safe.
- Surfactants derived from microorganisms are better than synthetic or chemical surfactants in several ways, including being able to handle higher temperatures, changes in pH, high salinity, faster degradation, lower toxicity, and better selectivity [9, 10].
- Functional at extreme conditions (change in pH, temperature, salinity).
- Characterized by a wide range of structural characteristics, strong emulsifying, and surface tension-reducing capabilities.
- Used in agriculture to enhance soil quality, pesticide delivery, pathogen control, and plant growth.
- Microbial and synthesized biosurfactants' low molecular weight reduces surface tension, while high-molecular-weight promotes stable emulsions [38, 39].

Some examples of biosurfactants include phospholipids, glycolipids, lipopeptides, fatty acids, and biopolymers with high molecular weights. They are more stable, selective, biodegradable, and environmentally friendly than their synthetic alternatives. Particularly in areas with harsh environmental circumstances and in emerging nations, their extensive use promotes sustainable development. These biosurfactants, which are naturally occurring, exhibit considerable potential as environmentally friendly substitutes for synthetic ones. Commercial production of these is on the rise due to their beneficial effects on human and environmental health, and they often beat synthetic surfactants in emulsifying and surface activity [17–20]. Additionally, microorganisms are a viable option for sustainable development and production due to their widespread presence, ease of cultivation, affordability, and minimal environmental impact [33].

An amphiphilic surface-active agent, biosurfactants can lower surface and interfacial tension by producing micelles above the Critical Micelle Concentration (CMC). They are distinguished by their capacity to reduce the interfacial surface tension between surfaces and solutions, as well as between immiscible phases such as the interfaces of water and oil. Many procedures, such as emulsification and solubilization, are made easier by this reduction. Both the formulation of pesticides and the solubilization of hydrophobic compounds are areas where these characteristics have proven advantageous [22, 25–27]. They are like synthetic surfactants in that they disperse and emulsify, but they are safer, more biodegradable, and have less toxicity [28–31]. These molecules contain hydrophilic polar head groups and hydrophobic fatty acid chains [32], and there are two types of these molecules, defined by their molecular weight and structure: low-molecular-weight (glycolipids, phospholipids, lipopeptides) and high-molecular-weight (polysaccharides, lipopolymers) [28, 33, 34]. When it comes to reducing surface tension, low-molecular-weight types really shine, whereas high-molecular-weight types are great at stabilizing emulsions. Their ability to self-assemble, driven by hydrophobic interactions, is crucial to their performance in various applications. Biosurfactants can partition between two immiscible liquids due to their amphiphilic character, which effectively lowers interfacial tension [40, 41]. Because of this characteristic, they are beneficial in solubilizing hydrophobic medications, increasing their bioavailability and therapeutic effectiveness [21].

Biosurfactants in agriculture serve to:

- Enhance soil quality and facilitate the mobilization of trace nutrients.
- Degrade hazardous contaminants.
- Optimize the delivery of agrochemicals.
- Inhibit diseases and augment plant immunity.
- Stimulate plant development and sustain beneficial microflora [22, 35–42].
- Biosurfactants have great promise for sustainable agriculture, particularly in underdeveloped nations, due to their non-toxic, biodegradable, and environmentally friendly characteristics.

These days, biosurfactants are used for a variety of purposes, including enhancing soil quality by increasing trace element concentrations, managing plant diseases, and mixing with pesticides [21]. As an antagonistic molecule against pests/pathogens or plant diseases, biosurfactants are commonly employed in sustainable agriculture to improve soil quality by degrading harmful pollutants or increasing soil trace element availability. The antibacterial properties of surfactants produced by certain microbial strains significantly hinder the growth of pathogens. It safeguards plants from harmful pathogens in multiple instances by enhancing their immune systems [42]. In addition, the application of biosurfactants has shown additional benefits in the plant by boosting the encouragement of growth. Furthermore, these biosurfactant molecules provide energy for the native microflora of the soil or plant system, which in turn regulates the physiological parameters of plants and ensures the system's health and quality.

2 Market Overview of Sustainable Agrochemicals

Growing environmental worries about conventional chemicals, improved farming techniques, and rising food demand are all driving agriculture toward more environmentally friendly methods. As a result, research into ecologically friendly agrochemicals to promote ethical farming practices has intensified. Table 2 reviews the market overview of sustainable agrochemicals.

2.1 Global Market Trends

The worldwide market for agrochemicals, which includes fertilizers and crop protection, was valued at \$271.4% of GDP in 2023 and is expected to reach \$390.2% of GDP by 2030, representing a compound annual growth rate of around 5.3% [43]. The fast expansion of bio-based and microbiological solutions like biopesticides and biofertilizers is being driven by increasing consumer and environmental concerns, stricter laws, and the desire for organic food. In the past few years, bio-based agrochemical products have accounted for around 30% of all new product launches, while slow-release or innovative formulations accounted for 20% [44]. The market for biofertilizers is expected to reach USD 3.2 billion by the year 2029, with a compound annual growth rate (CAGR) of around 11.5% [47]. In 2024, the Asia–Pacific region will account for more than 43% of the world's agrochemical revenue, strongly dominating the industry [48]. The region, comprising major economies

Table 2 Market overview of sustainable agrochemicals

Region	Market size (2023)	Projected size	CAGR (%)	Notes	References
Global	USD 271.4 B	USD 390.2 B (2030)	~ 5.3	Bio-based products ~ 30% of new launches	[43, 44]
India	USD 15.59 B	USD 24.68 B (2032)	~ 5.3	Increasing adoption of biopesticides/ biofertilizers	[45, 46]

such as India, China, and Japan, is experiencing several significant trends, including sustainability and integrated pest management (IPM) [49].

2.2 Indian Scenario: Agrochemicals and the Sustainable Shift

The agrochemical market in India was worth $15.59 billion in 2023 and is projected to reach $24.68 billion by 2032, as per Global Market Insights (CAGR ~ 5.3%) [45]. According to another analysis by Expert Market Research, the market is expected to reach $4.83 billion by 2034, an increase of $3.23 billion (CAGR ~ 4.1%), as stated in Ref. [46]. In terms of market value, fertilizers account for almost USD 10 billion, while crop-protection chemicals will bring in around USD 7.7 billion in 2023 [45]. With the help of government programs, an increasing number of people are utilizing organic amendments, biofertilizers, soil enhancers, and biopesticides [45, 50]. Optimal pesticide usage, waste reduction, and environmental impact are all outcomes of precision farming techniques enabled by AI, drones, and the IoT [45]. The Indian agrochemicals market is anticipated to reach USD 16.87 billion from 2024 to 2029, at a CAGR of 11.7%. Herbicides, integrated pest management (IPM) strategies, and environmentally friendly products are the main factors propelling this growth [51]. Sustainable agricultural inputs are gaining popularity, which is supported by government programs like PMKSY and PKVY, as well as subsidies and rising consumer awareness [45]. In fiscal year 22–23, India exported agrochemicals (pesticides) worth around ₹43,223 crore (~ USD 5.4 billion), supported by an export CAGR of around 13% [52]. This made India a top-four worldwide agrochemical market. There is a fast transition towards sustainable, bio-based solutions in the agrochemical sector, which is large (~ $271 b), growing (~ 5% CAGR), and on a global scale. With a strong tilt towards green agrochemicals, India is experiencing strong growth (also at a compound annual growth rate of approximately 5%). This growth is driven by policy, technology, and global demand. Sustainability in agrochemicals appears promising, given India's position as the world's second-largest exporter and the government's efforts to promote it through subsidies and the adoption of new technologies.

3 Biosurfactants and Their Fundamentals

3.1 Biosurfactants Produced by Soil Microorganisms

The multi-usefulness of microbial biosurfactants makes them powerful tools for ecological and agricultural management. They improve nutrient availability, control plant infections, stimulate biodegradation and bioremediation, and enable biofilm development. They also thrive under a variety of stressful circumstances. The low toxicity, excellent stability despite temperature fluctuations, and complete biodegradability of these chemicals make them ideal for decontaminating agricultural soils affected by organic contaminants and heavy metals. Biosurfactants enhance soil health and sustainability by improving nutrient mobilization, flushing, and solubilization through the formation of complexes with metals. Their advantageous qualities indicate that they will have more widespread, economical, and economically feasible uses [1].

Microorganisms associated with the rhizosphere, such as bacteria, yeasts, and fungi, are significant producers of biosurfactants. To shape plant–microbe interactions, suppress pathogens, and enhance beneficial microbial activity through improved nutrient uptake, these substances help microbial mobility, signaling, and biofilm growth. Additionally, they provide environmentally preferable substitutes for the commonly utilized synthetic surfactants in pesticide formulations [7].

Soil, freshwater, and the surfaces of leaves are just a few of the many environments where biosurfactant-producing bacteria can be found, according to the research [53]. Research has shown that certain microbes may synthesize biosurfactants; this has important implications for plant health, soil remediation, and other agricultural applications [54–60].

Bacteria are the predominant makers of biosurfactants among microorganisms, whereas fungi account for around 19% (12% from Ascomycota and 7% from Basidiomycota) of identified types [61]. Yeast species, such as *Saccharomyces lipolytica* and *Candida lipolytica*, are prominent producers, along with bacterial species including *Bacillus salmalaya*, *Pseudomonas aeruginosa*, *Bacillus subtilis*, and *Acinetobacter calcoaceticus*. Examples of chemically varied substances synthesized by these microbes include biodispersan and emulsan from *Acinetobacter*, liposan from *C. lipolytica*, and alasan from *A. radioresistens* [62]. *Pseudomonas rhamnolipids* and *Bacillus surfactin* are two notable biosurfactants [63]. Yeasts such as *Pseudozyma*, *Candida*, and *Yarrowia* can also create useful biosurfactants, but they have received less attention from researchers [64].

3.2 Chemistry and Mode of Action of Biosurfactants

Surface and interfacial tension are reduced by microbial biosurfactants, which are amphiphilic molecules that are either secreted into the environment or synthesized by cells. These molecules either form micelles or arrange themselves at fluid boundaries (Fig. 1). Hydrophobic components, such as long-chain fatty acids, can be attached to various substances, including carbohydrates, phosphates, carboxylic acids, alcohols, amino acids, and cyclic peptides. The polar (hydrophilic) section improves solubility in water, while the nonpolar section decreases surface tension by embedding within micelle cores or at air–water interfaces. Biosurfactants work primarily by creating micelles above a specific critical micelle concentration (CMC) and by reducing the

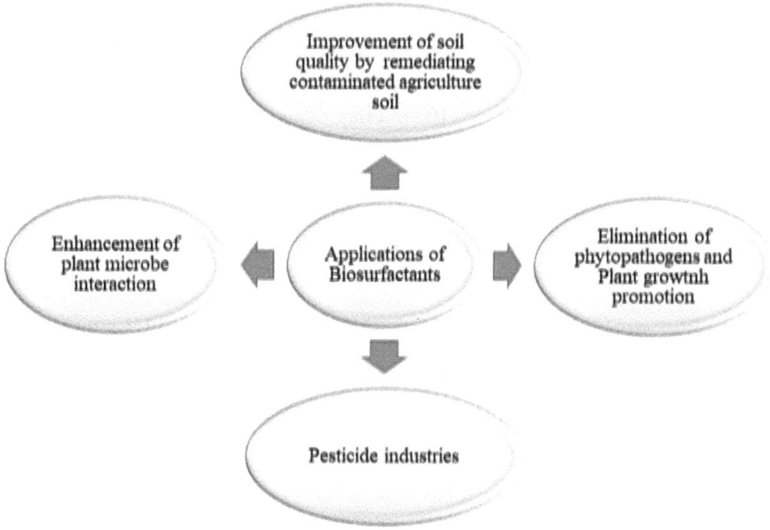

Fig. 1 Multifunctional prospects of biosurfactants in agriculture [7]

oil–water interfacial tension, which limits the mobility of the organic phase. Chemical molecular weight (CMC), chemical structure, hydrophilic-lipophilic balance (HLB), and microbial origin are essential properties that define their performance. Surfactants that are suitable for reducing interfacial tension and water surface tension can do so by solubilizing materials with high HLB values in water and solubilizing oils with low values [16, 17]. The lower CMCs of biosurfactants (usually between 0.15 and 30 mg/L) allow them to be more effective with less volume, making them superior to their synthetic analogues [65]. In addition, surface-active agents that are microbial are stable in a range of environments. One example is the biosurfactant produced by *Arthrobacter protophormiae*, which is effective from 30 to 100 °C and from 2 to 12 pH [17]. Their durability makes them ideal for use in environmental and industrial settings.

3.3 *Classification of Biosurfactants*

Fatty acids, glycolipids, peptides, lipopeptides, antibiotics, and phospholipids are some of the biosurfactants that fall into several categories, depending on their chemical composition and the species from which they originate. There are two categories for them based on molecular weight: low-molecular-weight and high-molecular-weight. Biosurfactants with low molecular weights, like phospholipids, glycolipids, and lipopeptides, reduce surface and interfacial tensions and usually have a molecular weight of 500–1500 Da [66]. Rhamnolipids, sophorolipids, and trehalose lipids are famous for their strong tension-reducing powers; they are members of the most investigated class of glycolipids. Polymeric bioemulsifiers, lipopolysaccharides, polysaccharides, and other high-molecular-weight surfactants mainly function as emulsifiers rather than surface tension reducers [67]. They consist of proteins and extracellular amphiphilic polysaccharides that are great at keeping oil-in-water emulsions stable. One way to categorize them is by cell type; prokaryotes are those that lack a cell wall, whereas eukaryotic organisms have a cell wall.

The following chemical categories apply to these surface-active substances, and the same has been mentioned in Table 3:

- **Fatty acids and phospholipids**, such as phosphatidylethanolamine and spiculisporic acid;
- **Glycolipids**, including rhamnolipids, sophorolipids, trehalose lipids, and mannosylerythritol lipids;
- **Lipopeptides**, such as surfactin, iturin, fengycin, lichenysin, gramicidin, and polymyxin;
- **Polymeric biosurfactants**, comprising emulsion, liposan, alasan, biodispersan, and complex protein–lipid–polysaccharide mixtures [68].
- For instance, rhamnolipids produced by *Pseudomonas* species are highly esteemed due to their antibacterial and hydrocarbon-emulsifying capabilities [68, 69]. Bioremediation, cosmetics, and agriculture all make use of sophorolipids because of their biodegradability, low toxicity, and dual lactonic/acidic forms [70]. Sophorolipids are compounds that are produced by yeasts such as *Candida* and *Starmerella. Rhodococcus* and *Arthrobacter* trehalose lipids are good hydrocarbon solubilizers and emulsifiers [71]. Notable lipopeptides with antibacterial, antiviral, and anticancer properties include surfactin (*Bacillus subtilis*) and lichenysin (*Bacillus licheniformis*) [72]. Their antimicrobial, anti-inflammatory, antiviral, anticancer, and thrombolytic actions are well-documented [73]. Microorganisms can oxidize alkanes to form fatty acids and phospholipids. The food sector makes extensive use of fatty acids, but phospholipids, as a component of membranes, have found utility in gene carrier systems [74]. Lastly, categorization can also be based on substrate choice. For example, some bacteria (like *Corynebacterium* and *Arthrobacter*) create biosurfactants when cultured on carbon sources that are insoluble in water (like *Bacillus*), whereas others (like *Pseudomonas*) grow on mixes of these types of substrates [1].

4 Surfactants in Agrochemicals

As surface-active agents, surfactants play an essential role in agrochemical formulations. They improve the dispersion, adhesion, solubilization, and penetration of agrochemicals by lowering surface and interfacial tension. Many agricultural products, including foliar fertilizers, insecticides, herbicides, and fungicides, contain

Table 3 Biosurfactants: types, examples, microbial sources, and their functions

Type of biosurfactant	Biosurfactant	Producing microorganisms	Agricultural function	References
Glycolipid	Rhamnolipids	*Pseudomonas aeruginosa*, *Burkholderia* spp.	Enhanced pesticide delivery, antifungal activity, hydrocarbon and heavy metal removal, and inhibition of zoospores Enhances wetting and penetration of foliar-applied pesticides Exhibits antifungal activity against pathogens like *Fusarium* and *Botrytis*	[7, 22, 76, 77]
Lipopeptides	Surfactin, iturin, fengycin	*Bacillus subtilis*	Biocontrol, biofilm formation, pathogen inhibition, soil remediation, and ISR induction Used in biocontrol formulations due to its potent antimicrobial activity Supports colonization of beneficial microbes in the rhizosphere	[7, 78]

(continued)

Table 3 (continued)

Type of biosurfactant	Biosurfactant	Producing microorganisms	Agricultural function	References
Glycolipids	Sophorolipids	*Candida bombicola*	Adjuvants in foliar sprays, cleaning agents, antimicrobials, and emulsifiers for agrochemicals Useful as adjuvants in foliar sprays and biodegradable cleaning agents for agricultural equipment Non-toxic to plants and beneficial insects	[7, 79]
Mannosylerythritol lipids (MELs)	MEL-A, MEL-B	*Pseudozyma* spp.	Pest control, emulsifiers, and seed coating agent Show promising properties in pest control and as emulsifiers in botanical pesticide formulations	[7, 22, 80]
	Viscosin	*Pseudomonas fluorescens*	Pathogen inhibition (e.g., *Rhizoctonia solani*)	[81]
Polymeric biosurfactants	Emulsan	*Acinetobacter calcoaceticus*	Emulsification of hydrophobic pollutants, soil washing	[23]
Phospholipids/ fatty acids	Various	*Rhodococcus* spp., *Corynebacterium* spp.	Metal chelation, hydrocarbon emulsification	[7]

surfactants. Surfactants can be roughly classified as either chemical (synthetic) or bio-based, depending on their origin and composition.

4.1 *Chemical Surfactants*

Chemical surfactants are man-made compounds typically derived from petroleum-based feedstocks. Agrochemical formulations rely on them heavily because of the positive effects they have on the wettability and solubility of active chemicals, which in turn improve their distribution and effectiveness [75].

Characteristics

- Generally categorized as anionic, cationic, non-ionic, or amphoteric based on the nature of their polar head groups.
- Common types include alkylphenol ethoxylates (APEs), sodium lauryl sulfate (SLS), and quaternary ammonium compounds.

Limitations

- Non-biodegradable or slow to degrade, leading to long-term environmental persistence.
- Associated with toxicity to aquatic life, soil microorganisms, and non-target organisms.
- They can contribute to endocrine disruption and bioaccumulation in food chains.

4.2 *Bio-Based Surfactants*

Biosurfactants, or bio-based surfactants, are surface-active chemicals that are produced by microbes. Biosurfactants are safer, less harmful, and more environmentally friendly than their synthetic counterparts because they are derived from renewable biological sources [7, 8].

Key Advantages

- Produced by microorganisms such as *Pseudomonas*, *Bacillus*, *Candida*, and *Acinetobacter* species.
- Function effectively under extreme pH, temperature, and salinity.
- Possess antimicrobial and antifungal properties, which are beneficial in pest and disease control.

Applications in Agrochemicals

- Improving pesticide delivery by reducing droplet size and increasing spreadability on leaf surfaces.
- Enhancing micronutrient solubilization for better uptake by plants.

- Supporting biocontrol agent formulation and soil remediation.

4.2.1 Examples of Biosurfactants Used in Agriculture

Several biosurfactants have shown promising applications in agriculture due to their surface-active properties and additional bioactivities.

Table 3 indicates examples of biosurfactants used in agriculture with their specific functions.

5 Advantages of Using Biosurfactants in Agriculture

Biosurfactants, which are microbe-produced amphiphilic surface-active molecules, are a potential replacement for synthetic surfactants and agrochemicals. A wide range of advantages, consistent with sustainability, safety, and ecological harmony, can be achieved through their use in farming.

5.1 Environmental Sustainability

The fact that biosurfactants are safe for the environment is one of their main selling points. Biosurfactants are greener than their synthetic counterparts since they break down naturally in the environment and are made from sustainable materials. Incorporating agro-industrial waste into their manufacturing process helps to reduce environmental burden and promotes a circular bioeconomy. Biosurfactants are designed to break down into harmless by-products when exposed to water or soil, preventing their accumulation. Because of their low critical micelle concentration (CMC), they are effective even at low dosages, which reduces their impact on the environment. Sustainable agriculture benefits from its use in soil remediation, improved nutrient solubilization, and integrated pest management [8, 15].

5.2 Low Toxicity

Organic biosurfactants have a lower toxicity level than their synthetic chemical counterparts and insecticides. Their low toxicity makes them an excellent choice for agricultural applications where protecting people, animals, and the environment from exposure is crucial. At amounts used in agriculture, rhamnolipids, sophorolipids, and surfactins are not very hazardous to cells, according to studies. They are perfect for use in biopesticides, organic fertilizers, and organic farming due to their low toxicity. Furthermore, unlike certain synthetic agrochemicals, biosurfactants do not display any carcinogenic or mutagenic properties [11, 12].

5.3 *Biocompatibility*

Soil ecosystems, beneficial microorganisms, and plants all tolerate biosurfactants quite well because of their high biocompatibility. Due to their ability to avoid interfering with native microbiomes, they find application in the development of biopesticides, biofertilizers, and microbial inoculants. Beneficial rhizobacteria (such as *Bacillus subtilis*) can improve plant–microbe interactions using biosurfactants, such as lipopeptides, which aid in biofilm development. Because they don't damage root-associated organisms or symbionts like mycorrhizae, they help biocontrol agents colonize and improve plant health. Agricultural personnel and post-harvest applications can safely use them because of their biocompatibility, which also includes their safe interaction with human skin and other biological tissues [7, 78].

6 Key Agricultural Applications of Biosurfactants

Biosurfactants, which are surface-active chemicals generated by microorganisms, provide a sustainable, biodegradable, and low-toxicity substitute for synthetic surfactants. Their amphiphilic characteristics facilitate several applications in agriculture, including pesticide formulation, micronutrient solubilization, soil remediation, and the delivery of biocontrol agents.

6.1 *Enhanced Pesticide Delivery*

Enhancing the wettability, spreadability, and penetration of active chemicals on hydrophobic surfaces, such as insect cuticles and plant leaves, biosurfactants improve the efficiency of pesticides. They reduce surface and interfacial tension, making weakly water-soluble insecticides more soluble.

For instance, research has demonstrated that rhamnolipids and sophorolipids can modify the permeability of insect cuticles and leaf surfaces, improving the contact efficiency and bioavailability of pesticides [39]. In addition to reducing pesticide dosage, biosurfactants help keep pest control efficacy while lowering environmental impact [7].

6.2 Micronutrient Solubilization

Essential micronutrients, such as copper (Cu), iron (Fe), and zinc (Zn), in soil can be chelated and dissolved by biosurfactants, thereby improving plant nutrient uptake. These nutrients are typically unavailable in calcareous or alkaline soils due to precipitation or adsorption, so this is especially helpful in those types of soils. Soil particles and mineral matrices can have their zinc (Zn) and iron (Fe) mobilized by microbial biosurfactants, such as glycolipids and lipopeptides. By utilizing these, the nutritional quality, yield, and plant growth could all be enhanced, all without the use of synthetic chelating agents.

6.3 Soil Remediation

Agricultural soils polluted with petroleum hydrocarbons, heavy metals, or pesticide residues are a common target for bioremediation efforts, and biosurfactants are finding increasing use in this field. For improved microbial degradation and plant uptake, they enhance the bioavailability, desorption, and mobilization of contaminants. Biosurfactants function by increasing the surface area and access for degrading microbes in hydrocarbon-contaminated soils, which in turn promotes emulsification and biodegradation [22]. They facilitate phytoextraction or microbial detoxification by chelating and desorbing metals from soils that are polluted with these elements.

6.4 Biocontrol Agent Delivery

Biosurfactants can act as effective carriers for biocontrol agents (e.g., *Trichoderma*, *Bacillus*, or *Pseudomonas* spp.), improving their adhesion, survival, and colonization on plant surfaces and rhizospheres. They also possess intrinsic antimicrobial properties that inhibit phytopathogens. For instance, surfactin, produced by *Bacillus subtilis*, not only facilitates biofilm formation for enhanced colonization but also inhibits the growth of pathogens such as Fusarium oxysporum and *Pythium* spp. [78]. Thus, biosurfactants can synergistically boost the effectiveness of microbial biopesticides and biofungicides.

7 Applications of Biosurfactants in Agriculture

The natural surfactants used by microbes living in soil as a source of carbon have the potential to replace chemical surfactants in pesticides. With their many functional characteristics, microbial biosurfactants have a wide range of potential applications, including but not limited to improving soil quality through improved remediation of pollutants, controlling phytopathogens, and enhancing soil micronutrient status [1].

Hydrophobic organic contaminants (HOCs) have their apparent solubility improved when surfactants, which act as mobilizing agents, increase the solubility of biohazardous chemical substances like polycyclic aromatic hydrocarbons (PAHs) [82]. Furthermore, they are essential because they help microbes adhere to pollutant-containing soil particles, which in turn shortens the diffusion distance between the absorption site and microbial uptake [82]. Hydrophilizing heavy soils with surfactants enhances wettability and ensures uniform fertilizer distribution in agriculture. On top of that, they stop fertilizer from caking while in storage and help pesticide toxicants disperse and penetrate [82]. With no known adverse effects on human or environmental health from cumulative exposure, the rhamnolipid biosurfactant, which the genus *Pseudomonas* mainly produces, demonstrates vigorous antibacterial activity [82]. One group of biosurfactants that has shown promise in biocontrol of plant diseases is fengycins, which are believed to have antifungal effects [83, 84].

Biosurfactants derived from microbes have dual hydrophobic and hydrophilic properties, which can be used in agricultural applications [85]. A Applications include increasing plant growth promotion due to their antimicrobial activity, improving soil quality by enhancing the biodegradation of pollutants, and increasing the plant–microbe interaction that is beneficial for plants [7]. These aspects are illustrated in Fig. 1, which shows the various applications. Also, Table 4 summarizes key agricultural applications of biosurfactants. These natural surfactants have been discovered to be utilized as a carbon source by soil-dwelling microbes [86–88], which is the reason for the biological removal of biosurfactants from agricultural soil. As a result, these biosurfactants have the potential to replace the harsh surfactants that are currently used in the pesticide industry.

7.1 *Improvement in the Soil Quality Through Enhancement of Soil Micronutrients*

Optimal soil micro- and macronutrient concentrations have a crucial role in crop growth and productivity. Reduced crop quality and illnesses can result from deficiencies or excesses of trace elements, which are essential for plant physiological activities. Reduced crop yields and food quality can result from alterations to the availability and efficacy of essential nutrients caused by climate change, increased temperatures, changes in soil pH, increases in salt, and the accumulation of environmental pollutants [97].

Micronutrient availability in nutrient-deficient soils can be enhanced by using biosurfactants. By interacting with metal ions through oxidation–reduction, adsorption, and desorption, these amphiphilic compounds form complexes, which increase their bioavailability [98]. A biosurfactant, such as rhamnolipid, can be introduced to soil to reduce the interfacial tension between solid particles and trace metal cations.

Table 4 Key agricultural applications of biosurfactants

Application area	Mechanism of action	Example biosurfactants	References
Pesticide delivery	Reduce surface/interfacial tension, increase spreadability	Rhamnolipids, sophorolipids	[7, 39]
Soil remediation (hydrocarbon and pesticide removal)	Desorption of hydrocarbons/heavy metals Micelle formation enhances the desorption and bioavailability of hydrophobic pollutants and accelerates biodegradation	Rhamnolipids, surfactin, sophorolipids	[7, 39, 89, 90]
Heavy metal removal	Chelation and mobilization of metal ions; foaming-surfactant technology for extraction	Rhamnolipids, surfactin	[7, 91]
Plant pathogen elimination	Disruption of pathogen membranes, inhibition of spore germination, and ISR induction in plants	Surfactin, iturin, rhamnolipids	[42, 92]
Pathogen inhibition	Antifungal/antibacterial action	Iturins, fengycins	[93, 94]
Nutrient mobilization/ solubilization	Chelation of micronutrients (Fe, Zn, Cu) for enhanced plant uptake	Glycolipids, lipopeptides	[7, 41, 95]
Agrochemical formulation enhancement	Improved solubility, spreadability, and penetration of active ingredients	Sophorolipids, MELs	[15]
Weed management	Enhancement of bioherbicide activity and penetration	Rhamnolipids with *Myrothecium verrucaria*	[96]
Biocontrol agent delivery	Improved adhesion/colonization	Surfactin, MELs	[78, 80]

This enhanced interaction improves the mobility and solubility of the metals. These biosurfactants are anionic, so they may readily form stable complexes with positively charged metal ions. This makes it easier for the ions to be released from soil particles and then taken up by plant roots. As a result of their sequestering properties, biosurfactants enhance nutrient availability and facilitate metal desorption [41]. On the other hand, biosurfactant-metal complexes can be mobilized and flushed out in polluted water or soil because of strong electrostatic interactions [23, 99].

Biosurfactants improve trace metal availability in nutrient-poor soils by chelating metals into stable biosurfactant complexes, which then undergo adsorption, desorption, and finally metal removal from soil interfaces. Nutrient mobility, bioavailability, and biodegradation are all enhanced when these chemicals tend to congregate at immiscible fluid boundaries, which increases the surface area for chemical interaction. Biosurfactants with low molecular weights mitigate phase-repulsive and interfacial pressures, whereas those with high molecular weights encourage emulsification and enable interactions between substances that would otherwise not mix.

One factor that contributes to micronutrient insufficiency is soil alkalinity, which reduces metal availability. Fortunately, biosurfactants can help dissolve these bound metals from soil complexes and even facilitate the creation of new complexes that are more accessible [100]. Applying ionic biosurfactants to soils deficient in micronutrients causes them to create stable metal complexes rather than metal-soil interactions. The desorption of metals is facilitated by the decrease in interfacial tension, which in turn increases their accessibility to plant roots [41]. Two methods exist for improving soil micronutrient uptake: bioaugmentation with microorganisms that produce biosurfactants and direct application of biosurfactants to infertile soils.

Soil alkalinity is a major contributor to micronutrient deficiency; biosurfactants work by creating stronger connections between metal and biosurfactant than between metal and soil. This improves the availability of metals to plant roots by allowing them to desorb from the soil matrix into the soil solution [86, 87]. Improved nutrient transfer to the root zone is achieved through biosurfactants by micelle formation and a reduction in interfacial tension. This strategy has been widely adopted in agriculture worldwide to enhance soil fertility sustainably [23, 101, 102].

Biosurfactants can increase the solubility of both organic and inorganic compounds, as well as decrease the interfacial tension between immiscible liquids, due to their amphiphilic character [103, 104]. By reducing interfacial tension between soil particles and metal cations, this feature improves the mobility and uptake of trace metals in agricultural soils [105]. Canola, maize, sudangrass, and tomato plants, among others, have benefited from *Bacillus* sp. J119's capacity to increase its uptake of trace metals and facilitate its growth [95]. According to Stacey et al. [106], the development of lipophilic metal-rhamnolipid complexes improved the absorption and translocation of Cu, Mn, and Zn in Brassica napus and Triticum durum.

Biosurfactants derived from microbes enhance plant development by stimulating the production of phytohormones and systemic resistance. Therefore, bioaugmentation of biosurfactant-producing bacteria may improve the efficacy and availability of soil micronutrients to plants [107]. They help promote the indigenous microflora, which aids in stress alleviation and pollutant degradation. Liao et al. noted that the utilization of rhamnolipid and lecithin in crude oil-contaminated soil augmented microbial populations, leading to enhanced hydrocarbon breakdown while not adversely affecting maize biomass [108].

Other studies further support these conclusions: Glycolipids synthesized by *Bacillus pumilus* 2A enhanced the growth of beans, radish, and beetroot [109]. *Pseudomonas aeruginosa* RTE4, isolated from the tea rhizosphere, demonstrated both plant growth-promoting and fungicidal characteristics [110]. Viscosin, produced by *Pseudomonas fluorescens* SBW25, improved motility and colonization, resulting in enhanced plant growth [111].

Strong interactions between biosurfactants and cationic metals produce complexes that are easily absorbed by plant roots, improving the mobility and solubility of micronutrients in soil columns. These molecules serve multiple purposes, including increasing nutrient availability, aiding in the dispersion of fertilizer, providing carbon, and protecting against pollutants. Rhizospheric microorganisms utilize plant exudates to synthesize biosurfactants and other chemical compounds, including Myc factors, Nod factors, volatile organic compounds (VOCs), and exopolysaccharides. They are essential in establishing symbiotic interactions with plants, especially when they associate with Rhizobium strains and mycorrhizal fungi [112]. The solubilization of micronutrients can be further improved by combining biosurfactants with inorganic ligands [103]. Soils low in nutrients can benefit from these natural agents' ability to increase nutrient uptake, whether administered topically or supplied by biosurfactant-producing bacteria.

Sheng et al. [95] looked at how the biosurfactant-producing *Bacillus* sp. strain J119 could boost plant growth by increasing the absorption of trace metals in crops such as canola, maize, sudangrass, and tomatoes. The development of lipophilic rhamnolipid-metal complexes was also suggested as the reason for the enhanced absorption of copper, manganese, and zinc in the roots of Brassica napus and Triticum durum. Soil flushing can eliminate these metal–surfactant complexes, as stated by Pacwa-Plociniczak et al. [23]. This interaction is facilitated by the strong electrostatic attraction between negatively charged biosurfactants and positively charged metal ions. So, biosurfactants applied to the rhizosphere are a green light for improving micronutrient availability and a step in the right direction for sustainable farming.

7.1.1 Organic Waste and Crop Residue Management

A growing environmental concern is the lack of treatment for agro-industrial waste, which is rich in organic matter and nutrients such as proteins, lignin, cellulose, starch, fibres, minerals, and vitamins. Waste production in Asia is around 4.4 billion tons per year, with 350 million tons of it coming from various sources in India alone [113]. There are significant environmental problems because much of this garbage is not handled correctly. On the other hand, a sustainable and environmentally friendly alternative is to produce biosurfactants using organic leftovers. One of the many interesting things about microbial biosurfactants is their capacity to increase microbial degrading activity, emulsify hydrophobic substances, and decrease surface and interfacial tension. A wide variety of microbes make these biosurfactants [114].

Biosurfactants have several benefits over synthetic surfactants, but they aren't always practical due to poor yields and high manufacturing costs. Making large-scale manufacturing more feasible has recently centered on finding cheaper alternatives to chemical inputs, such as agro-industrial wastes utilized as carbon sources. Biosurfactant synthesis using carbohydrate-rich wastewaters, such as those from rice and cereal processing, has demonstrated potential as an economically viable feedstock [115]. In a similar vein, bacteria grown on renewable resources such as spent cooking oil, potato peelings, and sunflower cake have been found to produce biosurfactants [116].

Soybean oil refinery waste and olive mill effluents are examples of residues that have been shown to generate glycolipids and rhamnolipids according to research effectively [117]. The high carbohydrate content of fruit and vegetable processing by-products, such as orange peels, cashew apples, and date molasses, makes them ideal for microbial fermentation. These by-products have been demonstrated to decrease surface tension when biosurfactants are being made in considerable amounts [118]. Both the commercial viability of biosurfactant applications and the proliferation of microbes are enhanced by these waste products.

Microbes that produce biosurfactants, such as *Candida* species, have also found promising carbon sources in the by-products of the dairy and oil processing industries [119]. Biosurfactant production and microbial metabolism are both enhanced by the lignocellulosic biomass that is prevalent in agricultural waste [120–122]. As an example, research has shown that *Penicillium citrinum* H9 can hydrolyze many materials, including hay and straw. One study found that it could produce as much as 209.25 g/mL of sugar from rice straw [123]. Additionally, surfactin can be produced by *Bacillus subtilis* 21332 from potato processing effluents [124, 125] Lipopeptide biosurfactants, such as fengycin and surfactin isoforms, can be synthesized from potato waste by *Bacillus pumilus* and *Bacillus mojavensis* A21 [126, 127]. The rhamnolipid synthesis by *Pseudomonas azotoformans* AJ15 has been demonstrated in studies using sugarcane bagasse and potato peels as substrates [121].

In addition to improving soil fertility, increasing microbial activity, and reducing the risk of harmful disposal techniques like open burning, using agricultural wastes to make biosurfactants helps farmers better manage their waste [119]. Research has demonstrated that biosurfactants, when added to compost systems, accelerate the decomposition of organic materials, reduce the time required for composting, and decrease the release of foul odors. Additionally, these chemicals aid in the breakdown of pesticide residues in agricultural waste, which reduces the negative impact of agriculture on the environment [98, 128, 129].

Biosurfactants also help microbes perform their functions more effectively, which in turn enhances the performance of bioconversion technologies such as anaerobic digestion and composting. In addition to producing organic fertilizers of superior quality, this process also lends credence to the ideas behind a regenerative agricultural economy [130]. In general, microbial biosurfactants are an eco-friendly, long-term solution for organic waste valorization that promotes soil health and sustainability in agriculture [117].

7.1.2 Improvement of Soil Quality Through Biosurfactant-Mediated Bioremediation

Micelles enhance pollutant solubilization, reduce toxicity, facilitate heavy metal chelation, and promote biodegradation of hydrophobic compounds through mechanisms such as micelle formation and microbial adherence. Figure 2 illustrates the key functional roles of biosurfactants and micelles in environmental remediation.

Role of Biosurfactants in Remediating Hydrocarbons and Organic Pollutants

Biosurfactants increase the mobilization, emulsification, and solubilization of hydrophobic organic contaminants, which makes them more degradable [89, 131].

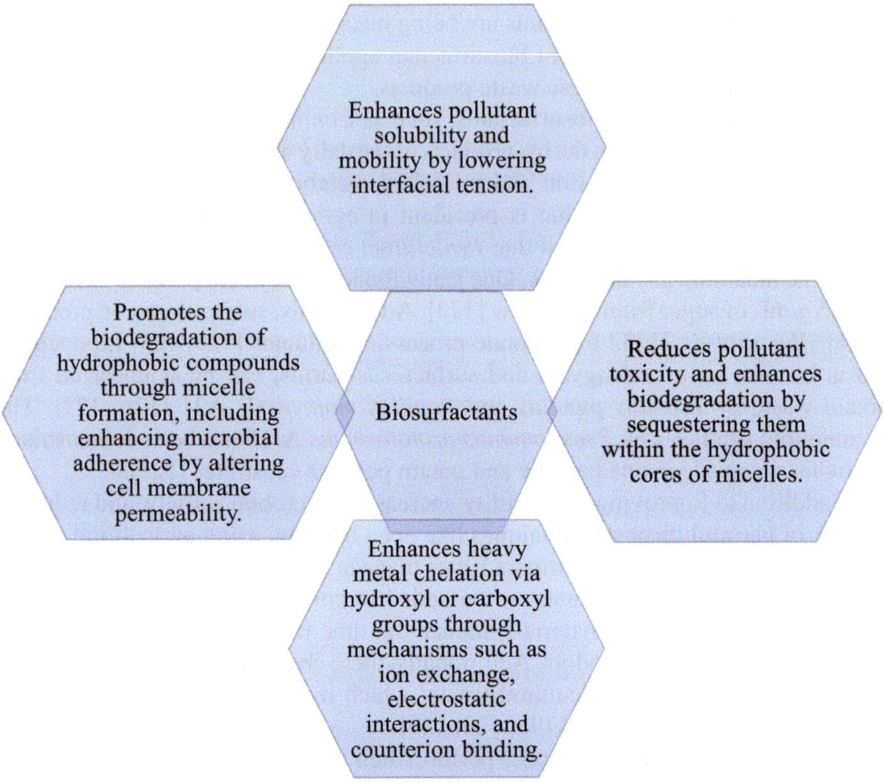

Fig. 2 Mode of action through which biosurfactants facilitate the bioremediation of pollutants

The mass transfer and availability of pollutants to bacteria that break them down are both enhanced by these procedures. Biosurfactants, for instance, improve solubilization by reducing soil-oil capillary forces and increasing contact angle, which in turn promotes hydrocarbon mobilization below the critical micelle concentration (CMC) and micelle formation above CMC [131]. Figure 3 illustrates the removal of soil hydrocarbons by Biosurfactants through desorption and emulsification.

Hydrocarbon removal effectiveness was 63% for rhamnolipids generated by *Pseudomonas aeruginosa*, 40% for Triton-100, 62% for surfactin, and 35% for Tween 80, according to comparative studies [132]. Similarly, pyrene, fluoranthene, and phenanthrene were up to 45-fold more soluble in lipopeptides derived from *P. aeruginosa* LBP9 [133]. Other strains, including *Chelidonium majus* endophytes and *Pseudomonas aeruginosa* PG1, also aided petroleum breakdown [134, 135].

In soils polluted with hydrocarbons, biosurfactants made by *Lactobacillus pentosus* decreased octane content by 58.6–62.8% [136]. Some research suggests that surfactin and glycolipids can break down pesticides and chlorinated hydrocarbons [131, 137].

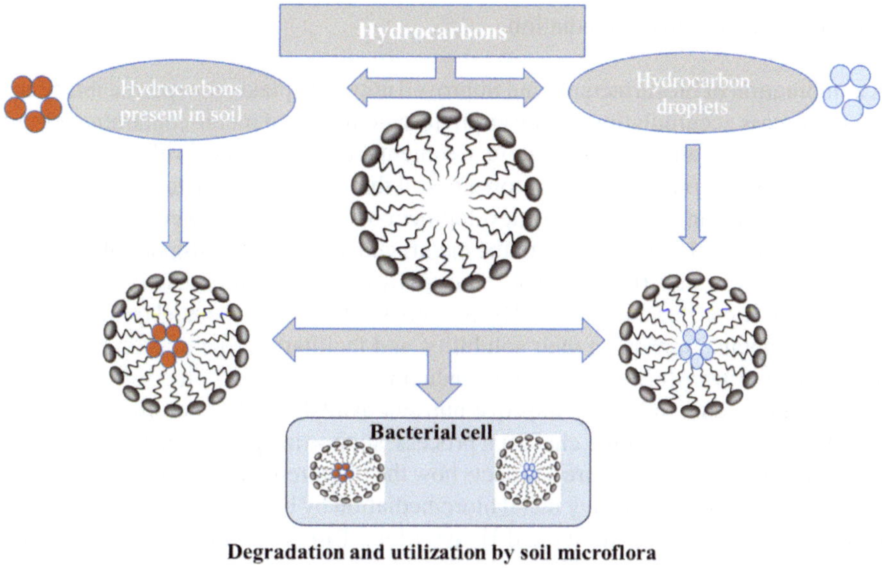

Fig. 3 Soil hydrocarbon removal by biosurfactant through desorption and emulsification

Bioremediation of Heavy Metals Using Biosurfactants

Soils can retain heavy metals like lead, cadmium, and mercury for a considerable period due to their low natural degradability. Biosurfactants are able to mobilize metals through a variety of mechanisms, such as sorption, desorption, emulsification, and complexation [138, 139]. Anionic rhamnolipids form compounds with cationic metal ions when they desorb from the soil matrix and integrate into micelles [140].

For instance, JBR-425, a biosurfactant, simultaneously mobilized Zn, Cu, Pb, and Cd while promoting earthworm development [141]. Sophorolipids were able to remove 83.6% of Cd and 44.5% of Pb [90]. Biosurfactants produced by *Bacillus* sp. MSI 54 was effective in removing 99.93% Cd, 89.5% Mn, 97.73% Pb, and 75.5% Hg from polluted solutions [142]. The following substances were removed from polluted soil: 69% lead, 54% cadmium, and 43% chromium by *Bacillus cereus* NWUAB01 [143], and around 96% zinc and copper were removed by yeast-derived biosurfactants [100].

Using precipitation or membrane filtering, biosurfactant-formed metal–surfactant complexes can be separated [144, 145]. With the use of rhamnolipids and surfactins utilizing foaming-surfactant technology, several metals, including Ni, Cd, Zn, Cu, Mn, and Ba, have been effectively removed [56, 139, 146].

Biosurfactants in Phytoremediation

A combination of biosurfactants and improved nutrient uptake, root permeability, and phytobiomass accumulation enhances phytoremediation of metal-contaminated soils [147, 148]. For example, sophorolipid treatment reduced proline levels and increased growth in Bidens pilosa that was grown in soil that was contaminated with Cd [148]. These biosurfactants facilitate the development of complexes between metals, which increases both the bioavailability of the metals and their toxicity. Reportedly, 41% Ni, 30% Cr, 29% Pb, and 20% Zn can be removed by biosurfactant-assisted phytoremediation [149]. Biosurfactants secreted by microbes enhance heavy metal remediation by chelating metals, improving their solubility, and facilitating root uptake. Plants then accumulate or transform these metals, reducing their toxicity in the soil. Alongside this, biosurfactants indirectly improve nutrient availability by mobilizing nutrients (like Fe, Zn, Mn) in a similar chelation process, supporting plant growth while detoxifying the environment. Figure 4 reflects how the phytoremediation process increases nutrient availability and heavy metal bioremediation by biosurfactant. The method of Increased Nutrient Availability and Heavy Metal Bioremediation by biosurfactants involves the following steps [1].

Step 1: Addition of Microbial Biosurfactants: Soil microorganisms release Biosurfactants, i.e., surface-active biomolecules and exopolysaccharides, which reduce surface tension, solubilize hydrophobic compounds, and interact with metal cations in the soil.

Step 2: Complexation/Chelation with Metal Cations: Biosurfactants and exopolysaccharides form complexes with metal ions (Pb^{2+}, Cd^{2+}, Cu^{2+}, etc.) through chelation. Chelation stabilizes the metals, preventing them from being free toxic ions in the soil solution, which reduces toxicity while also mobilizing them for plant uptake.

Step 3: Enhanced Bioavailability and Root Uptake: Once complexed, the metals become more bioavailable in the rhizosphere (root zone). Plant roots, aided by root exudates, absorb these metal-biosurfactant complexes more efficiently. Rhizosphere processes such as rhizofiltration (metal absorption by roots) and rhizoimmobilization (metal entrapment in soil-root complex) coincide.

Step 4: Precipitation of Biosurfactant: After metal uptake, some biosurfactants are precipitated out of the complex and recycled back into the soil environment. This ensures continuous interaction of microbial biosurfactants with available metal cations.

Step 5: Phytoaccumulation/Phyto-transformation: Inside the plant system, heavy metals undergo two key processes: (a) Phytoaccumulation: Metals are stored in plant tissues (roots, shoots, leaves) and (b) Phyto-transformation: Plants metabolically transform or detoxify metal ions and associated compounds. Some metals may also undergo phytovolatilization, being converted into volatile forms and released via leaves.

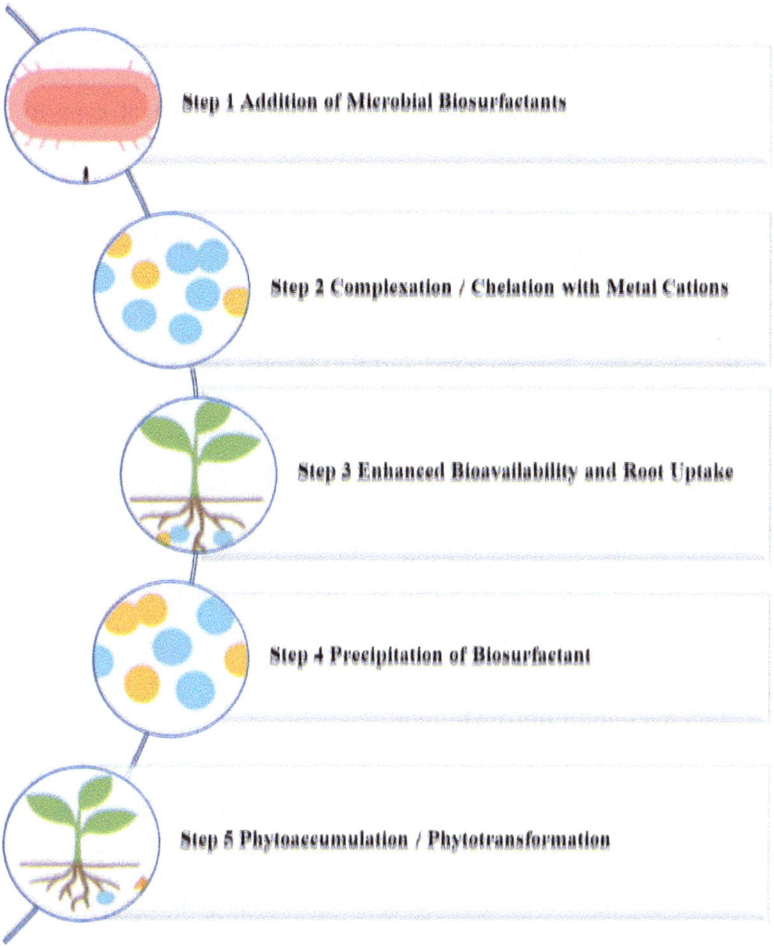

Fig. 4 Increased nutrient availability and heavy metal bioremediation by biosurfactant [1]

In addition, Rhizostabilisation (Roots and microbial products immobilize toxic metals, reducing leaching) and Rhizofiltration (Roots actively filter and concentrate metals from the soil solution) also play a vital role in the phytoremediation process.

Biosurfactants in Pesticide Bioremediation

It is estimated that only around 1% of pesticides really reach the pest species they were intended to kill, with the majority ending up in the environment due to their poor water solubility [150]. The use of biosurfactants aids in the remediation of these residues because they promote microbial decomposition and increase solubility [151, 152]. Rhamnolipids can dissolve pesticides like endosulfan and quinalphos [132, 133].

Some of the mechanisms involved are ion exchange, counter-ion binding, electrostatic interactions, and precipitation-dissolution [153, 154]. The rhamnolipids generated by *Pseudomonas rhoeophila* S211 helped solubilize pesticides because of their high emulsification index (90%) and oil displacement capacity [155]. Similarly, rhamnolipids isolated from *Lysinibacillus sphaericus* IITR51 successfully dissolved several pesticides and showed remarkable stability when exposed to different temperatures, salinities, and pH levels [155]. PGPR facilitates rhizoremediation and pesticide breakdown in polluted soils that can produce biosurfactants, such as *Burkholderia cenocepacia* BSP3 [156].

7.1.3 Organic Waste Management and Soil Health Improvement

The utilization of biosurfactants has improved the degradation of organic waste and contributed to the objectives of the circular economy in the treatment of municipal solid waste, crop residues, and composting processes [157, 158]. Because of their emulsifying properties, they are excellent biodegraders of oil and hydrocarbon residues found in organic waste streams. Additionally, biosurfactants enhance soil wettability, which in turn increases nutrient uptake by plants and helps fertilizers spread evenly [159, 160].

7.1.4 Challenges and Future Outlook

Soil remediation with microbial biosurfactants is successful, but the cost of producing these chemicals is high. One of the main obstacles preventing wider use in the field is the high production cost [120]. One way to cut expenses and make fermentation more sustainable is to use agro-industrial waste as a substrate [13]. Modern genetic engineering techniques, like CRISPR-based gene editing, have the potential to enhance the specificity and productivity of biosurfactants.

The rhizosphere may harbor new strains that produce biosurfactants; functional metagenomics and route engineering could aid in their discovery. Performing thorough Life Cycle Assessments (LCAs) and Life Cycle Sustainability Analyses (LCSAs) is crucial for confirming sustainability and financial viability [156]. For field applications to be effective, environmental parameters, including temperature, pH, and pollutant concentration, need to be optimized. The use of microbial biosurfactants in hydrocarbon, heavy metal, and pesticide remediation provides a more environmentally friendly alternative to synthetic surfactants. Sustainable agriculture relies on its multifunctional roles, which include degrading pollutants and promoting plant development.

7.2 Elimination of Phytopathogens and Plant Growth Promotion

A significant portion of the world's agricultural output, around 30%, is lost due to plant diseases [161]. These biosurfactants boost plant immunity and inhibit several plant diseases due to their antibacterial and antifungal properties. Biosurfactants derived from bacteria, such as lipopeptides and glycolipids, have antimicrobial properties that make them potential biocontrol agents for environmentally friendly farming [85]. Parasitis, antibiosis, competition, hypovirulence, and induced systemic resistance are some of the ways biosurfactants produced by rhizobacteria such as *Bacillus* and *Pseudomonas* promote plant growth. In agriculture, surfactants boost the antagonistic actions of microorganisms and microbial compounds that encourage plant development [162, 163]. Several systems have shown enhanced insecticidal activity when treated with biosurfactants [162, 164, 165]. A fungus called Myrothecium verrucaria is used in conjunction with biosurfactants to help eradicate weed species. However, this approach could have adverse effects on biodiversity and production [96]. Aspergillus species contaminate Cottonseed, peanuts, and corn, but they also prevent aflatoxin synthesis [166].

Some notable examples of biosurfactants in phytopathogen inhibition, directly or indirectly, at different processes related to agriculture are as follows:

- *Pseudomonas* and *Bacillus* biosurfactants control *Pectobacterium* and *Dickeya* spp. (soft rot pathogens) [167].
- Rhamnolipids are effective against fungi like *Alternaria*, *Botrytis*, *Fusarium*, *Rhizoctonia*, *Pythium*, *Phytophthora*, and *Plasmopara* [77].
- *P. putida* biosurfactants lyse *Phytophthora capsici* zoospores (damping-off in cucumber) [56].
- Lipopeptide biosurfactants from *Bacillus* spp. inhibit *Fusarium*, *Aspergillus*, and *Bipolaris sorokiniana* [168], can be considered as biocontrol agents.
- *Brevibacillus brevis* HOB1 produces surfactins with strong antibacterial and antifungal properties [169].
- *Pseudomonas fluorescens* strains display antifungal activity [57].
- Fluorescent pseudomonads inhibit *Pythium ultimum* (causative agent of damping-off and root rot of plants), *Fusarium oxysporum* (causes wilting in crop plants), and *Phytophthora cryptogea* (causes rotting of fruits and flowers) [170].
- Biosurfactants from *Pseudomonas* sp. suppress *Verticillium* spp. (potato wilt) [171].
- Viscosin and tensin (dual-function biosurfactants from *Pseudomonas*) inhibit *Rhizoctonia solani* and *Pythium ultimum* [81].
- *Bacillus subtilis* controls *Colletotrichum gloeosporioides* on papaya [172].
- *Staphylococcus* sp. biosurfactant inhibits *P. aeruginosa*, a plant pathogen [172].
- *B. subtilis* SPB1 biosurfactant reduced *Agrobacterium tumefaciens* disease by 83% in tomatoes [173].
- *B. subtilis* AKP produced surfactins, iturins, and fengycins, inhibiting *Colletotrichum capsici* (61.5%) [93].

- Rhamnolipids from *P. aeruginosa* JS29 inhibited *Alternaria solani* (73%) and *C. capsici* (97–98%) [174, 175].
- Rhamnolipids inhibit *Alternaria alternata, Verticillium dahliae, Mucor circinelloides, Aspergillus niger*, and *A. carbonarius* [94, 176].
- Rhamnolipids induce systemic resistance in *Arabidopsis* against *Botrytis cinerea* and *Hyaloperonospora arabidopsidis* [42, 177].
- Cyclic lipopeptides (CLPs) from *Bacillus* and *Pseudomonas* target diverse pathogens [77].
- *Bacillus methylotrophicus* OB9 inhibits *Xanthomonas campestris* [77].
- *P. aeruginosa* DR1 and *B. amyloliquefaciens* RHNK 22 produce rhamnolipids and iturins active against *Fusarium oxysporum* [146].
- Product Serenade ASO 6 (based on *B. subtilis* QST 713 + fengycin) is effective against fungal pathogens [156, 178].
- *P. aeruginosa* rhamnolipids enhance grapevine immunity against *B. cinerea* [179].
- *P. fluorescens* CLPs show antifungal activity [57].
- *P. aeruginosa* PNA1 and *Pseudomonas* CMR12a produce phenazines and CLPs, suppressing *Pythium* and *Rhizoctonia* spp. [180, 181].
- *B. licheniformis* biosurfactant controls *R. solani* in faba beans [144].
- *Acinetobacter* glycolipid biosurfactants inhibit *X. oryzae* [145].
- *Brevibacillus brevis* and *Burkholderia* spp. produce broad-spectrum antibacterial biosurfactants [169, 182].
- *B. subtilis* exhibits bio-nematicidal activity against *Meloidogyne incognita* [183].
- Biosurfactants reduce aflatoxin production by *Aspergillus* spp. [166].

7.3 Insect Pest Control and Pesticide Enhancement

The effectiveness of pesticides is reduced because insect cuticles are hydrophobic. Decreasing surface tension, surfactants aid in the dispersion and adhesion of pesticides. They can improve the biopesticide activity or remove the lipids from the insect's cuticle, which can cause dehydration and death [184, 185]. An environmentally preferable option is microbial biosurfactants.

Chemical pesticides are used extensively; however, they pose problems related to soil and water contamination, pest resistance, and human health concerns. The majority of pesticides pollute the environment, whereas just a small fraction, about 0.1%, actually reach their intended objectives [186, 187]. When developing pesticides, surfactants are often included to increase the pesticide's dispersion, spreading, and target reach [188, 189]. But using them too frequently can raise pesticide residues or damage the phytobiome [190, 191]. There are concerns about the toxicity of synthetic surfactants, such as Triton X-100, Tween 20, Silwet L-77, and others, although they increase delivery [7, 192]. Biosurfactants, on the other hand, are gaining popularity for direct application, are safer, and biodegrade [77, 193]. At the same time that they promote plant growth and the natural microbial flora, biosurfactants kill insect pests by destroying their cell walls and causing them to dehydrate [138, 194, 195].

These biosurfactants exhibit direct insecticidal activity or enhance biocontrol systems like entomopathogenic fungi:

- Rhamnolipids not only target fungi but also act against zoospore-forming pathogens and contribute to plant immunity. Their multifaceted role also allows for insecticidal action [42, 92, 170, 196].
- Biosurfactant from *Pseudomonas* demonstrated insecticidal activity against the green peach aphid (*Myzus persicae*) [92].
- *B. subtilis* shows bio-nematicidal activity against *Meloidogyne incognita* [183].
- Biosurfactants assist in weed eradication when used with *Myrothecium verrucaria* [96].

Lipopeptide biosurfactants from *Bacillus* spp. disrupt insect membranes via cytotoxic reactions through pore formation, cellular leakage, and interference with cellular signaling [197]. Examples:

- Lipopeptide biosurfactants from *B. amyloliquefaciens* AG1 show larvicidal activity against *Tuta absoluta*, *Spodoptera littoralis*, and *Myzus persicae*, and surfactins of *B. subtilis* showed larvicidal activity against *Spodoptera littoralis* [198, 199].
- *B. velezensis* PHP1601 biosurfactants exhibit antagonistic activity against *Lucilia cuprina* larvae [200].
- Rhamnolipids enhance the efficacy of entomopathogenic fungi (*Cordyceps javanica, Beauveria bassiana*) against *Bemisia tabaci* [201].
- *Pseudomonas* biosurfactants kill *M. persicae* [92].

Importantly, biosurfactants like those from *Candida sphaerica* show no phytotoxicity [202], indicating their environmental safety.

Environmentally preferable to chemical pesticides, these green surfactants have the added benefit of promoting composting through microbial development [203, 204]. However, to achieve specificity, it is crucial to adjust manufacturing tactics for biosurfactants to tailor their chemical composition. This is because biosurfactants can also affect non-target flora [7].

7.4 Enhancement of Geologic Carbon Storage and Environmental Sustainability

Injecting carbon dioxide into deep subterranean formations like saline aquifers or depleted gas reservoirs is known as Geological Carbon Storage (GCS), and it's a recognized approach for lowering atmospheric CO_2 levels [205]. Protecting the CO_2 immobilization for the long run is a big obstacle in this approach. To improve the efficiency of pore-scale sweeps in CO_2–water–rock systems, biosurfactants lower the capillary pressure and increase the efficacy of GCS.

Species of bacteria like *Bacillus* are capable of producing biosurfactants in subterranean environments like oil reservoirs and GCS sites, which makes them valuable for in-situ applications [206]. An economical and environmentally acceptable method for producing biosurfactants was suggested by Freitas et al. [207] it involves reusing bacteria from contaminated soils, effluents, and wastewater in conjunction with inexpensive agricultural waste products, including molasses, crop leftovers, and oils. Bioremediation, nutrient solubilization, and low toxicity are three key reasons why biosurfactants, such as arthrofactins, surfactins, and sophorolipids, are gaining popularity. With less hemolytic action on human red blood cells than their synthetic equivalents, these bio-based compounds are very biodegradable [208].

The effectiveness of biosurfactants in protecting the environment has been confirmed by experiments, such as the Modified MITI 301 C biodegradability assay. For example, sophorolipids began to break down shortly after cultivation and achieved a biodegradability (BOD/TOD ratio) of 61% within eight days, according to Hirata et al. [209]. Synthetic surfactants, on the other hand, exhibited zero degradation during that time. In a similar vein, Kłosowska-Chomiczewska et al. [210] found that biosurfactants degraded 60% in just 10 days, suggesting that they did not remain long in the environment. Even in agricultural settings, biosurfactants show minimal phytotoxicity. Cabbage had a germination index of more than 80% according to Jimoh and Lin [159], indicating that biosurfactants are not harmful to plants. For these reasons, biosurfactants are a more environmentally friendly and long-term viable option than synthetic surfactants for carbon sequestration and other uses.

7.5 Enhancement of Beneficial Plant–Microbe Interactions

It is crucial for rhizobacteria to successfully interface with plant root surfaces for them to have a favorable impact on plants. Microbial characteristics like motility, biofilm development, and the secretion of signaling molecules called quorum-sensing chemicals encourage this interaction [7]. Acyl homoserine lactones (AHLs) are essential for rhizobacteria to produce antifungal chemicals. The significance of AHLs in rhizosphere competence, or the capacity of beneficial microorganisms to efficiently colonize root zones, is supported by studies showing that their concentration is much higher in the rhizosphere than in bulk soil. In addition, AHLs have a role in controlling the formation of exopolysaccharides, which are essential for the steady growth of biofilms on the roots of plants [160, 211].

Research has shown that biosurfactants, specifically rhamnolipids produced by *Pseudomonas* spp., can influence the processes involved in quorum sensing [212]. Furthermore, these naturally occurring surfactants facilitate cellular communication and differentiation, biofilm formation, and microbial motility [16, 213–215]. Biosurfactants improve plant growth and health by allowing helpful bacteria to colonize roots more effectively.

Soil microbes can better metabolize and obtain hydrophobic nutrients when biosurfactants are present. In addition to helping plants grow, they improve soil wettability, which means water and chemical fertilizers are distributed more evenly. In sum, biosurfactants' many uses underscore their significance in promoting interactions between plants and microbes and their worth as environmentally benign instruments for promoting sustainable agriculture.

7.6 Potential of Biosurfactants in Pesticide Industries

When it comes to making insecticides, herbicides, and fungicides, surfactants are indispensable as adjuvants. Pesticide applications are made more effective by using these agents, which emulsify, disperse, moisten, and distribute the pesticide. The pesticide manufacturing business currently relies heavily on synthetic surfactants, which can be classified as either anionic, cationic, amphoteric, or non-ionic [216, 217]. Furthermore, these substances give new agricultural insecticidal formulations, defining capabilities [216].

The extensive usage of synthetic surfactants, however, brings up serious environmental issues. These surfactants tend to build up in agricultural soils when applied in excess, which has adverse effects on soil color, texture, and plant growth. Along with remaining on the produce, they contaminate groundwater via leaching [218, 219]. Because of their toxicity, resistance to degradation, and ability to contaminate soil, air, and water, as well as their persistence in the environment, these substances are classified as hazardous organic pollutants [220]. Considering these environmental dangers, the pesticide industry is increasingly turning to biodegradable, environmentally benign biosurfactants as alternatives to synthetic surfactants. Agrochemical formulations have the potential to diminish their ecological impact greatly; nevertheless, biosurfactants provide a sustainable alternative [221]. To create pesticides that are less harmful to the environment and more effective, the agrochemical industry is looking at new formulation methods, such as combining biosurfactants with polymers.

Soil bacteria with bioremediation capabilities offer a novel approach to reducing pollution from surfactants. Isolated from rice fields, bacterial strains including *Pseudomonas* spp. and *Burkholderia* spp. can use synthetic surfactants as carbon sources, allowing them to break them down [222]. Reducing pollutants and boosting soil health are two benefits of this biological technique.

Biosurfactants are gaining recognition as effective biopesticides, in addition to their adjuvant functions. Chemical pesticides with a wide range of applications have been used for a long time to control various pests, including weeds, nematodes, insects, flies, rodents, fungi, and harmful bacteria. However, this approach often has unintended consequences, including the development of pest resistance, residual toxicity, and increasing costs associated with new chemical formulations. Because of these restrictions, people are looking for long-term solutions to pest problems [84, 223].

Biosurfactants, particularly microbial lipopeptides, have antifungal and insecticidal effects. As an example, there are bacterial biosurfactants that have demonstrated efficacy against the fruit fly (*Drosophila melanogaster*) [224]. These metabolites produced by microbes have multiple uses: they reduce abiotic stress, solubilize nutrients, and limit the proliferation of pests [40, 156, 225].

Biosurfactants, which incorporate green chemistry principles, are being developed for use as pesticides. This is in line with worldwide initiatives to produce safer and more sustainable farming methods. Biosurfactants are a new and exciting development in the agrochemical sector because of their dual function as agents for formulation and active components for biocontrol.

8 Industrial Production of Biosurfactant-Based Agrochemicals

Environmentally preferable to synthetic surfactants, biosurfactants have recently become popular in the agricultural sector. An industrial-scale synthesis of biosurfactants is necessary for large-scale applications such as pesticide delivery, soil remediation, and plant growth encouragement. Steps include making sure the process is cost-effective and scalable, optimizing fermentation technologies, and choosing microbial strains that provide high yields.

8.1 Microorganisms Producing Biosurfactants

Biosurfactants can be produced by a wide range of microorganisms, including bacteria, yeasts, and fungi. Biosurfactant type (glycolipids, lipopeptides, phospholipids, etc.), agrochemical application, and substrate availability all play a role in the microbe selection process.

Common Biosurfactant-Producing Microorganisms

Pseudomonas aeruginosa—Produces *rhamnolipids*, effective in pest control and emulsification of agrochemicals [76].

Bacillus subtilis—Produces *surfactin*, a potent lipopeptide with antifungal and antibacterial activity, suitable for biocontrol formulations [78].

Candida bombicola—Yields *sophorolipids*, useful in biodegradable wetting agents and foliar sprays [226].

Acinetobacter calcoaceticus—Known for *emulsan*, used in soil remediation and oil-contaminated agricultural sites.

Pseudozyma spp.—Produce *mannosylerythritol lipids (MELs)*, with potential in pest control and fertilizer formulation [227].

8.2 Production Technologies

Industrial production of biosurfactants involves several fermentation strategies and process optimizations to reduce cost and improve yield [13, 228].

Fermentation Approaches

- Submerged Fermentation (SmF): The most used technique, where microorganisms grow in a liquid medium. Easier to scale but requires more downstream processing.
- Solid-State Fermentation (SSF): Microbes grow on moist solid substrates (e.g., bran, bagasse). More sustainable and cost-effective, especially for agro-industrial waste valorization [15].
- Co-culture Techniques: Using two or more strains for synergistic biosurfactant production, enhancing yield and bioactivity.
- Fed-Batch and Continuous Fermentation: Improve yield and productivity of biosurfactants like rhamnolipids and surfactin [229].

Substrate Utilization

- Utilization of molasses, whey, crude glycerol, used frying oil, and other agro-industrial by-products enhances sustainability and lowers costs [230].

Downstream Processing [231]

Involves cell separation, acid precipitation, solvent extraction, and purification.

Novel methods like ultrafiltration and foam fractionation are being integrated to reduce costs.

Cost Reduction Strategies

Use of agro-industrial wastes (molasses, whey, glycerol, etc.) as carbon sources.

Genetic engineering and metabolic pathway optimization to increase biosurfactant titers.

9 Intellectual Property and Technological Innovation

With the rise in sustainable agriculture, biosurfactants have emerged as a focus for technological innovation and patent protection. Patents play a critical role in encouraging investment in R&D, securing proprietary production methods, and facilitating the commercialization of biosurfactant-based agrochemicals [232].

9.1 Patents on Biosurfactants and Their Agricultural Applications

Biosurfactants are protected under patents in multiple domains:

1. Microbial Strain Patents

Patents have been filed for genetically engineered or wild-type strains capable of producing biosurfactants with high efficiency, thermal stability, or specific antimicrobial activity.

- Example: US Patent No. US8012645B2—*Recombinant Bacillus subtilis for enhanced surfactin production.*

2. Process and Formulation Patents

Innovations related to fermentation technologies, downstream processing, and formulations with biopesticides or fertilizers are widely patented.

- Example: Patent EP2609357B1—Process for producing sophorolipids using non-pathogenic yeast on renewable feedstocks.

3. Application-Specific Patents

Biosurfactants have been patented as components of fungicides, herbicide adjuvants, biostimulants, and biocontrol agents.

- Example: Indian Patent IN202021034376—*Use of biosurfactants in integrated pest management (IPM) systems.*

4. Green Innovations and Sustainable Packaging

New patents also address eco-friendly packaging, slow-release formulations, and synergistic combinations with nanomaterials or biofertilizers.

10 Challenges in Agricultural Applications of Biosurfactants

Several major obstacles prevent biosurfactants from being widely used, despite their considerable potential in sustainable agriculture. Production cost, environmental stability, and regulatory limitations are three of the most important of these concerns.

10.1 Production Cost

The high production cost of biosurfactants is one of the main obstacles to their use in agriculture. Producing biosurfactants can be 20–30% more expensive than synthetic surfactants [14]. This is mainly because of the intricate fermentation processes and the downstream processing costs involved. Costs associated with substrates alone can make up 10–50% of overall production costs [233]. Scalability and consistent quality are still challenges, even if agro-industrial wastes (such as molasses, bagasse, peels, and brewery waste) provide a low-cost alternative [234].

The costs associated with purification are even higher, accounting for 60–80% of the overall cost of producing biosurfactants [207, 235]. Sophorolipids and other biosurfactants can be mass-produced for $2–$5/kg, while iturin and surfactin, on the other hand, can command prices of $10,000–$15,000/kg owing to their scarcity [236].

Because of this price difference, biosurfactants are not as competitive as their synthetic counterparts in agricultural applications, where saving money is essential.

To address this:

- Technologies like in-situ foam fractionation and ultrafiltration can help reduce recovery costs.
- Use of genetically engineered strains and metabolic optimization may enhance yields and reduce overall costs.

10.2 Stability Concerns

The stability of biosurfactants, being biological molecules, might be a concern in agricultural settings with widely varying environmental factors. In real-world settings, biosurfactants may lose some of their efficacy due to deterioration caused by temperature sensitivity, pH fluctuations, and ultraviolet radiation. Storage and transportation over extended periods of time may change their surface-active characteristics, which could impact how well they work as soil conditioners, biopesticides, or adjuvants. Although biosurfactants are better for the environment than synthetic surfactants, their biodegradability means they don't last as long and may require stabilizing formulations or more frequent applications. This contrasts with synthetic surfactants, which are often made to be chemically stable.

To overcome these concerns:

- Research is focusing on custom-formulated biosurfactants, the use of polymer carriers, or encapsulation technologies to improve shelf-life and stability.
- Combining biosurfactants with nanotechnology (e.g., nanocarrier systems or "myconanotechnology") may enhance their stability and efficacy in field applications.

10.3 Regulatory Considerations

Another major obstacle to the widespread use of biosurfactants in agriculture is the absence of well-defined regulatory policies. The use of biosurfactants in agrochemicals is a recent development. Unlike synthetic fertilizers and pesticides, there are currently no established norms for the testing of safety, evaluation of risk, or recommendations for their application. Biosurfactants may not be ready for commercial usage until regulatory agencies review and approve them based on toxicological, ecotoxicological, and environmental impact data. Even though the end product is not hazardous, biosafety concerns might still arise when biosurfactants are sourced from harmful microbes like *Pseudomonas aeruginosa* [237]. Additionally, market fragmentation and delays in product commercialization are caused by differences in regulatory processes among countries.

Addressing regulatory issues involves:

- Comprehensive environmental fate studies, focusing on biodegradability, persistence, and ecological risks.
- Establishment of international standards and harmonized policies for biosurfactant registration and use in agriculture.
- Increased emphasis on non-pathogenic production strains and GRAS (Generally Recognized as Safe) organisms to streamline approval.

11 Conclusion and Future Perspectives

The microbial biosurfactants are a great alternative to the synthetic surfactants used in the agrochemical and agricultural sectors. They are sustainable, environmentally benign, and show great promise. Biodegradability, low toxicity, plant compatibility, stability under varied environmental conditions, and multifunctional capabilities are some of the promising characteristics of these naturally occurring surface-active chemicals, which are primarily produced by microbes such as *Bacillus* and *Pseudomonas*.

Their usefulness in farming extends to numerous areas:

Environmental Safety: Biosurfactants contribute to sustainability efforts by lessening the impact of agriculture on the environment.

Soil quality, nutrient availability, and absorption are all improved by enhanced agricultural practices, which in turn promote healthier and more productive crops by facilitating bioremediation of organic contaminants and heavy metals.

Biosurfactants are a promising alternative to traditional chemical pesticides due to their antibacterial and insecticidal properties, which contribute to their biocontrol capabilities. Biosurfactant production using agricultural and industrial waste as feedstock is a sustainable approach to waste management and a supporter of the circular economy concept. Biosurfactants have several potential applications in agriculture, but their high production costs, limited microbial diversity, and scaling issues have prevented their widespread adoption. A comprehensive strategy, including researchers from several fields, from basic science to industrial application, is needed to overcome these obstacles.

The following areas require joint investigation and concentrated effort to realize microbial biosurfactants' full potential:

- Discovery of Various Microbial Species: Novel biosurfactants with qualities suitable for specific agricultural uses can be discovered by expanding the search for microorganisms that produce biosurfactants.
- Production Process Optimization: To improve economic feasibility, it is vital to develop scalable, low-cost fermentation and extraction technologies using renewable and agro-industrial substrates.
- Progress in Genetic and Synthetic Biology: Biosurfactant yield, specificity, and performance can be improved by genetic engineering and route optimization using tools like CRISPR and functional metagenomics.

Effective integrated pest and nutrient management techniques should incorporate biosurfactants with organic fertilizers, biocontrol agents, and formulations based on nanotechnology. This will help to incorporate biosurfactants with other sustainable agricultural practices.

- Life Cycle and Sustainability Assessment: Important tools for assessing the social, economic, and environmental implications of biosurfactants across their whole lifecycle, from production to use, include Life Cycle Assessment (LCA) and Life Cycle Sustainability Analysis (LCSA).
- Ecological and Field Studies with Long Time Frames: To make sure biosurfactants are safe and sustainable to use, we need to know how they will affect soil health, microbial communities, and crop yield in the long run.

For biosurfactants to be more widely accepted and used responsibly, there must be a clear regulatory framework in place, and stakeholders must be educated.

To transform laboratory-scale triumphs into marketable commercial goods, it is essential to encourage cooperation between academic institutions, businesses, and government organizations.

Overall, microbial biosurfactants are more than just a replacement for synthetic surfactants; they herald a crucial step towards environmentally conscious and long-term sustainable farming. They play an essential role in lowering pesticide pollution, boosting crop health, and assisting with global sustainability objectives due to their environmental compatibility and multifunctionality. To fully utilize their potential in the agricultural and related industries, it is crucial to maintain investments in research, innovation, and governmental assistance.

References

1. Datta D, Ghosh S, Kumar S, Gangola S, Majumdar B, Saha R, Mazumdar SP, Singh SV, Kar G (2024) Microbial biosurfactants: multifarious applications in sustainable agriculture. Microbiol Res 279:127551
2. Vos R, Bellù LG (2019) Global trends and challenges to food and agriculture into the 21st century. In: Sustainable food and agriculture, p 11
3. Sunkara M (2023) Agrochemicals: an overview of their types and impact on agriculture. J Agric Sci Food Res 14(2):151
4. Anjaria P, Vaghela S (2024) Toxicity of agrochemicals: impact on environment and human health. J Toxicol Stud 2(1):250
5. Ganguly RK, Mukherjee A, Chakraborty SK, Verma JP (2021) Impact of agrochemical application in sustainable agriculture. In: New and future developments in microbial biotechnology and bioengineering. Elsevier, p 15
6. Sebastian A, Nangia A, Prasad MNV (2020) Advances in agrochemical remediation using nanoparticles. In: Agrochemicals detection, treatment and remediation. Elsevier, p 465
7. Sachdev DP, Cameotra SS (2013) Biosurfactants in agriculture. Appl Microbiol Biotechnol 97:1005
8. Marchant R, Banat IM (2012) Biosurfactants: a sustainable replacement for chemical surfactants? Biotechnol Lett 34(9):1597
9. Kuyukina MS, Ivshina IB, Makarov SO, Litvinenko LV, Cunningham CJ, Philp JC (2005) Effect of biosurfactants on crude oil desorption and mobilization in a soil system. Environ Int 31(2):155
10. Shekhar S, Sundaramanickam A, Balasubramanian T (2015) Biosurfactant producing microbes and their potential applications: a review. Crit Rev Environ Sci Technol 45(14):1522
11. Lang S, Wullbrandt D (1999) Rhamnose lipids–biosynthesis, microbial production and application potential. Appl Microbiol Biotechnol 51(1):22
12. Naughton P, Marchant R, Naughton V, Banat IM (2019) Microbial biosurfactants: current trends and applications in agricultural and biomedical industries. J Appl Microbiol 127(1):12
13. Satpute SK, Banpurkar AG, Dhakephalkar PK, Banat IM, Chopade BA (2010) Methods for investigating biosurfactants and bioemulsifiers: a review. Crit Rev Biotechnol 30(2):127
14. Hazra C, Kundu D, Ghosh P, Joshi S, Dandi N, Chaudhari A (2011) Screening and identification of *Pseudomonas aeruginosa* AB4 for improved production, characterization and application of a glycolipid biosurfactant using low-cost agro-based raw materials. J Chem Technol Biotechnol 86(2):185
15. Banat IM (1995) Biosurfactants production and possible uses in microbial enhanced oil recovery and oil pollution remediation: a review. Bioresour Technol 51(1):1
16. Van Hamme JD, Singh A, Ward OP (2006) Physiological aspects: part 1 in a series of papers devoted to surfactants in microbiology and biotechnology. Biotechnol Adv 24(6):604
17. Vijayakumar S, Saravanan V (2015) Biosurfactants-types, sources and applications. Res J Microbiol 10(5):181
18. Mehta S, Sharma S, Mehta N, Cameotra SS (2010) Biomimetic amphiphiles: properties and potential use. Biosurfactants 102
19. Zhong H, Liu G, Jiang Y, Yang J, Liu Y, Yang X, Liu Z, Zeng G (2017) Transport of bacteria in porous media and its enhancement by surfactants for bioaugmentation: a review. Biotechnol Adv 35(4):490
20. Kumar S, Choudhary AK, Suyal DC, Makarana G, Goel R (2022) Leveraging arsenic resistant plant growth-promoting rhizobacteria for arsenic abatement in crops. J Hazard Mater 425:127965
21. Kumar A, Singh SK, Kant C, Verma H, Kumar D, Singh PP, Modi A, Droby S, Kesawat MS, Alavilli H (2021) Microbial biosurfactant: a new frontier for sustainable agriculture and pharmaceutical industries. Antioxidants 10(9):1472

22. Banat IM, Franzetti A, Gandolfi I, Bestetti G, Martinotti MG, Fracchia L, Smyth TJ, Marchant R (2010) Microbial biosurfactants production, applications and future potential. Appl Microbiol Biotechnol 87:427
23. Pacwa-Płociniczak M, Płaza GA, Piotrowska-Seget Z, Cameotra SS (2011) Environmental applications of biosurfactants: recent advances. Int J Mol Sci 12(1):633
24. Cameotra SS, Makkar RS, Kaur J, Mehta SK (2010) Synthesis of biosurfactants and their advantages to microorganisms and mankind. Biosurfactants 261
25. Amaral PF, Coelho MAZ, Marrucho IM, Coutinho JA (2010) Biosurfactants from yeasts: characteristics, production and application. Biosurfactants 236
26. Koglin A, Doetsch V, Bernhard F (2010) Molecular engineering aspects for the production of new and modified biosurfactants. Biosurfactants 158
27. Xu Q, Nakajima M, Liu Z, Shiina T (2011) Biosurfactants for microbubble preparation and application. Int J Mol Sci 12(1):462
28. Lima TM, Procópio LC, Brandão FD, Carvalho AM, Tótola MR, Borges AC (2011) Biodegradability of bacterial surfactants. Biodegradation 22:585
29. Nitschke M, Costa S (2007) Biosurfactants in food industry. Trends Food Sci Technol 18(5):252
30. Das P, Mukherjee S, Sivapathasekaran C, Sen R (2010) Microbial surfactants of marine origin: potentials and prospects. Biosurfactants 88
31. Karnwal A, Shrivastava S, Al-Tawaha ARMS, Kumar G, Singh R, Kumar A, Mohan A, Yogita, Malik T (2023) Microbial biosurfactant as an alternate to chemical surfactants for application in cosmetics industries in personal and skin care products: a critical review. BioMed Res Int 2023(1):2375223
32. Rahman PK, Gakpe E (2008) Production, characterisation and applications of biosurfactants—review. Biotechnology 7(2):360
33. Varjani SJ, Upasani VN (2017) Critical review on biosurfactant analysis, purification and characterization using rhamnolipid as a model biosurfactant. Bioresour Technol 232:389
34. Dobler L, Vilela LF, Almeida RV, Neves BC (2016) Rhamnolipids in perspective: gene regulatory pathways, metabolic engineering, production and technological forecasting. New Biotechnol 33(1):123
35. Singh R, Rathore D (2019) Impact assessment of azulene and chromium on growth and metabolites of wheat and chilli cultivars under biosurfactant augmentation. Ecotoxicol Environ Saf 186:109789
36. Shao B, Liu Z, Zhong H, Zeng G, Liu G, Yu M, Liu Y, Yang X, Li Z, Fang Z (2017) Effects of rhamnolipids on microorganism characteristics and applications in composting: a review. Microbiol Res 200:33
37. De Almeida DG, Soares Da Silva RDCF, Luna JM, Rufino RD, Santos VA, Banat IM, Sarubbo LA (2016) Biosurfactants: promising molecules for petroleum biotechnology advances. Front Microbiol 7:1718
38. Mukherjee AK, Das K (2010) Microbial surfactants and their potential applications: an overview. Biosurfactants 54
39. Mulligan CN (2005) Environmental applications for biosurfactants. Environ Pollut 133(2):183
40. Akbari S, Abdurahman NH, Yunus RM, Fayaz F, Alara OR (2018) Biosurfactants—a new frontier for social and environmental safety: a mini review. Biotechnol Res Innov 2(1):81
41. Singh R, Glick BR, Rathore D (2018) Biosurfactants as a biological tool to increase micronutrient availability in soil: a review. Pedosphere 28(2):170
42. Vatsa P, Sanchez L, Clement C, Baillieul F, Dorey S (2010) Rhamnolipid biosurfactants as new players in animal and plant defense against microbes. Int J Mol Sci 11(12):5095
43. Global agrochemicals market size & outlook, 2023–2030. Available from: https://www.grandviewresearch.com/horizon/outlook/agrochemicals-market-size/global?utm_source=chatgpt.com
44. Agrochemical market size. Available from: https://www.globalgrowthinsights.com/market-reports/agrochemicals-market-108710?utm_source=chatgpt.com

45. India agrochemicals market, opportunity, growth drivers, industry trend analysis and forecast, 2024–2032. Available from: https://www.marketresearch.com/One-Off-Global-Market-Insights-v4130/India-Agrochemicals-Opportunity-Growth-Drivers-40939545/
46. India agrochemicals market growth analysis—forecast trends and outlook (2025–2034). Available from: https://www.expertmarketresearch.com/reports/india-agrochemicals-market?utm_source=chatgpt.com
47. Biofertilizers: nurturing crops, preserving the planet. Available from: https://www.reddit.com/r/Yield_Farming/comments/18xgtwk/biofertilizers_nurturing_crops_preserving_the/?utm_source=chatgpt.com
48. Agrochemicals market size, share, trends and forecast by fertilizer type, pesticide type, crop type, and region, 2025–2033. Available from: https://www.imarcgroup.com/agrochemicals-market?utm_source=chatgpt.com
49. GVR report coveragrochemicals market size, share & trends report agrochemicals market size, share & trends analysis report by product (fertilizers, crop protection chemicals), by application (cereal & grains, oilseeds & pulses, fruits & vegetables), by region, and segment forecasts, 2024–2030. Available from: https://www.grandviewresearch.com/industry-analysis/agrochemicals-market
50. Indian agrochemicals market analysis-industry size, share, research report, insights, COVID-19 impact, statistics, trends, growth and forecast 2025–2034. Available from: https://markwideresearch.com/india-agrochemicals-market/?utm_source=chatgpt.com
51. India agrochemicals market size 2025–2029. Available from: https://syovi.com/report/agrochemicals-market-in-india-2025-2029/
52. Economy of India. Available from: https://en.wikipedia.org/wiki/Economy_of_India?utm_source=chatgpt.com
53. Burch AY, Browne PJ, Dunlap CA, Price NP, Lindow SE (2011) Comparison of biosurfactant detection methods reveals hydrophobic surfactants and contact-regulated production. Environ Microbiol 13(10):2681
54. Amani H, Mehrnia MR, Sarrafzadeh MH, Haghighi M, Soudi MR (2010) Scale up and application of biosurfactant from *Bacillus subtilis* in enhanced oil recovery. Appl Biochem Biotechnol 162:510
55. Singh PB, Sharma S, Saini H, Chadha B (2009) Biosurfactant production by *Pseudomonas* sp. and its role in aqueous phase partitioning and biodegradation of chlorpyrifos. Lett Appl Microbiol 49(3):378
56. Kruijt M, Tran H, Raaijmakers JM (2009) Functional, genetic and chemical characterization of biosurfactants produced by plant growth-promoting *Pseudomonas putida* 267. J Appl Microbiol 107(2):546
57. Nielsen TH, Sørensen J (2003) Production of cyclic lipopeptides by *Pseudomonas fluorescens* strains in bulk soil and in the sugar beet rhizosphere. Appl Environ Microbiol 69(2):861
58. Bodour AA, Drees KP, Maier RM (2003) Distribution of biosurfactant-producing bacteria in undisturbed and contaminated arid southwestern soils. Appl Environ Microbiol 69(6):3280
59. Nielsen T, Sørensen D, Tobiasen C, Andersen JB, Christophersen C, Givskov M, Sørensen J (2002) Antibiotic and biosurfactant properties of cyclic lipopeptides produced by fluorescent *Pseudomonas* spp. from the sugar beet rhizosphere. Appl Environ Microbiol 68(7):3416
60. Takeyama H, Wada M, Matsunaga T (2002) Screening of soil bacteria for production of biocleaner. Appl Biochem Biotechnol 98:319
61. da Silva AF, Banat IM, Giachini AJ, Robl D (2021) Fungal biosurfactants, from nature to biotechnological product: bioprospection, production and potential applications. Bioprocess Biosyst Eng 44(10):2003
62. Sarubbo L, Rocha R Jr, Luna J, Rufino RD, Santos V, Banat IM (2015) Some aspects of heavy metals contamination remediation and role of biosurfactants. Chem Ecol 31(8):707
63. Kaawai Y, Yano I, Kaneda K, Yabuuchi E (1988) Ornithine-containing lipids of some *Pseudomonas* species. Eur J Biochem 175(3):633
64. Katemai W (2012) Biosurfactants from yeasts. Walailak J Sci Technol (WJST) 9(1):1

65. Pichot R, Watson RL, Norton IT (2013) Phospholipids at the interface: current trends and challenges. Int J Mol Sci 14(6):11767
66. Gayathiri E, Prakash P, Karmegam N, Varjani S, Awasthi MK, Ravindran B (2022) Biosurfactants: potential and eco-friendly material for sustainable agriculture and environmental safety—a review. Agronomy 12(3):662
67. Ron EZ, Rosenberg E (2002) Biosurfactants and oil bioremediation. Curr Opin Biotechnol 13(3):249
68. Li Z, Zhang Y, Lin J, Wang W, Li S (2019) High-yield di-rhamnolipid production by *Pseudomonas aeruginosa* YM4 and its potential application in MEOR. Molecules 24(7):1433
69. Liu G, Zhong H, Yang X, Liu Y, Shao B, Liu Z (2018) Advances in applications of rhamnolipids biosurfactant in environmental remediation: a review. Biotechnol Bioeng 115(4):796
70. Gaur VK, Regar RK, Dhiman N, Gautam K, Srivastava JK, Patnaik S, Kamthan M, Manickam N (2019) Biosynthesis and characterization of sophorolipid biosurfactant by *Candida* spp.: application as food emulsifier and antibacterial agent. Bioresour Technol 285:121314
71. Kuyukina MS, Ivshina IB, Baeva TA, Kochina OA, Gein SV, Chereshnev VA (2015) Trehalolipid biosurfactants from nonpathogenic *Rhodococcus* actinobacteria with diverse immunomodulatory activities. New Biotechnol 32(6):559
72. Caldeira AT, Arteiro JS, Coelho AV, Roseiro JC (2011) Combined use of LC–ESI-MS and antifungal tests for rapid identification of bioactive lipopeptides produced by *Bacillus amyloliquefaciens* CCMI 1051. Process Biochem 46(9):1738
73. Carrillo C, Teruel JA, Aranda FJ, Ortiz A (2003) Molecular mechanism of membrane permeabilization by the peptide antibiotic surfactin. Biochim Biophys Acta (BBA) Biomembr 1611(1–2):91
74. Sharma R, Singh J, Verma N (2018) Production, characterization and environmental applications of biosurfactants from *Bacillus amyloliquefaciens* and *Bacillus subtilis*. Biocatal Agric Biotechnol 16:132
75. Zhao N, Yang C (2024) Influence of nonionic surfactants on the adsorption and elution of atrazine in agriculturally modified soils of Hetao irrigation district
76. Abdel-Mawgoud AM, Lépine F, Déziel E (2010) Rhamnolipids: diversity of structures, microbial origins and roles. Appl Microbiol Biotechnol 86(5):1323
77. Crouzet J, Arguelles-Arias A, Dhondt-Cordelier S, Cordelier S, Pršić J, Hoff G, Mazeyrat-Gourbeyre F, Baillieul F, Clément C, Ongena M (2020) Biosurfactants in plant protection against diseases: rhamnolipids and lipopeptides case study. Front Bioeng Biotechnol 8:1014
78. Ongena M, Jacques P (2008) *Bacillus* lipopeptides: versatile weapons for plant disease biocontrol. Trends Microbiol 16(3):115
79. De Oliveira MR, Camilios-Neto D, Baldo C, Magri A, Celligoi M (2014) Biosynthesis and production of sophorolipids. Int J Sci Technol Res 3(11):133
80. Kitamoto D, Morita T, Fukuoka T, Konishi M-A, Imura T (2009) Self-assembling properties of glycolipid biosurfactants and their potential applications. Curr Opin Colloid Interface Sci 14(5):315
81. Andersen JB, Koch B, Nielsen TH, Sørensen D, Hansen M, Nybroe O, Christophersen C, Sørensen J, Molin S, Givskov M (2003) Surface motility in *Pseudomonas* sp. DSS73 is required for efficient biological containment of the root-pathogenic microfungi *Rhizoctonia solani* and *Pythium ultimum*. Microbiology 149(1):37
82. Makkar RS, Rockne KJ (2003) Comparison of synthetic surfactants and biosurfactants in enhancing biodegradation of polycyclic aromatic hydrocarbons. Environ Toxicol Chem 22(10):2280
83. Traudel K, Merten S (2017) Possible food and agricultural application of microbial surfactants: an assessment. In: Biosurfactants and biotechnology. Routledge, 183
84. Markam SS, Raj A, Kumar A, Khan ML (2024) Microbial biosurfactants: green alternatives and sustainable solution for augmenting pesticide remediation and management of organic waste. Curr Res Microb Sci 100266
85. Singh A, Van Hamme JD, Ward OP (2007) Surfactants in microbiology and biotechnology: part 2. Application aspects. Biotechnol Adv 25(1):99

86. Scott MJ, Jones MN (2000) The biodegradation of surfactants in the environment. Biochim Biophys Acta (BBA) Biomembr 1508(1–2):235
87. Takenaka S, Tonoki T, Taira K, Murakami S, Aoki K (2007) Adaptation of *Pseudomonas* sp. strain 7–6 to quaternary ammonium compounds and their degradation via dual pathways. Appl Environ Microbiol 73(6):1797
88. Lima TM, Procópio LC, Brandão FD, Leão BA, Tótola MR, Borges AC (2011) Evaluation of bacterial surfactant toxicity towards petroleum degrading microorganisms. Bioresour Technol 102(3):2957
89. Uzoigwe C, Burgess JG, Ennis CJ, Rahman PK (2015) Bioemulsifiers are not biosurfactants and require different screening approaches. Front Microbiol 245
90. Qi X, Xu X, Zhong C, Jiang T, Wei W, Song X (2018) Removal of cadmium and lead from contaminated soils using sophorolipids from fermentation culture of *Starmerella bombicola* CGMCC 1576 fermentation. Int J Environ Res Public Health 15(11):2334
91. Herman DC, Artiola JF, Miller RM (1995) Removal of cadmium, lead, and zinc from soil by a rhamnolipid biosurfactant. Environ Sci Technol 29(9):2280
92. Kim SK, Kim YC, Lee S, Kim JC, Yun MY, Kim IS (2011) Insecticidal activity of rhamnolipid isolated from *Pseudomonas* sp. EP-3 against green peach aphid (*Myzus persicae*). J Agric Food Chem 59(3):934
93. Kumar A, Rabha J, Jha DK (2021) Antagonistic activity of lipopeptide-biosurfactant producing *Bacillus subtilis* AKP, against *Colletotrichum capsici*, the causal organism of anthracnose disease of chilli. Biocatal Agric Biotechnol 36:102133
94. Thakur P, Saini NK, Thakur VK, Gupta VK, Saini RV, Saini AK (2021) Rhamnolipid the glycolipid biosurfactant: emerging trends and promising strategies in the field of biotechnology and biomedicine. Microb Cell Fact 20:1
95. Sheng X, He L, Wang Q, Ye H, Jiang C (2008) Effects of inoculation of biosurfactant-producing *Bacillus* sp. J119 on plant growth and cadmium uptake in a cadmium-amended soil. J Hazard Mater 155(1–2):17
96. Boyette CD, Walker HL, Abbas HK (2002) Biological control of kudzu (*Pueraria lobata*) with an isolate of *Myrothecium verrucaria*. Biocontrol Sci Technol 12(1):75
97. Robinson BH, Bañuelos G, Conesa HM, Evangelou MW, Schulin R (2009) The phytomanagement of trace elements in soil. Crit Rev Plant Sci 28(4):240
98. Roy A (2017) Review on the biosurfactants: properties, types and its applications. J Fundam Renew Energy Appl 8(2):104–172
99. Mnif I, Ghribi D (2015) Review lipopeptides biosurfactants: mean classes and new insights for industrial, biomedical, and environmental applications. Pept Sci 104(3):129
100. Rufino R, Rodrigues G, Campos-Takaki G, Sarubbo L, Ferreira S (2011) Application of a yeast biosurfactant in the removal of heavy metals and hydrophobic contaminant in a soil used as slurry barrier. Appl Environ Soil Sci 2011(1):939648
101. Rufino RD, Luna J, Sarubbo LA, Rodrigues LRM, Teixeira JAC, Campos-Takaki G (2011) Antimicrobial and anti-adhesive potential of a biosurfactant Rufisan produced by *Candida lipolytica* UCP 0988. Colloids Surf B Biointerfaces 84(1):1
102. Singh P, Cameotra SS (2004) Potential applications of microbial surfactants in biomedical sciences. Trends Biotechnol 22(3):142
103. Olaniran AO, Balgobind A, Pillay B (2013) Bioavailability of heavy metals in soil: impact on microbial biodegradation of organic compounds and possible improvement strategies. Int J Mol Sci 14(5):10197
104. Bustamante M, Duran N, Diez M (2012) Biosurfactants are useful tools for the bioremediation of contaminated soil: a review. J Soil Sci Plant Nutr 12(4):667
105. Wang S, Mulligan CN (2004) An evaluation of surfactant foam technology in remediation of contaminated soil. Chemosphere 57(9):1079
106. Stacey SP, McLaughlin MJ, Çakmak I, Hettiarachchi GM, Scheckel KG, Karkkainen M (2008) Root uptake of lipophilic zinc–rhamnolipid complexes. J Agric Food Chem 56(6):2112
107. Singha LP, Pandey P (2021) Rhizosphere assisted bioengineering approaches for the mitigation of petroleum hydrocarbons contamination in soil. Crit Rev Biotechnol 41(5):749

108. Liao C, Xu W, Lu G, Deng F, Liang X, Guo C, Dang Z (2016) Biosurfactant-enhanced phytoremediation of soils contaminated by crude oil using maize (*Zea mays* L.). Ecol Eng 92:10
109. Marchut-Mikołajczyk O, Drożdżyński P, Polewczyk A, Smułek W, Antczak T (2021) Biosurfactant from endophytic *Bacillus pumilus* 2A: physicochemical characterization, production and optimization and potential for plant growth promotion. Microb Cell Fact 20:1
110. Chopra A, Bobate S, Rahi P, Banpurkar A, Mazumder PB, Satpute S (2020) *Pseudomonas aeruginosa* RTE4: a tea rhizobacterium with potential for plant growth promotion and biosurfactant production. Front Bioeng Biotechnol 8:861
111. Alsohim AS, Taylor TB, Barrett GA, Gallie J, Zhang XX, Altamirano-Junqueira AE, Johnson LJ, Rainey PB, Jackson RW (2014) The biosurfactant viscosin produced by *Pseudomonas fluorescens* SBW 25 aids spreading motility and plant growth promotion. Environ Microbiol 16(7):2267
112. Upadhyay SK, Srivastava AK, Rajput VD, Chauhan PK, Bhojiya AA, Jain D, Chaubey G, Dwivedi P, Sharma B, Minkina T (2022) Root exudates: mechanistic insight of plant growth promoting rhizobacteria for sustainable crop production. Front Microbiol 13:916488
113. Madurwar MV, Ralegaonkar RV, Mandavgane SA (2013) Application of agro-waste for sustainable construction materials: a review. Constr Build Mater 38:872
114. Puyol McKenna P, Naughton PJ, Dooley JS, Ternan NG, Lemoine P, Banat IM (2024) Microbial biosurfactants: antimicrobial activity and potential biomedical and therapeutic exploits. Pharmaceuticals 17(1):138
115. Cerda A, Mejias L, Rodríguez P, Rodríguez A, Artola A, Font X, Gea T, Sánchez A (2019) Valorisation of digestate from biowaste through solid-state fermentation to obtain value added bioproducts: a first approach. Bioresour Technol 271:409
116. Biktasheva L, Gordeev A, Kirichenko A, Kuryntseva P, Selivanovskaya S (2024) Screening of microorganisms from wastes and identification of the optimal substrate for biosurfactant production. Microbiol Res 15(1):152
117. Eras-Muñoz E, Farré A, Sánchez A, Font X, Gea T (2022) Microbial biosurfactants: a review of recent environmental applications. Bioengineered 13(5):12365
118. Domínguez Rivera Á, Martínez Urbina MÁ, López y López VE (2019) Advances on research in the use of agro-industrial waste in biosurfactant production. World J Microbiol Biotechnol 35(10):155
119. Mohanty SS, Koul Y, Varjani S, Pandey A, Ngo HH, Chang J-S, Wong JW, Bui X-T (2021) A critical review on various feedstocks as sustainable substrates for biosurfactants production: a way towards cleaner production. Microb Cell Fact 20(1):120
120. Gudiña EJ, Teixeira JA, Rodrigues LR (2011) Biosurfactant-producing *Lactobacilli*: screening, production profiles, and effect of medium composition. Appl Environ Soil Sci 2011(1):201254
121. Das AJ, Kumar R (2018) Utilization of agro-industrial waste for biosurfactant production under submerged fermentation and its application in oil recovery from sand matrix. Bioresour Technol 260:233
122. Panadare D (2015) Applications of waste cooking oil other than biodiesel: a review. Iran J Chem Eng (IJChE) 12(3):55
123. Sari SK, Trikurniadewi N, Ibrahim SNMM, Khiftiyah AM, Abidin AZ, Nurhariyati T (2020) Bioconversion of agricultural waste hydrolysate from lignocellulolytic mold into biosurfactant by *Achromobacter* sp. BP(1)5. Biocatal Agric Biotechnol 24:101534
124. Begum W, Saha B, Mandal U (2023) A comprehensive review on production of bio-surfactants by bio-degradation of waste carbohydrate feedstocks: an approach towards sustainable development. RSC Adv 13(36):25599
125. Karnwal A (2021) Biosurfactant production using bioreactors from industrial byproducts. In: Biosurfactants for a sustainable future: production and applications in the environment and biomedicine, p 59
126. Sharma D, Ansari MJ, Gupta S, Al Ghamdi A, Pruthi P, Pruthi V (2015) Structural characterization and antimicrobial activity of a biosurfactant obtained from *Bacillus pumilus* DSVP18 grown on potato peels. Jundishapur J Microbiol 8(9):e21257

127. Ayed HB, Azabou MC, Hmidet N, Triki MA, Nasri M (2019) Economic production and biocontrol efficiency of lipopeptide biosurfactants from *Bacillus mojavenis* A21. Biodegradation 30:273
128. Sánchez C (2022) A review of the role of biosurfactants in the biodegradation of hydrophobic organopollutants: production, mode of action, biosynthesis and applications. World J Microbiol Biotechnol 38(11):216
129. Abraham P, Chinedu EO, Banaru AA, Dauda WP (2023) Biosurfactants: possible roles in environmental management—a review. Fudma J Sci 7(4):236
130. Kiruba NJM, Saeid A (2022) An insight into microbial inoculants for bioconversion of waste biomass into sustainable "bio-organic" fertilizers: a bibliometric analysis and systematic literature review. Int J Mol Sci 23(21):13049
131. Mata-Sandoval JC, Karns J, Torrents A (2001) Influence of rhamnolipids and Triton X-100 on the biodegradation of three pesticides in aqueous phase and soil slurries. J Agric Food Chem 49(7):3296
132. Lai C-C, Huang Y-C, Wei Y-H, Chang J-S (2009) Biosurfactant-enhanced removal of total petroleum hydrocarbons from contaminated soil. J Hazard Mater 167(1–3):609
133. Bezza FA, Chirwa EMN (2017) Biosurfactant-assisted bioremediation of polycyclic aromatic hydrocarbons (PAHs) in liquid culture system and substrate interactions. Polycycl Aromat Compd 37(5):375
134. Patowary K, Patowary R, Kalita MC, Deka S (2017) Characterization of biosurfactant produced during degradation of hydrocarbons using crude oil as sole source of carbon. Front Microbiol 8:279
135. Marchut-Mikolajczyk O, Drożdżyński P, Pietrzyk D, Antczak T (2018) Biosurfactant production and hydrocarbon degradation activity of endophytic bacteria isolated from *Chelidonium majus* L. Microb Cell Fact 17:1
136. Moldes AB, Paradelo R, Rubinos D, Devesa-Rey R, Cruz JM, Barral MT (2011) Ex situ treatment of hydrocarbon-contaminated soil using biosurfactants from *Lactobacillus pentosus*. J Agric Food Chem 59(17):9443
137. Awasthi N, Kumar A, Makkar R, Cameotra SS (1999) Biodegradation of soil-applied endosulfan in the presence of a biosurfactant. J Environ Sci Health B 34(5):793
138. Khedher SB, Boukedi H, Dammak M, Kilani-Feki O, Sellami-Boudawara T, Abdelkefi-Mesrati L, Tounsi S (2017) Combinatorial effect of *Bacillus amyloliquefaciens* AG1 biosurfactant and *Bacillus thuringiensis* Vip3Aa16 toxin on *Spodoptera littoralis* larvae. J Invertebr Pathol 144:11
139. Mishra S, Lin Z, Pang S, Zhang Y, Bhatt P, Chen S (2021) Biosurfactant is a powerful tool for the bioremediation of heavy metals from contaminated soils. J Hazard Mater 418:126253
140. Aşçı Y, Nurbaş M, Açıkel YS (2008) A comparative study for the sorption of Cd(II) by soils with different clay contents and mineralogy and the recovery of Cd(II) using rhamnolipid biosurfactant. J Hazard Mater 154(1–3):663
141. Slizovskiy IB, Kelsey JW, Hatzinger PB (2011) Surfactant-facilitated remediation of metal-contaminated soils: efficacy and toxicological consequences to earthworms. Environ Toxicol Chem 30(1):112
142. Ravindran A, Sajayan A, Priyadharshini GB, Selvin J, Kiran GS (2020) Revealing the efficacy of thermostable biosurfactant in heavy metal bioremediation and surface treatment in vegetables. Front Microbiol 11:222
143. Ayangbenro AS, Babalola OO (2020) Genomic analysis of *Bacillus cereus* NWUAB01 and its heavy metal removal from polluted soil. Sci Rep 10(1):19660
144. Akladious SA, Gomaa EZ, El-Mahdy OM (2019) Efficiency of bacterial biosurfactant for biocontrol of *Rhizoctonia solani* (AG-4) causing root rot in faba bean (*Vicia faba*) plants. Eur J Plant Pathol 153:1237
145. Shalini D, Benson A, Gomathi R, Henry AJ, Jerritta S, Joe MM (2017) Isolation, characterization of glycolipid type biosurfactant from endophytic *Acinetobacter* sp. ACMS25 and evaluation of its biocontrol efficiency against *Xanthomonas oryzae*. Biocatal Agric Biotechnol 11:252

146. Bee H, Khan MY, Sayyed R (2019) Microbial surfactants and their significance in agriculture. In: Plant growth promoting rhizobacteria (PGPR): prospects for sustainable agriculture, p 205
147. Fernandes PAV, Arruda IRD, Santos AFABD, Araújo AAD, Maior AMS, Ximenes EA (2007) Antimicrobial activity of surfactants produced by *Bacillus subtilis* R14 against multidrug-resistant bacteria. Braz J Microbiol 38:704
148. Shah V, Daverey A (2021) Effects of sophorolipids augmentation on the plant growth and phytoremediation of heavy metal contaminated soil. J Clean Prod 280:124406
149. Liduino VS, Servulo EF, Oliveira FJ (2018) Biosurfactant-assisted phytoremediation of multi-contaminated industrial soil using sunflower (*Helianthus annuus* L.). J Environ Sci Health Part A 53(7):609
150. Rawat G, Dhasmana A, Kumar V (2020) Biosurfactants: the next generation biomolecules for diverse applications. Environ Sustain 3:353
151. Sajadi Bami M, Raeisi Estabragh MA, Ohadi M, Banat IM, Dehghannoudeh G (2022) Biosurfactants aided bioremediation mechanisms: a mini-review. Soil Sediment Contam Int J 31(7):801
152. Rasheed T, Shafi S, Bilal M, Hussain T, Sher F, Rizwan K (2020) Surfactants-based remediation as an effective approach for removal of environmental pollutants—a review. J Mol Liq 318:113960
153. Pardhi DS, Panchal RR, Raval VH, Joshi RG, Poczai P, Almalki WH, Rajput KN (2022) Microbial surfactants: a journey from fundamentals to recent advances. Front Microbiol 13:982603
154. Wu Q, Zhi Y, Xu Y (2019) Systematically engineering the biosynthesis of a green biosurfactant surfactin by *Bacillus subtilis* 168. Metab Eng 52:87
155. Gaur VK, Bajaj A, Regar RK, Kamthan M, Jha RR, Srivastava JK, Manickam N (2019) Rhamnolipid from a *Lysinibacillus sphaericus* strain IITR51 and its potential application for dissolution of hydrophobic pesticides. Bioresour Technol 272:19
156. Karamchandani BM, Pawar AA, Pawar SS, Syed S, Mone NS, Dalvi SG, Rahman PK, Banat IM, Satpute SK (2022) Biosurfactants' multifarious functional potential for sustainable agricultural practices. Front Bioeng Biotechnol 10:1047279
157. Bilal M, Iqbal HM (2020) Microbial bioremediation as a robust process to mitigate pollutants of environmental concern. Case Stud Chem Environ Eng 2:100011
158. Azubuike CC, Chikere CB, Okpokwasili GC (2016) Bioremediation techniques–classification based on site of application: principles, advantages, limitations and prospects. World J Microbiol Biotechnol 32:1
159. Jimoh AA, Lin J (2020) Biotechnological applications of *Paenibacillus* sp. D9 lipopeptide biosurfactant produced in low-cost substrates. Appl Biochem Biotechnol 191:921
160. Newton J, Fray R (2004) Integration of environmental and host-derived signals with quorum sensing during plant–microbe interactions. Cell Microbiol 6(3):213
161. Kumar A, Droby S (2020) Food security and plant disease management. Woodhead Publishing
162. Jazzar C, Hammad EA-F (2003) The efficacy of enhanced aqueous extracts of *Melia azedarach* leaves and fruits integrated with the *Camptotylus reuteri* releases against the sweetpotato whitefly nymphs. Bull Insectol 56:269
163. Kim P, Bai H, Bai D, Chae H, Chung S, Kim Y, Park R, Chi YT (2004) Purification and characterization of a lipopeptide produced by *Bacillus thuringiensis* CMB26. J Appl Microbiol 97(5):942
164. Gronwald JW, Plaisance KL, Ide DA, Wyse DL (2002) Assessment of *Pseudomonas syringae* pv. *tagetis* as a biocontrol agent for Canada thistle. Weed Sci 50(3):397
165. Krishnayyaand P, Grewal P (2002) Effect of neem and selected fungicides on viability and virulence of the entomopathogenic nematode *Steinernema feltiae*. Biocontrol Sci Technol 12(2):259
166. Rodriguez SB, Mahoney NE (1994) Inhibition of aflatoxin production by surfactants. Appl Environ Microbiol 60(1):106
167. Krzyzanowska D, Potrykus M, Golanowska M, Polonis K, Gwizdek-Wisniewska A, Lojkowska E, Jafra S (2012) Rhizosphere bacteria as potential biocontrol agents against soft rot caused by various *Pectobacterium* and *Dickeya* spp. strains. J Plant Pathol 367

168. Velho R, Medina L, Segalin J, Brandelli A (2011) Production of lipopeptides among *Bacillus* strains showing growth inhibition of phytopathogenic fungi. Folia Microbiol 56:297
169. Haddad NI, Wang J, Mu B (2008) Isolation and characterization of a biosurfactant producing strain, *Brevibacilis brevis* HOB1. J Ind Microbiol Biotechnol 35(12):1597
170. Hultberg M, Bergstrand K-J, Khalil S, Alsanius B (2008) Characterization of biosurfactant-producing strains of fluorescent pseudomonads in a soilless cultivation system. Anton Van Leeuw 94:329
171. Debode J, Maeyer KD, Perneel M, Pannecoucque J, Backer GD, Höfte M (2007) Biosurfactants are involved in the biological control of *Verticillium microsclerotia* by *Pseudomonas* spp. J Appl Microbiol 103(4):1184
172. Kim P-I, Ryu J-W, Kim Y-H, Chi Y-T (2010) Production of biosurfactant lipopeptides iturin A, fengycin, and surfactin A from *Bacillus subtilis* CMB32 for control of *Colletotrichum gloeosporioides*. J Microbiol Biotechnol 20(1):138
173. Bouassida M, Mnif I, Hammami I, Triki M-A, Ghribi D (2023) *Bacillus subtilis* SPB1 lipopeptide biosurfactant: antibacterial efficiency against the phytopathogenic bacteria *Agrobacterium tumefaciens* and compared production in submerged and solid state fermentation systems. Food Sci Biotechnol 32(11):1595
174. Lahkar J, Borah SN, Deka S, Ahmed G (2015) Biosurfactant of *Pseudomonas aeruginosa* JS29 against *Alternaria solani*: the causal organism of early blight of tomato. Biocontrol 60:401
175. Lahkar J, Goswami D, Deka S, Ahmed G (2018) Novel approaches for application of biosurfactant produced by *Pseudomonas aeruginosa* for biocontrol of *Colletotrichum capsici* responsible for anthracnose disease in chilli. Eur J Plant Pathol 150:57
176. Sha R, Meng Q (2016) Antifungal activity of rhamnolipids against dimorphic fungi. J Gen Appl Microbiol 62(5):233
177. Sanchez L, Courteaux B, Hubert J, Kauffmann S, Renault J-H, Clément C, Baillieul F, Dorey S (2012) Rhamnolipids elicit defense responses and induce disease resistance against biotrophic, hemibiotrophic, and necrotrophic pathogens that require different signaling pathways in *Arabidopsis* and highlight a central role for salicylic acid. Plant Physiol 160(3):1630
178. Patel H, Tscheka C, Edwards K, Karlsson G, Heerklotz H (2011) All-or-none membrane permeabilization by fengycin-type lipopeptides from *Bacillus subtilis* QST713. Biochim Biophys Acta (BBA) Biomembr 1808(8):2000
179. Varnier AL, Sanchez L, Vatsa P, Boudesocque L, Garcia-Brugger A, Rabenoelina F, Sorokin A, Renault JH, Kauffmann S, Pugin A (2009) Bacterial rhamnolipids are novel MAMPs conferring resistance to *Botrytis cinerea* in grapevine. Plant Cell Environ 32(2):178
180. Perneel M, D'Hondt L, De Maeyer K, Adiobo A, Rabaey K, Höfte M (2008) Phenazines and biosurfactants interact in the biological control of soil-borne diseases caused by *Pythium* spp. Environ Microbiol 10(3):778
181. D'aes J, Hua GKH, De Maeyer K, Pannecoucque J, Forrez I, Ongena M, Dietrich LE, Thomashow LS, Mavrodi DV, Höfte M (2011) Biological control of *Rhizoctonia* root rot on bean by phenazine- and cyclic lipopeptide-producing *Pseudomonas* CMR12a. Phytopathology 101(8):996
182. Radhakrishnan EK, Mathew J (2020) Characterization of biosurfactant produced by the endophyte *Burkholderia* sp. WYAT7 and evaluation of its antibacterial and antibiofilm potentials. J Biotechnol 313:1
183. Hussain T, Haris M, Shakeel A, Ahmad G, Ahmad Khan A, Khan MA (2020) Bio-nematicidal activities by culture filtrate of *Bacillus subtilis* HussainT-AMU: new promising biosurfactant bioagent for the management of root galling caused by *Meloidogyne incognita*. Vegetos 33:229
184. Mascarin GM, Kobori NN, Quintela ED, Arthurs SP, Júnior ÍD (2014) Toxicity of non-ionic surfactants and interactions with fungal entomopathogens toward *Bemisia tabaci* biotype B. Biocontrol 59:111
185. Liu TX, Stansly PA (2000) Insecticidal activity of surfactants and oils against silverleaf whitefly (*Bemisia argentifolii*) nymphs (Homoptera: Aleyrodidae) on collards and tomato. Pest Manage Sci 56(10):861

186. Gill HK, Garg H (2014) Pesticide: environmental impacts and management strategies. Pesticides Toxic Asp 8(187):10
187. Mostafalou S, Abdollahi M (2013) Pesticides and human chronic diseases: evidences, mechanisms, and perspectives. Toxicol Appl Pharmacol 268(2):157
188. Huang H, Yu Q, Ren H, Geng J, Xu K, Zhang Y, Ding L (2018) Towards physicochemical and biological effects on detachment and activity recovery of aging biofilm by enzyme and surfactant treatments. Bioresour Technol 247:319
189. Palma-Bautista C, Tataridas A, Kanatas P, Travlos IS, Bastida F, Domínguez-Valenzuela JA, De Prado R (2020) Can control of glyphosate susceptible and resistant *Conyza sumatrensis* populations be dependent on the herbicide formulation or adjuvants? Agronomy 10(10):1599
190. Rebello S, Asok AK, Mundayoor S, Jisha M (2013) Surfactants: chemistry, toxicity and remediation. In: Pollutant diseases, remediation and recycling, p 277
191. Ryckaert B, Spanoghe P, Haesaert G, Heremans B, Isebaert S, Steurbaut W (2007) Quantitative determination of the influence of adjuvants on foliar fungicide residues. Crop Prot 26(10):1589
192. Jibrin MO, Liu Q, Jones JB, Zhang S (2021) Surfactants in plant disease management: a brief review and case studies. Plant Pathol 70(3):495
193. Mizumoto S, Shoda M (2007) Medium optimization of antifungal lipopeptide, iturin A, production by *Bacillus subtilis* in solid-state fermentation by response surface methodology. Appl Microbiol Biotechnol 76:101
194. Edosa TT, Jo YH, Keshavarz M, Han YS (2018) Biosurfactants: production and potential application in insect pest management. Trends Entomol 14(79):79
195. Yun DC, Yang SY, Kim YC, Kim IS, Kim YH (2013) Identification of surfactin as an aphicidal metabolite produced by *Bacillus amyloliquefaciens* G1. J Korean Soc Appl Biol Chem 56:751
196. Sha R, Jiang L, Meng Q, Zhang G, Song Z (2012) Producing cell-free culture broth of rhamnolipids as a cost-effective fungicide against plant pathogens. J Basic Microbiol 52(4):458
197. Cossus L, Roux-Dalvai F, Kelly I, Nguyen TTA, Antoun H, Droit A, Tweddell RJ (2021) Interactions with plant pathogens influence lipopeptides production and antimicrobial activity of *Bacillus subtilis* strain PTB185. Biol Control 154:104497
198. Khedher SB, Boukedi H, Kilani-Feki O, Chaib I, Laarif A, Abdelkefi-Mesrati L, Tounsi S (2015) *Bacillus amyloliquefaciens* AG1 biosurfactant: putative receptor diversity and histopathological effects on *Tuta absoluta* midgut. J Invertebr Pathol 132:42
199. Ghribi D, Abdelkefi-Mesrati L, Boukedi H, Elleuch M, Ellouze-Chaabouni S, Tounsi S (2012) The impact of the *Bacillus subtilis* SPB1 biosurfactant on the midgut histology of *Spodoptera littoralis* (Lepidoptera: Noctuidae) and determination of its putative receptor. J Invertebr Pathol 109(2):183
200. Ramesar DR, Hunter CH (2023) Elucidation of lipopeptide biosurfactants responsible for the larvicidal activity of *Bacillus velezensis* PHP1601 towards *Lucilia cuprina* larvae. Biocontrol Sci Technol 33(8):772
201. do Nascimento Silva J, Mascarin GM, de Castro RPV, Castilho LR, Freire DM (2019) Novel combination of a biosurfactant with entomopathogenic fungi enhances efficacy against *Bemisia* whitefly. Pest Manage Sci 75:2882
202. Sobrinho HB, Luna JM, Rufino RD, Porto ALF, Sarubbo LA (2013) Biosurfactants: classification, properties and environmental applications. Recent Dev Biotechnol 11(14):1
203. Zhang C, Wang S, Yan Y (2011) Isomerization and biodegradation of beta-cypermethrin by *Pseudomonas aeruginosa* CH7 with biosurfactant production. Bioresour Technol 102(14):7139
204. Zhang F, Gu W, Xu P, Tang S, Xie K, Huang X, Huang Q (2011) Effects of alkyl polyglycoside (APG) on composting of agricultural wastes. Waste Manage 31(6):1333
205. Stevens SH, Kuuskraa VA, Gale J, Beecy D (2001) CO_2 injection and sequestration in depleted oil and gas fields and deep coal seams: worldwide potential and costs. Environ Geosci 8(3):200
206. Peet KC, Freedman AJ, Hernandez HH, Britto V, Boreham C, Ajo-Franklin JB, Thompson JR (2015) Microbial growth under supercritical CO_2. Appl Environ Microbiol 81(8):2881

207. Freitas BG, Brito JG, Brasileiro PP, Rufino RD, Luna JM, Santos VA, Sarubbo LA (2016) Formulation of a commercial biosurfactant for application as a dispersant of petroleum and by-products spilled in oceans. Front Microbiol 7:1646
208. Shin K-H, Kim K-W, Ahn Y (2006) Use of biosurfactant to remediate phenanthrene-contaminated soil by the combined solubilization–biodegradation process. J Hazard Mater 137(3):1831
209. Hirata Y, Ryu M, Oda Y, Igarashi K, Nagatsuka A, Furuta T, Sugiura M (2009) Novel characteristics of sophorolipids, yeast glycolipid biosurfactants, as biodegradable low-foaming surfactants. J Biosci Bioeng 108(2):142
210. Klosowska-Chomiczewska I, Medrzycka K, Karpenko E (2011) Biosurfactants–biodegradability, toxicity, efficiency in comparison with synthetic surfactants. Adv Chem Mech Eng 2:1
211. Loh J, Pierson EA, Pierson LS III, Stacey G, Chatterjee A (2002) Quorum sensing in plant-associated bacteria. Curr Opin Plant Biol 5(4):285
212. Dusane DH, Zinjarde SS, Venugopalan VP, Mclean RJ, Weber MM, Rahman PK (2010) Quorum sensing: implications on rhamnolipid biosurfactant production. Biotechnol Genet Eng Rev 27(1):159
213. Ron EZ, Rosenberg E (2001) Natural roles of biosurfactants: minireview. Environ Microbiol 3(4):229
214. Berti AD, Greve NJ, Christensen QH, Thomas MG (2007) Identification of a biosynthetic gene cluster and the six associated lipopeptides involved in swarming motility of *Pseudomonas syringae* pv. tomato DC3000. J Bacteriol 189(17):6312
215. Kearns DB, Losick R (2003) Swarming motility in undomesticated *Bacillus subtilis*. Mol Microbiol 49(3):581
216. Rostás M, Blassmann K (2009) Insects had it first: surfactants as a defence against predators. Proc R Soc B Biol Sci 276(1657):633
217. Mulqueen P (2003) Recent advances in agrochemical formulation. Adv Coll Interface Sci 106(1–3):83
218. Blackwell P (2000) Management of water repellency in Australia, and risks associated with preferential flow, pesticide concentration and leaching. J Hydrol 231:384
219. Street J (1969) Methods of removal of pesticide residues. Can Med Assoc J 100(4):154
220. Petrović M, Barceló D (2004) Analysis and fate of surfactants in sludge and sludge-amended soils. TrAC Trends Anal Chem 23(10–11):762
221. Hopkinson M, Collins H, Goss G (1997) Pesticide formulations and application systems: ASTM Committee E-35 on Pesticides. ASTM Int 17(1328):1
222. Nishio E, Ichiki Y, Tamura H, Morita S, Watanabe K, Yoshikawa H (2002) Isolation of bacterial strains that produce the endocrine disruptor, octylphenol diethoxylates, in paddy fields. Biosci Biotechnol Biochem 66(9):1792
223. Adetunji AI, Olaniran AO (2021) Production and potential biotechnological applications of microbial surfactants: an overview. Saudi J Biol Sci 28(1):669
224. Moldes AB, Vecino X, Rodríguez-López L, Rincón-Fontán M, Cruz JM (2019) Microbial glycoprotein and lipopeptide biosurfactants: production, properties and applications. In: Microbial biosurfactants and their environmental and industrial applications. CRC Press, p 106
225. Chaprão MJ, Ferreira IN, Correa PF, Rufino RD, Luna JM, Silva EJ, Sarubbo LA (2015) Application of bacterial and yeast biosurfactants for enhanced removal and biodegradation of motor oil from contaminated sand. Electron J Biotechnol 18(6):471
226. Van Bogaert IN, Saerens K, De Muynck C, Develter D, Soetaert W, Vandamme EJ (2007) Microbial production and application of sophorolipids. Appl Microbiol Biotechnol 76(1):23
227. Morita T, Fukuoka T, Imura T, Kitamoto D (2015) Mannosylerythritol lipids: production and applications. J Oleo Sci 64(2):133
228. Adetunji CO (2022) Applications of biosurfactant in agriculture. Academic Press
229. Öztürk S, Calik P, Özdamar TH (2016) Fed-batch biomolecule production by *Bacillus subtilis*: a state of the art review. Trends Biotechnol 34(4):329

230. Banat IM, Satpute SK, Cameotra SS, Patil R, Nyayanit NV (2014) Cost effective technologies and renewable substrates for biosurfactants' production. Front Microbiol 5:697
231. Smyth TJ, Perfumo A, McClean S, Marchant R (2010) Isolation and analysis of lipopeptides and high molecular weight biosurfactants. In: Handbook of hydrocarbon and lipid microbiology. Springer, p 3687
232. Kesarwani S, Tripathy DB, Mishra A, Gupta A (2021) Surfactants: patent landscape of the most versatile class of materials. In: Surfactants from renewable raw materials, p 227
233. Luna JM, Rufino RD, Jara AMA, Brasileiro PP, Sarubbo LA (2015) Environmental applications of the biosurfactant produced by *Candida sphaerica* cultivated in low-cost substrates. Colloids Surf A Physicochem Eng Asp 480:413
234. Rane AN, Baikar VV, Ravi Kumar V, Deopurkar RL (2017) Agro-industrial wastes for production of biosurfactant by *Bacillus subtilis* ANR 88 and its application in synthesis of silver and gold nanoparticles. Front Microbiol 8:492
235. Santos DKF, Rufino RD, Luna JM, Santos VA, Sarubbo LA (2016) Biosurfactants: multifunctional biomolecules of the 21st century. Int J Mol Sci 17(3):401
236. Dhanarajan G, Sen R (2014) Cost analysis of biosurfactant production from a scientist's perspective. Biosurfactants 159:153
237. Lan G, Fan Q, Liu Y, Chen C, Li G, Liu Y, Yin X (2015) Rhamnolipid production from waste cooking oil using *Pseudomonas* SWP-4. Biochem Eng J 101:44

Surfactants: Basic Principles and Applications in Biomedical Sciences

Sandeep R. Patil and Dharmesh S. Varade

Abstract Surface active agents or surfactants have a unique molecular structure, wherein a non-polar hydrophobic part, typically a straight or branched hydrocarbon chain containing 8–18 carbon atoms referred to as a tail is attached to a polar or ionic part or hydrophilic group, commonly known as a headgroup. The different interactions of these two groups in aqueous media are an important factor resulting in self-aggregation of surface-active agents into different morphological forms, viz. micelles, lyotropic liquid crystalline phases, vesicles and bilayers, etc. Surfactants sometimes are also referred as amphiphiles which in broader terms includes surface active substance like polar lipids as well. The term surfactant generally refers to man-made substance even though it contains bio-sourced molecules. Surfactants find applications in diverse aspects of chemical industries, including agrochemicals, cosmetics, detergents, dyestuffs, emulsifiers, water treatment, drug delivery systems, tertiary oil recovery, firefighting, nanotechnology, catalysis (specifically micellar catalysis) and biochemical formulations. A comprehensive account of applications of surfactants in biomedical sciences, emphasizing their advantages and disadvantages are detailed herein. Surfactants enhance solubilization of hydrophobic compounds and drugs, thereby improving bioavailability, and subsequently ensure controlled release of drug. Surfactants exhibit antimicrobial action and their antimicrobial efficacy stems from their ability to disrupt of microbial membranes. Surfactants improve stabilization of lipid nanoparticles thereby ensuring their suitability in gene and protein delivery, specifically mRNA vaccines and CRISPR-based therapeutics. Additionally, their utility in cancer diagnostics, fluorescence microscopy for biomedical imaging, enhancement of MRI contrast has attracted interest of researchers to explore their potential for theranostic applications. Surfactants facilitate scaffold fabrication and find use in biomaterial coating thereby playing a critical role in tissue engineering. However, their clinical translation is hindered by cytotoxicity, environmental concerns, and regulatory limitations.

S. R. Patil (✉) · D. S. Varade
School of Science, Navrachana University, Vadodara, Gujarat, India
e-mail: sandeepp@nuv.ac.in

M. Abdul Rub et al. (eds.), *Industrial Applications of Individual and Mixed Surfactant Systems—Volume 2*, Sustainable Materials and Technology,
https://doi.org/10.1007/978-981-95-9195-4_4

Keywords Drug delivery systems · Gene delivery · Tissue engineering · Theranostics · Regenerative medicine · Biosurfactants

1 Introduction

Surface active agents or surfactants have a unique molecular structure [1], wherein a non-polar hydrophobic part, typically a straight or branched hydrocarbon chain containing 8–18 carbon atoms referred to as a tail is attached to a polar or ionic part or hydrophilic group, commonly known as a headgroup [1]. The different interactions of these two groups in aqueous media [2, 3] are an important factor resulting in self-aggregation of surface-active agents into different morphological forms, viz. micelles, lyotropic liquid crystalline phases, vesicles, and bilayers, etc. [4]. Surfactants enhance solubilization of hydrophobic compounds and drugs, thereby improving bioavailability, and subsequently ensure controlled release of drug. Surfactants exhibit antimicrobial action and their antimicrobial efficacy stems from their ability to disrupt of microbial membranes. Surfactants improve stabilization of lipid nanoparticles thereby ensuring their suitability in gene and protein delivery, specifically mRNA vaccines and CRISPR-based therapeutics. The unique molecular structure of surfactants, their ability to reduce surface tension, even when present in very low concentration, ability to self-aggregate into different morphological forms, hydrophilic-lipophilic balance (HLB), makes them a viable choice for medicinal formulations [5]. Surfactants do play a vital role in pharmaceutical formulations such as ointments, creams, lubricants, and gels. They enhance drug absorption and penetration and help stabilize semisolid preparations. Additionally, surfactants exhibit various medicinal properties including antimicrobial, anti-adhesive, antibiofilm, anti-inflammatory, and antioxidant effects. Some studies also highlight their spermicidal and laxative activities [6–8].

1.1 Classification of Surface-Active Agents/Surfactants

Surfactants are primarily classified based on the nature/charge of hydrophilic portion, referred to as a head group [2–4]. The four broad classes of surfactants are, therefore, anionic, cationic, non-ionic and zwitterionic. If the charge on the polar head group is positive, the surfactant is classified as cationic; if the charge is negative, it is designated as anionic. Whereas, if a surfactant contains a polar head with two oppositely charged groups, it is known as zwitterionic/amphoteric [2–4].

The important characteristics of four classes of surfactants are detailed below:

- o The anionics and cationics, which dissociate in water into two oppositely charged species (the surfactant ion and its counterion) [2–4],

- o The non-ionics based on ethylene oxide, referred to as ethoxylated surfactants [2–4] include a highly polar moiety (without charge), such as polyoxyethylene group. Different types of non-ionic surfactants are alcohol ethoxylates, alkyl phenol ethoxylates, fatty acid ethoxylates, sorbitan ester ethoxylates, fatty amine ethoxylates and ethylene oxide–propylene oxide copolymers (block copolymer/ pluronic surfactants), polyglucoside, glycol esters etc.
- o The zwitterionics (or amphoterics), contain both anionic and cationic groups. They exhibit stability in acids and alkali, additionally they form mixed micellar aggregates with other classes of surfactants (Table 1).

Moreover, new classes of surfactants have been developed by researchers which exhibit synergistic properties resulting in improved surface activity and self-aggregation behaviour. Some representative surfactants include catanionic surfactants, bolaform surfactants, gemini (or dimeric) surfactants, N-Acyl L-glutamates

Table 1 Classification of surfactants [2–4]

Type	Examples
Anionic	Ammonium dodecyl sulfate, Sodium dodecyl sulfate Sodium lauryl ether sulfate, Sodium myreth sulfate N-Acyl-n-Alkyltaurates, Arylalkanesulfonates, Sulfosuccinate Esters Alkyldiphenylether(di)sulfonates, Alkylnaphthalenesulfonates Dioctyl sodium sulfosuccinate, Perfluorooctanesulfonate Perfluorobutanesulfonate, Linear alkylbenzenesulfonates Sodium stearate, Sodium lauroylsarcosinate, Fluorinated Polyoxetanes Fatty Acid Monoethanolamide Sulfates, Perfluorononanoate, Perfluorooctanoate
Cationic	N-Alkyltrimethylammonium chlorides: Hexadecyltrimethylammonium bromide, Cetrimonium bromide, Hexadecyltrimethylammonium chloride N-Benzyl-N-alkyldimethylammonium halides: Dioctadecyldimethylammonium bromide, Dimethyldioctadecylammonium chloride Cetylpyridinium bromide, Benzethonium chloride, Benzalkonium chloride
Nonionic surfactants	Alkyl phenol Ethoxylates, Alcohol Ethoxylates, Tertiary Acetylenic Glycols Polyoxypropylene Glycols, Polyoxyethylene Silicones, Polyoxyethylene Glycol Esters, Polyoxyethylene Mercaptans, Alkylpolyglycosides
Zwitterionic/Amphoteric surfactants	β-N-Alkylaminopropionic Acids, N-Alkyl-β-iminodipropionic Acids, Imidazoline Carboxylates, N-Alkylbetaines, Amidoamines, Amidobetaines, Amine Oxides, Sulfobetaines

Table 2 Structural characteristics and examples of new classes of surfactants [2–4]

Class	Structural characteristics	Example
Catanionic	Prepared by mixing a cationic surfactant and an anionic surfactant	Decyltrimethylammonium decylsulfonate, mixtures of cetyltrimethylammonium bromide and sodium octyl sulfate, mixtures of cetyltrimethylammonium tosylate and negatively charged sodium dodecylbenzene sulfonate
Bolaform	Made up of two charged groups connected by a long linear polymethylene chain	Hexadecanediyl-l,16-bis (trimethylammonium bromide), Potassium hexadecanedioate, Dodecanediyl-l,12-bis (tetrakis (oxyethylene) monoether), Demecarium bromide
Gemini (or dimeric)	Two monomers linked by a short spacer between the polar headgroups	$[C_{10}H_{21}OCH_2CH(OCH_2COONa^+)\ CH_2]_2O$, $[C_{12}H_{25}N^+(CH_3)_2CH_2CHOH]_2\ 2Br^-$, $[C_8H_{17}N^+(CH_3)_2CH_2CHOH]_2\ 2Br^-$
N-Acyl L-Glutamates (AG)	Prepared by N-acylation of L-glutamic acid with fatty acid chloride in a mixed solvent of water and water-miscible organic solvent	$RCONHCH(COO^-M^+)\ CH_2CH_2COO^-M^+$; $M = H^+$ or Cation
Polymerisable	Surfactant capable of undergoing homo polymerisation or copolymerisation with other components of the system	11-(acryloyloxy)undecyltrimethyl ammonium bromide 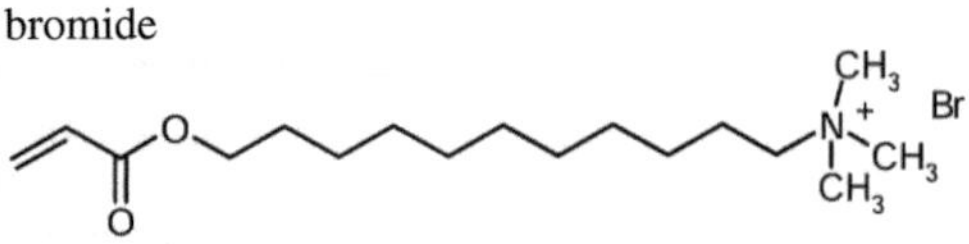

and polymerisable surfactants. The general characteristic features are enlisted in Table 2.

Chemical and biological surfactants are extensively utilized for several fundamental and industrial applications [9]. The applications of surfactants in diverse subject area are collated in Table 3.

1.2 *Biomedical Applications of Surfactants: Significance of Unique Molecular Structure and Physicochemical Properties*

The ability to adsorb at surfaces/interfaces, micelle formation, emulsifying ability/hydrophilic-lipophilic balance (HLB) are the most important characteristic properties of a surfactant vis-à-vis biomedical formulation [2–5]. At low concentrations, the

Table 3 Applications of surface-active agents/surfactants [9]*

Industry/Subject area	Application	Role of surfactants
Petroleum	Enhanced oil recovery	Improving oil drainage into well bore; stimulating release of oil entrapped by capillaries; wetting of solid surfaces; reduction of oil viscosity and oil pour point; lowering of interfacial tension; dissolving of oil
	De-emulsification	De-emulsification of oil emulsions; oil solubilization; viscosity reduction, wetting agent
Environmental	Bioremediation	Emulsification of hydrocarbons; lowering of interfacial tension; metal sequestration
	Soil remediation and flushing	Emulsification through adherence to hydrocarbons; dispersion; foaming agent; detergent; soil flushing
Food	Emulsification and de-emulsification	Emulsifier; solubilizer; demulsifier; suspension, wetting, foaming, defoaming, thickener, lubricating agent
	Functional ingredient	Interaction with lipids, proteins and carbohydrates, protecting agents
Biology	Microbiological	Physiological behaviour such as cell mobility, cell communication, nutrient accession, cell–cell competition, plant and animal pathogenesis
	Pharmaceuticals and therapeutics	Antibacterial, antifungal, antiviral agents; adhesive agents; immunomodulatory molecules; vaccines; gene therapy
Agriculture	Biocontrol	Facilitation of biocontrol mechanisms of microbes such as parasitism, antibiosis, competition, induced systemic resistance and hypovirulence
Bioprocessing	Downstream processing	Biocatalysis in aqueous two-phase systems and microemulsions; biotransformation; recovery of intracellular products; enhanced production of extracellular enzymes and fermentation products
Cosmetics	Health and beauty products	Emulsifiers, foaming agents, solubilizers, wetting agents, cleansers, antimicrobial agent, mediators of enzyme action

* Reproduced with permission from Singha, A.; Van Hammeb, J. D.; Ward, O. P. Surfactants in microbiology and biotechnology: Part 2. Application aspects, Biotechnology Advances 2007, 25, 99

properties of surfactants are like an electrolyte. However, above a particular concentration, surfactant molecules exhibit abrupt change in interfacial and bulk properties, e.g. spontaneous self-aggregation into aggregates (micelles). The concentration above which a surfactant self-assembles into colloidal aggregates is known as the critical micelle concentration (CMC) [2–4]. Micelles enhance solubility of otherwise insoluble drugs aiding in improved bioavailability drugs. The interfacial properties of surfactants owing to their orientation at interface are important with reference to stability in biological systems, drug solubilization, drug encapsulation, and drug release kinetics. HLB (hydrophile–lipophile balance) a number (0–40) indicative of emulsification behavior and related to the balance between the hydrophilic and lipophilic (hydrophobic) portions of the surfactant molecule [1–3] is a vital parameter that ensures judicious choice of a surfactant for improving solubility, stability and bio-availability of hydrophobic drug molecule in a pharmaceutical formulation [10, 11].

The mixed micellar systems aid drug solubility and binding effectiveness. This subsequently helps in development and formulation of effective drug delivery systems. The drug delivery efficiency relies majorly on the micelle formation, especially at low concentration of surfactant, ensuring solubilization of drug [12–14], encapsulation of a drug and imparting stability to the overall formulation. Surfactants that exhibit micelle formation at low critical micelle concentration (CMC), viz. gemini surfactants (unique type of surface-active molecules constituting of two hydrophobic tails and two polar head groups connected by a spacer) ensure good loading efficiency and bioavailability of drugs. Similar properties are exhibited by benzalkonium chloride which forms micellar aggregates at low concentration. The critical micelle concentration (CMC) is a key determinant in identifying and selecting the most suitable surfactant for pharmaceutical and biomedical applications [12–14].

Surfactants reduce surface tension substantially even at very low concentration and ensure uniform drug dispersion. This is an important parameter of interest for utilization of a surfactant with enhanced drug delivery potential in pharmaceutical/biomedical applications [15–18]. Higher surface activity exhibited by cetyltrimethylammonium bromide (CTAB)-a cationic surfactant and double-chain cationic surfactants facilitate DNA-binding and antimicrobial applications. Block copolymer/polymeric surfactants, pluronic surfactants owing to their mild nature and Polysorbate 80 (Tween 80) are ideal choice for enhanced stability and decreased toxicity of a formulation. Zwitterionic/Amphoteric surfactants are well-suited for cosmetics, regenerative medicine and tissue engineering owing to their mild nature and biocompatibility [15–18].

HLB (hydrophile–lipophile balance) a number (0–40) indicative of emulsification behavior is a deciding factor for use of surfactants in medication formulations as it guarantees enhanced solubilization of a hydrophobic drug and its targeted delivery thereby improving therapeutic efficacy. Low HLB surfactants (HLB 4.7) form persistent lipid-based vesicles like niosomes, whereas high HLB surfactants have high hydrophilicity. The high HLB surfactants are suitable for water-soluble drug formulations. Biosurfactants (sophorolipids) with a HLB in the range of 7–13 and lipid-based surfactants (HLB 8–16) have attracted interest of researchers owing to their promise as a biocompatible alternative for design and development of drug delivery systems [19–22].

The comprehensive list/collation of different applications of surfactants in pharmaceutical and biomedical sciences [5], emphasizing their advantages and disadvantages are detailed herein (Fig. 1).

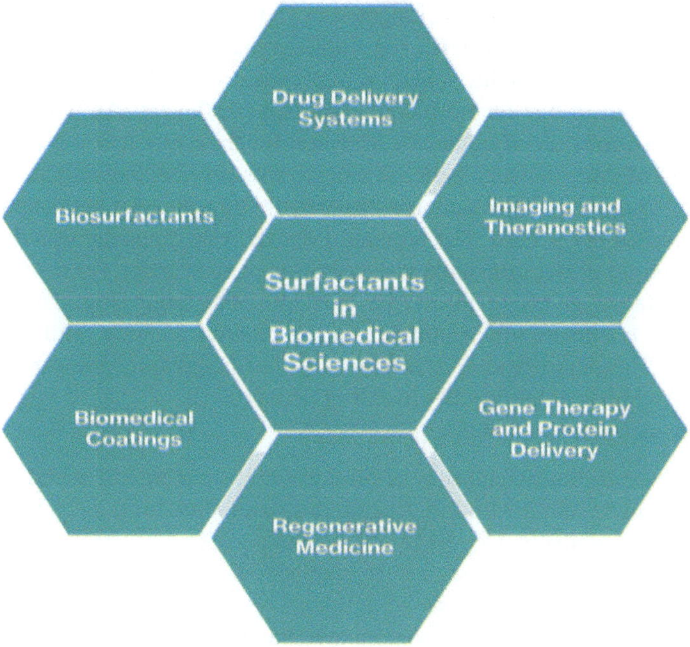

Fig. 1 Applications of surfactants in biomedical sciences [5]

Surfactants improve solubilization, facilitate targeted delivery and enable controlled release of a hydrophobic drug. These traits are important for applications in biomedical sciences and pharmaceutical industry. The ability of surfactants to disrupt membranes ensures their antimicrobial efficacy. Surfactants stabilize lipid nanoparticles, thereby enhancing their utility in vaccines and therapeutic applications. Additionally, their utility in cancer diagnostics, fluorescence microscopy for biomedical imaging, enhancement of MRI contrast has attracted interest of researchers to explore their potential for theranostic applications. Surfactants facilitate scaffold fabrication and find use in biomaterial coating thereby playing a critical role in tissue engineering. However, their clinical translation is substantially dependent and limited owing to cytotoxicity, environmental concerns, and regulatory constraints [5].

2 Applications of Surfactants in Biomedical Sciences

2.1 Drug Delivery Systems

The shape of the micelle formed by a surfactant in aqueous media is of importance in determining various properties of the surfactant solution. Above critical micelle concentration, surfactant self-aggregates into (1) relatively small, spherical micelles (aggregation number < 100), (2) elongated cylindrical, rodlike micelles with hemispherical ends (prolate ellipsoids), (3) liquid crystalline hexagonal, lamellar, and cubic phases and (4) vesicles-spherical aggregates consisting of bilayer lamellar micelles arranged in one or more concentric spheres/layers. An understanding of the phase transitions of surfactants is essential for their efficient use in design and development of drug delivery systems. Surfactants serve either as passive carriers, transporting drugs without influencing their distribution or targeting systems, guiding drugs to specific body sites. The targeted drug delivery happens in two ways; (a) passive targeting: utilizes the natural distribution of the carrier, (b) active targeting: Employs surface ligands to direct the carrier to the target site. These methods enhance drug concentration/availability at the desired location while minimizing side effects. The detailed understanding of phase behaviour, phase transitions of surfactant systems is vital for effective drug delivery.

Key considerations include:

- Dilution effects: Surfactants may be diluted below their critical micelle concentration (cmc) upon administration, risking drug precipitation, especially dangerous in intravenous delivery.
- Preferred surfactants: Non-ionic surfactants are favoured due to their lower cmc, minimizing precipitation risks.
- Phase transitions: Dilution or temperature changes (e.g., body temperature being ~ 12 °C above ambient) can shift surfactant phases (e.g., hexagonal to micellar), potentially altering drug release and causing dose dumping.
- Hysteresis: Cubic phases may remain pseudo-stable at unexpected temperatures, affecting drug delivery reliability.
- Biological interactions: In vivo agents like bile salts can disrupt surfactant aggregates, such as solubilizing liposomes and releasing their contents.

The afore-mentioned factors must be carefully managed to ensure safe and effective drug delivery using surfactant systems. With an aim to improve the oral bioavailability of hydrophobic drugs (limited solubility in water) surfactants were utilized in self-emulsifying drug delivery systems (SEDDS), thereby facilitating microemulsion formation in the gastrointestinal tract. Additionally, surfactants have transformed design of drug delivery systems and are instrumental in forming liposomes and micelles for efficient drug diffusion and transport [23–27]. Surfactants ensure targeted drug delivery by forming micelles and liposomes (Fig. 2).

Micelles, composed of amphiphilic molecules, encapsulate hydrophobic drugs to improve solubility and stability. Liposomes, made of phospholipid bilayers, enable precise delivery to specific cells or tissues. Utilization of liposome as a drug delivery system and routes of drug administration to the human body is mentioned in Fig. 3 [28].

Surface modification such as receptor-specific ligands or stimuli-responsive surfactants significantly improve drug release control, minimize side effects, and boost therapeutic efficacy [27, 28]. Stimuli-responsive nanoparticles enable site-directed drug delivery in response to changes in pH and other variables of the system.

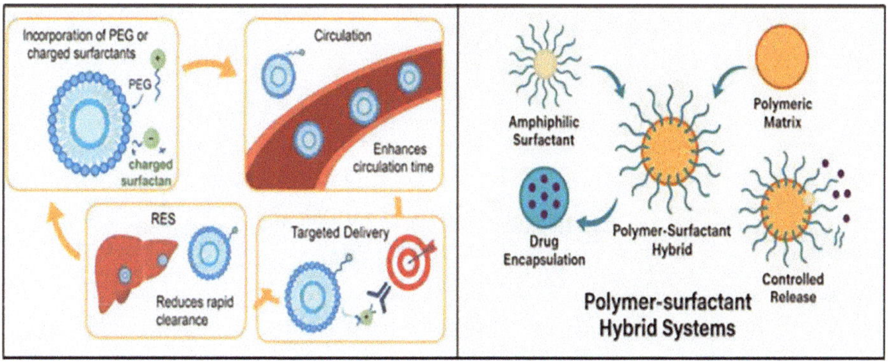

Fig. 2 Surfactant-based nanocarriers and liposomes for targeted drug delivery. Reproduced with permission from Reference [27]

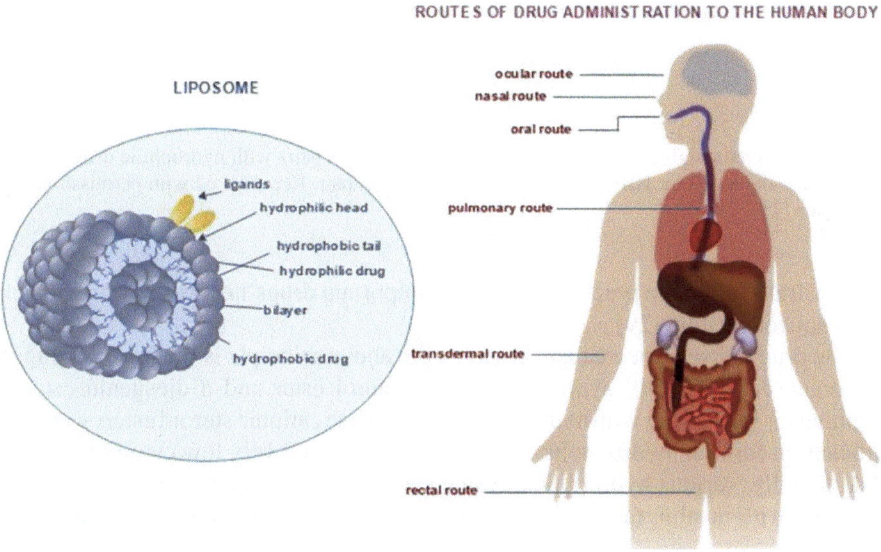

Fig. 3 Liposomal drug delivery system and routes of drug administration into the human body. Reproduced with permission from Reference [28]

In oncology, surfactants ensure controlled drug release, achieving high tumor-specific deposition while minimizing systemic toxicity.

Owing to their natural origin, easy degradability and low cytotoxicity; amino acid surfactants are an inevitable choice for pharmaceutical formulations, specifically drug delivery systems. Ion pairing of surfactants with hydrophilic macromolecular drugs, specifically arginine ester-based dicationic surfactants has been reported as a viable strategy to fine-tune the solubility of charged hydrophilic drugs, encompassing small molecules, peptides, and protein-based agents. Notably, arginine ester

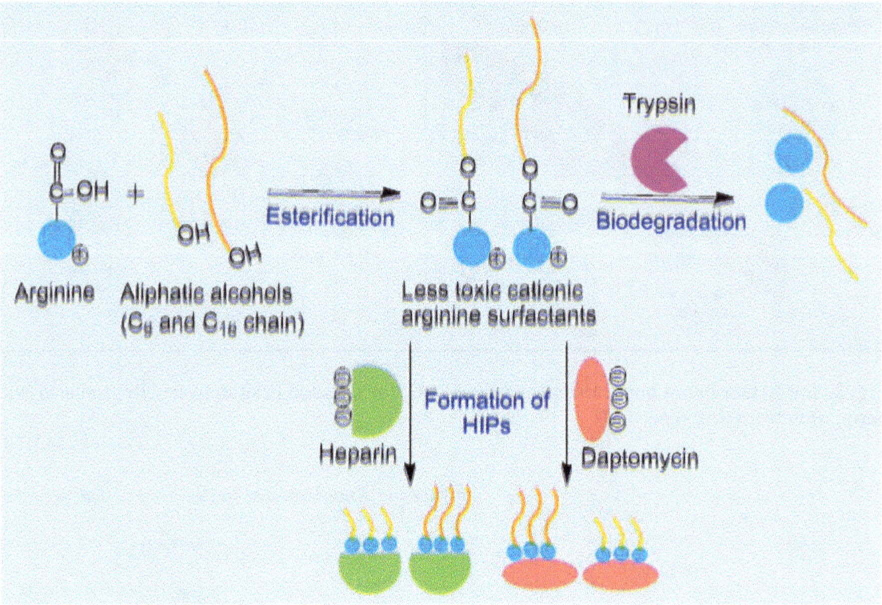

Fig. 4 Long-chain arginine esters can form hydrophobic ion pairs with hydrophilic drugs such as heparin and daptomycin. HIP stands for hydrophobic ion pair. Reproduced with permission from Reference [29]

demonstrated effective precipitation of two important drugs-heparin and daptomycin from aqueous media [29].

The drug delivery mechanism based on the above principle is illustrated in Fig. 4.

Jorgensen et al. [30] synthesized a cholesterol ester and a diosgenin ester of arginine for ion pairing with hydrophilic drugs. These cationic steroid esters were also efficient in forming hydrophobic ion pairs and exhibited very low cytotoxicity [30]. Additionally, various hydrophobic lysine esters were utilized to form hydrophobic ion pairs with insulin, facilitating complex formation through nonpolar interactions [31].

Surfactants enhance ophthalmic formulations by increasing the solubility and uniform dispersion of hydrophobic drugs. The cleansing action of surfactants in contact lens solution is attributed to their ability to prolong drug residence time by reducing tear film washout. Additionally, surfactants aid transdermal drug delivery by facilitating drug penetration through the skin's stratum corneum. Surfactants improve drug permeation through the skin by modifying the lipid barrier, allowing both hydrophilic and hydrophobic drugs to pass more effectively [29, 30]. Surfactants enable effective delivery of anaesthetics, antibiotics, and anticancer agents via lipid-based intravenous formulations [32–35]. In oral drug delivery, surfactants aid in absorption of polypeptides/proteins and peptides by forming protective micelles or liposomes, shielding them from enzymatic degradation. This strategy has expanded the oral delivery potential for drugs such as insulin and vaccines [36–38].

Surfactants play a vital role in the formulation of nasal sprays/inhalers designed for chronic respiratory conditions, wherein they improve dispersibility of aerosolized particles and therapeutic efficacy [39, 40]. In addition to their role as excipients, natural lung surfactants are employed as active therapeutic agents in conditions like neonatal respiratory distress syndrome, where they help restore alveolar stability and improve pulmonary function. This dual functionality both as delivery enhancers and bioactive agents position surfactants as indispensable components in respiratory medicine.

The morphological aggregates formed by surfactants are efficient delivery systems for several drugs; though some, like fibre gels, are not viable for current biomedical/pharmaceutical applications. Most surfactant systems are not ideal for targeted delivery, except ligand-modified vesicles. Recognizing the inherent limitations of each delivery system is essential, as attempting to enhance the aqueous solubility of an already water-soluble hydrophilic drug using an aqueous-based surfactant system is often infructuous. Nonetheless, solubilizing such drugs into surfactant-based formulations may prove advantageous when aiming to achieve sustained drug release.

2.2 *Antimicrobial and Antiviral Applications*

Considerable work has been dedicated to designing several cationic amphiphilic peptides, surfactants and polymers as antimicrobials [41–54]. Surfactants disrupt microbial membranes owing to their amphiphilic nature, inducing membrane disruption and enhanced permeability and cell rupture [55–57]. Cetyltrimethylammonium bromide and other cationic surfactants exhibit potent bactericidal and antiviral action by destabilizing membranes and impairing biological functions. Surfactants are extensively used as antimicrobial agents, skin sanitizers, cleansing solutions, and germicidal preparations, as well as in personal hygiene products like facial cleansers, body washes, shampoos, intimate care formulations, and oral rinses. They are routinely combined with alcohols or hydrogen peroxide for enhanced efficacy. All the three major types of surfactants, i.e. cationic, anionic and non-ionic surfactants (particularly polysorbates) are often useful as disinfectants. They exhibit potent antimicrobial activity and effectively eliminate different pathogens thereby helping in mitigation of hospital-acquired infections (HAIs) [58].

Polymyxin and amino glycoside micellar formulations have exhibited efficacy against multidrug-resistant (MDR) bacteria [59] owing to their low toxicity and efficient bacterial clearance. Surfactants have emerged as potent agents in antiviral formulations [60]. Their mechanism involves disrupting the lipid bilayer of viral envelopes, thereby neutralizing infectivity and curbing transmission [59, 61]. Surfactants are increasingly utilized in antiviral nasal sprays and inhalation therapies to combat respiratory infections such as influenza and COVID-19 [62]. Surfactants reduce viral adhesion to epithelial surfaces and facilitate deeper penetration of therapeutic agents into the lungs. Recent studies have explored the prophylactic potential of surfactant-based nasal formulations, reporting promising reductions in viral load within the upper respiratory tract. These findings suggest that surfactants may serve as both delivery enhancers and direct antiviral agents in respiratory care. Anionic surfactants like sodium lauryl sulfate (SLS) are widely incorporated into nasal inhalants, germicidal preparations, and mouthwashes to reduce viral load on mucosal surfaces and skin. Beyond topical applications, lipid-based surfactants play a critical role in vaccine adjuvants, where they enhance antigen stability and immunogenicity, contributing to more robust and sustained immune responses [63]. In addition to antiviral applications, surfactants do play a crucial role in dermal bandages and topical disinfectant systems. Surfactants infiltrate and destabilize biofilms, structured microbial communities that confer antibiotic resistance and impede healing. Surfactant-based products have demonstrated efficacy in dispersing biofilms formed by *Pseudomonas aeruginosa* and *Staphylococcus epidermidis*, two common pathogens in chronic wounds. This disruption enhances the penetration of antimicrobial agents and accelerates tissue regeneration, making surfactants valuable adjuncts in wound management strategies [64]. Overall, surfactants are important in mitigating microbial contamination and environmental sanitation systems. Cationic surfactants are routinely employed in wastewater remediation to inactivate pathogenic bacteria and viruses within sewage containing streams. Additionally, during bioremediation, surfactants infiltrate and destabilize membranes and colonies of microbes and microbial colonies in industrial pipelines and medical devices [65].

The broad-spectrum antimicrobial activity of surfactants combined with the physicochemical properties that enhance dispersion and penetration, makes them indispensable in public health preservation. From food safety and pharmaceutical formulations to healthcare hygiene and environmental decontamination, surfactants serve as versatile agents, disrupting microbial membranes, improving drug delivery, and functioning as effective disinfectants.

2.3 Theranostics and Imaging

Theranostics refers to the combination of diagnosis and therapy (treatment) of disease in a single medical intervention. This technique utilizes radioisotopes to visualize a patient's tumor for diagnostic purposes and subsequently deliver targeted therapeutic treatment to the same site. Due to their capacity to enhance the stability, distribution, and performance of diagnostic agents and therapeutic nanocarriers, surfactants are important in biomedical imaging and theranostic applications (Fig. 5) [5, 66]. Surfactants are well suited for maintaining the stability of contrast materials used for magnetic resonance imaging (MRI), ultrasound diagnostics, and fluorescence-guided bioimaging. Surfactants enhance MRI contrast agent performance by improving solubility, stability, and biocompatibility of SPIONs and gadolinium compounds. These characteristics enhance the suitability of surfactants for precise targeted imaging in oncology [67]. In gadolinium-based contrast agents (GBCAs), surfactants improve dispersion during formulation, facilitating uniform distribution of GBCAs

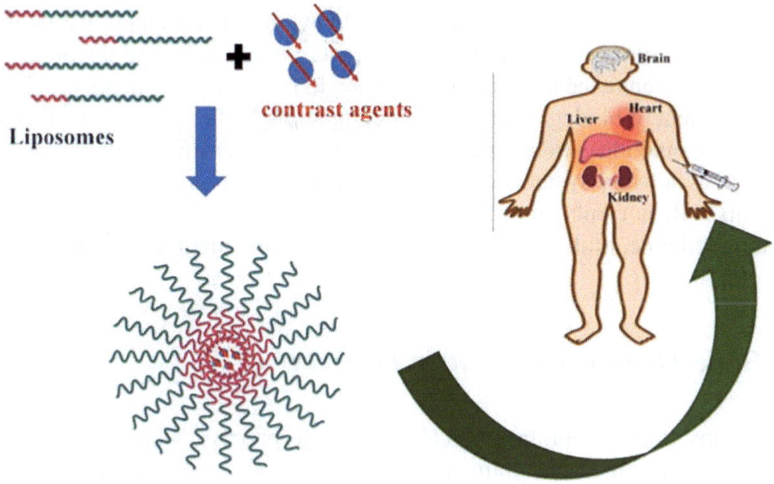

Fig. 5 Application of surfactants in biomedical imaging and oncology. Reproduced with permission from Reference [5]

in injectable solutions. Surfactants reduce viscosity, thereby enhancing injectability and tissue penetration. They also impart stability to nano formulations.

Surfactants like dipalmitoylphosphatidylcholine (DPPC) and distearoylphosphatidylcholine (DSPC) are essential in ultrasound imaging as they stabilize contrast agents, improve echogenicity, and prolong in vivo circulation. In ultrasound imaging, microbubble contrast agents are gas-filled spheres encapsulated by surfactant shells that dramatically improve the visualization of blood flow and tissue perfusion. These agents oscillate in response to ultrasound waves, producing strong acoustic signals that enable real-time imaging [68]. Surfactants are essential for dispersing and stabilizing fluorescent micelles and quantum dots (QDs) in bioimaging, enhancing solubility, biocompatibility, and reducing toxicity. In the field of fluorescence-based bioimaging, the use of surfactant-stabilized nanoparticles especially fluorescent dye-loaded micelles and quantum dots (QDs) has revolutionized cellular and molecular imaging. These nanostructures offer high brightness, photostability, and tunable emission spectra [69]. Moreover, surfactants play a transformative role in cancer theranostics, specifically nanomedicines by stabilizing liposomes and micelles, two of the most effective nanocarriers for drug delivery and imaging. Their amphiphilic nature allows them to encapsulate hydrophobic drugs and imaging agents while enhancing biocompatibility, circulation time, and targeting precision. Photoacoustic imaging (PAI) is an integrated technique that merges the fine spatial resolution of optical imaging with the deep tissue accessibility of ultrasound. It relies on the generation of acoustic waves from light-absorbing agents exposed to pulsed laser light. Surfactant-modified nano formulations enhance photoacoustic imaging by improving light absorption, signal contrast, and stability, with promising applications in vascular imaging, hypoxic tumor detection, and brain-targeted drug delivery [69].

Surfactants have emerged as indispensable agents in contemporary biomedical imaging, MRI, ultrasound imaging, fluorescence microscopy and oncology, owing to their ability to enhance the stability of dispersion and targeting precision. In future, the development of biodegradable and biocompatible surfactant systems is poised to revolutionize imaging-guided therapies, particularly in sensitive applications such as cerebral drug delivery systems and hypoxic tumor detection. These next-generation surfactants will not only improve safety profiles but also unlock new avenues in precision medicine, multimodal imaging, and real-time therapeutic monitoring.

2.4 Gene Therapy and Protein Delivery

Gene therapy is an important clinical therapeutic process, wherein, the genes are deliberately introduced into human somatic cells for stimulating/restoring a gene function. This results in production of a therapeutic protein exhibiting desired action [70]. The success of this therapeutic procedure depends on several parameters: (i) convenient and patient-friendly route of gene administration, (ii) precise control over the location, (iii) functioning and expression of the administered gene, and (iv) formulation of a robust gene product with favourable cost/risk–benefit ratio [70]. The ability of surfactants to interact with both hydrophilic and hydrophobic components make them suitable for stabilizing, solubilizing and transporting biological molecules like DNA, RNA, and proteins. Surfactants are essential in lipid nanoparticle (LNP) formulations for vaccine formulations, CRISPR gene editing, and therapeutic applications. Surfactants impart stability, ensure encapsulation, and intracellular delivery of medicines [71–73]. Surfactant-based systems are vital in the field of nucleic acid and protein delivery, because of their integration into lipid nanoparticles (LNPs) [74]. Surfactants are vital for enhancing the delivery of gene-editing reagents because they ensure efficient encapsulation of these gene-editing reagents into lipid vesicles or polymeric nanoparticles. Cationic surfactants form stable electrostatic complexes with negatively charged nucleic acids, thereby enabling precise gene editing along with minimal adverse effects [74].

Surfactants are important adjuvants that promote immunogenicity and antigen stability thereby enhancing their potential utility in vaccine formulations. Niosomes, non-ionic surfactant-based vesicles have gained significant attention as efficient carriers for protein and peptide drugs, viz. insulin and bacitracin, which are otherwise limited by poor bioavailability, instability in-vivo and ex-vivo, and potential side effects. Niosomes are classified into small unilamellar vesicles (SUV), large unilamellar vesicles (LUV), and multilamellar vesicles (MLV) based on the size and number of layers (Fig. 6) [75].

Small unilamellar vesicles (SUVs), typically ranging from 10 to 100 nm in diameter, are commonly produced through techniques such as sonication, high-pressure extrusion, or high-shear homogenization. SUVs are thermodynamically unstable and therefore prone to aggregation. They exhibit limited encapsulation efficiency for hydrophilic compounds due to their constrained aqueous core.

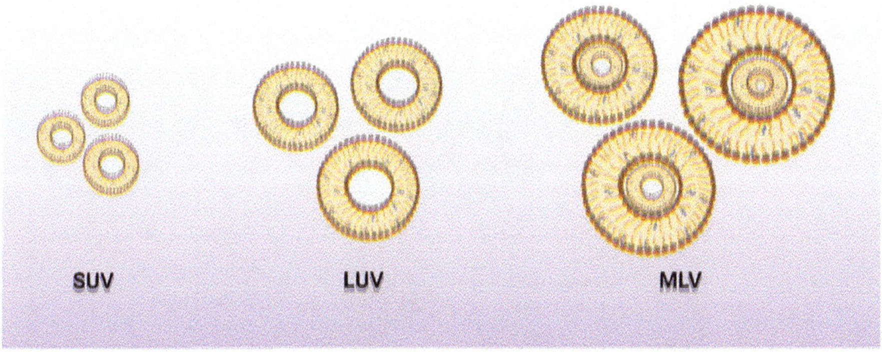

Fig. 6 Types of niosomes based on the size and number of lamellar. SUV: small unilamellar vesicles, LUV: large unilamellar vesicles, MLV: multilamellar vesicles. Reproduced from Reference [75]

Large unilamellar vesicles (LUVs), typically measure between 100 and 1000 nm in diameter. They can be synthesized using various techniques such as the transmembrane pH gradient method (also known as remote loading), reverse-phase evaporation, solvent injection, thermal processing, and dehydration–rehydration cycles.

In contrast, multilamellar vesicles (MLVs) range in size from approximately 0.5–10 μm and are characterized by multiple concentric bilayers, each encapsulating its own aqueous compartment. This layered structure contributes to their high encapsulation capacity and structural complexity, making them suitable for sustained-release applications.

Compared to living organism-based vaccines, niosome-based vaccine formulations demonstrate superior efficiency and reduced adverse reactions. The bilayer structure of niosomes facilitates the encapsulation and controlled release of therapeutic agents, protecting them from enzymatic degradation and improving pharmacokinetics. Moreover, niosomes exhibit excellent potential in the development of protein subunit vaccines, offering targeted delivery, enhanced immunogenicity, and scalable production. This makes them a valuable platform for treating chronic diseases like diabetes and for advancing next-generation vaccine technologies [72–76].

Surfactants are extremely beneficial for pulmonary and ophthalmic delivery of gene and protein therapeutics, overcoming biological barriers and enhancing therapeutic efficacy. A schematic illustration depicting the design of Theranostic Niosomes (iSPN) for intracellular delivery of siRNA/miRNA is presented in Fig. 7 [76]. This results in enhancement of gene transfer efficiency in diseases such as cystic fibrosis and lung cancer [76]. Moreover, for pulmonary applications, phophatidylcholine-based natural surfactant [77], ensures dispersion of nanoparticles within the respiratory tract of lungs, thereby enhancing the delivery of genetic materials like mRNA and siRNA, which are otherwise hindered by mucosal barrier.

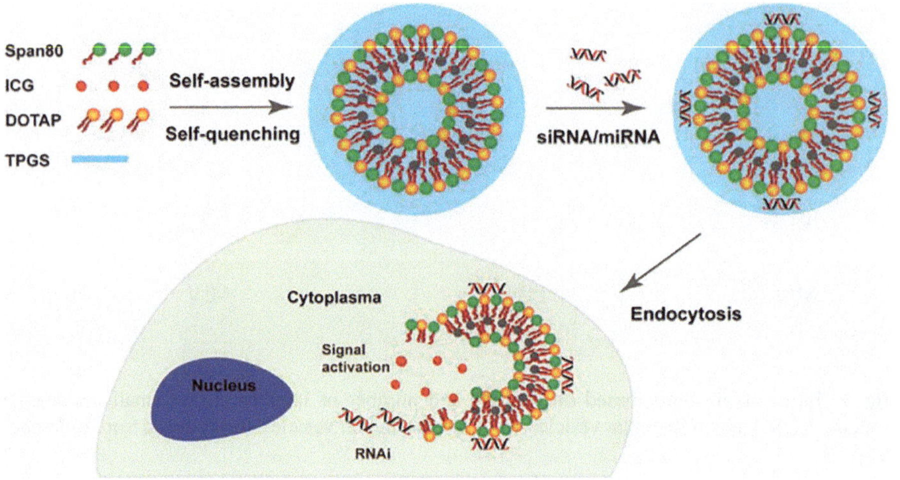

Fig. 7 Illustration showing the design of theranostic niosomes (iSPN) for intracellular delivery of siRNA/miRNA and activatable labelling of cells upon dequenching. Reproduced with permission from Reference [76]

Whereas, in ophthalmology, surfactants are used to solubilize therapeutic proteins in intravitreal injections, increasing drug retention and preventing agglomeration [78]. The targeted gene delivery to retinal cells by surfactants is a promising therapeutic strategy, particularly for the treatment of Leber congenital amaurosis [79]. These applications underscore the versatility of surfactants in advancing localized, non-invasive, and high-precision delivery systems for complex biologics. The potential of surfactants is increasingly recognized in the development of oral formulations for protein and gene-based therapeutics [79, 80]. Surfactants like bile salts and PEGylated lipids enhance the permeability of biologics across the intestinal epithelium despite the harsh gastrointestinal environment (prone to enzymatic degradation), thereby advancing oral delivery systems for treatments such as insulin and gene therapies targeting gut-related diseases [80]. Overall, surfactants play a central role in maintaining the stability, bioavailability, and therapeutic performance of genetic and protein-based drugs. Their integration into platforms like mRNA vaccines, CRISPR delivery systems, and protein stabilization underlines their growing significance in modern medicine. As research progresses, innovative surfactant-based delivery technologies are expected to further elevate the precision and effectiveness of personalized therapeutic interventions.

2.5 Tissue Engineering and Regenerative Medicine

Tissue engineering and regenerative medicine cater to the replacement or regeneration of cells, tissues or organs to restore normal biological function. Surfactants are vital in tissue engineering and regenerative medicine as they promote scaffold formation, cell membrane permeability, and biomaterial delivery in stem cell therapy. Surfactants ensure stability of emulsions, influence surface properties, and interact with cellular membranes, thereby facilitating development of biomaterials for tissue regeneration [81]. During scaffold fabrication, surfactants are utilized to fine tune porosity, mechanical integrity, and biocompatibility. These scaffolds act as structural frameworks that support cell growth and tissue formation, with their physical and chemical attributes significantly influencing regenerative outcomes [82, 83]. Surfactants also influence surface wettability, a key factor in promoting cell adhesion and differentiation. Hydrophilic agents like sodium dodecyl sulfate (SDS) can be employed to increase the hydrophilicity of polymeric scaffolds, improving their compatibility with cells. In bone and cartilage tissue engineering, where effective cell attachment to synthetic matrices underpins successful tissue formation, the aforementioned properties play a significant role. Additionally, surfactants aid in the deposition of bioactive molecules such as growth factors and extracellular matrix (ECM) proteins by modulating surface tension, thereby enhancing the bioactivity of the scaffold [81]. Non-ionic surfactants are particularly effective in emulsifying polymers like polycaprolactone (PCL) and polylactic acid (PLA), thereby improving their processability and facilitating uniform pore formation [84]. Incorporating surfactants during scaffold fabrication enhances the uniformity of porous architectures, thereby facilitating optimal diffusion of nutrients and oxygen, which is critical for cell viability and proliferation. Surfactants exhibit a significant role in decellularization protocols used to generate acellular tissue matrices. These matrices retain the extracellular matrix (ECM) while removing cellular components from donor tissues, ensuring biocompatibility for organ transplantation [81]. Surfactants are commonly used to lyse cells and dissolve membrane lipids without significantly compromising ECM integrity. This approach has proven effective in engineering tissues like heart valves, liver, and cornea [85]. Surfactant-coated nanoparticles and hydrogels have been utilized for effective delivery of carriers for immunomodulatory agents, helping to regulate inflammation and promote tissue regeneration [86]. Recently, surfactant-functionalized liposomes have been utilized to deliver anti-inflammatory therapeutics in wound healing, reducing scar formation and accelerating tissue repair [87]. Surfactants exhibit versatility in regenerative medicine applications spanning from scaffold fabrication and stem cell delivery to surface functionalization and immune modulation. The ability of surfactant to alter surface properties, enhance cellular uptake, and stabilization of bioactive compounds underscores their indispensability in the development of advanced biomaterials.

2.6 Biomedical Coatings

Biocompatible coatings refer to the surface modifications or layers applied to a material to make it compatible with biological systems, they are a crucial component in the development of medical devices, implants, and other biomedical applications. These coatings are designed to interact with biological systems without causing adverse reactions (viz. inflammation, toxicity or rejection, when the material is in contact with living tissues or bodily fluids, ensuring the safety, longevity and efficacy of the medical devices or implants. The integration of surfactants into biomaterial coatings has emerged as a promising strategy to improve the functionality of medical implants and prosthetics. Hydrophilic surfactants, especially PEGylated variants, are employed to tailor surface properties, thereby reducing bacterial adhesion and enhancing biocompatibility [88, 89]. These surfactants effectively inhibit protein -adsorption and biofilm formation, mitigating infection risks associated with long-term implantation. Moreover, surfactant-coated titanium and hydroxyapatite substrates demonstrate improved osteointegration, a key factor in bone tissue engineering and orthopaedic applications [90]. Such surface modifications not only promote cellular compatibility but also support tissue regeneration and implant longevity.

Surfactant-based coatings are pivotal in advancing the safety and efficacy of medical implants, particularly those deployed in cardiovascular interventions [91]. The direct interaction of biomaterials with body fluids and tissue can result in thrombosis, inflammation, and immune activation. These complications are specifically pronounced in devices like stents and catheters. To mitigate these risks, surfactant coatings are engineered to modulate surface properties, reducing protein adsorption, platelet activation, and inflammatory signalling [91]. By enhancing compatibility with the biological milieu, these coatings significantly improve clinical outcomes and device longevity in high-contact vascular environments. The development of surfactant coatings for cardiovascular stents has been developed with an aim to prevent clot formation and platelet adhesion [91]. Surfactants form a functional barrier that prevents direct contact between the biomaterial and blood components. The dual affinity of a surfactant molecule owing to the unique structure is vital for mitigating thrombus formation. Clinically, stents coated with such surfactants have demonstrated significantly lower incidences of thrombosis and restenosis, underscoring their potential to enhance vascular implant performance and patient outcomes [92].

Surfactant coatings that mitigate thrombogenicity offer significant advantages for vascular grafts, particularly those employed in coronary artery bypass procedures. Non-ionic surfactants such as polysorbate 80 exhibited excellent performance in reducing clot formation and improving graft patency over time. These coatings not only suppress thrombus development by minimizing platelet adhesion and protein adsorption but also facilitate endothelialisation, a critical process for graft integration and long-term viability of implants [93–95].

Self-healing and intelligent surfactant coatings represent a transformative advancement in biomaterials, particularly for medical devices and implants. These dynamic coatings are engineered to respond to external stimuli, viz. pH changes, temperature changes, light exposure, or mechanical stress by autonomously repairing micro-damage and restoring surface integrity [96]. Through reversible molecular interactions and adaptive self-assembly, these surfactants enhance the longevity, functionality, and safety of biomedical interfaces. Their ability to maintain biocompatibility while progressively reinforcing structural resilience positions them as key enablers of next-generation implantable technologies [97]. Surfactant-based self-healing systems offer a novel approach to repairing mechanical damage in biomedical coatings. These systems rely on dynamic reforming of the surfactant layer, triggered by external stimuli or surface disruption. Surfactants featuring reversible covalent bonds such as disulfide linkages can autonomously reconstruct damaged regions when the coating is scraped or fractured. This molecular adaptability enables the restoration of surface integrity without external intervention, preserving biocompatibility and functional performance. Such smart coatings are particularly valuable for implants and devices exposed to mechanical stress, where sustained surface resilience is critical for long-term clinical success [97, 98]. Incorporating stimuli-responsive surfactants into biomaterial coatings offers a versatile platform for targeted and controlled drug release. This strategy is particularly valuable in managing chronic or complex diseases like cancer, where synergistic drug action at precise locations and time points is critical for therapeutic efficacy. By integrating responsiveness with biocompatibility, surfactant-based coatings pave the way for next-generation implantable drug delivery systems [99].

2.7 *Biosurfactants*

Biosurfactants are microbial biosurfactants, which are synthesized as metabolic byproducts during the fermentation of bacteria or fungi. These compounds are naturally occurring, sustainably produced, and environmentally friendly. Compared to conventional synthetic surfactants, biosurfactants exhibit significantly lower toxicity. Structurally diverse, biosurfactants mirror the complexity of synthetic variants and are generated by microorganisms grown on either insoluble substrates, such as oils, hydrocarbons, and industrial residues or soluble sources like carbohydrates [100–102]. Biosurfactants are classified into five broad groups, viz. glycolipids, lipopeptides and lipoproteins, phospholipids, hydroxylated and crossed-linked and fatty acids, polymeric surfactants and particulate surfactants [100–102]. This classification is based on chemical constitution and types of microbes from which a respective biosurfactant is produced. Biosurfactants have been extensively utilized in the medicinal and pharmaceutical industries in different therapeutic applications due to their antimicrobial, antifungal, anticancer, anti-adhesive, immunomodulatory and enzyme-inhibiting characteristics [102]. Applications of biosurfactants as antimicrobial and antibiofilm agents have been illustrated in Fig. 8 [101].

Surfactin is known for its antibacterial, antiviral, antitumor, and hemolytic activities. Structurally, it is an anionic cyclic lipopeptide, comprising a heptapeptide ring

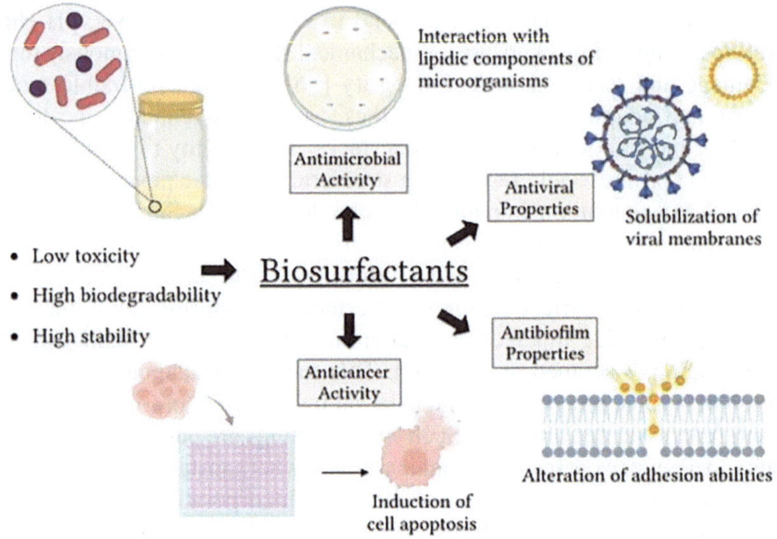

Fig. 8 Applications of biosurfactants as antimicrobial and antibiofilm agents. Reproduced from Reference [101]

linked to a β-hydroxy fatty acid. Surfactin integrates into phospholipid membranes, where it disrupts membrane integrity and induces structural perturbation of target cells [103]. Surfactin A, a biosurfactant, serves as a representative micellar system and was employed as a nanocarrier for Itraconazole (ITZ), a broad-spectrum antifungal agent. While ITZ is therapeutically potent, its high hemolytic activity poses notable safety concerns. In contrast, surfactin A exhibits minimal hemolytic potential, making it a suitable carrier to alleviate this issue. Encapsulation of ITZ within surfactin A micelles effectively limits direct interaction between the drug and red blood cells, thereby reducing cytotoxic effects. This micellar formulation achieved a $53 \pm 2.63\%$ decrease in hemolytic activity compared to an equivalent concentration of ITZ in solution, highlighting its promise for safer antifungal delivery. The micellar formulation demonstrated notable antimicrobial efficacy against Aspergillus fumigatus, Aspergillus niger, and Candida albicans. This highlights the potential of biosurfactants as nanocarrier components, especially for hydrophobic drugs [104].

Wound healing is a complex biological process that restores damaged tissue through three sequential phases: inflammation, proliferation, and remodelling. Effective healing relies heavily on both the modulation of inflammation and protection against oxidative stress. Suppressing inflammation helps reduce swelling, cleanse the wound, and limit tissue damage, while antioxidants shield cells from oxidative injury, thereby promoting cell growth and tissue regeneration. Consequently, therapeutic agents that possess both anti-inflammatory and antioxidant capabilities are especially valuable. During the inflammatory phase, biosurfactants recognized for their anti-inflammatory and antimicrobial properties can attenuate the immune response by downregulating intracellular inflammatory mediators. Notably, the ceramide-like structure of MEL-A, a biosurfactant variant, enhances water permeability across intercellular junctions, enabling it to crystallize within and maintain intracellular hydration. MEL-A has also shown promise in repairing SDS-induced cellular damage at a concentration of 1 wt% [105]. Additionally, crude biosurfactants derived from *Lactobacillus casei* strains exhibit antioxidant activity, offering a protective mechanism against oxidative stress throughout the wound healing cascade [106]. Biosurfactants modulate the immune system by influencing immune cell interactions, regulating cytokine balance, and orchestrating overall immune responses. It has been reported that MEL-A liposomes exhibit targeted binding to macrophage receptors and show strong affinity for immunoglobulins, facilitating the precise delivery of immune-modulating agents to relevant immune cells [107]. Additionally, sophorolipids have demonstrated immunoregulatory effects by suppressing IgE production in the U266 B lymphocyte cell line. This is achieved through the downregulation of key genes associated with IgE biosynthesis and pathobiology, highlighting their potential in managing hypersensitivity and allergic conditions [108].

A new class of biosurfactants-Rhamnolipids exhibit excellent physicochemical properties. They are highly surface active, impart stability and have excellent emulsification characteristics. Rhamnolipid biosurfactants exhibit potent antibacterial and antifungal action [109].

With the advent of personalized medicine, biosurfactants are being custom engineered to align with individual patient profiles. These tailored formulations consider factors such as genetic predisposition, metabolic activity, and disease-specific characteristics to optimize drug delivery and therapeutic outcomes [110]. Biosurfactants can be functionalized to recognize and bind specific biomarkers, enabling precise targeting of therapeutic agents. This approach is particularly valuable in oncology, where tumor heterogeneity demands individualized drug delivery strategies [111]. Furthermore, personalized biosurfactants show promise in regenerative medicine, enhancing cell adhesion and tissue repair in accordance with patient-specific physiological conditions.

Emerging research signifies the potential role of biosurfactants in vaccine formulations and adjuvant therapies [112–114]. Their ability to modulate immune responses makes them ideal candidates for enhancing vaccine stability and efficacy. Specifically, biosurfactants can improve the solubility of hydrophobic components, thereby facilitating more efficient antigen delivery and immune system activation. The immunomodulatory properties of biosurfactants also open avenues for designing novel adjuvants that boost vaccine-induced protection against infectious diseases [115]. Looking ahead, biosurfactants are poised to transform medicine through advancements in sustainable formulation strategies, intelligent drug delivery systems, and AI-driven optimization [116]. Applications in nanomedicine, antimicrobial therapies, and personalized treatment approaches continue to expand, addressing critical challenges in pharmaceutical research. Although industrial-scale production remains a hurdle, ongoing progress in biotechnology and synthetic biology is expected to accelerate their commercialization, positioning biosurfactants as key components of next-generation medical solutions.

3 Conclusions and Prospects

Surfactants are extremely useful with reference to several applications in biomedical sciences: specifically, drug delivery systems, antimicrobial activity, gene and protein delivery, imaging, biomedical coatings, biosurfactants, tissue engineering and regenerative medicine.

- The unique structural features of different classes of surfactants, viz. anionic, cationic, non-ionic, and zwitterionic surfactants and their self-aggregation behaviour ensure enhancement of solubility, stability, and bioavailability of the therapeutic drugs thereby facilitating pharmaceutical drug formulations and biomedical applications. However, biocompatibility and cytotoxicity are critical factors with reference to their clinical applications.
- In drug delivery systems, surfactants exhibit significant therapeutic efficacy. Micelle and liposome-based systems ensure precise drug delivery, sustained release along with regulatory compliance.
- For antimicrobial and antiviral uses, surfactants disrupt microbial membranes, exhibit germicidal properties, and facilitate the creation of surfactant-modified antibiotics and antiviral products, such as against COVID-19 and influenza.
- Surfactants play an important role in gene and protein delivery by stabilizing lipid-based nanoparticles for mRNA vaccines. They facilitate CRISPR-based gene editing, and protein stabilization in therapeutics.
- Surfactants have emerged as indispensable agents in contemporary biomedical imaging, MRI, ultrasound imaging, fluorescence microscopy and oncology, owing to their ability to enhance the stability of dispersion and targeting precision.
- Surfactants facilitate scaffold fabrication and find use in biomaterial coating thereby playing a critical role in tissue engineering and regenerative medicine.
- However, challenges vis-a-vis long-term in vivo stability, cytotoxicity, biodegradability, scalability and regulatory approvals are still significant hurdles for clinical translation, thereby impending surfactant-based medical product approval.
- The advent of machine learning (ML) and artificial intelligence (AI) is expected to expedite design of surfactant formulations capable of targeted drug delivery with enhanced therapeutic efficacy and minimal adverse effects.
- The design of surfactants or utilization of biosurfactants that are specifically designed for diverse biomedical applications is a crucial determinant for prospects.

Acknowledgements Authors thank Navrachana University for Seed Grant (NUV_Seed Grant_2025-26_07) for providing the financial support. Authors declare that they have no conflict of interest.

References

1. Sharma KS, Rodgers C, Palepu RM, Rakshit AK (2003) Studies of mixed surfactant solutions of cationic dimeric (gemini) surfactant with nonionic surfactant $C_{12}E_6$ in aqueous medium. J Colloid Interface Sci 268(2):482
2. Holmberg K, Jönsson B, Kronberg B, Lindman B (2002) Surfactants and polymers in aqueous solution. Wiley, Chichester, U.K.
3. Rosen MJ, Kunjappu JT (2012) Surfactants and interfacial phenomena, 4th edn. Wiley, Hoboken, NJ
4. Eastoe J (2005) Surfactant aggregation and adsorption at interfaces, colloid science; Cosgrove T, (ed) Blackwell Publishing Ltd., Oxford, p 50
5. Nagaraj K, Kamalesu S (2025) State-of-the-art surfactants as biomedical game changers: unlocking their potential in drug delivery, diagnostics, and tissue engineering. Int J Pharm 676:12559
6. Rodrigues LR (2015) Microbial surfactants: fundamentals and applicability in the formulation of nano-sized drug delivery vectors. J Colloid Interface Sci 449:304
7. Suhail M, Janakiraman AK, Khan A, Naeem A, Badshah SF (2019) Surfactants and their role in pharmaceutical product development: an overview. J Pharma Pharm 6:72
8. Zhou P, Yang XL, Wang XG, Hu B, Zhang L, Zhang W, Si HR, Zhu Y, Li B, Huang CL, Chen HD, Chen J, Luo Y, Guo H, Jiang RD, Liu MQ, Chen Y, Shen XR, Wang X, Zheng XS, Shi ZL (2020) A pneumonia outbreak associated with a new coronavirus of probable bat origin. Nature 579(7798):270
9. Singha A, Van Hammeb JD, Ward OP (2007) Surfactants in microbiology and biotechnology: part 2. Appl Aspects Biotechnol Adv 25:99
10. (a) Srivastava A, Tiwari S, Khan JM, Deb DK, Ullah MW (2025) Investigating the binding interactions of cetirizine and diphenhydramine in SDS-SDBS mixed micelles. Colloid Polym Sci, 4, 1. (b) Srivastava A, Thapa U, Saha M, Jalees M (2019) Aggregation behaviour of tetracaine hydrochloride with Gemini surfactants and the formation of silver nanoparticles using drug-Gemini surfactants mixture, J Mol Liq, 276, 399
11. Srivastava A, Kumar M, Singh RP, Khan JM, Singh SK (2024) Effect of the mixed micelles of zwitterionic-anionic surfactant on efficiency of antibiotic azithromycin dihydrate. J Mol Liq 408:125317
12. Srivastava A, Kumar M, Ahmed MZ (2024) Investigation of the binding and partition properties of azithromycin drug using drug-surfactant mixed micelles in the presence of trisodium citrate electrolyte. J Mol Liq 414:126114
13. Kumar M, Khushi K, Singh SK, Deb DK, Khan JM, Ahmad A, Singh OG, Singh N, Srivastava A (2024) Binding, and occupancy properties of gabapentin in mixed surfactant systems. J Surfactant Deterg 27(3):433
14. Kumar M, Elahi D, Bhardwaj A, Sharma S, Khushi K, Singh E, Singh N, Srivastava A (2022) Physiochemical investigation of the excipients mixed micelles for improvement of encapsulation and controlled release of antihistamine drugs. J Mol Liq 364:119971
15. Srivastava A, Kumar M, Deb DK, Muzaffar F, Singh S (2022) Utilization of amphiphilic antihistamines drugs to enhance micellization of anionic surfactant and improve the binding and solubility of Itraconazole drug. J Mol Liq 348:118018
16. Kumar M, Singh V, Choudhary R, Singh S, Srivastava A (2021) Mixed micellization of drug-excipients and its application to enhance the binding and encapsulation efficacy of ibuprofen in aqueous media. Colloids Surf A Physicochem Eng Asp 628:127268
17. Masjedi M, Montahaei T (2021) An illustrated review on nonionic surfactant vesicles (niosomes) as an approach in modern drug delivery: fabrication, characterization, pharmaceutical, and cosmetic applications. J Drug Deliv Sci Technol 61:102234
18. Sarkar R, Pal A, Rakshit A, Saha B (2021) Properties, and applications of amphoteric surfactant: a concise review. J Surfactant Deterg 24:709
19. Gursoy RN, Benita S (2004) Self-emulsifying drug delivery systems (SEDDS) for improved oral delivery of lipophilic drugs. Biomed Pharmacother 58:173

20. Kohli K, Chopra S, Dhar D, Arora S, Khar RK (2010) Self-emulsifying drug delivery systems: an approach to enhance oral bioavailability. Drug Discov Today 15:958
21. Saha U, De R, Das B (2023) Interactions between loaded drugs and surfactant molecules in micellar drug delivery systems: a critical review. J Mol Liq 382:121906
22. Desai SA, Patel VP, Nagare S, Bhosle K (2025) In: Biosurfactants as Potential Drug Delivery Systems with Special Reference to Nanovesicle Liposomes. In: Microbial surfactants in pharmaceuticals and cosmetics. CRC Press, pp 230–254
23. (a) Jacob S, Kather FS, Boddu SHS, Attimarad M, Nair AB (2025) Nanosuspension innovations: expanding horizons in drug delivery techniques. Pharmaceutics, 17, 136. (b) Jayne Lawrence M (1994) Surfactant systems: their use in drug delivery, Chem Soc Rev, 23, 417
24. Attwood D (2012) Surfactant systems: their chemistry, pharmacy and biology. Springer Science & Business Media
25. Lawrence MJ, Rees GD (2000) Microemulsion-based media as novel drug delivery systems. Adv Drug Deliv Rev 45(1):89–121
26. Hiromasa U, Srivastava A, Fujimori A, Tomoo K, Nakanishi A, Tandia M, Kadota K, Tozuka Y (2018) Investigation of physiological properties of transglycosylated stevia with cationic surfactant and its application to enhance the solubility of rebamipide. J Phys Chem B 122(43):10051
27. Nagaraj K, Anbazhagan GK, Govindasamy RS, Muthu V, Rajkumar S, Senthilnathan S, Muddana S, Esanakula RR, Sivakumar D, Sri CD, Ravi R (2025) Biomimetic surfactants for tunable interfacial properties in drug delivery, biomedical coatings and tissue engineering. Int J Pharm 677:125658
28. Maja L, Željko K, Mateja P (2020) Sustainable technologies for liposome preparation. J Supercrit Fluids 165:104984
29. Shahzadi I, Asim MH, Dizdarevic A, Wolf JD, Kurpiers M, Matuszczak B, Bernkop-Schnürch A (2019) Arginine-based cationic surfactants: biodegradable auxiliary agents for the formation of hydrophobic ion pairs with hydrophilic macromolecular drugs. J Colloid Interface Sci 552:287
30. Matteo Jörgensen A, Knoll P, Haddadzadegan S, Fabian H, Hupfauf A, Gust R, Georg Jörgensen R, Bernkop-Schnürch A (2023) Biodegradable arginine-based steroid-surfactants: cationic green agents for hydrophobic ion-pairing. Int J Pharm 630:122438
31. Kurpiers M, Wolf JD, Spleis H, Steinbring C, Jörgensen AM, Matuszczak B, Bernkop-Schnürch A (2021) Lysine-based biodegrad able surfactants: increasing the lipophilicity of insulin by hydrophobic ion paring. J Pharmaceut. Sci 110:124
32. Sun T, Zhang YS, Pang B, Hyun DC, Yang M, Xia Y (2014) Engineered nanoparticles for drug delivery in cancer therapy. Angew Chem Int Ed Engl 53(46):12320
33. Ghafourian T, Nokhodchi A, Kaialy W (2015) Surfactants as Penetration Enhancers for Dermal and Transdermal Drug Delivery. In: Dragicevic N, Maibach H (eds) Percutaneous penetration enhancers chemical methods in penetration enhancement. Springer, Berlin, Heidelberg, p 207
34. Maulvi FA, Parmar MB, Shetty KH, Patel AR, Desai BV, Vyas BA, Desai DT, Kalaiselvan P, Masoudi M, Shah DO, Willcox MDP (2024) Role of micelle dynamics in enhancing cyclosporine uptake in hyaluronic acid-contact lenses for improved critical lens properties in dry eye management. Colloids Surf A: Physicochem Eng Aspects 688:133550
35. Maya SA, Alam MM, Khan JM, Anis-Ul-Haque KM, Rana S, Hasan K, Posa M, Kumar D, Rahman MM, Hoque MA (2025) The aggregation behaviour of tetradecyltrimethylammonium bromide in aqueous solution of an antidiabetic drug at variable temperatures: influences of dihydroxy organic compounds and temperature. Colloids Surf A: Physicochem Eng Aspects 714:136552
36. Srivastava A, Kumar M, Ahmed MZ (2024) Investigation the binding and partition properties of azithromycin drug using drug-surfactant mixed micelles in the presence of trisodium citrate electrolyte. J Mol Liq 414:126114
37. Kumar M, Singh V, Choudhary R, Deb DK, Singh S, Srivastava A (2021) Mixed Micellization of drug-excipients and its application to enhance the binding and encapsulation efficacy of ibuprofen in aqueous media. Colloids Surf A: Physicochem Eng Aspects 628:127268

38. Li G, Srivastava A, Liu C, Qiao W (2020) Interaction of doxorubicin hydrochloride in the presence of, mixed aggregate of ibuprofen sodium and cationic lipid. J Mol Liq 313:113451
39. Bastani MN, Jalilian S (2024) Unraveling the enigma: the emerging significance of pulmonary surfactant proteins in predicting, diagnosing, and managing COVID-19. Immun Inflamm Dis 12:1302
40. Cojocaru E, Petriş OR, Cojocaru C (2024) Nanoparticle-based drug delivery systems in inhaled therapy: improving respiratory medicine. Pharmaceuticals 17:1059
41. Inacio AS, Domingues NS, Nunes A, Martins PT, Moreno MJ, Estronca LM, Fernandes R, Moreno AJ, Borrego MJ, Gomes JP, Vaz WL (2016) Quaternary ammonium surfactant structure determines selective toxicity towards bacteria: mechanisms of action and clinical implications in antibacterial prophylaxis. J Antimicrob Chemother 71:641
42. Zhou C, Wang F, Chen H, Li M, Qiao F, Liu Z, Hou Y, Wu C, Fan Y, Liu L, Wang S (2016) Selective antimicrobial activities and action mechanism of micelles self-assembled by cationic oligomeric surfactants. ACS Appl Mater Interfaces 8:4242
43. Bai H, Zhang H, Hu R, Chen H, Liu F, Lv L, Wang S (2017) Supramolecular conjugated polymer systems with controlled antibacterial activity. Langmuir 33:1116
44. Pinazo A, Manresa MA, Marques AM, Bustelo M, Espuny MJ, Pérez L (2016) Amino acid–based surfactants: new antimicrobial agents. Adv Colloid Interface Sci 228:17
45. Herzog IM, Fridman M (2014) Design and synthesis of membrane-targeting antibiotics: from peptides-to aminosugar-based antimicrobial cationic amphiphiles. Med Chem Comm 5:1014
46. Hoque J, Akkapeddi P, Yarlagadda V, Uppu DS, Kumar P, Haldar J (2012) Cleavable cationic antibacterial amphiphiles: synthesis, mechanism of action, and cytotoxicities. Langmuir 28:12225
47. El Hage S, Lajoie B, Stigliani JL, Furiga-Chusseau A, Roques C, Baziard G (2014) Synthesis, antimicrobial activity and physico-chemical properties of some n-alkyldimethylbenzylammonium halides. J Appl Biomed 12:245
48. Muthukumarasamyvel T, Baskar R, Chandirasekar S, Umamaheswari K, Rajendiran N (2016) Hierarchical self-assembly of bile-acid-derived dicationic amphiphiles and their toxicity assessment on microbial and mammalian systems. ACS Appl Mater Interfaces 8:25111
49. Nagamune H, Maeda T, Ohkura K, Yamamoto K, Nakajima M, Kourai H (2000) Evaluation of the cytotoxic effects of bis-quaternary ammonium antimicrobial reagents on human cells. Toxicol in Vitro 14:139
50. Kimura R, Shibata M, Koeda S, Miyagawa A, Yamamura H, Mizuno T (2018) Development of new antimicrobial agents from cationic PG-surfactants containing oligo-lys peptides. Bioconjug Chem 29(12):4072
51. Jennings MC, Minbiole KP, Wuest WM (2015) Quaternary ammonium compounds: an antimicrobial mainstay and platform for innovation to address bacterial resistance. ACS Infect Dis 1:288
52. Zhou C, Wang D, Cao M, Chen Y, Liu Z, Wu C, Xu H, Wang S, Wang Y (2016) Self-aggregation, antibacterial activity, and mildness of cyclodextrin/cationic trimeric surfactant complexes. ACS Appl Mater Interfaces 8:30811
53. Qi R, Zhang P, Liu J, Zhou L, Zhou C, Zhang N, Han Y, Wang S, Wang Y (2018) Peptide amphiphiles with distinct supramolecular nanostructures for controlled antibacterial activities. ACS Appl Bio Mater 1:21
54. Zhou C, Wang Y (2020) Structure-activity relationship of cationic surfactants as antimicrobial agents. Curr Opin Colloid Interface Sci 45:2843
55. Kaur G, Mehta SK (2017) Developments of Polysorbate (Tween) based microemulsions: preclinical drug delivery, toxicity and antimicrobial applications. Int J Pharm 529(1):134
56. Sultan F, Maji D, Phatake RS, Kumar K (2025) Pharmaceutical applications of microbial biosurfactants. Int J Pharm 681:125887
57. Kelkar SS, Bhojwani KR (2025) Kamble. CRC Press, E.E. In Biosurfactants as Antimicrobial Agents, pp 256–273
58. Thomas RE, Thomas BC, Conly J, Lorenzetti D (2022) Cleaning and disinfecting surfaces in hospitals and long-term care facilities for reducing hospital- and facility-acquired bacterial and viral infections: a systematic review. J Hosp Infect 122:9

59. Kirtane AR, Verma M, Karandikar P, Furin J, Langer R, Traverso G (2021) Nanotechnology approaches for global infectious diseases. Nat Nanotechnol 16:369
60. Sarangi MK, Padhi S, Patel LD, Rath G, Nanda SS, Yi DK (2022) Theranostic efficiency of biosurfactants against COVID-19 and similar viruses—a review. J Drug Deliv Sci Technol 76:103764
61. Lorin S, Dhanakotti R, Selvam S, Jaganathan R, Kumaradhas P, Nagaraj K, Raja K (2024) Synthesis, DFT, in-silico molecular docking, molecular dynamic simulation and ADMET studies of (Z)-2,6-bis(4-bromophenyl)-3,3-dimethyl-4-(2-(2,4,6-trichlorophenyl) hydrazono) piperidine derivatives against the SARS-CoV-2 main-protease. Z Phys Chem 238(4):729
62. Pramod K, Kotta S, Jijith US, Aravind A, Abu Tahir M, Manju CS (2020) Gangadharappa, H.V. Surfactant-based prophylaxis and therapy against COVID-19: a possibility. Med Hypotheses 143:110081
63. Bolhassani A (2023) Lipid-based delivery systems in development of genetic and subunit vaccines. Mol Biotechnol 65:669
64. Ceresa C, Fracchia L, Fedeli E, Porta C, Banat IM (2021) Recent advances in biomedical. Ther Pharm Appl Microb Surf Pharm 13:466
65. Sharma G, Shyam K, Thakur S, Yadav M, Nair S, Kumar N, Chandel H, Saxena G (2023) Metagenomic characterization and applications of microbial surfactants in remediation of potentially toxic heavy metals for environmental safety. In: Bilal M, Ferreira LFR, Iqbal HMN (eds) Genomics approach to bioremediation: principles, tools, and emerging technologies, vol Kumar. Wiley, New York, pp 89–124
66. Astafyeva K, Somaglino L, Desgranges S, Berti R, Patinote C, Langevin D, Lazeyras F, Salomir R, Polidori A, Contino-P´ epin C, Urbach W (2015) Perfluorocarbon nanodroplets stabilized by fluorinated surfactants: characterization and potentiality as theranostic agents. J Mater Chem B, (14), 2892
67. Raychaudhuri R, Pandey A, Hegde A, Abdul Fayaz SM, Chellappan DK, Dua K, Mutalik S (2020) Factors affecting the morphology of some organic and inorganic nanostructures for drug delivery: characterization, modifications, and toxicological perspectives. Expert Opin Drug Deliv 17(12):1737–1765
68. Krafft MP (2012) Controlling phospholipid self-assembly and film properties using highly fluorinated components–fluorinated monolayers, vesicles, emulsions and microbubbles. Biochimie 94(1):11
69. Salabat A, Mirhoseini F (2022) Polymer-based nanocomposites fabricated by microemulsion method. Polym Compos 43(3):1282
70. Shahiwala A, Surti A (2025) Barriers to gene and protein delivery, In: (eds) Shahiwala A, Surti N, Challenges in delivery of therapeutic genomics and proteomics. 2nd edn, Academic Press, pp 1–34
71. Ahmady AR, Hosseinzadeh P, Solouk A, Akbari S, Szulc AM, Brycki BE (2022) Cationic gemini surfactant properties, its potential as a promising bioapplication candidate, and strategies for improving its biocompatibility: a review. Adv Colloid Interface Sci 299:102581
72. Dhayalan M, Wang W, Riyaz SM, Dinesh RA, Shanmugam J, Irudayaraj SS, Stalin A, Giri J, Mallik S, Hu R (2024) Advances in functional lipid nanoparticles: from drug delivery platforms to clinical applications. Biotech 14(2):57
73. Grijalvo S, Puras G, Zárate J, Sainz-Ramos M, Qtaish NA, López T, Mashal M, Attia N, Diaz Diaz D, Pons R, Fernández E (2019) Cationic niosomes as non-viral vehicles for nucleic acids: challenges and opportunities in gene delivery. Pharmaceutics 11(2):50
74. (a) Thirumal V, Aravindhan V, Ac N (2023) Liposomal carriers: development, evaluation, applications with its regulatory aspects, World J Adv Res Rev, 20 (3), 324. (b) Khaksa G, D'Souza R, Lewis S, Udupa N (2000) Pharmacokinetic study of niosome encapsulated insulin. Ind J Exp Bio, 38(9), 901
75. Moammeri A, Chegeni MM, Sahrayi H, Ghafelehbashi R, Memarzadeh F, Mansouri A, Akbarzadeh I, Abtahi MS, Hejabi F, Ren Q (2023) Current advances in niosomes applications for drug delivery and cancer treatment. Mater Today Bio 23:100837

76. Yang C, Gao S, Song P, Dagnæs-Hansen F, Jakobsen M, Kjems J (2018) Theranostic niosomes for efficient siRNA/MicroRNA delivery and activatable near-infrared fluorescent tracking of stem cells. ACS Appl Mater Interfaces 10(23):19494
77. (a) Vangala AK, Perrie Y (2003) Preparation and characterisation of DNA loaded non-ionic surfactant vesicles. J Pharm Pharmacol, 55, 101. (b) Alwani S, Wasan EK, Badea I (2024) Solid lipid nanoparticles for pulmonary delivery of biopharmaceuticals: a review of opportunities, challenges, and delivery applications. Mol Pharm, 21 (7), 3084
78. Jiao J (2008) Polyoxyethylated nonionic surfactants and their applications in topical ocular drug delivery. Adv Drug Deliv Rev 60(15):1663
79. Abdulsalam L, Mordecai J, Ahmad I (2025) Non-viral gene therapy for Leber's congenital amaurosis: progress and possibilities. Nanomedicine 20(3):291
80. Desai SA, Patel VP, Nagare S, Bhosle K (2025) Biosurfactants as potential drug delivery systems with special reference to nanovesicle liposomes. In: Microbial surfactants in pharmaceuticals and cosmetics. CRC Press, p 230
81. Beygi M, Oroojalian F, Azizi-Arani S, Hosseini SS, Mokhtarzadeh A, Kesharwani P, Sahebkar A (2024) Multifunctional nanotheranostics for overcoming the blood–brain barrier. Adv Funct Mater 34(19):2310881
82. Suamte L, Tirkey A, Barman J, Babu PJ (2023) Various manufacturing methods and ideal properties of scaffolds for tissue engineering applications. Smart Mater Manuf 1:100011
83. Wei W, Bai F, Fan H (2019) Surfactant-assisted cooperative self-assembly of nanoparticles into active nanostructures. Science 11:272
84. Singla P, Garg S, McClements J, Jamieson O, Peeters M, Mahajan RK (2022) Advances in the therapeutic delivery and applications of functionalized Pluronics: a critical review. Adv Colloid Interface Sci 299:102563
85. Otzen DE (2017) Biosurfactants and surfactants interacting with membranes and proteins: same but different? Biochim Biophys Acta Biomembr 1859(4):639
86. Qin XS, Wang M, Li W, Zhang YS (2022) Biosurfactant-stabilized micropore-forming GelMA inks enable improved usability for 3D printing applications. Regenerative Translational Med 8:471
87. Foldvari M, Chen DW, Nafissi N, Calderon D, Narsineni L, Rafiee A (2016) Non-viral gene therapy: gains and challenges of non-invasive administration methods. J Control Release 240:165
88. Nagaraj K (2025) State-of-the-art surfactant-based coatings for biomaterials: Innovations, applications, and future prospects in biomedical engineering. Int J Pharm 681:125879
89. Priya S, Desai VM, Singhvi G (2022) Surface modification of lipid-based nanocarriers: a potential approach to enhance targeted drug delivery. ACS Omega 8(1):74
90. Prasad PS, Marupalli BC, Das S, Das K (2023) Surfactant-assisted synthesis of hydroxyapatite particles: a comprehensive review. J Mater Sci 58(14):6076
91. Devi LA, Selvam IK, Nagaraj K, Arun M, Senthilkumar N, Saravanan K, Maity P, Renuga V, Bremkumar KS, Manimegalai T (2024) Potassium silicate applications in Abelmoschus esculentus (okra) cultivation: a comprehensive review on mitigating abiotic and biotic stresses. Physiol Mol Plant Pathol 1:102437
92. Kuehl B, Raman S, Becker A, Garg V, Roberts-Dobie J, McCaslin A, Brensdal J, Attinger J, Burton L, Forrester M, Hohmann A (2024) Fully atom-efficient olvent-mediated biopolymer manufacturing: a base case illustrated with macromolecular surfactants tailored to stable polymer–water interfaces. ACS Appl Mater Interfaces 16:59280
93. Torchilin VP (2001) Structure and design of polymeric surfactant-based drug delivery systems. J Control Release 73:137
94. Tranquillo E, Bollino F (2020) Surface modifications for implants lifetime extension: an overview of sol-gel coatings. Coatings 10:589
95. Siafaka PI, Üstündağ Okur N, Karavas E, Bikiaris DN (2016) Surface modified multifunctional and stimuli responsive nanoparticles for drug targeting: current status and uses. Int J Mol Sci 17(9):1440

96. Jamrozy M, Kudłacik-Kramarczyk S, Drabczyk A, Krzan M (2024) Advanced drug carriers: a review of selected protein, polysaccharide, and lipid drug delivery platforms. Int J Mol Sci 25:786
97. Wu G, Li P, Feng H, Zhang X, Chu PK (2024) Engineering, and functionalization of biomaterials via surface modification. J Mater Chem B 2015:3
98. Wu W, Cheng R, das Neves J, Tang J, Xiao J, Ni Q, Liu X, Pan G, Li D, Cui W, Sarmento B (2017) Advances in biomaterials for preventing tissue adhesion. J Controlled Release 261:318–336
99. Anestopoulos I, Kiousi DE, Klavaris A, Galanis A, Salek K, Euston SR, Pappa A, Panayiotidis MI (2020) Surface active agents and their health-promoting properties: molecules of multifunctional significance. Pharmaceutics 12:688
100. Markande A, Patel D, Varjani S (2021) A review on biosurfactants: properties, applications, and current developments. Bioresour Technol 330:124963
101. Puyol McKenna P, Naughton PJ, Dooley JSG, Ternan NG, Lemoine P, Banat IM (2024) Microbial biosurfactants: antimicrobial activity and potential biomedical and therapeutic exploits. Pharmaceuticals 17:138
102. Jemil N, Ben Ayed H, Manresa A, Nasri M, Hmidet N (2017) Antioxidant properties, antimicrobial and anti-adhesive activities of DCS1 lipopeptides from Bacillus methylotrophicus DCS1. BMC Microbiol 17:144
103. Tsan P, Volpon L, Besson F, Lancelin JM (1968) Structure and dynamics of surfactin studied by NMR in micellar media. J Am Chem Soc 2007:129
104. Usman F, Farooq M, Wani TA, Ahmad H, Javed I, Iqbal M, Sheikh FA, Siddique F, Zargar S, Sheikh S (2023) Itraconazole loaded biosurfactin micelles with enhanced antifungal activity: fabrication. Eval Mol Simul Antibiot 12:1550
105. Morita T, Kitagawa M, Suzuki M, Yamamoto S, Sogabe A, Yanagidani S, Imura T, Fukuoka T, Kitamoto D (2009) A yeast glycolipid biosurfactant, mannosylerythritol lipid, shows potential moisturizing activity toward cultured human skin cells: the recovery effect of MEL-A on the SDS-damaged human skin cells. J Oleo Sci 58:639
106. Merghni A, Dallel I, Noumi E, Kadmi Y, Hentati H, Tobji S, Ben Amor A, Mastouri M (2017) Antioxidant and antiproliferative potential of biosurfactants isolated from lactobacillus casei and their anti-biofilm effect in oral staphylococcus aureus strains. Microb Pathog 104:84
107. Im JH, Nakane T, Yanagishita H, Ikegami T, Kitamoto D (2001) Mannosylerythritol lipid, a yeast extracellular glycolipid, shows high binding affinity towards human immunoglobulin G. BMC Biotechnol 1:5
108. Hagler M, Smith-Norowitz TA, Chice S, Wallner SR, Viterbo D, Mueller CM, Gross R, Nowakowski M, Schulze R, Zenilman ME (2007) Sophorolipids decrease IgE production in U266 cells by downregulation of BSAP (Pax5), TLR-2, STAT3 and IL-6. J Allergy Clin Immunol 119:S263
109. Puyol McKenna P, Naughton PJ, Dooley JS, Ternan NG, Lemoine P, Banat IM (2024) Microbial biosurfactants: antimicrobial activity and potential biomedical and therapeutic exploits. Pharmaceuticals 17(1):138
110. Purohit GK, Dubey D, Rath S, Subhadarsini SS, Tripathi D, Behera S, Sahu LS (2024) Biosurfactants: moving towards healthcare applications. Int J Adv Life Sci Res 7:24
111. Karnwal A, Shrivastava S, Al-Tawaha ARMS, Kumar G, Singh R, Kumar A, Mohan A, Yogita Malik T (2023) Microbial biosurfactant as an alternate to chemical surfactants for application in cosmetics industries in personal and skin care products: a critical review. Biomed Res Int 1:2375223
112. Gaur VK, Sharma P, Gupta S, Varjani S, Srivastava JK, Wong JW, Ngo HH (2022) Opportunities and challenges in omics approaches for biosurfactant production and feasibility of site remediation: strategies and advancements. Environ Technol Innov 25:102132
113. Sajid M, Khan MSA, Cameotra SS, Al-Thubiani AS (2020) Biosurfactants: potential applications as immunomodulator drugs. Immunol Lett 223:71
114. Joshi R, Jadhao M (2022) Application of biosurfactant as an adjuvant in medicine. In: Green sustainable process for chemical and environmental engineering and science. Academic Press, p 61

115. Fan J, Jin S, Gilmartin L, Toth I, Hussein WM, Stephenson RJ (2022) Advances in infectious disease vaccine adjuvants. Vaccines 10(7):1120
116. Bharadwaj E, Deepika K, Kumar A, Jaiswal S, Miglani S, Singh D, Fartyal P, Kumar R, Singh S, Singh MP, Gaidhane AM (2024) Exploring the artificial intelligence and its impact in pharmaceutical sciences: insights toward the horizons where technology meets tradition. Chem Biol Drug Des 104(4):14639

Applications of Microbial Surfactants in Biotechnology

Ninad Mhatre, Fatema Saiger, and Amit P. Pratap

Abstract Biosurfactants are surface-active biomolecules synthesized by microorganisms. In recent years, there has been increasing interest in biosurfactants as sustainable and eco-friendly alternatives to conventional petrochemical surfactants. Due to their amphiphilic nature, biosurfactants can effectively reduce surface and interfacial tension. Biosurfactants demonstrate high biodegradability, low toxicity, and environmental compatibility. This chapter comprehensively discusses classification, physicochemical properties, and production strategies for biosurfactants, including submerged, solid-state, and co-culture fermentation systems. Special emphasis has been given to cost-effective production using renewable substrates such as agro-industrial wastes and optimization through genetic and process engineering. Among biosurfactants, the following types are most studied: rhamnolipids, sophorolipids, and surfactin. These biosurfactants have found versatile uses in such industry sectors as pharmaceutical, food, cosmetic, agriculture, petroleum recovery, and bioremediation. Their multifunctional roles include emulsification, antimicrobial activity, disruption of biofilm, and enhancement of pollutant degradation, hence making biosurfactants attract attention in circular and green biotechnological processes. In addition to many advantageous characteristics, major challenges to commercialization at large scale involve high production costs and complex downstream processing. However, recent advances in synthetic biology, improvement of strains, and valorisation of wastes open a way to large-scale commercialization. Overall, biosurfactants represent a promising class of bio-based materials with many applications that align the global drive toward sustainable industrial biotechnology and environmental stewardship.

Keywords Biosurfactants · Microbial surfactants · Bioremediation · Genetic engineering · Sustainable biotechnology · Industrial microbiology · Fermentation optimization · Renewable substrates · Environmental applications · Green

N. Mhatre · F. Saiger · A. P. Pratap (✉)
Department of Oils, Oleochemicals and Surfactants Technology, Institute of Chemical Technology, Nathalal Parekh Marg, Matunga (East), Mumbai, India
e-mail: ap.pratap@ictmumbai.edu.in
URL: https://www.ictmumbai.edu.in

M. Abdul Rub et al. (eds.), *Industrial Applications of Individual and Mixed Surfactant Systems—Volume 2*, Sustainable Materials and Technology,
https://doi.org/10.1007/978-981-95-9195-4_5

chemistry · Bioprocess engineering · Sophorolipids · Rhamnolipids · Surface-active agents · Eco-friendly surfactants

1 Introduction

Our primary energy source is fossil fuels, such as coal, oil, and natural gas. There are worries that the world's fossil fuel supplies may soon run out as a result of rising energy consumption brought on by the global population growth. Researchers are working to develop sustainable and cutting-edge biocatalytic techniques to transform biodegradable resources, such as trash and biomass, into chemicals, fuels, and energy to meet the expanding needs of contemporary society. The utilization of these renewable resources to switch from petrochemical plants to biorefineries is currently encouraged by government legislation and policies, which benefits communities and the environment. Bio-based chemicals and materials made with microbial catalysts (like yeast, bacteria, and fungi) are environmentally benign, biodegradable, and biocompatible, in contrast to those derived from fossil fuels. Their production costs are still greater than those of conventional petrochemical alternatives, nonetheless, because of continuous research and technological developments. The biotechnological synthesis of biosurfactants is one prominent example. The potential of mostly fat-rich and hydrophobic organic wastes as raw materials for the fermentation and enzymatic processes that provide these highly prized compounds has been demonstrated by numerous research. Their fatty acid-based triglycerides can be used as precursors by a range of microorganisms or as initial reactants in lipase-catalysed processes to create biosurfactants. Carbohydrate-based biosurfactants, which are made from renewable resources like sugars and starches, have drawn a lot of interest because of their potential to solve environmental issues while still fulfilling the demands of different applications. This study aims to investigate the use and prospects of biosurfactants. The investigation of bio-based substitutes has become crucial in a number of industries as they strive for environmentally conscious practices and sustainable development. Surface-active compounds called biosurfactants are produced by microorganisms and provide a sustainable substitute for traditional surfactants. They are a more environmentally friendly alternative to petroleum-based surfactant products because of their well-known biodegradability, low toxicity, and renewable sources. Growing worries about the environmental effects of conventional surfactants highlight the need for sustainable surfactants. Conventional surfactants, which are mostly obtained from non-renewable petrochemical sources, pollute the environment, degrade soil, and contaminate water. For companies that depend on surfactant technologies, biosurfactants present a feasible substitute that is more environmentally conscious and sustainable. Understanding the environmental effects of biosurfactants becomes not only a scientific endeavour but also an essential part of making responsible industrial decisions as we traverse an era characterized by increased environmental consciousness and a global need for sustainable practices. The future of biosurfactants and their environmentally responsible and sustainable development in surfactant technology are the main topics of this chapter.

2 Surfactant

Surfactants are molecules that can gather at the interfaces between liquids of different polarity, such as oil–water or air–water, since they have both hydrophobic (which repels water) and hydrophilic (which attracts water) regions. They interact with surfaces and interfaces due to their dual nature, which effectively reduces the surface tension of liquids. The hydrophilic head group might be cationic, anionic, non-ionic, or amphoteric, whereas the hydrophobic portion is usually a hydrocarbon chain. The cost and complexity of creating or joining hydrophilic groups, as well as the availability, variety, and affordability of hydrophobic components, all have a significant impact on the dynamics of the surfactant sector. The ability of a surfactant to lower surface tension—the energy required to increase a liquid's surface area—is a measure of its efficacy. Surfactants reduce tension at the oil–water and air–water interfaces by making hydrophobic materials more soluble in water. High-performance surfactants may reduce the interfacial tension of water and n-hexadecane from 40 mN/m to barely 1 mN/m, as well as the surface tension of water from 72 mN/m to 35 mN/m. Surface tension falls with increasing surfactant concentration in water, leading to the formation of micelles, which are spherical aggregates with hydrophilic shells and hydrophobic cores. The smallest quantity of surfactant required for maximal tension reduction is known as the critical micelle concentration (CMC), which indicates the point at which surface tension stabilizes at its lowest level. Micelle formation becomes commonplace beyond the CMC, as shown in Fig. 1. Since they are typically less hazardous and more biodegradable than cationic or amphoteric versions, the majority of commercial surfactants are either anionic (negatively charged) or non-ionic (uncharged). Depending on the source, there are two types of surfactants that are produced: petrochemical and natural or renewable [9].

2.1 Effect of Synthetic Surfactant on the Environment

2.1.1 Release into the Aquatic Environment

Since many consumer products that include surfactants are made to be used just once, a sizable amount of these substances ends up in the environment. According to research, about 60% (by weight) of all manufactured surfactants eventually make their way into aquatic environments; this amounts to an estimated 7 million tons annually worldwide. Both direct routes—such as spills, industrial discharges, and

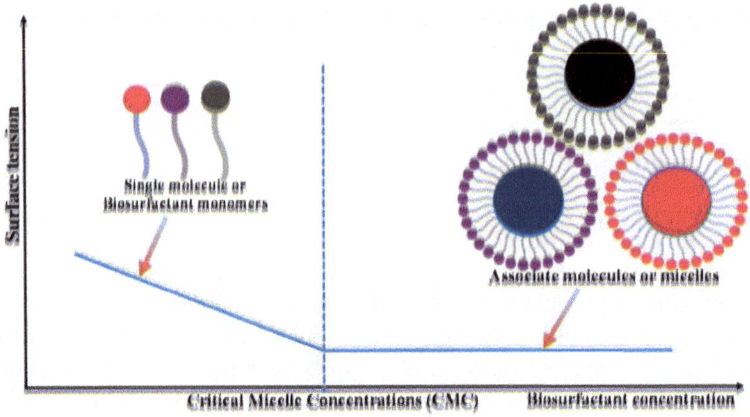

Fig. 1 Critical micelle concentration (CMC) and micelle formation of biosurfactant monomers [15]

landfill leaks—and indirect ones—such as treated wastewater discharged from treatment facilities—allow these substances to enter the environment. For treatment facilities, the presence of surfactants in wastewater presents a significant difficulty. They are difficult to totally eradicate due to their distinct molecular makeup, leaving behind remnants even after processing. Other very water-soluble substances have comparable removal challenges; therefore, this problem is not just with surfactants. Furthermore, by preventing microbiological activity, surfactants might interfere with wastewater treatment. Their antimicrobial qualities exacerbate environmental contamination by hindering the good bacteria that break down toxins and sewage sludge. Synthetic surfactants can interfere with vital biological processes and have hazardous effects on living things in addition to making wastewater treatment more difficult. Additionally, they worsen pollution by making other pollutants, like heavy metals, more mobile, endangering ecosystems.

2.1.2 Effect on Micro-Organisms

The potential for certain synthetic surfactants to be hazardous to wildlife and disrupt normal biological processes is one of the main environmental concerns. The impact of detergents on heterotrophic bacterial communities in an estuary was noted by Effendi et al. [7], as the concentration of surfactant increased, the number of water bacteria dramatically declined. The chemical characteristics of surfactants determine their antibacterial action. Anionic surfactants interfere with DNA and enzymes, impairing cellular functioning, whereas cationic surfactants mainly damage bacterial cell membranes [7]. Alkylphenol ethoxylates (APEOs) and other non-ionic surfactants work against bacteria by denaturing vital proteins, which eventually causes bacterial cell damage. The investigation's main focus was on APEOs, which make up around 7% of the world's surfactant usage [16]. Due to their sluggish rate of degradation, these compounds are persistent organic contaminants in the environment. Aquatic ecosystems deteriorate and the water cycle may be disrupted when surfactants are introduced into the environment because they create foam in rivers, which reduces the water's oxygen content. Similar to the phenomena seen with antibiotics, prolonged exposure to sublethal surfactant doses in the environment may cause bacterial resistance. Over time, this adaptive reaction may progressively reduce the antibacterial effectiveness of products based on surfactants.

2.1.3 Effect on Fauna

In contaminated water systems, aquatic organisms are extremely vulnerable to synthetic surfactants. These substances weaken fish's defence mucus layers, making them more vulnerable to infections and seriously harming their gills. At detergent concentrations of about 15 ppm, experimental tests show > 50% fish mortality; even lower concentrations ($\leq$ 5 ppm) are fatal to fish eggs. According to Ostroumov's research on oysters and clams, bivalves that are exposed to surfactants have a reduced ability to filter-feed. Standardized toxicity testing shows concentration-dependent mortality across species, while zebrafish studies show developmental defects and disturbed metamorphosis. Fish embryos, algae, and Daphnia are among the aquatic biota that are acutely harmed by firefighting foams that contain surfactants. Beyond aquatic environments, cationic surfactants such as methyl pyridinium bromide have been shown to be cytotoxic in mammalian models, harming rabbit corneal cells and human lymphocytes. Additionally, agricultural surfactants have been shown to negatively impact pollinators, especially bees. R. M. Goodwin showed that three common surfactants were oral noxious to bees, as evidenced by their detrimental impacts on survival. They also mention additional research that revealed negative impacts on bees' ability to learn by smell. Food shortages may result from falling bee populations, which is a major issue because bees are essential for pollination plants. As a result of chemical exposures, some nations have now designated bees as endangered, demonstrating the growing governmental acknowledgment of surfactant consequences. Given the considerable toxicity and irritancy potential of some surfactant classes, the threats to human health are as urgent. Cleaning products without surfactants, especially those with alcohol for sensitive skin, are among the market's reactions. Studies on toxicity show a distinct order.

1. Cationic surfactants: Most hazardous, causing dermal/ocular burns and respiratory irritation
2. Anionic surfactants: Intermediate toxicity
3. Non-ionic varieties: Least toxic

2.1.4 Effect on Flora

Surfactants have detrimental effects on plant systems in addition to animals. According to research by Masakorala et al. these substances reduce the efficiency of photochemical energy conversion, which is essential for plant survival. Greywater reuse in agriculture has both financial and environmental advantages, but research shows that surfactants can cause acute toxicity in lettuce (Lactuca sativa). Many surfactants remain in greywater, creating persistent phytotoxic hazards that require thorough toxicity assessment, even if some can be broken down by bioremediation. Moreover, stable surface foams with significant liquid fractions are produced by aquatic surfactant pollution. Because of their restricted photic zone, these optically dense structures greatly reduce light penetration, which may limit photosynthetic activity in benthic and submerged aquatic plants. depth [20].

2.1.5 Persistence, Accumulation, and Secondary Effects

Non-biodegradable components are frequently found in the molecular architecture of synthetic surfactants, which results in prolonged environmental retention. In low-energy depositional environments, such as sedimentary matrices, lentic systems (lakes), and riverine zones with lower flow velocity, Ferrer's research shows that these compounds have a considerable capacity for accumulation. Because of structural characteristics that prevent microbial breakdown, its persistence leads to long-term environmental sequestration as opposed to degradation. Ferrer et al. [10] documented seasonal variations in linear alkylbenzene sulphate (LAS) dispersion, noting rapid dilution during high-flow winter conditions versus prolonged persistence in summer low-flow periods. Although sub-toxic surfactant concentrations are present in industrial and consumer formulations (usually 10–15% in detergents), environmental accumulation presents serious risks due to: (A) the potential for bioaccumulation, where concentrations gradually build up in sediment beds and low-energy aquatic systems; (B) the effects of chronic exposure, such as bacterial resistance and endocrine disruption from degradation products (especially alkylphenol polyethoxylates/APEOs that exhibit estrogenic activity linked to reproductive disorders); and (C) secondary contamination due to increased mobilization of co-pollutants such as heavy metals and hydrophobic organics, as well as increased bioavailability of soil-bound toxins. Current risk assessment paradigms frequently neglect critical considerations including (1) long-term consequences of low-dose exposure, (2) the formation of toxic intermediates through partial biodegradation, and (3) the gradual development of antimicrobial resistance—knowledge gaps that may lead to underestimation of ecological threats, particularly in vulnerable ecosystems where these compounds persist and interact with other environmental contaminants. Heavy metal contamination and surfactant pollution coexist in emerging nations, raising concerns. These discoveries have encouraged interest in biosurfactants as sustainable alternatives, exhibiting greater biodegradability and reduced ecotoxicity while keeping surface-active qualities [20].

3 Biosurfactants

Biosurfactants are surface-active molecules synthesized by microorganisms and plants [17]. Biosurfactants can also be defined as surface-active biomolecules produced by microorganisms with a wide range of applications. Vijayakumar and Saravanan [32]. Because of its many benefits, including low toxicity, excellent selectivity, and easy production methods, biosurfactants have become increasingly popular in industry. Across a variety of industries, these biomolecules exhibit extraordinary functional diversity by acting as food additives, cleaning agents, emulsifiers, demulsifiers, wetting agents, foaming agents, and dispersants. Petroleum processing, petrochemicals, mining (especially bioleaching), metallurgy, agricultural chemicals, food production, cosmetics, and pharmaceuticals are just a few of the many sectors that use them. Their capacity to significantly reduce interfacial tension is one of their most advantageous qualities, which makes them highly beneficial for increased oil recovery processes and the bioremediation of heavy crude oil contamination [32]. Using a variety of substrates, including waste materials, a wide range of microorganisms, including bacteria, archaea, algae, and fungi, create biosurfactants, which are amphiphilic microbial metabolites, extracellularly. These substances can be classified as glycolipids, lipopeptides, lipopolysaccharides, glycoproteins, and polysaccharides because they contain hydrophilic components (such amino acids, peptides, carbohydrates, or phosphates) and hydrophobic moieties (usually long-chain fatty acids). They are divided into two groups according to their molecular weight: low molecular weight types (like surfactin and rhamnolipids) that efficiently lower surface tension and high molecular weight types (like polysaccharides and glycoproteins) that are excellent bio emulsifiers. High MW biosurfactants mainly stabilize emulsions, even though they are less effective in reducing interfacial tension. Low MW biosurfactants, such as glycolipids and lipopeptides (e.g., trehalose lipids, lichenysin), frequently exhibit dual functionality in both tension reduction and emulsion stabilization. Their diverse spectrum of biological activities and industrial applications are supported by this structural diversity.

4 The Physiochemical Properties of Biosurfactants

Biosurfactants are very important for industrial application because of their unique properties that set them apart from synthetic surfactants. These microbially derived surfactants exhibit exceptional environmental compatibility and functional adaptability in contrast to chemically manufactured alternatives. This includes extraordinary stability under high temperatures, variable pH conditions, and fluctuating salt concentrations. Their exceptional surface-active qualities, biodegradability, low toxicity, and dual use as emulsifiers and demulsifiers are some of the main factors that contribute to their economic success. Furthermore, a lot of biosurfactants have antibacterial qualities by nature. For large-scale applications in a variety of industries, these combined qualities, together with their capacity to be manufactured from a variety of frequently affordable substrates, provide substantial benefits. The major distinctive features of each property of biosurfactant are discussed below:

(1) Surface and interface activity—Biosurfactants have exceptional surface-active qualities that greatly lower interfacial and surface tensions. Notably, Bacillus subtilis surfactin reduces the interfacial tension of water and hexadecane to less than 1 mN/m and the surface tension of water to 25 mN/m. Likewise, rhamnolipids generated from Pseudomonas aeruginosa reduce the surface tension of water to 26 mN/m and attain a similar reduction in interfacial tension (< 1 mN/m). Because of their significantly lower Critical Micelle Concentrations (CMC), frequently by an order of magnitude, these microbial surfactants work more efficiently than their synthetic equivalents. Because of their improved efficacy, biosurfactants exhibit superior surface activity and can reduce tension to their maximum at far lower concentrations than chemical surfactants [33].
(2) Temperature, pH tolerance, and ionic strength tolerance—Biosurfactants exhibit remarkable environmental stability, continuing to function in harsh environments that would render their synthetic counterparts inactive. Many biosurfactants are active over a wide range of operating conditions, such as temperatures up to 120 °C, pH extremes (3–12), and saline concentrations up to 10%, in contrast to chemical surfactants that lose their efficacy at salt concentrations exceeding 2% NaCl [24]. Significant commercial interest has recently been sparked by this exceptional stability, especially in biosurfactants generated from extremophiles. These biological surfactants are better than conventional surfactants for applications needing performance in demanding environments because of their structural robustness, which allows consistent surface activity despite variations in environmental circumstances [32].
(3) Specificity:—The structural diversity of biosurfactants is astounding, and the distinct composition and functional groups of each molecule dictate its particular bioactive characteristics. Similar to synthetic surfactants, these biological substances exhibit the ability to self-assemble, generating micellar structures on their own that improve their functional specificity and allow for a variety of morphological configurations. Their ability to produce a variety of nanostructures, such as vesicular formations, rod-shaped aggregates, and spherical micelles, has attracted a lot of attention from a variety of industrial sectors, especially in the fields of pharmaceutical research, food processing, and cosmetic formulation. Beyond their structural adaptability, biosurfactants exhibit encouraging uses as efficient emulsion breakers in industrial processes and in environmental remediation through pollutant detoxification [24].
(4) Biodegradability—Microbial biosurfactants are especially useful for environmentally friendly applications like bioremediation because of their higher biodegradability as compared to synthetic surfactants. The hunt for sustainable substitutes has accelerated due to growing environmental consciousness, and biosurfactants have emerged as viable options. Marine-derived biosurfactants

provide a green alternative to persistent synthetic surfactants that are harmful to the environment, particularly when it comes to treating hydrophobic pollutants like polycyclic aromatic hydrocarbons (PAHs). In particular, recent studies have shown how well they work to improve the biosorption of phenanthrene, a poorly soluble PAH that commonly contaminates aquatic environments. Vijayakumar and Saravanan [32]. Because of their makeup, biosurfactants are more biodegradable and biocompatible than their chemical counterparts when temperature, pH, and breakdown time vary [25].

(5) Low toxicity—There is limited research on the toxicity of biosurfactants, although they are generally deemed low or non-toxic and suitable for medicinal, cosmetic, and food applications [32]
(6) Emulsion forming and emulsion breaking—By coating oil droplets, Lipson, a water-soluble emulsifier produced by Candida lipolytic, has been used to emulsify edible oils and create stable emulsions. These liposomes were frequently utilized in the food and cosmetics sectors to create stable oil/water emulsions [32].
(7) Antiadhesive agents—Because they can change a material's hydrophobicity, biosurfactants show great promise for surface modification applications by affecting the dynamics of microbial adherence. Certain biosurfactants have been found to be useful in preventing bacterial colonization on industrial surfaces. For example, a surfactant generated from Streptococcus thermophilus significantly slows down the surface fouling that other thermophilic Streptococcus strains cause on steel substrates. Similarly, it has been demonstrated that Pseudomonas fluorescens produces biosurfactants that successfully stop harmful Listeria monocytogenes from adhering to steel surfaces. These results demonstrate how biosurfactants may be used in industrial and food processing environments as environmentally friendly substitutes for traditional antimicrobial surface treatments [22].

5 Classification of Biosurfactants

Biosurfactants are categorized based on their molecular weight, chemical composition, and microbiological origin, whereas chemically produced surfactants are categorized based on their polarity. Since biosurfactants typically have molecular masses between 500 and 1500 Da, they can be generically classified as either low molecular weight or high molecular weight. Further, the biosurfactants can classified based on molecular weight:

(1) **Low-molecular-weight biosurfactants**—Low molecular weight biosurfactants are among them. While larger molecular weight biosurfactants are best suited for stabilizing oil-in-water emulsions, low molecular weight biosurfactants are better at lowering the interfacial tension at the oil–water interface and the surface tension at the air–water interface. Fatty acids, phospholipids, lipopeptides, and glycolipids are a few examples [22].

(2) **High-molecular-weight biosurfactants**- Among them are high molecular weight biosurfactants. Examples of high molecular weight biosurfactants include lipoproteins, lipopolysaccharides, polymer biosurfactants, and particle biosurfactants [22].

The classification of Biosurfactants is as follows:

(1) **Glycolipid**: An ester group binds these carbohydrates to long-chain aliphatic or hydroxyaliphatic acids. Glycolipids make up the majority of biosurfactants. Rhamnolipids, trehalolipids, and sophorolipids are the three most well-known types of glycolipids. The sources and properties of several glycolipids are explored below.

(a) **Rhamnolipids** -Glycolipids made of rhamnose and hydroxydecanoic acid are known as rhamnolipids. The biosurfactants that have been studied in great detail are the main glycolipids produced by *P. aeruginosa.*

(b) **Trehalolipids**—These are typical of species of *Corynebacterium, Nocardia, and Mycobacterium.* Trehalose lipids from *Arthrobacter* species and *Rhodococcus erythropolis.* decreased the culture broth's surface and interfacial tension by 25–40 and 1–5 mNm, respectively.

(c) **Sophorolipids**—Glycolipids are produced by yeasts and consist of a long-chain hydroxyl fatty acid joined by a glycosidic bond with the dimeric carbohydrate sophorose. Six to nine hydrophobic sophorolipids are frequently combined to form sophorolipids. For numerous applications, the lactone form of sophorolipids is recommended [32].

(d) **Mannosylerythritol lipids (MEL)**—High amounts of glycolipids known as mannosylerythritol lipids (MEL) are produced by the non-pathogenic yeast *Pseudozyma antarctica.* The structural components of MEL are a hydrophilic 4-O-b-D-mannopyranosyl D-erythritol moiety, and a hydrophobic group made up of a combination of C6–C18 fatty acyl chains at the mannose moiety's C2' and C3' (Mawani et al. n.d.).

(2) **Lipopeptides and lipoproteins**:—These molecules consist of a polypeptide chain connected to a lipid [21]. Numerous biosurfactants have shown antibacterial efficacy against viruses, fungus, bacteria, and algae [5]. Asserted that Turin, a lipopeptide produced by *Bacillus subtilis*, has antifungal qualities, while Singh and Camerota [26] found that Turin also had antibacterial qualities. B. subtilis turin was found to be active at pH 5–11 even after autoclaving, and it had a 6-month shelf life at − 18 °C [32].

(a) **Surfactin:**—A powerful biosurfactant, surfactin is a cyclic lipopeptide with a seven-amino acid ring structure connected to a fatty acid chain by lactone bonding [2]. The physicochemical characteristics of *B. subtilis* surfactin were described in a prior study. It has been demonstrated that surfactin reduces the surface and interfacial tension of water. It was also discovered that surfactin inhibited the replication of retroviruses and herpes [32].

(b) **Lichenysin**:—Like surfactin, *Bacillus licheniformis* produces biosurfactants that are extremely stable under extremes of pH, temperature, and salt. *B. licheniformis* lichenysin has the potential to lower water's surface and interfacial tensions to 27 and 0.36 mN/m, respectively.

(3) **Fatty acids, phospholipids, and neutral lipids**:—As they feed on n-alkanes, bacteria and yeast generate large amounts of fatty acids and phospholipid surfactants. Phosphatidyl ethanolamine-rich vesicles, which are produced by Acinetobacter spp. 1-N, create optically transparent alkane micro-emulsions in water. These biosurfactants are essential in medicine.

(4) **Polymeric biosurfactants**:—The most researched polymeric biosurfactants include several polysaccharide-protein complexes, emulsan, liposan, alasan, and lipomanan. Even in water with hydrocarbon concentrations as low as 0.001–0.01 percent, emulsan works wonders as an emulsifying agent. Liposan, an extracellular water-soluble emulsifier composed of 17% protein and 83% carbohydrates, is produced by *Candida lipolytica*. The food and cosmetic industries' usage of liposan, a polymeric biosurfactant, as an emulsifier.

(5) **Particulate biosurfactants**:—The microemulsion produced by extracellular membrane vesicles aids in the absorption of alkanes by microbial cells. Protein, phospholipids, and lipopolysaccharides make up the 20–50 nm diameter, 1.158 cubic g cm buoyant density vesicles of *Acinetobacter spp.* strain HO1-N.

6 Advantages of Biosurfactants Over Synthetic Biosurfactants

Over the synthetic biosurfactants, the biosurfactant shows the following advantages

(1) Low toxicity
(2) Biodegradability
(3) Eco friendly
(4) Temperature tolerance
(5) pH tolerance
(6) Excellent selectivity
(7) Biocompatibility
(8) Bioavailability
(9) Ecological acceptability, and
(10) Enhanced efficiency salt concentrations

Sr. No.	Advantage of biosurfactants	Description
1	Biodegradability	Biosurfactants degrade naturally, reducing environmental pollution compared to synthetic surfactants
2	Eco-friendliness	Produced from renewable and sustainable sources, reducing reliance on petrochemicals

(continued)

(continued)

Sr. No.	Advantage of biosurfactants	Description
3	Low toxicity	They are less harmful to humans and ecosystems, making them safer for widespread use
4	Antimicrobial property	Some biosurfactants show antibacterial, antifungal, and antiviral properties, enhancing their utility
5	Stability under extreme conditions	Remain functional in extreme pH, salinity, and temperature, unlike many synthetic alternatives
6	Carbon footprint reduction	Production from renewable resources and lower environmental impacts reduce their carbon footprint
7	Biocompatibility	Non-irritating and compatible with biological systems, making them suitable for pharmaceutical and personal care applications

7 Disadvantages of Biosurfactants

Biosurfactants have many advantages there are some disadvantages are there. Some of the disadvantages are as follows.

(1) High production cost—Because biosurfactant manufacture is so expensive, the products' prices are unaffordable. 90% pure rhamnolipids cost about $1250 per kilogram, while high-purity synthetic surfactants, like 99% (SDS), cost $10 to $20 per kilogram. About 60% of the cost of producing biosurfactants is incurred during downstream processing, whereas up to 30% of the total cost is incurred during raw material procurement (Mawani et al. n.d.). The production of biosurfactants requires suitable raw materials with high carbohydrate and lipid content, which makes them commercially unviable.

(2) Low yield and productivity of biosurfactants—The fermentation substrate is a crucial determinant of inadequate production yields and process efficiency, which are a major constraint in the synthesis of biosurfactants. The output concentration and structural properties of biosurfactant molecules are directly influenced by the choice and metabolic conversion efficiency of carbon sources. Reduced product titres and impaired molecular functioning are two common consequences of improper substrate selection or poor usage. Therefore, a key prerequisite for process intensification in the synthesis of biosurfactants is strategic substrate optimization. Two significant financial limitations in commercial-scale biosurfactant manufacturing can be addressed by production systems that simultaneously improve product quality and volumetric productivity through logical substrate engineering that takes into account both biochemical suitability and cost-effectiveness [29].

(3) Difficulty in the downstream process—Biosurfactant purification is complicated and necessitates extra procedures, which raises production costs and duration.

(4) Substrate dependents—The quality and cost of substrates for microbial growth significantly affect biosurfactant production [4].

8 Comparison Between Synthetic and Biosurfactants

Property	Synthetic surfactants	Biosurfactants
Source	Chemically synthesized, often from petrochemical derivatives	Naturally produced by microorganisms (bacteria, fungi)
Cost	Generally lower due to large-scale industrial production	Typically, higher due to more complex production processes
Environmental impact	Can be toxic, non-biodegradable, and persistent in the environment	Biodegradable, less toxic, and environmentally friendly
Biodegradability	Often non-biodegradable or slow to degrade	Highly biodegradable and environmentally safe
Applications	Cleaning products, detergents, cosmetics, oil recovery, etc	Bioremediation, food, pharmaceuticals, cosmetics, agriculture
Efficiency	High performance, especially in harsh conditions	Generally effective but may not perform as well in extreme conditions
Surface activity	High surface tension reduction; effective at lower concentrations	Lower surface tension and good performance at low concentrations
Toxicity	Can be toxic to aquatic life and humans	Non-toxic, often regarded as safe for humans and animals
Chemical structure	Typically consists of a hydrophilic head and a hydrophobic tail	Can be glycolipids, lipopeptides, phospholipids, etc
Stability	Stable under harsh chemical and physical conditions	May have lower stability under extreme conditions but can be engineered for better stability
Renewability	Non-renewable as derived from petroleum	Renewable as derived from renewable resources (microorganisms)

8.1 Method of Production of Biosurfactants

The molecular structure of biosurfactants is amphipathic, containing both hydrophilic and hydrophobic domains that need different biosynthesis processes. The hydrophobic components usually come from relatively simple synthetic processes that represent basic lipid metabolic pathways. Hydrophilic moieties, on the other hand, exhibit far more structural variation, requiring more intricate and diverse biosynthetic processes. According to this structural dichotomy, the synthesis of biosurfactants entails at least two distinct metabolic pathways working together: one that is responsible for the synthesis of hydrophobic tails and another that regulates the development of various hydrophilic head groups. The vast variety of potential polar functional groups (such as phosphate esters, peptides, or polysaccharides) that can make up the hydrophilic domain of the latter contributes to its synthetic complexity. According to Syldatk and Wagner, four principal possibilities of biosynthesis of such amphiphilic molecules are

(a) Both the hydrophilic and hydrophobic moieties are synthesized de novo by two separate pathways.
(b) The hydrophilic moiety is synthesized de novo whereas the synthesis of the hydrophobic moiety is induced by substrate.
(c) The hydrophobic moiety is synthesized de novo whereas the synthesis of the hydrophilic moiety is substrate-dependent.
(d) The synthesis of both the hydrophobic and hydrophilic moieties depends on the substrate used.

Microbial Biosurfactants Producers:

Biosurfactants are made by microbes using strains of bacteria, fungi, and yeast. Bacteria are currently the most studied source of biosurfactant. The most common genera that create biosurfactants are Rhodococcus, Bacillus, Candida, and Pseudomonas. The ability to produce biosurfactants is well known for bacterial species as *Pseudomonas, Actinobacteria, Sphingomonas, Halomonas, Bacillus, Mycobacterium, Alcanivorax, Rhodococcus, Pseudoalteromonas, and Arthrobacter. The genera Aspergillusustus, Trichosporonashii, Candida lipolytica, Aspergillusustus, Candida batistae, Candida ishiwadae*, and *Candida bombicola* are among the fungus that produce biosurfactants. Formerly known as *Candida antarctica*, Moesziomyces antarcticus is a highly prized strain of the Antarctic bacterium that secretes mannosylerythritol lipids, a biosurfactant that is extensively utilized in the cosmetics industry. This strain can use vegetable oil and intermittent bioreactor feeding to produce up to 100 g/L of biosurfactant [30].

Fermentation

The process of fermentation for different biosurfactants can be produced by a variety of microorganisms, such as yeasts, fungus, and bacteria. The bacteria that produce them determine their nature, chemical makeup, and amount. These organisms are able to exploit potentially harmful substrates because they are not exposed to contaminated soil, sewage, or sewage sources. Because of this, the possible substrates are varied and can be broadly categorized as industrial wastes from agriculture, plants and animals, and other sources. Wastewater from oil mills, vegetable oils and oily wastes, and wastes containing starch, such as wastewater from potato processing, are examples of agricultural industrial wastes. Whey, animal fat, molasses, and soy molasses are examples of dairy industry waste that is classified as industrial waste of vegetable and animal origin. Soap scum is associated with other industrial waste. Biosurfactants are derived from a variety of biological substrates. Up to 60% of the overall production cost is incurred as a result of this additional processing. Because of this, producing biosurfactants commercially is highly costly. Therefore, in order to overcome high manufacturing costs, the focus nowadays is on manufacture from inexpensive and wasteful substrates. The world produces a lot of regular and hazardous waste each year. We desperately need to control and use them properly. The generation of biosurfactants has thus been claimed to be facilitated by the use of wastes from tropical crops, fruit processing industries, petroleum refineries, coffee industries, etc.

Some fermentation strategies used are

(1) Submerged fermentation: A common bioprocessing method for producing microbial metabolites is submerged fermentation (SmF), in which microorganisms are grown in an aqueous suspension solution. This method involves the direct secretion of target molecules into the nutrient-rich liquid medium. SmF offers unique technological benefits, especially its ability to precisely control process parameters (such as pH, aeration, and nutrient feeding) and its proven effectiveness in turning liquid waste streams into fermentation substrates.
(2) Solid-state fermentation: A new bioprocessing method for the synthesis of biosurfactants is solid-state fermentation (SSF), which is defined by the culture of microorganisms on moist solid substrates without the presence of a free-flowing liquid. This method successfully avoids the troublesome foam generation connected to liquid-phase systems and provides a competitive substitute for submerged fermentation (SmF). Microorganisms thrive in SSF systems on hydrated solid matrices, which can be either naturally water-retentive or nutrient-impregnated materials. This results in a triphasic environment that consists of solid (substrate), liquid (bound water), and gaseous (interstitial air) phases. Microbial development at the solid–liquid interfaces of these wet particulate systems is what gives SSF its heterogeneous character.
(3) Co-culture fermentation: Co-culture fermentation has revolutionary possibilities for environmental cleaning applications and signifies a paradigm leap in biosurfactant biotechnology. In comparison to traditional single-strain fermentation systems, this novel method, which uses mixed microbial consortia for biosurfactant synthesis, exhibits superior environmental remediation performance through three main advantages: (1) increased contaminant degradation efficiency, (2) microbial synergistic interactions that enhance functionality, and (3) improved environmental adaptability. In these co-culture systems, interspecies collaboration frequently results in metabolic complementarity that greatly increases the capacity to break down contaminants and produce biosurfactants [29].

Strategies for Low-Cost Production

Different Types of Substrates Are Used to Reduce Carbon Cost in Fermentation:

(1) Oil wastes: Industrial waste products from the production of food and vegetable oils are becoming more and more affordable substrates for the biosynthesis

of biosurfactants. Studies show that oleic acid and sunflower seed oil can be efficiently converted into rhamnolipid biosurfactants by the thermophilic bacterium *Thermus thermophilus* HB8. One of the main financial obstacles in the commercial-scale manufacture of biosurfactants may be overcome by the clever use of such low-value oil wastes in production media, which offers a promising way to drastically cut manufacturing costs. In addition to reducing the cost of raw materials, this strategy helps the food processing industry achieve sustainable waste valorization.

(2) Dairy whey: About 6 L of whey are produced for every kilogram of paneer (cheese) produced by the dairy sector, providing a stable and affordable substrate for the fermentation of biosurfactants. Through an improved two-stage cultivation technique, *Cryptococcus curvatus* ATCC 20,509 produces large quantities of sophorolipids, demonstrating the effectiveness of using dairy wastes in microbial biosynthesis. By turning dairy effluents into valuable biosurfactants, this method not only improves process sustainability but also lowers production costs by reusing waste.

(3) Starchy waste: For the synthesis of biosurfactants, starchy waste products from agro-industrial processes—in particular, potato-processing by-products—represent inexpensive and nutritionally ideal substrates. These carbohydrate-rich residues are rich in carbon sources, such as fermentable sugars and starch polymers, as well as nitrogen/sulphur compounds from leftover proteins and vital micronutrients, including vitamins that support microbial metabolism, trace elements (iron, zinc), and inorganic minerals (phosphorus, potassium). Because of their full nutritional profile, these starchy wastes are especially well-suited for growing microorganisms that produce biosurfactants. They also offer a cost-effective and environmentally friendly substitute for synthetic culture media components in industrial fermentation processes.

(4) Molasses: Because of its high fermentable sugar content, molasses, a nutrient-rich byproduct of sugar production from sugarcane or sugar beet, is a great carbon source for microbial fermentation. Patel and Desai showed how *Pseudomonas aeruginosa* GS3 may efficiently use molasses and corn-steep liquor as dual carbon and nitrogen sources for the synthesis of rhamnolipid biosurfactants. By using industrial leftovers to make a cost-effective culture medium, this method addresses waste valorization and lowers the cost of producing biosurfactants while preserving high product yields [19].

(5) Animal fat: Animal fat and tallow, which are produced in large quantities by the meat processing industry, are suitable substrates for the synthesis of biosurfactants. Deshpande and Daniels' research showed that the yeast *Starmerella bombicola* (previously *Candida bombicola*) could successfully convert animal fat into sophorolipids. In addition to making use of inexpensive byproducts, this strategy supports sustainable waste management techniques in the meat industry and increases the economic viability of biosurfactant synthesis.

(6) Soap stock: A lipid-rich residue of refining oilseeds, soap stock serves as a sustainable substrate for the synthesis of biosurfactants. Studies show that *Pseudomonas aeruginosa* LB1, a strain obtained from soil contaminated by

petroleum, may effectively produce surface-active rhamnolipids (referred to as RLLBI) by batch fermentation. This procedure highlights the strain's metabolic adaptability in turning industrial wastes into value-added biosurfactants while promoting circular economy concepts in the oil processing industry. It uses a minimum mineral salts medium with soap stock as the only carbon source [6].

Approaches to Maximize Biosurfactant Yield:

(1) Genetic engineering

Recombinant DNA (rDNA) technology is one possible strategy for boosting microbial strain improvement in biosurfactant production. One advantage is that it makes it possible to tailor biosurfactants to specific industrial and environmental needs. Recent research has shown that scientists can optimize biosurfactant outputs, reduce manufacturing costs, and enhance stability at high temperature, salt concentration, and pH levels by altering microbial genetics. Effective optimization of cloned microbial strains is necessary to scale up the production of biosurfactants while accounting for various environmental conditions. To get beyond the production limitations, new biosurfactants with improved properties can be made using genetic engineering. This entails using state-of-the-art genetic engineering techniques to enhance the recovery and fermentation procedures. Genetic engineering and omics methods including transcriptomics, proteomics, metabolomics, and genomics can be combined to better tailor biosurfactants for specific applications, revolutionizing the environmental and biotechnological sectors. These advancements demonstrate how genetic engineering has the potential to transform the market and advance environmental initiatives. For example, Zhang et al. modified strain Aureobasidium V33 [35] produced 25.10 g/L of cell dry weight and 55.38 g/L of limonin, indicating significant advancements in biosurfactant production. The wild-type strain 9-1, on the other hand, generated 21.61 g/L of cell dry weight and 30.61 g/L of limousin when grown in the same circumstances. This striking contrast demonstrates the substantial improvement made possible by genetic engineering. Similarly increase to rhamnolipids production using a recombinant *Saccharomyces cerevisiae* strain, designated as Generally Recognized as Safe (GRAS), expressing enzymes sourced from *P. putida* [3]. These findings underscore the pivotal role of genetic and metabolic engineering tools in boosting yields [29].

(2) Process optimization:

Research into using inexpensive industrial waste streams as fermentation substrates has been prompted by the increasing need for commercially viable biosurfactants. Advanced optimization techniques, especially statistical experimental designs, are essential for improving yield and functionality to enhance cost efficiency. Among them, factorial designs and Response Surface Methodology (RSM) are frequently used; factorial designs assess multifactor linear interactions, while RSM combines statistical and mathematical modelling to create empirical connections. Although studies indicate that hybrid optimization strategies incorporating various designs may outperform single-method approaches due to inherent constraints in individual models, Central Composite Design (CCD) and Box-Behnken Design (BBD) are commonly utilized RSM methodologies. Membrane aeration methods, which minimize excessive foaming, a typical problem in biosurfactant fermentation, while preserving foam-adhered cells to increase yield, can further improve large-scale production. Developing competitive, commercially viable biosurfactant technologies requires combining these optimization and scale-up techniques [29]

(3) Downstream processing:

A crucial and expensive stage in the synthesis of biosurfactants is downstream processing, which entails a series of unit operations for product isolation and purification. In the first step of cell separation, microbial cells are separated from the fermentation broth via flocculation (chemically induced aggregation), filtration (membrane-based size exclusion), or centrifugation (high-speed rotating separation). While the final product recovery makes use of methods such as solvent extraction (for hydrophobic compounds), pH-dependent precipitation, or chromatographic techniques (e.g., ion-exchange, affinity) tailored to the biosurfactant's physicochemical properties, subsequent biomass removal uses secondary filtration or centrifugation to remove any remaining cellular debris. The properties of the target molecule essentially dictate the choice and improvement of these purification techniques, necessitating a careful balancing act between yield, purity, and economic feasibility for large-scale applications. Acid precipitation was used to partially purify the biosurfactants generated by *Bacillus sp.* HIP3. Using 3 M HCl to acidify the cell-free supernatant to pH 2.0 caused biosurfactant precipitation to occur overnight. Centrifugation was used to gather the resultant precipitate, and the pellet was then extracted using a solvent. Rotary evaporation was then used to concentrate the purified biosurfactant [13]. Advanced purification methods combined with synthetic biology tools (like CRISPR-Cas9 genome editing) and multi-omics analyses (like proteomics, metabolomics, and genomics) have the potential to revolutionize the production of biosurfactants by enabling precise molecular customization for specific industrial applications as well as a significant increase in yield. This multidisciplinary approach facilitates: (1) optimized metabolic pathways through targeted genetic modifications, (2) tailored surface-active properties via structure-function characterization, and (3) streamlined downstream processing through biomarker-assisted purification strategies—collectively representing a next-generation framework for biosurfactant development that combines precision engineering with scalable manufacturing [29].

9 Application of Biosurfactants

Biosurfactants have diverse applications in various fields such as bioremediation, petroleum, textiles, cosmetics, food, pharmaceuticals, biomedicine, and nanotechnology.

Industries in which biosurfactants are used are mentioned below and shown in Fig. 2.

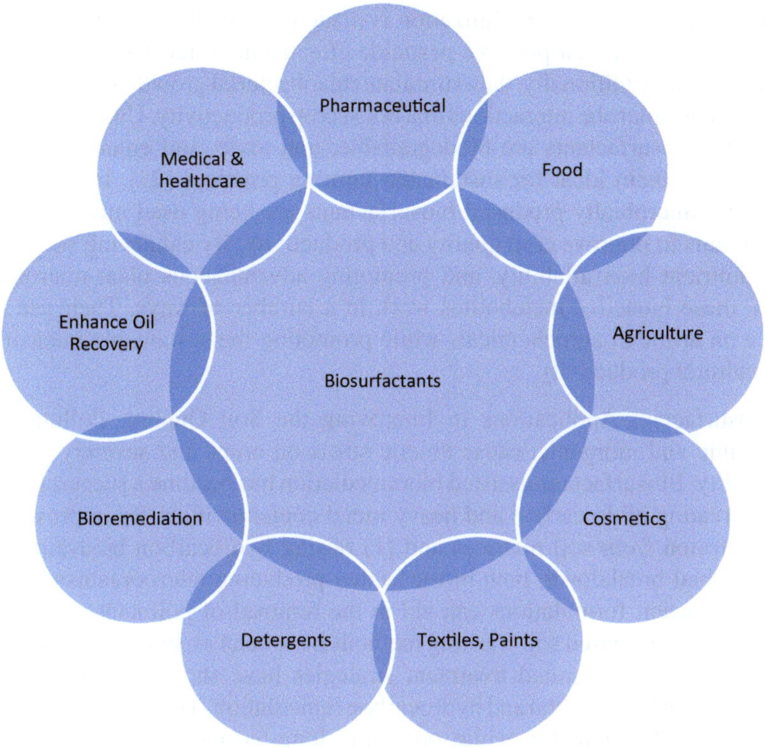

Fig. 2 Schematic presentation of various application fields of Biosurfactants

(a) Pharmaceutical Industry
(b) Food
(c) Agriculture
(d) Cosmetics
(e) Textile and paint, detergents
(f) Bioremediation
(g) Enhance oil recovery

9.1 Application in the Agriculture

Biosurfactants have become essential agricultural tools, providing sustainable substitutes for synthetic surfactants to satisfy the needs of an expanding population. These green surfactants, which are produced by bacteria, yeasts, and fungi, boost nutrient availability for helpful rhizobacteria, fight plant diseases, and promote soil health. Biosurfactants offer an economical and environmentally beneficial alternative to the 200,000 tons of surfactants used annually in insecticides and plant protection products. Superior soil hydrophilization is made possible by their special qualities, which guarantee the best possible pesticide effectiveness, fertilizer dispersion, and water retention. Additionally, they stimulate rhizobacterial growth in the rhizosphere, fostering plant-microbe interactions that boost crop productivity. Unlike conventional surfactants, biosurfactants are biodegradable, non-toxic, and enhance soil microbiota, making them ideal for sustainable farming practices [25]. In contemporary agriculture, microbially produced biosurfactants are being used more and more as biostimulants to improve crop quality and productivity. By enhancing soil structure, raising nutrient bioavailability, and promoting advantageous plant-microbe interactions, these bioactive metabolites work in a number of ways. Their use lessens reliance on artificial agrochemicals while promoting the sustainable intensification of agricultural productivity.

(a) Biosurfactant Applications in Improving the Soil Quality: Pollutants, both organic and inorganic, cause abiotic stress on crops and severely reduce soil fertility. Biosurfactant-assisted bioremediation has become a successful method for treating hydrocarbon and heavy metal contamination. By improving metal desorption from soil matrices and (1) raising hydrocarbon bioavailability for microbial breakdown, both biosurfactant-producing microorganisms and pure biosurfactant formulations can aid in the removal of pollutants. In particular, biosurfactant-based soil washing methods have been especially effective in field settings, and integrated treatment strategies have shown positive results for concurrent heavy metal and hydrocarbon remediation. These biological methods restore soil production while providing long-term substitutes for traditional chemical treatments.
(b) In seed germination and Plant growth: Biosurfactants have been demonstrated to efficiently encourage seed germination in agricultural procedures generally. While the majority of biosurfactants have been shown to promote plant growth, some research has also shown that they can also have inhibitory effects [25]
(c) Biosurfactant Application in Plant Disease Management: An alternative to conventional pesticides, biosurfactants can reduce the risk to both human and environmental health. Chemical pesticides pose a serious risk to both people and the environment.

9.2 *Application in the Food Industry*

In the food business, biosurfactants are multipurpose additives that play important secondary functions in food processing and quality improvement in addition to being efficient emulsifiers (such as glycerol esters, lecithin derivatives, and ethoxylated monoglycerides). From fat stabilization in edible oils and foam control in sugar production to solubility enhancement in instant foods and starch modification in dehydrated products, these microbially derived compounds enhance texture, flavor, and shelf-life while satisfying clean-label requirements. In addition to their main function of lowering interfacial tension, biosurfactants can help stabilize aerated systems in baked goods, regulate fat globule aggregation in dairy products, alter dough rheology for better baking results, and improve texture in fat-based spreads. Their versatility is further demonstrated by their capacity to create protective coatings on fresh produce, which makes biosurfactants essential for contemporary food manufacturing by concurrently meeting the needs of consumer safety, product consistency, and processing efficiency [14]. Biosurfactants have several uses in the manufacturing of ice cream and baked goods, including improving flavor oil dispersion, reducing starch retrogradation (stalling), and adjusting product uniformity. In particular, these biomolecules operate exceptionally well as fat emulsifiers when heated. Specific quality characteristics in baked goods, such as dough stability, loaf volume, crumb texture, and shelf-life extension through moisture retention, are improved by the use of rhamnolipid-based surfactants. When taken as a whole, these impacts preserve clean-label appeal while producing notable gains in ultimate product quality [31]. Additionally, the study recommended using rhamnolipids to enhance the qualities of frozen confections and buttercream. There is evidence for L-Rhamnose. In food processing, biofilms provide serious contamination issues, however biosurfactants provide a workaround by preventing microorganisms from sticking to surfaces. Rhamnolipids and surfactin dramatically lower bacterial attachment to polypropylene, reducing the formation of biofilms, according to studies by Huang et al. Because of these characteristics, biosurfactants are useful as anti-fouling coatings for surfaces that come into touch with food and processing machinery, especially in dairy applications. Their probiotic potential, including with Lactobacillus acidophilus, was further highlighted by Rodrigues et al. In addition to controlling biofilms, biosurfactants improve the stability, texture, and consistency of food emulsions, which are mixes of immiscible liquids. In goods like sauces, dressings, and drinks, they act as emulsifiers to reduce interfacial tension and stop droplets from aggregating. Biosurfactants perform important tasks including thickening, gelling, and stabilizing while offering clean-label alternatives to synthetic additives (such glyceryl monostearate). They have recently been used in cookie formulations, as demonstrated by Ribeiro et al. who experimented with various dough preparations. With growing interest in sustainable ingredients, research is shifting toward microbial-derived biosurfactants over genetically modified plant sources, reinforcing their role in safer food production and preservation [1].

9.3 Application in the Pharmaceutical Industry

(a) As antioxidants: Because of their established antioxidant properties, certain biosurfactants have been incorporated into the pharmaceutical formulation process. Mannosyl erythritol lipids, or MELs, are multipurpose biosurfactants with interfacial and biochemical properties. MEL-C may scavenge free radical ions and has the highest antioxidant activity. It is therefore commonly used in the manufacturing of skin care products and provides the finest protection.

Similar activity has been shown by RW1, a biosurfactant derived from Bacillus subtilis, which shows a natural alternative to antioxidants [1].

(b) As a drug delivery agent: Biosurfactants are widely used as therapeutic agents and drug delivery systems in the pharmaceutical sector. The study by Baferani et al. claims that biosurfactants exhibit emulsification, foaming, detergency, and scattering. Sophorolipids and rhamnolipids can be combined with lecithin to create biocompatible microemulsions. Because of their robust phase behaviour to temperature and salt changes, they are consequently ideal for use in the administration of medications. Furthermore, a recent study by Sadeghi et al. confirmed that the isotropic frameworks of the microemulsions created with biosurfactants develop spontaneously and that they are thermodynamically stable. The high solubilization limit, ease of preparation, and long-term stability of biosurfactant-assisted micro-emulsions make them extremely intriguing fluid carriers for future drug delivery systems. Furthermore, in contrast to conventional methods of drug distribution. Biosurfactants are widely used as therapeutic agents and drug delivery systems in the pharmaceutical sector. The study by Baferani et al. claims that biosurfactants exhibit emulsification, foaming, detergency, and scattering. Sophorolipids and rhamnolipids can be combined with lecithin to create biocompatible microemulsions. Because of their robust phase behaviour to temperature and salt changes, they are consequently ideal for use in the administration of medications. Furthermore, a recent study by Sadeghi et al. confirmed that the isotropic frameworks of the microemulsions created with biosurfactants develop spontaneously and that they are thermodynamically stable. The high solubilization limit, ease of preparation, and long-term stability of biosurfactant-assisted micro-emulsions make them extremely intriguing fluid carriers for future drug delivery systems. Furthermore, unlike conventional medication delivery methods [1]

(c) Gene transfection: A key tool in clinical therapies is gene transfection, and lipofection with cationic liposomes is becoming a popular non-viral delivery system because of its easy formulation, minimal cytotoxicity, and high transfection effectiveness. The physicochemical properties of cationic liposomes, such as their lipid packing density, surface charge (zeta potential), and structural shape, have a significant impact on the transfection effectiveness. As shown by Madihalli et al. who showed better transfection efficacy in monolayers combining MEL-A (a mannosylerythritol lipid) with DPPC compared to DPPC alone, biosurfactants improve this process by enabling membrane fusion between liposomes and target cells. Because of their distinct self-assembly characteristics, the four structural variations of mannosylerythritol lipids (MEL-A to D) are especially useful for applications involving drug encapsulation and controlled release. Biosurfactants are becoming more and more popular in genetic medicine as environmental and global health issues worsen. This is causing current research to focus on improving new non-viral delivery systems as well as viral vectors for precision gene editing applications.

(d) In preparation of micro-emulsion system and nanoparticles: Feng et al. showed that four crucial elements are needed for microemulsion-based drug delivery systems: an aqueous phase, an oil phase, a surfactant, and a co-surfactant. These thermodynamically stable colloidal systems efficiently separate hydrophilic and hydrophobic medications from the continuous phase by encapsulating them in their dispersed phase core. Three essential characteristics—inherent thermodynamic stability, ultralow oil–water interfacial tension, and improved solubilization capacity—are responsible for their remarkable drug delivery capability. In support of this, Nguyen et al. used rhamnolipid biosurfactant formulations to create a novel alcohol-free microemulsion system. Phase diagram analysis offered crucial information on the best surfactant ratios and system selection for upcoming uses. All of this research demonstrates the increasing pharmacological importance of biosurfactants in cutting-edge drug delivery systems.

(e) Removing the biofilm: As intricate, diverse structures, biofilms show considerable resistance to both biocides and traditional antibiotics. By causing steric repulsion upon surface contact, rhamnolipid biosurfactants improve bacterial mobility through porous matrices and consistently disrupt intercellular adhesion across a variety of bacterial strains and environmental conditions, thereby exhibiting a unique dual mechanism in the fight against biofilm persistence. Rhamnolipids are positioned as prospective adjuvants to overcome biofilm-mediated antimicrobial resistance because of their capacity to destroy biofilms, which has been confirmed by numerous experimental studies.

(f) As an anti-adhesive agent: In addition to their antibacterial properties, several biosurfactants have the capacity to inhibit microbial adherence. Particularly, on polystyrene surfaces, rhamnolipids and surfactin exhibit strong anti-adhesive properties against harmful bacteria like *Listeria monocytogenes* and *Staphylococcus aureus*, making them viable options for biofilm prevention in medical and food processing applications [1].

9.4 Application in the Medical and Healthcare Industry

Biosurfactants exhibit unique structural and functional properties that make them valuable for developing advanced therapies, either alone or in synergistic combinations. Their antimicrobial activity targets common pathogens, including multidrug-resistant strains such as *Acinetobacter baumannii* and *Staphylococcus aureus*, with biofilm inhibition observed at 25–50 mg/mL and bactericidal effects at 50–100 mg/mL. Additionally, biosurfactants from *Lactobacillus rhamnosus* and *Lactobacillus jensenii* demonstrate anti-adhesive and antimicrobial effects at concentrations of 25–100 mg/mL. Beyond their use in antimicrobial applications, biosurfactants exhibit intriguing anticancer effects by promoting the generation of tumour necrosis factor-α (TNF-α) in immune cells and causing apoptosis by membrane breakdown, chromatin condensation, and loss of cell adhesion. They may be used as adjuvants to increase antimicrobial activity because of their method of action, which involves changing phospholipid membranes, increasing permeability, and boosting drug bioavailability [28].

9.5 Application in the Cosmetic Industry

While providing better interfacial activity, biosurfactants outperform synthetic surfactants in terms of skin mildness, biocompatibility, and biodegradability. Mannosylerythritol lipids (MELs), rhamnolipids, and sophorolipids (from Starmerella bombicola) are the main variations that are used in cosmetics. MELs are excellent in anti-aging treatments, rhamnolipids are used in oral care and anti-wrinkle products, and sophorolipids are moisturizers and antimicrobials for acne and dandruff. Their eco-friendly profile satisfies the needs of contemporary skincare [11]. Ingredients for contemporary cosmetics, such as body creams, shampoos, soaps, and makeup, must be both environmentally friendly and compatible with surfaces. Because of their ecological compatibility and non-toxicity, biosurfactants have become excellent options. These macromolecules have several advantages for dermatology. Their fatty acid constituents support the preservation of the skin's natural acidic pH (4–6), which promotes the formation of good microbiota while preventing the growth of harmful microorganisms. Furthermore, they protect oxidative damage by neutralizing free radicals produced by UV light thanks to their antioxidant qualities. Biosurfactants' distinct amphiphilic structure increases their permeability across the stratum corneum, facilitating the efficient transport of active ingredients for hair restoration and skin protection. Because of their safety, practicality, and environmental friendliness, biosurfactants are becoming more and more useful in personal care products [12]. Important physicochemical factors such as ionic strength, balanced hydrophilic-lipophilic characteristics, and critical micelle concentration (CMC) affect the self-assembly behaviour of biosurfactant-containing products and determine their functional performance. Biosurfactants exhibit better environmental and dermatological profiles than traditional synthetic surfactants. They are distinguished by increased biodegradability, decreased cellular toxicity, and inherent antibacterial activity that helps treat cutaneous conditions. These characteristics, which result from their natural origin and structural complexity, allow for superior surface-active performance while allowing for softer but more focused interactions with biological surfaces.

9.6 Application in the Laundry

Because of their greater interfacial activity and natural nature, biosurfactants show great promise for detergent compositions. Biosurfactants provide an environmentally friendly substitute with similar efficacy to traditional synthetic surfactants, which are the most common in commercial laundry products but endanger aquatic habitats. Effective surface tension reduction (25–35 mN/m), strong emulsification ability with plant-derived oils, and exceptional stability in challenging environments—maintaining functionality across a wide pH spectrum (7.0–12.0) and at high temperatures (up to 121 °C)—are some of its key benefits. Because of these characteristics, as well as their innate biodegradability and low ecotoxicity, biosurfactants are excellent choices for creating more environmentally friendly laundry detergent formulations that satisfy sustainability and performance standards [8].

9.7 Application in the Hygiene and Sanitation

Because of their antibacterial, anti-adhesive, and low-toxicity qualities, biosurfactants are suitable for use in both oral and cutaneous applications and provide efficient, environmentally friendly alternatives for hygiene and sanitation. Their distinct amphiphilic structure facilitates the elimination of microorganisms by two different mechanisms: hydrophilic association with water molecules, which aids in the emulsification and detachment of pollutants, and hydrophobic interactions between fatty acid chains and microbial membranes. Because biosurfactants stay active on treated surfaces and inhibit germ reattachment and biofilm formation, this dynamic process guarantees long-lasting cleaning efficacy. Research demonstrates their wide range of applications in home, industrial, and medical cleaning solutions by confirming their capacity to prevent bacterial colonization on a variety of materials, such as silicone and polystyrene.

9.8 Application in the Bioremediation of Toxic Pollutants

By maximizing environmental factors (such as pH, moisture, and aeration) and microbial activity—often by adding bacterial inoculants and nitrogen/phosphorus fertilizers—bioremediation speeds up the natural degradation of pollutants. When low temperatures, high pollutant loads, or persistent substances like polycyclic aromatic hydrocarbons (PAHs) prevent decomposition, emulsifiers are essential. While bioemulsifiers can be directly applied to accelerate cleanup, recent genetic engineering technologies now permit the employment of microorganisms genetically modified to overproduce these surface-active chemicals. This innovative strategy has proven effective in practical applications, such as oil pipeline cleaning, demonstrating its potential for large-scale environmental remediation. To improve oil bioremediation, cultures of A. radioresistant [18] were introduced to a combination of bacteria that break down oil. These bacteria create the bio emulsifier alasan, but they cannot use hydrocarbons as a carbon source. The significant hydrophobicity of polycyclic aromatic hydrocarbons (PAHs), including phenanthrene and chrysene, which are common in sediments and wastewater contaminated by oil, is correlated with their molecular complexity; the more aromatic rings there are in the PAH, the less soluble it is in water. This structural feature significantly restricts the bioavailability of PAHs, which is a significant barrier to microbial breakdown. This problem is addressed by biosurfactants, which have the ability to improve the aqueous solubility of PAHs by up to 1–2 orders of magnitude due to their amphipathic character. In order to establish a molecular bridge between pollutants and microbial degraders in aqueous environments, the technique entails the development of micellar structures that contain hydrophobic PAH molecules within their core while preserving water compatibility through outer hydrophilic groups [34]. Low mobility states are common for soil-adsorbed hydrocarbons, which severely restricts their mass transfer and remedial effectiveness. A kinetic barrier to degradation is produced by this adsorption process, especially for extremely hydrophobic substances. By lowering the interfacial tension between hydrocarbons and aqueous phases, creating micellar structures that contain hydrophobic molecules, and increasing the rates of desorption from soil matrices, biosurfactants can be introduced as solubilizing agents and get around these restrictions. These mechanisms collectively increase the apparent solubility and bioavailability of otherwise recalcitrant contaminants, thereby improving their accessibility to microbial degraders and overall remediation potential [23].

9.9 Application in the Bioremediation of Heavy Metals

Soil contamination by heavy metals poses a serious risk to humans and other ecosystem inhabitants. Even trace levels of heavy metals in soil have been discovered to pose a significant risk because of their severe toxicity. Metals can be cleaned by a variety of biosurfactants, and heavy metals can be eliminated by rhamnolipids [27].

10 Economical Important

10.1 Overview

According to recent studies, the worldwide market supports new initiatives and is searching for biological alternatives to synthetic surfactants, which saw sales of over $1.74 billion in 2011. Global biosurfactant production was predicted to be over 344 thousand tons in 2013, while sales of biosurfactants exceeded $1.8 billion in 2016. About 442 thousand tonnes and $2.21 billion were estimated for 2018, while 2.31 billion and roughly 462 thousand tonnes were projected for 2020. Between 2014 and 2020, this market was predicted to increase at an annual pace of 4.3%. In 2023, biosurfactant sales are anticipated to reach 2.6 billion, with sophorolipids and rhamnolipids seeing an 8% increase in sales. According to another industry study, the global biosurfactant market is expected to reach a value of over $5.52 billion by 2022, with a compound annual growth rate (CAGR) of 5.6% between 2017 and 2022. However, the expensive cost of biosurfactants is one of the main barriers preventing their broad usage in industry. The cost of sophorolipids, the most viable microbial biosurfactants, is $34 per kilogram, whereas the average cost of synthetic surfactants, such as sodium dodecyl sulphate and plant-based amino acid surfactants, ranges from $1 to $4 per kilogram. Lower yields, longer processing times, higher downstream costs, energy requirements for sterilisation, and maintenance of biological culture are some of the production aspects that contribute to a biosurfactant's higher price [9].

10.2 Companies Producing Biosurfactants Commercially with Industrial Applications

Sr. No	Company name	Product name	Type of surfactant	Substrate used	Source organism	Applications
1	EcoChem Organic Ltd., Canada	EC601	Rhamnolipid	–	Bacteria	Dispersant of insoluble hydrocarbons in water
2	Cleveland Biotech Ltd., Teesside	Citrasolv	–	Orange peel	–	
3	Bio Future Ltd. Dublin, Ireland	Bio Future	Rhamnolipid	–	Bacteria	Washing of fuel tanks
4	Saraya Co. Ltd (Osaka, Japan)	Sophoron TM	Sophorolipids	Palm oil	*Candida spp.*	Cleaning products, hygiene eroducts
5	Soliance (Pomacle, France)	Sopholiance S	Sophorolipids	Rapeseed fermentation	*C. bombicola*	Cosmetics and pharmaceuticals
6	Sigma-Aldrich	Surfactin	Surfactin	–	*Bacillus subtilis*	
7	Petroferm research Inc	RAG-1 emulsan		–	–	
8	Sigma-Aldrich	R-90	Rhamnolipid	–	*P. aeruginosa*	
9	Fraunhofer IGB–Germany	–	Glycolipid and cellobiose, lipid biosurfactants	Mono- and disaccharides, vegetable oils or residues	*Moesziomyces and Ustilago*	Used glycolipids for cleaning products, dish washing liquids. And Lipids as pharmaceutical products (bioactive properties)
10	AGAE Technologies– USA	R90 AND R95 Rhamnolipids	Rhamnolipids	–	–	Pharmaceuticals, cosmetics, personal care products, bioremediation (in situ and ex situ), enhanced oil recovery (EOR)
11	TeeGene Biotech–UK	–	Rhamnolipids and lipopeptides	–	–	Pharmaceutical products, cosmetics, antimicrobials and anticarcinogen ingredients

(continued)

(continued)

Sr. No	Company name	Product name	Type of surfactant	Substrate used	Source organism	Applications
12	Jeneil Biosurfactant—USA	JBR 320, JBR 425, ZONIX BIOFUNGICIDE™	Rhamnolipids	–	–	Cleaning and oil recovery from storage tanks, EOR
13	Allied Carbon Solutions (ACS) Ltd–Japan	Sophomore®, BioToLife®	Sophorolipids	Vegetable oil, sugar	Yeasts *Starmerella bombicola*	Agricultural products, ecological research

11 Challenges and Future Opportunities

The future of biosurfactants lies in their broader adoption across diverse industries driven by technological advancements, regulatory shifts toward environmentally friendly products, and increased consumer demand for sustainable alternatives. Some key future prospects for biosurfactants include:

1. **Cost-Effective Production**: Innovations in fermentation technology, genetic engineering, and the use of industrial and agricultural waste as low-cost substrates will significantly reduce production costs, making biosurfactants more competitive with synthetic surfactants.
2. **Synthetic Biology and Strain Improvement**: Leveraging synthetic biology tools to develop engineered microbial strains with enhanced biosurfactant production capacity and efficiency will boost commercial viability.
3. **Enhanced Oil Recovery (EOR) and Bioremediation**: The petroleum industry will benefit from biosurfactants for EOR and remediation of oil spills, providing eco-friendly solutions to reduce environmental impact while improving resource recovery.
4. **Food and Beverage Industry**: Increased use of biosurfactants as natural emulsifiers and preservatives to improve product quality, stability, and shelf life is anticipated, driven by the growing trend for clean-label products.
5. **Cosmetic and Pharmaceutical Applications**: Their antimicrobial, anti-inflammatory, and skin-compatible properties will expand their use in personal care, dermatological products, and as carriers for drug delivery systems.
6. **Agriculture and Biotechnology**: Biosurfactants will play a critical role in sustainable agriculture, including applications in biopesticides, biofertilizers, and soil conditioning agents to improve crop yields while reducing chemical inputs.
7. **Regulatory and Market Growth**: As regulations increasingly favour biodegradable and non-toxic products, biosurfactants are poised for greater market penetration. Strategic investments, partnerships, and policy support will further drive innovation and adoption.

Overall, the continued research and commercialization efforts will enable biosurfactants to transform various sectors, contributing to more sustainable, eco-friendly, and circular industrial processes globally.

12 Conclusion

Finally, the topic of biosurfactants and their possible uses is covered. Because of their decreased toxicity, biodegradability, and capacity to be made from renewable resources, biosurfactants have become a viable substitute for synthetic surfactants. Research on biosurfactants is currently focused on increasing production yields, cost-effectiveness, and investigating various microbial strains and genetic engineering methods to boost productivity. Commercialization initiatives are also being propelled by developments in downstream processing and bioprocess optimization. With their growing applications in a variety of industries, biosurfactants have bright futures. They present opportunities for improved oil recovery and bioremediation of hydrocarbon-contaminated areas in the petroleum industry. Biosurfactants enhance food texture, stability, and emulsification in the food and beverage sector. Like this, their antibacterial and skin-compatible qualities make them perfect components for medication delivery systems and personal care items in the pharmaceutical and cosmetic industries. Additionally, biosurfactants have potential for improving plant growth and protecting plants in agriculture. Notwithstanding notable advancements, issues including exorbitant production costs and constraints on scalability still need to be resolved. These obstacles should be removed by continued research and advancements, such as the utilization of synthetic biology and waste-based substrates. To fully utilize biosurfactants and promote a shift to more environmentally friendly and sustainable industrial processes, cooperation between academia, business, and legislators will be essential.

References

1. Abbot V, Paliwal D, Sharma A, Sharma P (2022) A review on the physicochemical and biological applications of biosurfactants in biotechnology and pharmaceuticals. Heliyon 8(8):e10149. https://doi.org/10.1016/j.heliyon.2022.e10149
2. Arima K, Kakinuma A, Tamura G (1968) Surfactin, a crystalline peptidelipid surfactant produced by Bacillus subtilis: Isolation, characterization and its inhibition of fibrin clot formation. Biochem Biophys Res Commun 31(3):488–494. https://doi.org/10.1016/0006-291X(68)90503-2
3. Bahia R, dos Santos JC, de Almeida DG, Sarubbo LA, Banat IM (2018) Biosurfactant production by Saccharomyces cerevisiae expressing Pseudomonas putida enzymes for rhamnolipid synthesis. Biotechnol Rep 20:e00288. https://doi.org/10.1016/j.btre.2018.e00288
4. Banat IM, Satpute SK, Cameotra SS, Patil R, Nyayanit NV (2019) Cost effective technologies and renewable substrates for biosurfactant production. Front Microbiol 10:2026. https://doi.org/10.3389/fmicb.2019.02026
5. Besson F, Michel G, Lelièvre V (1976) Antifungal properties of iturin, a peptidolipid antibiotic produced by Bacillus subtilis. Biotech Lett 8(5):347–350. https://doi.org/10.1007/BF01026623
6. Deshpande M, Daniels L (1995) Evaluation of sophorolipid biosurfactant production by Candida bombicola using animal fat. Bioresource Technol 54:143–150
7. Effendi H, Yulianto E, Pratiwi A (2017) Toxicity of synthetic surfactants on estuarine bacterial communities. Mar Pollut Bull 118(1–2):111–118. https://doi.org/10.1016/j.marpolbul.2017.02.043
8. Enush O, Peled Y, Bar N (2021) Sustainable biosurfactant-based detergent formulations for environmental safety. Clean Chem Eng 2(1):100015. https://doi.org/10.1016/j.clce.2021.100015
9. Farias CB, Almeida FC, Silva IA, Souza TC, Meira HM, Luna JM, Santos VA, Converti A, Banat IM, Sarubbo LA (2021) Production of green surfactants: market prospects. Electron J Biotechnol 51:28–39. https://doi.org/10.1016/j.ejbt.2021.02.002
10. Ferrer I, Ginebreda A, Barceló D (1997) Occurrence and distribution of linear alkylbenzene sulfonates in riverine sediments. Environ Sci Technol 31(4):1103–1109. https://doi.org/10.1021/es960506u

11. Gayathri S, Anjana S, Surendran S (2023) Biosurfactants in cosmetic formulations: trends, benefits and challenges. Int J Cosmet Sci 45(2):220–233. https://doi.org/10.1111/ics.12812
12. Henkel M, Müller MM, Kügler JH, Lovaglio RB, Contiero J, Syldatk C, Hausmann R (2012) Rhamnolipids as biosurfactants from renewable resources: concepts for next-generation rhamnolipid production. Process Biochem 47(8):1207–1219. https://doi.org/10.1016/j.procbio.2012.04.018
13. Hisham NZ, Ibrahim ASS (2019) Partial purification and characterization of biosurfactants produced by Bacillus sp. HIP3. Biocatalysis Agric Biotechnol 18:101050. https://doi.org/10.1016/j.bcab.2019.101050
14. Kachholz T, Schlingmann M (1987) Biosurfactants in food processing and technology. Food Biotechnol 1(3):215–227. https://doi.org/10.1080/08905438709549785
15. Kumar A, Singh SK, Kant C, Verma H, Kumar D, Singh PP, Modi A, Droby S, Kesawat MS, Alavilli H, Bhatia SK, Saratale GD, Saratale RG, Chung SM, Kumar M (2021) Microbial biosurfactant: a new frontier for sustainable agriculture and pharmaceutical industries. Antioxidants 10(9):1473. https://doi.org/10.3390/antiox10091473
16. Mahalakshmi M, Pugazhendhi A (2020) Environmental impact and bioremediation of alkylphenol ethoxylates. J Hazard Mater 392:122479. https://doi.org/10.1016/j.jhazmat.2020.122479
17. Manga EB, Celik PA, Cabuk A, Banat IM (2021) Biosurfactants: opportunities for the development of a sustainable future. Curr Opin Colloid Interface Sci 56:101514. https://doi.org/10.1016/j.cocis.2021.101514
18. Navon-Venezia S, Zosim Z, Gottlieb A, Legmann R, Carmeli S, Ron EZ, Rosenberg E (1995) Alasan, a new bioemulsifier from Acinetobacter radioresistens. Appl Environ Microbiol 61(9):3240–3244
19. Patel RM, Desai AJ (1997) Surface-active properties of rhamnolipids from Pseudomonas aeruginosa GS3. J Basic Microbiol 37(4):281–286
20. Philip L, Kumar S, Ramaswamy S (2020) Environmental persistence and phytotoxicity of synthetic surfactants in soil and aquatic systems. Environ Technol Innov 19:100875. https://doi.org/10.1016/j.eti.2020.100875
21. Rosenberg E, Ron EZ (1999) High- and low-molecular-mass microbial surfactants. Appl Microbiol Biotechnol 52(2):154–162
22. Sarubbo LA, da Silva MGC, Durval IJB, Bezerra KGO, Ribeiro BG, Silva IA, Twigg MS, Banat IM (2022) Biosurfactants: production and applications in the twenty-first century. Trends Biotechnol 40(9):1134–1150. https://doi.org/10.1016/j.tibtech.2022.04.003
23. Shin KH, Kim HS, Kang ST (2004) Biosurfactant-enhanced bioremediation of hydrophobic organic contaminants. J Hazard Mater 115(1–3):111–119. https://doi.org/10.1016/j.jhazmat.2004.05.014
24. Silva IA, Farias CBB, Almeida FCG, Sarubbo LA (2024) Functional stability of microbial biosurfactants under extreme physicochemical conditions. Process Biochem 133:35–46. https://doi.org/10.1016/j.procbio.2024.03.002
25. Silva IA, Meira HM, Luna JM, Sarubbo LA (2024) Role of biosurfactants in sustainable agriculture and plant protection. Front Sustain Food Syst 8:1427351. https://doi.org/10.3389/fsufs.2024.1427351
26. Singh P, Cameotra SS (2004) Potential applications of microbial surfactants in biomedical sciences. Trends Biotechnol 22(3):142–146
27. Sourav A, Dutta J, Ghosh A (2015) Rhamnolipid-mediated removal of heavy metals from contaminated soil and water. J Environ Chem Eng 3(1):812–820. https://doi.org/10.1016/j.jece.2015.01.023
28. Thakur R, Jadhav J, Pratap A (2021) Biomedical and therapeutic potential of microbial biosurfactants. Curr Opin Biotechnol 72:207–215. https://doi.org/10.1016/j.copbio.2021.09.013
29. Thakur R, Jadhav J, Pratap A (2024) Statistical and bioprocess optimization for cost-effective biosurfactant production. Biochem Eng J 208:109680. https://doi.org/10.1016/j.bej.2024.109680

30. Thakur R, Jadhav J, Pratap A (2024) Co-culture fermentation strategies for enhanced biosurfactant yield. Bioresour Technol Rep 23:101012. https://doi.org/10.1016/j.biteb.2024.101012
31. Van Haesendonck I, Vanzeveren E (2004) Application of rhamnolipids in bakery and frozen confectionery products. Food Hydrocolloids 18(5):903–909. https://doi.org/10.1016/j.foodhyd.2003.12.001
32. Vijayakumar S, Saravanan V (2015) Biosurfactants—Types, sources and applications. Res J Microbiol 10(5):181–192. https://doi.org/10.3923/jm.2015.181.192
33. Vijayakumar S, Saravanan V (2015) Physicochemical properties of microbial biosurfactants. J Surf Sci Technol 31(3):239–252. https://doi.org/10.18311/jsst/2015/1329
34. Yin H, Qiang J, Jia Y, Ye J, Peng H, Qin H, Zhang N, He B (2009) Characteristics of biosurfactant produced by Pseudomonas aeruginosa and its application in oil recovery. Biochem Eng J 43(2):85–91. https://doi.org/10.1016/j.bej.2008.09.009
35. Zhou Y, Zhang Y, Li X (2023) Genetic modification of Aureobasidium pullulans for enhanced limonin biosurfactant production. Microb Cell Fact 22(1):189. https://doi.org/10.1186/s12934-023-02018-3

Sustainable Biosurfactants Examining for Replacement of Chemical Surfactants

Jagruti Barot, Neha Mistry, Urjita Sheth, Vijay Patel, Paresh Parekh, and Mehul Khimani

Abstract Surfactants adhere to the interface between two different phases, like liquid–gas or liquid–solid. They self-organize into micelles in a solution (oil or water) beyond the "critical micelle concentration" (CMC). Under different stimuli, their amphiphilic properties promote the formation of various geometric structures in solution. Thus, they have gained extensive application in consumer and commercial products, including food, healthcare, cosmetics, paints, and many more industrial processes such as oil recovery and printing. The global surfactant market was valued at $ 39,901 million in 2019 and is expected to reach $52,417 million by 2025. Many commercial goods that contain surfactants are discharged in native form in the ecosystem. Current estimates indicate that the quantity of surfactants in global aquatic habitats is approximately 7 million metric tons (60% by weight) annually. Nonetheless, their ecological ramifications have frequently been disregarded. Compared to synthetic counterparts, biosurfactants offer superior environmental compatibility. They have been extensively researched as alternatives to synthetic surfactants due to their versatile characteristics, including low-cost production, easy degradability,

J. Barot (✉)
Department of Biosciences, Veer Narmad South Gujarat University, Surat, Gujarat, India
e-mail: jkbarot@vnsgu.ac.in

N. Mistry
Shree Jalaram International School, Surat, Gujarat, India

U. Sheth
Department of Biotechnology, Faculty of Science, Vanita Vishram Women's University, Surat, Gujarat, India

V. Patel
Department of Chemistry, Navyug Science College, Surat, Gujarat, India
e-mail: vijaypatel4686@gmail.com

P. Parekh
Department of Chemistry, Veer Narmad South Gujarat University, Surat, Gujarat, India
e-mail: pyparekh@vnsgu.ac.in

M. Khimani (✉)
Chemical Division, P T H Consultancy Chemical LLP, Surat, Gujarat, India
e-mail: khimani_mehul@yahoo.com

M. Abdul Rub et al. (eds.), *Industrial Applications of Individual and Mixed Surfactant Systems—Volume 2*, Sustainable Materials and Technology,
https://doi.org/10.1007/978-981-95-9195-4_6

less toxicity, biofilm formation, and activity in extreme environmental conditions. They are produced by many microorganisms, including bacteria, fungi, and yeasts, as secondary metabolites. They are one of the best alternatives to synthetic surfactants to be used in cosmetic, food, and pharmaceutical fields. This chapter covers topics such as biosurfactant production techniques, various types of biosurfactants and their properties, and the challenges encountered during commercialization and large-scale production. The potential benefits of biosurfactant include bioremediation, healthcare, cosmetics, genetic engineering, bioprocess, etc., encouraging sustainable surfactant technology innovation and a greener industrial approach.

Keywords Synthetic surfactants · Industrial applications · Nonbiodegradable · Biosurfactants · An environment friendly

1 Introduction

1.1 Introduction of Chemical Surfactants

Surfactants, or surface-active agents, are multifunctional compounds that significantly reduce surface tension at the interfaces of distinct phases, including two immiscible liquids, [1] a gas and a liquid, [2] or a liquid and a solid [3]. These molecules exhibit unique amphiphilic characteristics, consisting of both hydrophobic (water-repellent) and hydrophilic (water-attracting) components. The hydrophobic component, typically a long hydrocarbon chain, is designated as the "tail," and the hydrophilic component, often an ionic or polar group, is called the "head". Owing to their amphiphilic characteristics, surfactant molecules self-organize into diverse forms in aqueous solutions. The hydrophilic heads orient toward the aqueous interface, and the hydrophobic tails align toward oil or air, depending on the system. This molecular arrangement allows surfactants to bridge the gap between water-insoluble and water-soluble components, making them invaluable in numerous applications. When surfactants mix with water, they disperse and adsorb at interfaces, forming a monolayer that significantly alters the surface properties. This adsorption process leads to a decrease in surface tension, which is the primary mechanism behind their functionality. The ability of surfactants to alter interfacial properties enables them to achieve several functions:

1. Detergents: They facilitate the removal of oily contaminants and particulates from surfaces.
2. Wetting agents: They enhance the dispersion and diffusion of liquids on surfaces.
3. Emulsifiers: They stabilize mixtures of immiscible liquids by reducing interfacial tension.
4. Foaming agents: They enhance the formation and stabilization of foam.
5. Dispersants: They facilitate the suspension of solid particles within a liquid medium.

The diverse applications of surfactants stem from their capacity to modify interfacial properties, making them essential in sectors from household cleaning products to advanced materials science and technology.

When it comes to surfactants, the most common kind has one hydrocarbon chain, while the tail usually has two. Surfactants having dual chains exhibit superior performance at the interface and can form more stable structures, such as bilayers and vesicles [4]. Surfactants are typically classified into four types based on the chemical properties of their head groups: anionic, cationic, non-ionic, and zwitterionic (also known as amphoteric). Anionic surfactants possess negatively charged head groups, whereas cationic surfactants exhibit a positive charge. Non-ionic surfactants lack charged groups; their hydrophilicity is derived from polyether chains, generally composed of ethoxylated sequences resembling polyethylene oxide. Zwitterionic surfactants possess both cationic and anionic groups within a single molecule, resulting in a net neutral charge. The structural diversity of surfactants markedly affects their physicochemical properties and uses in numerous industries.

1.2 *Types of Surfactants*

1.2.1 Anionic Surfactants

Anionic surfactants contain negatively charged functional groups at their hydrophilic head, such as phosphate ($-OPO_3^{2-}$), sulfate ($-OSO_3^-$), carboxylates ($-COO^-$), and sulfonate ($-SO_3^-$) moieties [5]. Commercial and industrial uses abound for anionic surfactants because of the exceptional detergency and surface activity imparted by these groups. Linear alkylbenzene sulfonates (LAS) and the alkyl phenol ethoxylates (APE) are among the most notable anionic surfactants. Cleaning products in both the domestic and commercial sectors frequently employ LAS compounds, which are made by sulfonating linear alkylbenzene precursors, due to their potent cleaning capabilities and comparatively high biodegradability [6]. Alkyl sulfates comprise another primary class of anionic surfactants, which includes sodium lauryl sulfate (SLS) and alkyl-ether sulfates, such as sodium lauryl ether sulfate (SLES), also known as sodium laureth sulfate, and sodium myreth sulfate. They are frequently spotted in toothpaste, body washes, and shampoos and are renowned for their emulsifying and foaming abilities. Sodium stearate is a prominent anionic surfactant and is commonly used in conventional bars (soaps). The stearates account for more than 50% of the global surfactant usage [7]. Other specialized anionic surfactants, such as dioctyl sodium sulfosuccinate (DOSS), are used as wetting agents and dispersants, especially in pharmaceutical and cosmetic formulations. Personal care, cleaning agents, agrochemicals, and industrial processes are just a few of the potential applications for anionic surfactants due to their performance, affordability, and adaptability (Fig. 1).

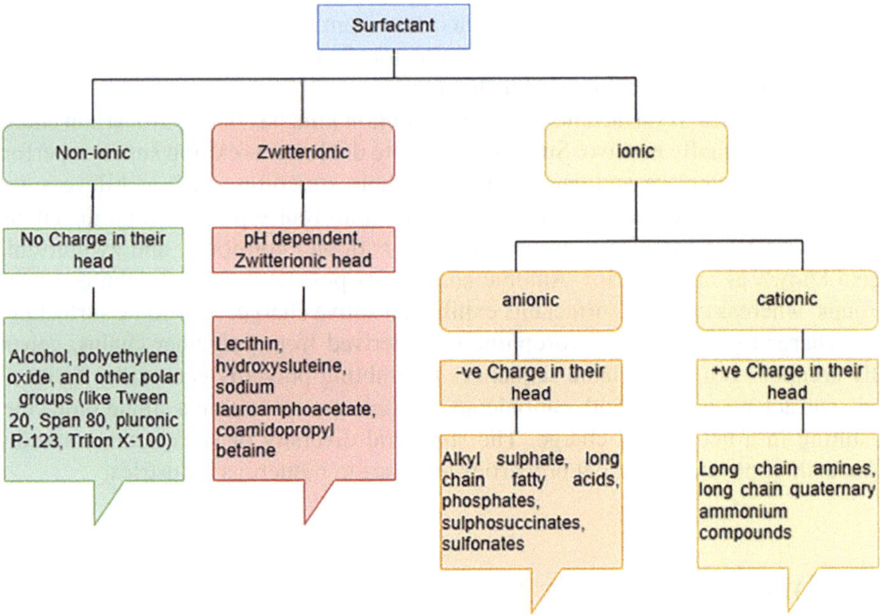

Fig. 1 Classification of surfactants

1.2.2 Cationic Surfactants

Cationic surfactants have a positively charged amine head group, which enables them to form unique electrostatic interactions with negatively charged surfaces, such as microbial membranes and other negatively charged surfactants. Depending on the number of alkyl groups attached to the nitrogen atom, an amino group is classified as primary, secondary, or tertiary. If the nitrogen atom is directly bonded to the fourth carbon atom, the molecular structure has a quaternary ammonium functional group with a positive charge [8]. They are frequently employed in health and industrial applications due to their antibacterial and antifungal properties. Hexadecyl trimethylammonium bromide, also known as cetyltrimethylammonium bromide (CTAB), is a quaternary ammonium compound and one of the most widely recognized cationic surfactants. It has a water-repelling tail made of a long 16-carbon chain (the cetyl group) and a water-attracting head made of a quaternary ammonium group, which is balanced by a bromide ion. Cetrimonium chloride is another type of cationic surfactant used as a conditioning agent in personal care products.

Alkyl dimethyl benzyl ammonium chloride (ADBAC), also known as benzalkonium chloride (BZK, BKC, BAK, BAC), is a well-known cationic surfactant that comes in different concentrations, like BKC-50 and BKC-80. The manufacturing of cationic surfactants is not a single-step reaction, but rather a series of selective chemical processes; therefore, the cost of these surfactants is higher than that of other types of surfactants. They are widely utilized in disinfectants due to their ability to disrupt microbial cell membranes, leading to cell lysis and subsequent cell death. This method is efficacious against a wide range of bacteria, viruses, and fungi. Cationic surfactants have a crucial role in textile processing, corrosion inhibition, and emulsification, particularly in their interactions with negatively charged substances such as clays, proteins, DNA, or synthetic polymers [9].

1.2.3 Non-ionic Surfactants

In an aqueous solution, non-ionic surfactants do not form ions. Instead, they have oxygen-rich groups that are water-attracting, such as ether (–CH_2CH_2O–), hydroxyl (–OH), or amide (–$CONH_2$), connected to hydrophobic components like fatty alcohols or long-chain hydrocarbons. These surfactants exhibit stability throughout an extensive pH spectrum and demonstrate compatibility with anionic, cationic, and amphoteric surfactants. Their solubility in water is affected by hydrogen bonding, which decreases with rising temperature and the introduction of chaotropic salts, resulting in the cloud point—defined as the temperature at which the surfactant no longer dissolves and begins to separate from the solution [10]. Non-ionic surfactants function well in hard water, produce less foam, and are less irritating to the skin than anionic surfactants.

Non-ionic surfactants can be classified structurally into several categories, including ester types (e.g., polyoxyethylene fatty acid esters), ether types (e.g., polyoxyethylene alkyl or alkylphenol ethers), amine types, amide types, and mixed kinds such as polysorbates and sorbitan esters. Both polar (hydrophilic) and non-polar (lipophilic) segments contribute to their surface activity. These surfactants are extensively employed in cosmetics, food formulations, and oilfield applications for functions including emulsification, corrosion inhibition, wax prevention, and improved oil recovery [11]. Although most non-ionic and anionic surfactants are generally non-toxic (with LD_{50} values equal to that of table salt), prolonged contact with surfactants can cause skin irritation owing to disruption of the lipid barrier. In general, skin irritation increases in order: non-ionic < amphoteric < anionic < cationic.

1.2.4 Amphoteric (Zwitterionic) Surfactants

A distinct category of surfactants, referred to as zwitterionic or amphoteric surfactants, possesses an electrically neutral configuration due to the presence of both negative and positive charged functional groups within a single molecule [12]. This dual-charge alignment bestows exceptional adaptability and sensitivity to environmental factors such as pH. The positive component of zwitterionic surfactants typically originates from the first, second, or third amines, with the group of quaternary ammoniums serving as the predominant source. The anionic component can vary and may include carboxylates, sulfonates, phosphates, or sulfates. For instance, betaines, such as cocamidopropyl betaine (CAPB), consist of a quaternary ammonium cation paired with a carboxylate anion. Therefore, the surface behavior of these surfactants is highly dependent on pH, as they are ampholytic. At low pH, the molecule exhibits cationic character due to protonation, while at high pH, the anionic form prevails. This pH sensitivity enables applications that require switching. However, the ability of zwitterionic surfactants to maintain stability across a wide range of pH and temperature, along with their compatibility with anionic, cationic, and non-ionic surfactants, renders them exceptionally versatile [13]. Consequently, they are valued in skincare products—such as shampoos, facial cleansers, and baby items—owing to their remarkable gentleness, non-irritating properties, and skin compatibility. They are also employed in textile and leather processing as dye-leveling and wetting agents, in industrial cleaners for their compatibility with various water conditions, and in biomedical applications due to their low cytotoxicity and potential as drug delivery agents.

1.3 Importance of Synthetic Surfactants in Industrial Applications

Synthetic surfactants are extremely useful across a broad range of industrial applications because of their tailored chemical structures and the predictable characteristics of their performance. From domestic cleaning products to sophisticated medications, they are essential in improving product efficacy and performance. Their distinctive amphiphilic chemical structure facilitates interaction at molecular interfaces, serving as a bridge between water-soluble (polar) and oil-soluble (non-polar) compounds. Their superior emulsifying and foaming capabilities make them ideal for use in cleaning products and detergents. In the textile sector, they facilitate dyeing, scouring, and finishing activities. In enhanced oil recovery, they assist in diminishing interfacial tension to optimize oil displacement. Synthetic surfactants are crucial in pharmaceuticals for drug solubilization and administration. In agrochemical formulations, they also serve to stabilize suspensions and emulsions. In paints and coatings, they enhance dispersion and flow characteristics. Cleaning, emulsification, and conditioning are some of their functions in cosmetics. When processing food, surfactants aid in emulsification and control texture. Due to their versatile structural composition, they are very crucial in industrial applications.

1.3.1 Household and Industrial Cleaning

Alcohol ethoxysulfates (AES), alkyl sulfates (AS), linear alkylbenzene sulfonates (LAS), [14] and secondary alkyl sulfonates (SAS) [15] widely used anionic surfactants are in the industrial and domestic sectors. They are frequently utilized in formulating dishwashing and laundry detergents, personal hygiene items, household cleaning goods, and soil conditioning agents for excavation or automated tunneling [14]. Another significant example of an anionic surfactant is GenX (2,3,3,3-tetrafluoro-2-(heptafluoropropoxy)propanoic acid), which has been extensively utilized since 2010 as a substitute for perfluorooctanoic acid (PFOA) in the production of fluoropolymers such as polytetrafluoroethylene (PTFE) [16]. GenX exhibits high water solubility with a pK_a value of 2.84 [16, 17]. In terms of consumption, DuPont consumes up to 100 tons of GenX annually [18].

1.3.2 Personal Care and Cosmetics

The multibillion-dollar worldwide cosmetic industry is built on the foundation of healthy, glowing skin and lustrous, well-groomed hair, which are considered to be key markers of overall well-being and confidence. Personal care products rely significantly on surfactants, which serve as conditioners, emulsifiers, and agents for foaming. A good example of this would be the use of sodium dodecyl sulfate (SDS) in shampoos for cleansing and lathering, but conditioners often contain cationic surfactants such as polyquaternium-10 to improve the softness of the hair [19]. Milder non-ionic and amphoteric surfactants, such as cocoamidopropyl betaine (CAPB) and sodium lauryl ether sulfates (SLES), are preferred for sensitive skin. SLES containing a higher number of ethylene oxide units (e.g., SLES-3EO) is milder and appropriate for infant care. Still, CAPB and alkylpolyglucosides (APGs) are considered some of the most delicate surfactants available. Infant shampoos often include SLES and CAPB in a 2:1 ratio, while high-end formulas may include CAPB, APG, and glycerin to optimize gentleness. Additionally, on a scale from 1 to 100, with 1 being the least intense and 100 the most intense, the subsequent order is as follows: Water < APG < SLES + CAPB (2:1) < syndet < SCI < soap < SLS [20].

1.3.3 Pharmaceuticals and Medicine

Micellar systems enhance drug solubility by encapsulating hydrophobic active pharmaceutical ingredients (APIs) within the surfactant's hydrophobic core, enabling dispersion in aqueous media [21, 22]. Surfactants serve as solubilizers, emulsifiers, and delivery agents in pharmaceutical formulations like creams and ointments. They enhance the bio-availability and stability of less water-soluble drugs. One example is Doxil®, a PEGylated liposomal formulation of doxorubicin that exhibits reduced cardiotoxicity when used to treat multiple myeloma, Kaposi's sarcoma, and ovarian cancer [23, 24]. Surfactant-based DDS exploits the EPR effect to enhance tumor targeting and reduce off-target toxicity [24, 25]. Beyond liposomes, surfactant-based micellar drug carriers have shown clinical success for cancers such as breast and lung cancers. Surfactants also support nebulized drug delivery, especially in infants with respiratory distress. Trials using vibrating mesh nebulizers demonstrated clinical safety and improved outcomes. These systems reflect the expanding therapeutic utility of surfactant-based nanocarriers [26–29].

1.3.4 Food Industry

By enhancing the consistency, stability, and shelf life of products such as mayonnaise, ice cream, and sauces within the food industry, surfactants serve a crucial function, particularly as emulsifiers in formulations. These additives help prevent ingredient separation, thereby improving product appearance and uniformity. In recent decades, lecithin has gained prominence as a naturally derived emulsifier, serving as both a functional food ingredient and a nutraceutical supplement. Extracted from sources like soybeans and egg yolks, it is widely used in the production of chocolates, baked goods, and dairy products due to its remarkable ability to blend ingredients effectively [30, 31]. In chocolate production, lecithin significantly alters the rheology and polymorphic behavior of cocoa butter, enhancing the flow properties and overall quality of the final product [31, 32]. It ensures even distribution of ingredients, contributing to a consistent texture and prolonged shelf life. Lecithin is also commonly incorporated into salad dressings, margarine, and confectionery items to ensure a uniform blend of oil- and water-based components. Beyond its technological applications, lecithin is valued for its potential health benefits. It is a rich source of choline, which is essential for maintaining healthy liver function, nerve signaling, and muscle activity. Emerging research suggests that it may also help lower cholesterol levels and support cognitive and neurological health. Thus, lecithin is an indispensable ingredient in both industrial and dietary contexts due to its versatility and its role in bridging food technology with human wellness [33].

1.3.5 Agriculture

When it comes to crop protection, agricultural science is dependent on agrochemical formulations, and surfactants are crucial components in these kinds of products (e.g., suspensions, emulsions, and microcapsules). They improve formulation stability and biological effectiveness during application. The diffusion rate of surfactants and micelle dynamics determine how quickly they adsorb to the air–liquid interface when spraying, reducing dynamic surface tension. They enhance droplet adhesion by diminishing surface tension and contact angle, so assuring superior retention on plant surfaces. On inclined surfaces, surfactants elevate contact angle hysteresis, diminishing droplet mobility and improving spray retention. Furthermore, they enhance wetting and spreading properties, ensuring uniform application of agrochemicals. Moreover, surfactants can directly enhance pesticide penetration through plant barriers by solubilizing active components and increasing concentration gradients, thereby improving performance. In addition to having a strong resistance to saline solutions, short-chain APGs exhibit respectable wetting and penetrating agent qualities [34].

1.3.6 Textile Industry

The traditional purpose of textiles is to manufacture products, primarily for esthetic purposes. The processing of fibers and fabrics using physical and chemical treatments imparts various qualities that enhance their performance. The three main steps in textile wet processing are pretreatment, coloring, and finishing. Finishing, being the final step in fabric manufacture, enhances both performance and value. This stage is divided into two primary types: chemical and mechanical. Chemical finishing imparts desired qualities to fabrics without altering their appearance. This process includes de-sizing, scouring, bleaching, dyeing, printing, and functional finishes such as softening, crease resistance, and water repellency. Surfactants play a crucial role as active components in finishing formulations for textile wet processing.

Non-ionic and anionic surfactants are utilized in pretreatment, dyeing, finishing, and post-treatment processes for functions like wetting, spreading, leveling, emulsifying, dispersing, solubilizing, cleaning (scouring), and foaming. Cationic surfactants are predominantly employed in the finishing process due to their beneficial properties, which include softening, lubrication, antistatic effects, antibacterial qualities, and the ability to repel water and oil. They also act as dye fixatives and leveling agents during dyeing. Amphoteric surfactants, while less commonly used, offer distinct advantages in specific applications because of their unique dual-charge structure. The development of multifunctional specialized surfactants aims to enhance sustainability and efficiency in textile processing. These innovative molecules are ideal for modern textile production as they reduce processing times, minimize water usage, and make operation at lower temperatures feasible by integrating two or more processing functions into a single compound [35].

1.3.7 Oil and Gas Industry

Improving oil recovery from depleting reservoirs is of the utmost importance due to the increasing energy demand and diminishing oil stocks [36]. The process of oil recovery can be broken down into three distinct stages: primary, secondary (conventional), and tertiary, which is also known as enhanced oil recovery (EOR) [37]. Surfactants are crucial for enhanced oil recovery (EOR) [38] as they modify the properties of the rock surface and reduce interfacial tension [39]. However, their efficacy is limited by high surfactant adsorption and instability in difficult reservoir conditions (such as high salinity or temperature). It is crucial to optimize surfactant compositions to decrease adsorption while maintaining optimal performance. Nanoparticles can improve rheological characteristics and foam stability, which in turn leads to an increase in recovery efficiency [40, 41]. For surfactants to be cost-effective, they should be able to reduce IFT to 0.01–0.001 dyn/cm, operate at concentrations of 0.1–0.3%, and demonstrate adsorption below 1 mg/g of rock [42] (Fig. 2).

Fig. 2 Industrial applications of surfactants

1.4 Environmental Impact of Chemical Surfactants

Domestic, industrial, and agricultural sectors extensively utilize surfactants due to their surface-active properties. They penetrate the environment via multiple pathways, such as agrochemicals (e.g., herbicides and insecticides), industrial uses (such as emulsifiers, coatings, and cleaners), and household activities (such as laundry, disinfection, and personal care). Therefore, significant amounts of surfactants are present in wastewater, ranging from 1 to 10 mg/L in household wastewater to as high as 300 mg/L in industrial effluents [43]. These surfactants frequently enter municipal wastewater treatment facilities (MWTPs) or are discharged directly into the environment [44, 45]. Even though secondary treatment techniques are employed, research has shown that residual surfactants and decomposition products may still be present in the treated effluent. This inadequate removal raises concerns about the contamination of drinking water, as well as potential risks to human health and ecosystems. Moreover, the quality of treated effluent is predominantly influenced by the efficacy of MWTPs and the chemical characteristics of the surfactants utilized. It is particularly challenging to deal with persistent and non-biodegradable surfactants because they tend to resist traditional treatment techniques, which can contribute to environmental contamination over an extended period.

Several investigations have revealed significant ecological and health effects of toxic surfactants on humans as well as on flora and fauna of the ecosystem, including microbes [46]. For example, anionic surfactants, especially linear alkylbenzene sulfonates (LAS), have demonstrated considerable detrimental impacts on both aquatic and terrestrial ecosystems. LAS can interfere with physiological and metabolic processes in marine organisms by hindering metabolism and development, weakening cell membrane integrity, and facilitating the destruction of chlorophyll-protein complexes [47]. Furthermore, LAS has been associated with several biochemical, clinical, and physiological abnormalities [45]. It diminishes the resilience of aquatic species to external stressors, adversely impacting their reproductive and developmental abilities [45, 48]. Surfactants generally increase the solubility of pollutants, therefore promoting eutrophication [49, 50]. The toxicity of surfactants is directly related to their hydrophobicity [51]. Nonylphenol ethoxylates (NPEOs) and octylphenol ethoxylates, together with their decomposition products, including nonylphenol carboxylic acids, exhibit greater toxicity than their parent chemicals. These pollutants can harm humans, soil organisms, aquatic creatures, and flora. The significant effects of these chemicals, therefore, elevate public health and environmental concerns over the elevated quantities of surfactants [45].

The REACH (Registration, Evaluation, Authorization, and Restriction of Chemicals) Regulation supports the utilization of *in-silico* tools, such as quantitative structure–activity relationship (QSAR) models and the ECOSAR program, [52] in conjunction with laboratory-based assays, such as the Daphtox Kit and the Algal Toxkit, to evaluate the ecotoxicity of surfactants. Factors such as water pH, dissolved oxygen levels, surfactant concentration, and species-specific sensitivity all influence the toxicity of these chemicals. It is common practice to assess environmental risk using important toxicological metrics, including the median inhibitory concentration (IC_{50}) and the median effective concentration (EC_{50}). For instance, the EC_{50} value for acetyl trimethyl ammonium chloride in relation to the green microalga *Chlorella vulgaris* was obtained at 145 ± 13.35 μg/L over a period of 96 h of exposure. Furthermore, surfactants can enter the bloodstream through the food chain and bind to proteins, which can affect human hormones and metabolic problems [53].

1.5 Need for Sustainable Alternatives

The majority of consumer products containing surfactants are intended for disposal after use, resulting in their extensive release into the environment through wastewater systems [54]. Approximately 60% (by weight) of the total surfactant produced, which is equivalent to around 7 million tons of worldwide production, ends up in aquatic habitats [55]. Sustainable alternatives to conventional chemical surfactants are in high demand due to increasing concerns about pollution, biodegradability, and potential health risks to humans. Petrochemicals and petroleum oil distillates, including paraffin, benzene, and other aromatic compounds, are derived from crude petroleum to produce chemical surfactants. These processes generally take place in oil refineries, which are significant contributors to pollution and can lead to aquatic toxicity and bio-accumulation [56].

Biosurfactants are a type of surfactant derived from living organisms, including plants, animals, and microbes such as yeasts, fungi, and bacteria. Their production provides eco-friendly and biodegradable alternatives for various applications [57]. Additionally, they exhibit equal or superior emulsifying properties and reduce surface tension more effectively than synthetic surfactants, even under challenging conditions such as temperature, salinity, and pH variations. However, their low toxicity, renewable production sources, and biodegradability make them strong candidates from the perspectives of sustainable development and green chemistry. This suitability extends to a wide range of industrial applications, including environmental remediation, pharmaceuticals, food processing, cosmetics, and other related fields [58]. Although, the costs associated with large-scale production and downstream processing currently hinder widespread adoption [59]. Advancements in microbial engineering and bioprocess optimization are promptly bridging the gap, positioning biosurfactants as a promising alternative to traditional chemical surfactants [60].

2 Biosurfactants: An Overview

Living cells produce biosurfactants, which have similar characteristics to synthetic surfactants in terms of reducing surface tension and interfacial tension [61]. Emulsion, dispersion, solubilization, foaming, wetting, and detergent capacity are some of the primary activities of biosurfactants [62]. They are popular for their antibacterial properties. Biosurfactants are produced by microorganisms such as yeast, bacteria, and fungi from diverse sources, such as waste materials, oils, alkanes, and sugars. Studies have demonstrated that Pseudomonas aeruginosa synthesizes rhamnolipids from C_{11} and C_{12} longer chain length alkanes, succinate, pyruvate, citrate, fructose, glycerol, olive oil, glucose, and mannitol [63]. Biosurfactants are considered environmentally friendly because they are non-toxic or less toxic and biodegradable. As a result, biosurfactants have gained considerable interest from the scientific and industrial sectors.

2.1 Definition and Characteristics

Surfactants are amphiphiles that possess both hydrophilic and hydrophobic constituents. The capacity of surfactants to lower surface and interfacial tension is what allows them to enhance the bio-availability and solubility of non-polar organic molecules [64]. The critical micelle concentration (CMC) is the concentration at which biosurfactant monomers begin to self-assemble into an organized structure known as micelles. When the concentration exceeds the CMC, these monomers spontaneously assemble into various geometric structures, such as spherical micelles, bilayers, or vesicles, significantly modifying the solution's physicochemical properties. As a general rule, surfactant efficiency is measured using the CMC. According to Desai and Banat [65], more effective biosurfactants have a lower CMC, meaning that the efficiency of the biosurfactants is inversely proportional to the amount of biosurfactants, CMC. In other words, to reduce surface tension, a minimal amount of biosurfactant is required. HLB value, which stands for hydrophilic-lipophilic balance, is a measurement that can be used to determine the kind of emulsion, such as oil-in-water or water-in-oil. Emulsifiers with a low HLB stabilize water-in-oil emulsions,

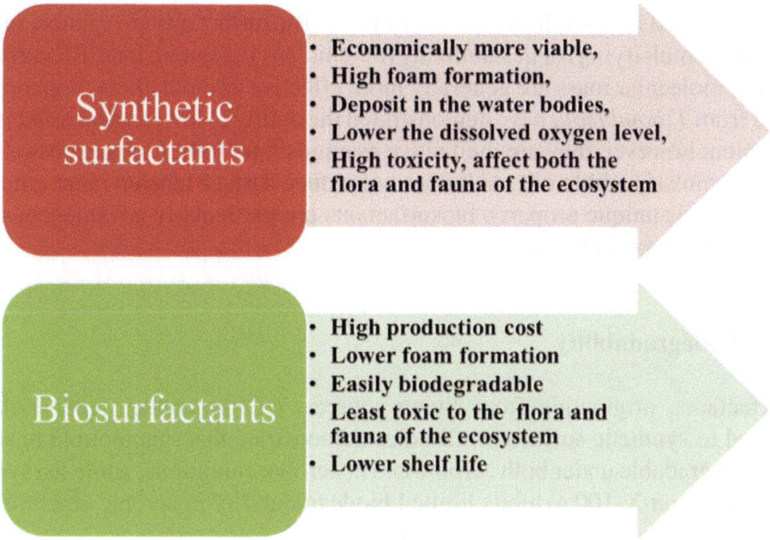

Fig. 3 Synthetic surfactants versus biosurfactants

while those with a high HLB perform the inverse function. Research indicates that the surface activity of biosurfactants is identical to that of synthetic surfactants. Biosurfactants can reduce the surface tension of water to 29.0 mN/m at the CMC, in comparison with Pluronic® F-68 at 42.8 mN/m and sodium dodecyl sulfate (SDS) at 28.6 mN/m [66] (Fig. 3).

The key unique characteristics of biosurfactants are.

2.1.1 Surface Activity

Surface tension of water is ~ 72 mN/m (at 20 °C). It decreases on the addition of a minor volume of surfactant. The interfacial tension of water against n-hexadecane can be reduced from 40 to 1 mN/m, and the surface tension of water can be reduced from 72 to 35 mN/m by an efficient biosurfactant [67]. The surface tension of water can be reduced to 25 mN/m, and the interfacial tension between water and n-hexadecane can be reduced to less than 1 mN/m by surfactin produced by B. subtilis. In comparison with synthetic surfactants, biosurfactants demonstrate superior efficiency and efficacy, as evidenced by their 10–40 times lower CMC values.

2.1.2 Emulsification and De-emulsification

Emulsification occurs when a liquid disperses into another as tiny droplets, combining two immiscible liquids. Using biosurfactants, more stable emulsions can be produced with a prolonged lifespan from months to years. Biosurfactants are capable of either stabilizing (emulsifying) or destabilizing (de-emulsifying) emulsions. Biosurfactants with high molecular mass are generally more effective as emulsifiers. Sophorolipids derived from *T. bombicola* have demonstrated the ability to lower surface and interfacial tension; however, they are ineffective as emulsifiers [68]. Liposan possesses the ability to emulsify edible oils, as it does not reduce surface tension more efficiently. Because of this unique property, biosurfactants are particularly advantageous to the cosmetics and food industries.

2.1.3 Biodegradability

Biosurfactants, originating from living sources, exhibit greater biodegradability compared to synthetic surfactants. Studies demonstrate that rhamnolipid biosurfactants are degradable under both aerobic and anaerobic conditions, while the synthetic surfactant Triton X-100 exhibits limited biodegradability in aerobic conditions and is non-biodegradable in anaerobic situations [69].

2.1.4 Temperature, pH, and Ionic Strength Tolerance

Biosurfactants demonstrate stability across various temperatures and pH ranges. Lichenysin, sourced from B. licheniformis JF-2, demonstrates stability at temperatures reaching 50 °C and within a pH range of 4.5–9.0. It has commendable stability with elevated doses of sodium chloride (50 g/L) and calcium (25 g/L), respectively [70].

2.2 Types of Biosurfactants

Chemically produced surfactants possess polar head groups that can be categorized into four types: cationic, anionic, zwitterionic, and non-ionic. Biosurfactants possess a unique structure comprising both hydrophilic and hydrophobic components. The hydrophilic component, which influences solubility in aqueous environments, may consist of organic acids, alcohols, peptides, or sugars. The lipophilic part, which helps it interact with oily surfaces and liquids, usually consists of fatty acids or long chains of hydrocarbons. These two parts are linked together by different kinds of chemical bonds: ester bonds, which include lactones, with organic or inorganic acids; amide bonds, which can be single amide or peptide connections; or glycosidic bonds that include sugar units.

Based on their molecular mass, biosurfactants are classified into two primary categories: those of Rosenberg and those of Ron [71].

i. **Low-Molecular-Weight Surface-Active Agents**: Biosurfactants typically possess a molecular weight of less than one kilodalton (kDa) and are noted for their potent capacity to reduce the surface tension of immiscible fluids, such as oil and water. This property enhances the solubility and dispersion of hydrophobic compounds, facilitating their microbial degradation. Major classes in this category include glycolipids, lipopeptides, and phospholipids.
ii. **High-Molecular-Weight Surface-Active Agents**: These biosurfactants usually have molecular weights over one kDa and are very good at keeping oil-in-water mixtures stable. They play a crucial role in stabilizing complex systems through steric or electrostatic mechanisms. Major classes in this category include polymeric biosurfactants and particulate biosurfactants.

2.3 *Key Properties*

2.3.1 Biodegradability

As they are of biological origin, they exhibit easy chemical structure and low toxicity. They do not survive in nature for an extended period and are readily decomposed by microorganisms.

2.3.2 Biocompatibility and Digestibility

Their biological nature ensures inherent compatibility, enabling their ongoing use in cosmetics, medicines, agriculture, and as functional food additives.

2.3.3 Low Toxicity

Biosurfactants have reduced toxicity in comparison with chemically generated surfactants. In naphthalene solubilization experiments, glycolipids from Rhodococcus species 413A showed a 50% reduction in toxicity relative to Tween 80 [72].

2.3.4 Surface and Interface Activity

According to Mulligan, an efficient surfactant may lower the interfacial tension of water and hexadecane from 40 to 1 mN/m and the surface tension of water from 72 to 35 mN/m. Surfactins can lower the interfacial tension between water and hexadecane to less than 1 mN/m and the surface tension of water to 25 mN/m [73].

2.3.5 Availability of Raw Materials

A broad array of raw materials is accessible to produce biosurfactants at a comparatively inexpensive cost—carbon sources such as hydrocarbons, carbohydrates, and lipids may be utilized separately or in combination [74].

2.3.6 Physical Factors

Numerous biosurfactants can endure harsh environments such as temperature, pH, and salt.

2.3.7 Specificity

Complex organic biosurfactants with functional groups frequently have unique actions. This feature is essential for pollutant detoxification, cosmetic development, industrial emulsion de-emulsification, and medicinal and food applications. Many biosurfactants work in severe temperatures, pH, and salinity.

2.3.8 Environmental Control

Biosurfactants have proven to be effective in managing oil spills, handling industrial emulsions, detoxifying and biodegrading industrial effluents, and bioremediation of contaminated soil.

3 Mechanisms of Biosurfactant Production

3.1 Microbial Sources

Many bacteria, yeasts, and fungi create surface-active substances during growth [75]. For instance,

Biosurfactants Producing Bacteria: Rhanolipids from *Pseudomonas aeruginosa* and surfactin from *Bacillus Subtilis,*

Biosurfactants Producing Yeasts: Sophorolipids from *Torulopsis bombicola* and peptidolipid from *Candida petrophilum,*

Biosurfactants Producing Fungi: Sophorolipids from *Candida batistae* and *Candida bombocola.*

Biosurfactants are biologically produced from diverse substrates. Downstream processing fees may constitute up to 60% of overall production costs, rendering the

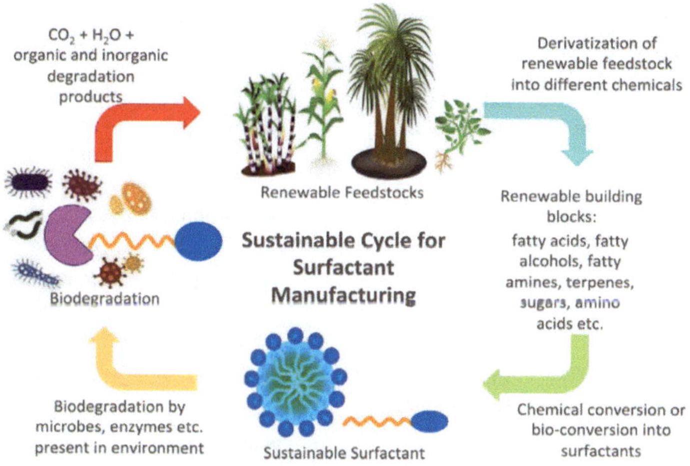

Fig. 4 Sustainable ideal cycle for producing biosurfactants [77]

commercial manufacture of biosurfactants considerably more expensive. To mitigate these high costs, there is a growing focus on utilizing low-cost and waste substrates for production. Each year, a considerable amount of both hazardous and non-hazardous waste is generated globally, necessitating effective management and utilization strategies. Substrates recognized for biosurfactant production encompass residues from tropical crops, outputs of the fruit processing sector, trash from oil processing mills, and remnants from coffee processing [76]. These are explained in detail (Fig. 4 and Table 1).

3.1.1 Oil Wastes

Biosurfactant manufacture can benefit significantly from waste oils from the food industry and vegetable oil refineries. The bacteria Thermus thermophilus HB8 can be used to manufacture rhamnolipids from oleic acid and sunflower seed oil [78]. The total cost of producing biosurfactants might be lowered by using these cheaper oils and oil wastes in industrial manufacturing media.

3.1.2 Dairy Whey

The amount of whey produced by one kilogram of paneer is around six liters. A stable and cost-effective substrate for biosurfactant fermentation is whey, a by-product of the dairy industry. A two-stage cultivation process utilizing dairy waste as a substrate for the yeast *Cryptococcus curvatus* ATCC 20,509 successfully produced high concentrations of sophorolipids [79].

Table 1 Significant types of biosurfactants and their respective microbial producers

Biosurfactant			Microorganisms
Group	Subgroup	Class	
Low-molecular-weight surface-active agents	Glycolipids	Rhamnolipids	*Pseudomonas aeruginosa*, *Pseudomonas sp.*
		Sophorolipids	*T. bombicola*, *T. apicola*
		Trehalolipids	*R. erythropolis*, *Microbacterium sp.*
		Cellobiolipids	*U. Zeae*, *U. maydis*
	Lipopeptides and Lipoprotiens	Surfactin	*B. subtilis*
		Viscisin	*P. fluorescense*
		Peptide lipid	*B. licheniformis*
	Fatty acids, neutral lipids, and phospholipids	Fatty acids	*C. lepus*
		Neutral lipids	*N. erythropolis*
		Phospholipids	*Thiobacillus thiooxidans*
High-molecular-weight surface-active agents	Polymeric surfactants	Emulsion	*A. calcoaceticus*
		Biodisperson	*A. calcoaceticus*
		Alasan	*A. radioresistens*
		Liposan	*C. lipolytica*
	Particulate surfactants	Vescicle and Fimbriae	*A. calcoaceticus*
		Whole cells	Variety of bacteria

3.1.3 Starchy Waste

For the production of biosurfactants, starchy waste materials serve as both suitable and inexpensive raw materials. The potato processing sector is one of the primary producers of these starchy by-products. These waste products are rich in various components, including sugars, starch, nitrogen, sulfur, inorganic minerals, trace elements, and vitamins.

3.1.4 Molasses

The sugar industry produces molasses as a by-product for processing sugarcane or sugar beets. Molasses is an excellent source of recyclable carbon. Patel and Desai employed P. aeruginosa GS3 for the manufacture of rhamnolipid biosurfactant, utilizing molasses and corn-steep liquor as the principal carbon and nitrogen sources [80].

4 Applications for Biosurfactants

4.1 Agriculture

The widespread application of chemicals and pesticides in agriculture to manage arthropods frequently leads to detrimental consequences for both the environment and human health. In contrast, biosurfactants present a safer alternative due to their natural antimicrobial and insecticidal properties, which do not pose significant environmental or health risks. This quality makes them highly promising for various agricultural applications, such as enhancing soil quality, promoting plant growth, and facilitating bioremediation. Several factors, including soil pH, temperature, and heavy metal concentration, affect microbial surfactant-mediated bioremediation under natural conditions. Therefore, optimizing processes is crucial for maximizing the effectiveness of biosurfactants to remove heavy metals from contaminated soils. Microorganisms play an essential role by either releasing chemicals that modify metal mobility or indirectly adjusting soil pH [81]. For example, research has demonstrated that a mixture of rhamnolipids and biosurfactants called JBR-425 can help remove heavy metals like Zn, Cu, Pb, and Cd from soil while simultaneously encouraging the growth of earthworms like Eisenia fetida and Lumbricus terrestris [82]. The finding that lipopeptide biosurfactants made by different bacterial strains can kill fruit flies (Drosophila melanogaster) further supports their use as a green biopesticide [67].

4.2 Cosmetics and Personal Care

A compound's ability to pass through skin membranes depends on its lipophilicity and surface activity. Biosurfactants possess unique properties that enable them to easily penetrate skin cell membranes, regulate protein and skin barrier functions, and promote beneficial effects related to hair repair and skin protection [83–85]. Several in vitro studies have demonstrated compatibility between human skin and rhamnolipids, sophorolipids, mannosylerythritol lipids (MELs), and surfactin [84, 86]. The skin-friendly properties of biosurfactants make them ideal for the healthcare and cosmetic industries. As skin moisturizers, sophorolipids can be esterified or blended with ethylene oxide. Tokyo-based Kao Corporation uses them in its personal care products. An example of a biosurfactant's application includes foaming agents, emulsifiers, solubilizers, wetting agents, cleansers, antimicrobial agents, enzyme mediators, insect repellents, acne pads, antacids, bath products, anti-dandruff products, contact lens solutions, baby products, mascara, lipsticks, and toothpaste [87]. Numerous commercial cosmetic and skincare products presently incorporate microbial biosurfactants, such as: RelipidiumTM (body and face moisturizer, manufactured by BASF, Monheim, Germany), SopholianceTM S (deodorant, facial cleanser, and shower gel, produced by Givaudan Active Beauty, Paris, France), and Kanebo skincare (moisturizer, cleanser, and UV filter, created by Kanebo Cosmetics, Tokyo,

Japan), among others [88]. Additionally, the action of Cutibacterium acnes (formerly known as Propionibacterium acnes) on microbial biosurfactants through triglyceride hydrolysis aids in maintaining the skin's acidic pH. This process not only promotes the adhesion of beneficial resident flora but also inhibits the growth of pathogenic microbes, contributing to a balanced and healthy skin microbiome [88].

4.3 Pharmaceuticals

Biological surfactants have significant applications in biomedical fields. Numerous biosurfactants exhibit substantial antibacterial, antifungal, and antiviral properties. Surfactants serve as anti-adhesive agents against microorganisms, rendering them advantageous for the treatment of multiple disorders. They may function as medicinal and probiotic agents. Rhamnolipids from *P. aeruginosa*, mannosylerythritol lipids from *Candida antarctica*, and lipopeptides from *Bacillus subtilis* and *Bacillus licheniformis* demonstrate antibacterial properties [89]. A rhamnolipid mixture from *P. aeruginosa* AT10 inhibits various bacteria, including *E. coli, S. marcescens, M. luteus, A. faecalis, M. phlei, and S. epidermidis.* It also shows vigorous antifungal activity against *Aspergillus niger, Penicillium chrysogenum, Chaetomium globosum, Aureobasidium pullulans*, and plant pathogens like *Rhizoctonia solani* and *Botrytis cinerea* [90]. Sophorolipids obtained from *C. bombicola* demonstrate virucidal efficacy against human semen [74]. Further benefits and uses of bio-surfactants in medicine encompass gene transport, immunomodulation, wound healing, insecticidal capabilities, and anticancer activity.

4.4 Environmental Remediation

4.4.1 Microbial Enhanced Oil Recovery (MEOR)

MEOR is a potent technique that is used to recover oil that is still present in reservoirs that have a low permeability or high viscosity. The residual oil in these reservoirs is typically located in inaccessible places, and it gets stuck in the pores because of capillary pressure [91]. Biosurfactants are an essential component in the process of reducing the interfacial tension that exists between oil and water, as well as between oil and rock. The decrease in interfacial tension diminishes the capillary forces that impede oil flow. Furthermore, biosurfactants can produce emulsions by tightly binding to the oil–water interface. This procedure stabilizes the desorbed oil in water and aids in the extraction of oil alongside the injected water [92]. *Bacillus subtilis* at 45 °C has excellent sand-pack oil-recovering ability [93].

4.4.2 Soil Washing Technology and Bioremediation of Crude Oil-Contaminated Environments

Extremely hydrophobic pollutants can adhere strongly to soil, rendering them inaccessible to biodegradation. Surfactants may facilitate the desorption of contaminants from soil. Rhamnolipids are efficient biosurfactants for the extraction of polycyclic aromatic hydrocarbons (PAHs) from soil, with a removal effectiveness of 60–80%.

Garcia-Junco et al. [94] crude oils exhibit minimal water solubility and significant adsorption to the soil matrix. Oil-contaminated soil poses challenges for bioremediation, as surplus oil creates droplets or films on soil particles. It constitutes a formidable impediment to microbial breakdown. Biological surfactants diminish the tension at the hydrocarbon–water interface, facilitating the mobility of hydrocarbons, enhancing their bio-availability, and subsequently promoting degradation. Consequently, biosurfactants are highly advantageous for applications within the oil industry, and their significance is evident in the market, where the vast majority of produced biosurfactants are intended for petroleum-related uses. BS29 bio-emulsions derived from Gordonia sp. are effective washing agents for the bioremediation of hydrocarbon-contaminated soil [95].

4.4.3 Metals Remediation

Soil contamination with heavy metals poses significant risks to humans and other species within this ecosystem. The existence of even minimal levels of heavy metals in the soil presents a considerable risk owing to their highly poisonous characteristics. Numerous biosurfactants possess the capability to remediate metals. Rhamnolipids possess the ability to eliminate heavy metals [23].

5 Challenges and Limitations

Biosurfactants present significant potential as eco-friendly alternatives to synthetic surfactants; however, their widespread use is hindered by several challenges. These challenges include (a) high costs for making and extracting them, often because expensive materials are used; (b) low amounts that can be extracted, needing large quantities or long processing times; (c) impurities and similar substances that make purification more complicated; (d) too much foam during production, which can block equipment and reduce efficiency; and (e) not enough information about how they interact with the human gut, which raises safety concerns for use in food and healthcare [96–98]. In combination, these issues pose significant challenges to the widespread commercial use of biosurfactants.

5.1 Economic Barriers

Biosurfactants are not widely utilized in industry due to their high production costs, despite being more environmentally friendly and biodegrading faster than conventional surfactants. Currently, surfactin and rhamnolipids are the only biosurfactants available on the market; the United States Environmental Protection Agency (US EPA) has authorized rhamnolipids for use in food, cosmetics, and pharmaceuticals [99]. The primary challenge is the high production cost, which can account for up to 80% of manufacturing expenses. Complex fermentation processes, costly downstream processing, and expensive raw materials contribute to this expense. Additionally, sterilization and yield optimization contribute to increasing operational expenses [100]. While the current price of synthetic surfactants is approximately USD 2/kg, [101] the cost of biosurfactants varies significantly, ranging from USD 1/kg to over USD 10,000,000/kg, depending on factors such as purity and production methods [59]. For biosurfactants to be commercially competitive, their prices need to be at least as low as those of synthetic surfactants. Rhamnolipids are priced between USD 1.5/g and USD 1500/g, while lipopeptides used in cosmetics and pharmaceuticals range from USD 20 to USD 130/mg [102]. Other biosurfactants, such as Fengycin (> 90%) and Iturin A (> 95%), have total production costs of USD 114/mg and USD 115/mg, respectively. In contrast, synthetic surfactants, such as sodium dodecyl sulfate, are priced between USD 1.46 and USD 1.59 per gram. The significant cost gap, along with inadequate infrastructure and stringent regulatory restrictions, hampers the commercialization of biosurfactants in competitive markets.

5.2 Scalability Issues

Biosurfactant production primarily occurs through batch and fed-batch fermentations, which often yield low results, mainly when cost-effective substrates are used. Although some studies have shown moderate success, such as surfactin production from fish waste in a 100 L pilot reactor yielding 0.274 g/L, the productivity remains insufficient for economically viable industrial-scale applications [103]. Similar efforts with various microorganisms, including *Candida tropicalis* [104], *Aureobasidium pullulans* [105], and *Gordonia amarae* [106], have reported biosurfactant concentrations below 10 g/L, limiting their commercial feasibility.

Another major problem with biosurfactant fermentation is foaming, which is brought on by surface-active substances in the broth. The process is interrupted when there is an excess of foam, which entraps cells, lowers the effective microbial concentration, increases the risk of contamination, and makes process management more difficult [107]. Chemical antifoams can minimize foam, [108] but they are not ideal for use in the creation of biosurfactants since there is a possibility that they will contaminate the final product.

Downstream processing, which contributes significantly to production costs, is another key challenge. Even within the same microbial strain, biosurfactants present a wide range of molecular compositions, which necessitate the implementation of complex and frequently economically unfeasible purification procedures. High purity is crucial for applications in food, cosmetics, and pharmaceuticals; however, it is less critical in industrial sectors such as oil recovery, where the purity requirements are comparatively lower. Biosurfactants are not yet commercially viable on a broad scale because of the high expense and technical complexity of the numerous purification procedures required from diluted fermentation broths [109].

6 Conclusion

Biosurfactants are appealing alternatives to conventional synthetic surfactants because they are biodegradable, have low toxicity, and are derived from renewable sources. They offer significant environmental benefits, making them suitable for sustainable applications in various sectors, including agriculture, cosmetics, food, and bioremediation. Their unique properties, such as emulsification and surface activity, often resemble those of synthetic surfactants. Biosurfactants can be made from industrial waste and renewable materials, which supports a circular economy and sustainable development goals. Despite numerous benefits, biosurfactants have not yet been widely used in industry because of several obstacles. These include low production yields, quality issues, complicated downstream processing, and higher manufacturing costs than synthetic counterparts. Biosurfactant composition can vary based on microbial strain and substrate, which adds another layer of complexity to the commercialization process in areas like pharmaceuticals and cosmetics, where high-purity products are required. Their ability to compete in the market is severely limited due to a lack of standardized regulatory frameworks and an inadequate industrial infrastructure for producing biosurfactants on a considerable scale. As a result, despite the fact that biosurfactants have a great deal of promise as sustainable and environmentally friendly alternatives, their extensive acceptance in industrial applications will necessitate a coordinated effort to make scientific advances, implement methods to reduce costs, standardize regulations, and establish infrastructure.

References

1. Zhang X-J, Zhou Z-H, Han L, Zhang Y-Q, Zhang Q, Ma D-S, Ma W-J, Zhang L, Zhang L (2022) Mechanism responsible for the reduction of interfacial tension by extended surfactants. Colloids Surf, A: Physchemical Eng Asp 634:128013
2. Khimani M, Vora S (2011) Effect of inorganic additives on a conventional anionic–nonionic mixed surfactants system in aqueous solution. J Surfactants Deterg 14(4):545
3. Howard SC, Craig VS (2009) Very slow surfactant adsorption at the solid–liquid interface is due to long lived surface aggregates. Soft Matter 5(16):3061

4. Krylov DI, Arslanov VV (2025) Static and dynamic surface properties of mixed non-ionic surfactant/cholesterol monolayers as niosome shell models. Colloids Surf, A: Physchemical Eng Asp 717:136771
5. Tamjidi S, Moghadas BK, Esmaeili H, Khoo FS, Gholami G, Ghasemi M (2021) Improving the surface properties of adsorbents by surfactants and their role in the removal of toxic metals from wastewater: a review study. Process Saf Environ Prot 148:775
6. Costa MF, de Oliveira AM, de Oliveira Junior EN (2020) Biodegradation of linear alkylbenzene sulfonate (LAS) by Penicillium chrysogenum. Bioresour Technol Rep 9:100363
7. Rieger MM (2020) Surfactants, In: Pharmaceutical dosage forms. CRC Press, p 211
8. Ahmadi M, Aliabadian E, Liu B, Lei X, Khalilpoorkordi P, Hou Q, Wang Y, Chen Z (2022) Comprehensive review of the interfacial behavior of water/oil/surfactant systems using dissipative particle dynamics simulation. Adv Coll Interface Sci 309:102774
9. Arnold WA, Blum A, Branyan J, Bruton TA, Carignan CC, Cortopassi G, Datta S, DeWitt J, Doherty A-C, Halden RU (2023) Quaternary ammonium compounds: a chemical class of emerging concern. Environ Sci Technol 57(20):7645
10. Cortés H, Hernández-Parra H, Bernal-Chávez SA, Del Prado-Audelo ML, Caballero-Florán IH, Borbolla-Jiménez FV, González-Torres M, Magaña JJ, Leyva-Gómez G (2021) Non-ionic surfactants for stabilization of polymeric nanoparticles for biomedical uses. Materials 14(12):3197
11. Hassabo AG, Hegazy BM, Sediek A, Saad F, Ghazal H (2023) The use of non-ionic surfactants in the textiles industry. J Text, Color Polym Sci 20(2):217
12. Mansour AS, Abo-Aly M, Rizk S, Abd El Rahman AM, Ghaly N, Ragab A, Alsabagh A (2025) Synthesis of new zwitterionic surfactants and investigation of their surface active and thermodynamic properties. Sci Rep 15(1):1
13. Gerola AP, Costa PF, Nome F, Quina F (2017) Micellization and adsorption of zwitterionic surfactants at the air/water interface. Curr Opin Colloid Interface Sci 32:48
14. Sasi S, Rayaroth MP, Aravindakumar CT, Aravind UK (2021) Alcohol ethoxysulfates (AES) in environmental matrices. Environ Sci Pollut Res 28(26):34167
15. Baena-Nogueras RM, González-Mazo E, Lara-Martín PA (2013) Determination and occurrence of secondary alkane sulfonates (SAS) in aquatic environments. Environ Pollut 176:151
16. Gannon SA, Fasano WJ, Mawn MP, Nabb DL, Buck RC, Buxton LW, Jepson GW, Frame SR (2016) Absorption, distribution, metabolism, excretion, and kinetics of 2, 3, 3, 3-tetrafluoro-2-(heptafluoropropoxy) propanoic acid ammonium salt following a single dose in rat, mouse, and cynomolgus monkey. Toxicology 340:1
17. Chelcea IC, Ahrens L, Örn S, Mucs D, Andersson PL (2020) Investigating the OECD database of per-and polyfluoroalkyl substances–chemical variation and applicability of current fate models. Environ Chem 17(7):498
18. Wang Z, Cousins IT, Scheringer M, Hungerbuehler K (2015) Hazard assessment of fluorinated alternatives to long-chain perfluoroalkyl acids (PFAAs) and their precursors: status quo, ongoing challenges and possible solutions. Environ Int 75:172
19. Johnson P, Trybala A, Starov V, Pinfield VJ (2021) Effect of synthetic surfactants on the environment and the potential for substitution by biosurfactants. Adv Coll Interface Sci 288:102340
20. Ananthapadmanabhan KP, Subramanyan K, Rattinger GB (2002) Moisturizing cleansers. In: James AVR, Leyden J (eds) Skin moisturization, CRC Press, p 405
21. Maity B, Chatterjee A, Ahmed SA, Seth D (2015) Interaction of the nonsteroidal anti-inflammatory drug indomethacin with micelles and its release. J Phys Chem B 119(9):3776
22. Lu X, Li M, Arce FA, Ling J, Setiawan N, Wang Y, Shi X, Campbell HR, Nethercott MJ, Xu W (2021) Mechanistic investigation of drug supersaturation in the presence of polysorbates as solubilizing additives by solution nuclear magnetic resonance spectroscopy. Mol Pharm 18(12):4310
23. Guo W, Zhang X, Wan L, Wang Z, Han M, Yan Z, Li J, Deng R, Li S, Mao Y (2024) β-glucan-modified nanoparticles with different particle sizes exhibit different lymphatic targeting efficiencies and adjuvant effects. J Pharm Anal 14(12):100953

24. Kang Y, Guan Y, Li S (2024) Innovative hydrogel solutions for articular cartilage regeneration: a comprehensive review. Int J Surg 110(12):7984
25. Guan Y, Zhang W, Mao Y, Li S (2024) Nanoparticles and bone microenvironment: a comprehensive review for malignant bone tumor diagnosis and treatment. Mol Cancer 23(1):246
26. Ellyett K, Broadbent R, Fawcett E, Campbell A (1996) Surfactant aerosol treatment of respiratory distress syndrome in the spontaneously breathing premature rabbit. Pediatr Res 39(6):953
27. Arrøe M, Pedersen-Bjergaard L, Albertsen P, Greisen G, Lundstrøm K, Struck J (1998) Inhalation of aerosolized surfactant (Exosurf®) to neonates treated with nasal continuous positive airway pressure. Prenat Neonatal Med 3(3):346
28. Berggren E, Liljedahl M, Winbladh B, Andreasson B, Curstedt T, Robertson B, Schollin J (2000) Pilot study of nebulized surfactant therapy for neonatal respiratory distress syndrome. Acta Paediatr 89(4):460
29. Finer NN, Merritt TA, Bernstein G, Job L, Mazela J, Segal R (2010) An open label, pilot study of Aerosurf® combined with nCPAP to prevent RDS in preterm neonates. J Aerosol Med Pulm Drug Deliv 23(5):303
30. Silva TJ, Okuro PK, da Silva MG, Ribeiro APB, da Cunha RL (2023) Lecithin. In: Fat mimetics for food applications, p 57
31. Gutiérrez-Méndez N, Chavez-Garay DR, Leal-Ramos MY (2022) Lecithins: a comprehensive review of their properties and their use in formulating microemulsions. J Food Biochem 46(7):e14157
32. Svanberg L, Ahrné L, Lorén N, Windhab E (2011) Effect of sugar, cocoa particles and lecithin on cocoa butter crystallisation in seeded and non-seeded chocolate model systems. J Food Eng 104(1):70
33. Harasym J, Banaś K (2024) Lecithin's roles in oleogelation. Gels 10(3): 169
34. Castro MJ, Ojeda C, Cirelli AF (2013) Surfactants in agriculture. In: Green materials for energy, products and depollution, 287
35. Sivaramakrishnan C (2013) The use of surfactants in the finishing of technical textiles. In: Advances in the dyeing and finishing of technical textiles. Elsevier, 199
36. Joshi SJ, Geetha S, Desai AJ (2015) Characterization and application of biosurfactant produced by Bacillus licheniformis R2. Appl Biochem Biotechnol 177:346
37. Pogaku R, Mohd Fuat NH, Sakar S, Cha ZW, Musa N, Awang Tajudin DNA, Morris LO (2018) Polymer flooding and its combinations with other chemical injection methods in enhanced oil recovery. Polym Bull 75: 1753
38. Sheng JJ (2015) Status of surfactant EOR technology. Petroleum 1(2): 97
39. Zulkifli NN, Mahmood SM, Akbari S, Manap AAA, Kechut NI, Elrais KA (2020) Evaluation of new surfactants for enhanced oil recovery applications in high-temperature reservoirs. J Pet Explor Prod Technol 10:283
40. Almahfood M, Bai B (2018) The synergistic effects of nanoparticle-surfactant nanofluids in EOR applications. J Petrol Sci Eng 171:196
41. Guo F, He J, Johnson PA, Aryana SA (2017) Stabilization of CO_2 foam using by-product fly ash and recyclable iron oxide nanoparticles to improve carbon utilization in EOR processes. Sustain Energy Fuels 1(4):814
42. Massarweh O, Abushaikha AS (2020) The use of surfactants in enhanced oil recovery: a review of recent advances. Energy Rep 6:3150
43. Jardak K, Drogui P, Daghrir R (2016) Surfactants in aquatic and terrestrial environment: occurrence, behavior, and treatment processes. Environ Sci Pollut Res 23:3195
44. Kulkarni D, Jaspal D (2023) Surfactants in waste water: development, current status and associated challenges. Mater Today: Proc 1
45. Badmus SO, Amusa HK, Oyehan TA, Saleh TA (2021) Environmental risks and toxicity of surfactants: overview of analysis, assessment, and remediation techniques. Environ Sci Pollut Res 28(44):62085

46. Arora J, Ranjan A, Chauhan A, Biswas R, Rajput VD, Sushkova S, Mandzhieva S, Minkina T, Jindal T (2022) Surfactant pollution, an emerging threat to ecosystem: approaches for effective bacterial degradation. J Appl Microbiol 133(3):1229
47. Rebello S, Asok AK, Mundayoor S, Jisha M (2014) Surfactants: toxicity, remediation and green surfactants. Environ Chem Lett 12:275
48. Christoffersen K, Hansen BW, Johansson LS, Krog E (2003) Influence of LAS on marine calanoid copepod population dynamics and potential reproduction. Aquat Toxicol 63(4):405
49. Skei J, Larsson P, Rosenberg R, Jonsson P, Olsson M, Broman D (2000) Eutrophication and contaminants in aquatic ecosystems. AMBIO: J Hum Environ 29(4): 184
50. Bolan S, Padhye LP, Mulligan CN, Alonso ER, Saint-Fort R, Jasemizad T, Wang C, Zhang T, Rinklebe J, Wang H (2023) Surfactant-enhanced mobilization of persistent organic pollutants: potential for soil and sediment remediation and unintended consequences. J Hazard Mater 443:130189
51. Lémery E, Briançon S, Chevalier Y, Bordes C, Oddos T, Gohier A, Bolzinger M-A (2015) Skin toxicity of surfactants: structure/toxicity relationships. Colloids Surf, A 469:166
52. Toma C, Cappelli CI, Manganaro A, Lombardo A, Arning J, Benfenati E (2021) New models to predict the acute and chronic toxicities of representative species of the main trophic levels of aquatic environments. Molecules 26(22):6983
53. Stoica C, Stanescu E (2013) Surfactants (Biodegradation, toxicity and risk assessment). In: Biodegradation: life of science, p 83
54. Zembrzuska J, Budnik I, Lukaszewski Z (2016) Monitoring of selected non-ionic surfactants in river water by liquid chromatography–tandem mass spectrometry. J Environ Manage 169:247
55. Pradhan A, Bhattacharyya A (2017) Quest for an eco-friendly alternative surfactant: surface and foam characteristics of natural surfactants. J Clean Prod 150:127
56. Zargar AN, Srivastava P (2024) Biosurfactants: sustainable alternatives to chemical surfactants. In: Industrial applications of biosurfactants and microorganisms, Elsevier. p 425
57. Xu Q, Nakajima M, Liu Z, Shiina T (2011) Biosurfactants for microbubble preparation and application. Int J Mol Sci 12(1):462
58. Varvaresou A, Iakovou K (2015) Biosurfactants in cosmetics and biopharmaceuticals. Lett Appl Microbiol 61(3):214
59. Zargar AN, Kumar M, Srivastava P (2023) Biosurfactants: challenges and future outlooks. In: Advancements in biosurfactants research, Springer, p 551
60. Perfumo A, Banat IM, Marchant R (2018) Going green and cold: biosurfactants from low-temperature environments to biotechnology applications. Trends Biotechnol 36(3):277
61. Singh A, Van Hamme JD, Ward OP (2007) Surfactants in microbiology and biotechnology: Part 2 application aspects. Biotechnol Adv 25(1): 99
62. Mukherjee S, Das P, Sen R (2006) Towards commercial production of microbial surfactants. Trends Biotechnol 24(11):509
63. Mulligan C, Gibbs B (2004) Types, production and applications of biosurfactants. Proc Indian Natl Sci Academy Part B, Rev Tracts-Biol Sci 70(1): 31
64. Whang L-M, Liu P-WG, Ma C-C, Cheng S-S (2008) Application of biosurfactants, rhamnolipid, and surfactin, for enhanced biodegradation of diesel-contaminated water and soil. J Hazard Mater 151(1):155
65. Desai JD, Banat IM (1997) Microbial production of surfactants and their commercial potential. Microbiol Mol Biol Rev 61(1):47
66. Pornsunthorntawee O, Wongpanit P, Chavadej S, Abe M, Rujiravanit R (2008) Structural and physicochemical characterization of crude biosurfactant produced by Pseudomonas aeruginosa SP4 isolated from petroleum-contaminated soil. Biores Technol 99(6):1589
67. Mulligan CN (2005) Environmental applications for biosurfactants. Environ Pollut 133(2):183
68. Cavalero DA, Cooper DG (2003) The effect of medium composition on the structure and physical state of sophorolipids produced by Candida bombicola ATCC 22214. J Biotechnol 103(1):31
69. Mohan PK, Nakhla G, Yanful EK (2006) Biokinetics of biodegradation of surfactants under aerobic, anoxic and anaerobic conditions. Water Res 40(3):533

70. McInerney MJ, Javaheri M, Nagle DP Jr (1990) Properties of the biosurfactant produced by Bacillus licheniformis strain JF-2. J Ind Microbiol Biotechnol 5(2–3):95
71. Rosenberg E, Ron EZ (1999) High-and low-molecular-mass microbial surfactants. Appl Microbiol Biotechnol 52:154
72. Kanga SA, Bonner JS, Page CA, Mills MA, Autenrieth RL (1997) Solubilization of naphthalene and methyl-substituted naphthalenes from crude oil using biosurfactants. Environ Sci Technol 31(2):556
73. Krishnaswamy K (2008) Traditional Indian spices and their health significance. Asia Pac J Clin Nutr 17(S1):265
74. Kosaric N (2001) Biosurfactants and their application for soil bioremediation. Food Technol Biotechnol 39(4):295
75. Bhardwaj G, Cameotra SS, Chopra HK (2013) Biosurfactants from fungi: a review. J Pet Environ Biotechnol 4(06):1
76. De Lima C, Ribeiro E, Servulo E, Resende M, Cardoso V (2009) Biosurfactant production by Pseudomonas aeruginosa grown in residual soybean oil. Appl Biochem Biotechnol 152:156
77. Bhadani A, Kafle A, Ogura T, Akamatsu M, Sakai K, Sakai H, Abe M (2020) Current perspective of sustainable surfactants based on renewable building blocks. Curr Opin Colloid Interface Sci 45:124
78. Pantazaki AA, Dimopoulou MI, Simou OM, Pritsa AA (2010) Sunflower seed oil and oleic acid utilization for the production of rhamnolipids by Thermus thermophilus HB8. Appl Microbiol Biotechnol 88:939
79. Daniel H-J, Reuss M, Syldatk C (1998) Production of sophorolipids in high concentration from deproteinized whey and rapeseed oil in a two stage fed batch process using Candida bombicola ATCC 22214 and Cryptococcus curvatus ATCC 20509. Biotech Lett 20:1153
80. Patel R, Desai A (1997) Biosurfactant production by Pseudomonas aeruginosaGS3 from molasses. Lett Appl Microbiol 25(2):91
81. Datta D, Ghosh S, Kumar S, Gangola S, Majumdar B, Saha R, Mazumdar SP, Singh SV, Kar G (2024) Microbial biosurfactants: multifarious applications in sustainable agriculture. Microbiol Res 279:127551
82. Slizovskiy IB, Kelsey JW, Hatzinger PB (2011) Surfactant-facilitated remediation of metal-contaminated soils: efficacy and toxicological consequences to earthworms. Environ Toxicol Chem 30(1):112
83. Heinrich K, Heinrich U, Tronnier H (2014) Influence of different cosmetic formulations on the human skin barrier. Ski Pharmacol Physiol 27(3):141
84. Vecino X, Cruz J, Moldes A, Rodrigues L (2017) Biosurfactants in cosmetic formulations: trends and challenges. Crit Rev Biotechnol 37(7):911
85. Alonso C, Lucas R, Barba C, Marti M, Rubio L, Comelles F, Morales JC, Coderch L, Parra JL (2015) Skin delivery of antioxidant surfactants based on gallic acid and hydroxytyrosol. J Pharm Pharmacol 67(7):900
86. Sen S, Borah SN, Kandimalla R, Bora A, Deka S (2020) Sophorolipid biosurfactant can control cutaneous dermatophytosis caused by Trichophyton mentagrophytes. Front Microbiol 11:329
87. Gharaei-Fathabad E (2011) Biosurfactants in pharmaceutical industry: a mini-review. Am J Drug Discov Dev 1(1):58
88. Adu SA, Naughton PJ, Marchant R, Banat IM (2020) Microbial biosurfactants in cosmetic and personal skincare pharmaceutical formulations. Pharmaceutics 12(11):1099
89. Rodrigues L, Banat IM, Teixeira J, Oliveira R (2006) Biosurfactants: potential applications in medicine. J Antimicrob Chemother 57(4):609
90. Abalos A, Pinazo A, Infante MR, Casals M, Garcia F, Manresa A (2001) Physicochemical and antimicrobial properties of new rhamnolipids produced by Pseudomonas a eruginosa AT10 from soybean oil refinery wastes. Langmuir 17(5):1367
91. Sen R (2008) Biotechnology in petroleum recovery: the microbial EOR. Prog Energy Combust Sci 34(6):714
92. Suthar H, Hingurao K, Desai A, Nerurkar A (2008) Evaluation of bioemulsifier mediated microbial enhanced oil recovery using sand pack column. J Microbiol Methods 75(2):225

93. Makkar R, Cameotra SS (1997) Biosurfactant production by a thermophilic Bacillus subtilis strain. J Ind Microbiol Biotechnol 18(1):37
94. Garcia-Junco M, Gomez-Lahoz C, Niqui-Arroyo J-L, Ortega-Calvo J-J (2003) Biosurfactant- and biodegradation-enhanced partitioning of polycyclic aromatic hydrocarbons from nonaqueous-phase liquids. Environ Sci Technol 37(13):2988
95. Franzetti A, Bestetti G, Caredda P, La Colla P, Tamburini E (2008) Surface-active compounds and their role in the access to hydrocarbons in Gordonia strains. FEMS Microbiol Ecol 63(2):238
96. Drakontis CE, Amin S (2020) Biosurfactants: formulations, properties, and applications. Curr Opin Colloid Interface Sci 48:77
97. Naughton P, Marchant R, Naughton V, Banat IM (2019) Microbial biosurfactants: current trends and applications in agricultural and biomedical industries. J Appl Microbiol 127(1):12
98. Nurfarahin AH, Mohamed MS, Phang LY (2018) Culture medium development for microbial-derived surfactants production—an overview. Molecules 23(5):1049
99. Nitschke M, Costa S (2007) Biosurfactants in food industry. Trends Food Sci Technol 18(5):252
100. Petrides D (2000) Bioprocess design and economics. In: Bioseparations science and engineering, p 1
101. Santos DKF, Rufino RD, Luna JM, Santos VA, Sarubbo LA (2016) Biosurfactants: multifunctional biomolecules of the 21st century. Int J Mol Sci 17(3):401
102. Luna J, Rufino R, Campos G, Sarubbo L (2012) Properties of the biosurfactant produced by Candida sphaerica cultivated in low-cost substrates. Chem Eng 27:67
103. Hu J, Luo J, Zhu Z, Chen B, Ye X, Zhu P, Zhang B (2021) Multi-scale biosurfactant production by Bacillus subtilis using tuna fish waste as substrate. Catalysts 11(4):456
104. De Almeida DG, Soares Da Silva RDCF, Luna JM, Rufino RD, Santos VA, Banat IM, Sarubbo LA (2016) Biosurfactants: promising molecules for petroleum biotechnology advances. Front Microbiol 7:1718
105. Brumano LP, Antunes FAF, Souto SG, Dos Santos JC, Venus J, Schneider R, da Silva SS (2017) Biosurfactant production by Aureobasidium pullulans in stirred tank bioreactor: new approach to understand the influence of important variables in the process. Biores Technol 243:264
106. Pagilla K, Sood A, Kim H (2002) Gordonia (Nocardia) amarae foaming due to biosurfactant production. Water Sci Technol 46(1–2):519
107. Winterburn J, Martin P (2012) Foam mitigation and exploitation in biosurfactant production. Biotech Lett 34:187
108. Miller CA (2008) Antifoaming in aqueous foams. Curr Opin Colloid Interface Sci 13(3):177
109. Marchant R, Banat IM (2012) Microbial biosurfactants: challenges and opportunities for future exploitation. Trends Biotechnol 30(11):558

An Overview of the Role of Biosurfactant in Industrial Agriculture

Nitin Srivastava, Braj Kishore Rathour, and Abhishek Srivastava

Abstract Industrial agriculture is trying to fulfill the growing food demand of the fast-increasing population. In pursuit of this, synthetic chemicals are extensively used in terms of pesticides, insecticides, fertilizers, hydrocarbons which employ harsh synthetic surfactants having harmful effects on the microenvironment of soil and ecosystem due issues of non-biodegradability and toxicity associated with them. Biosurfactants of microbial, animal or plant origin are eco-friendly and may replace synthetic and chemical surfactants. Biosurfactants are significant in formation of biofilm, dealing with the plant pathogen, bioremediation, biodegradation, and enhancing the bioavailability of nutrients in the soil for the plants. At present, potential biosurfactants are basically surfactin, rhamnolipid and sophorolipids derived from *B. subtilis*, *P. aeruginosa*, *T. bombicola*, respectively. Biosurfactants are extremely helpful with bioremediation of heavy metals by enhancing their mobility and facilitating complex formation through emulsification and adsorption. Biosurfactants enhance the bioavailability of various elements and micronutrients to the soil due to formation of cation–anion complexes and solubilization facilitating their absorption by roots of the plants. Biosurfactants cause wetting and thus find promising applications as emulsifiers which are extremely useful in the pesticide industry. They also antagonize plant pathogens naturally and thus can be developed as biopesticides to promote pest management in a sustainable way. Biosurfactants which are in their infancy need to be carried to the next level in terms of their production, extraction in an economically viable way for the maximum benefit in the era of industrial agriculture. Genetic engineering may help in modifying biosurfactants extraction befitting the specific requirements and exploring new horizons. There is also a need for evaluating the repercussions of biosurfactants on

N. Srivastava (✉)
Department of Chemistry, Amity University Uttar Pradesh Lucknow Campus, Lucknow, India
e-mail: nsrivastava1@lko.amity.edu

B. K. Rathour
Department of Chemistry, FAA Govt PG College, Mahmudabad, Sitapur, India

A. Srivastava (✉)
Department of Chemistry, GLA University Mathura, Mathura, India
e-mail: aabhichem@gla.ac.in

M. Abdul Rub et al. (eds.), *Industrial Applications of Individual and Mixed Surfactant Systems—Volume 2*, Sustainable Materials and Technology,
https://doi.org/10.1007/978-981-95-9195-4_7

soil and crops in long run, in terms of microbiome, their capacity for nutrient absorption, utilization for agricultural sustainability. Biosurfactants are associated with the bioremediation of the heavy metals and hydrocarbon pollutants in the soil or the oil spill. The biosurfactants are less toxic, biodegradable, highly active and stable in the temperature and varying environmental conditions. These biosurfactants are a promising help in promoting green technology in the era of industrial agricultural. Soil quality is improved by flushing, mobilization, and solubilization of nutrients. Biosurfactant may play a key role in sustainable development in next generation of agriculture which may be achieved by extensive commercialization, uses, enhancing their affordability.

Keywords Industrial agriculture · Biosurfactants · Emulsion · Plant pathogen · Pesticides · Environment remediation · Soil quality

1 Introduction

The rising population and urbanization have put extreme pressure on agriculture to fulfill the demand for food and agricultural products for both humankind and industry. The United Nations projects the average population to reach around 10 billion people by 2050 [1, 2]. This has led to the development of intensive farming or industrial agriculture, a modern farming system focusing on maximization of crops and animal produce through large-scale, mechanized processes using chemical inputs in the form of pesticides, insecticides, fertilizers, and Genetically Modified Organisms (GMOs) [3]. These organisms are basically plants, with genetically engineered DNA to alter or modify their traits and are known as genetically modified (GM) foods. Promotion of monoculture, synthetic fertilizers and pesticides, and modified animal feeding are the main characteristics of industrial agriculture, which aims to enhance yield and efficiency to meet the demands of a growing global population. The key features of industrial agriculture are shown in Fig. 1.

Despite its advantages, industrial agriculture is also associated with many limitations [4], like degrading the environment, bad health effects of GM crops, loss of biodiversity, zoonotic diseases, antimicrobial resistance, promotion of social

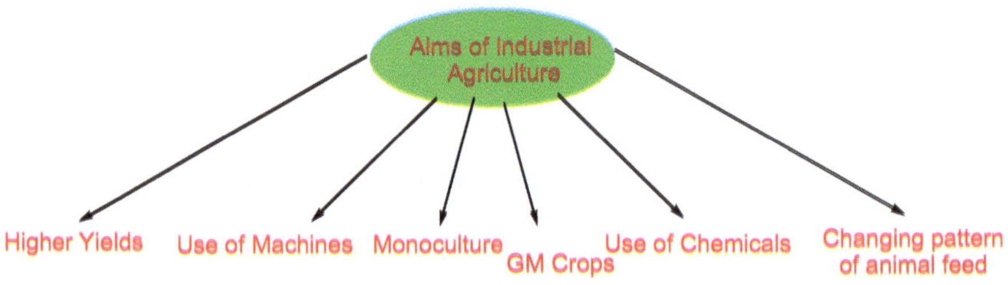

Fig. 1 Key features of industrial agriculture

inequality and injustice to animals [5]. So, we need to mitigate the shortcomings with advances in the field of agrochemicals. Agrochemicals have proved their significance as herbicides [6], fungicides [7], pesticides [8], etc. Herbicide glyphosate has been banned in Germany and Australia as it is associated with cancer among children [9]. Agrochemicals are not easily decomposed and pose threat to the ecosystems by contaminating waterbodies that lead to bioaccumulation in plants and animals [6, 10, 11]. Agrochemicals are also linked to occupational health hazards [12].

Industrial agriculture has paved the way to the development of genetically modified (GM) crops for dwindling use of agrochemicals in agriculture. GM crops are modified genetically to develop pathogen resistance for increased productivity [13] but require enhanced application of agrochemicals. Such crops, although resistant to some pathogens, are prone to other diseases and thus need large amounts of agrochemicals [14]. The GM crops are in high global demand, so we may work on environmental protection in some other ways for research in the field of natural agricultural defense. Agriculture is the primary sector which provides us with food and raw materials for industries. Agrochemicals used in industrial agriculture are highly toxic and degrade our ecosystem. Thus, research on green agrochemicals has become inevitable for the reduction of toxic agrochemicals in industrial agriculture [10]. Our group is working in the field of sustainable chemical synthesis and identification and removal of the heavy metals from the ecosystem [15–27]. Natural agrochemicals are one such type of chemical which are biologically obtained from bacteria, fungi, or plants, etc. Their biocompatibility and biodegradability help in the preservation of the ecosystem [28–30].

2 Significance of Biosurfactant

Surfactants play a significant role in the preparation, stability, and biological performance of agrochemicals. Agrochemicals employ chemical surfactants in their production and use in agriculture [31, 32]. Now, for the sake of the protection of the ecosystem and sustainable development, chemical surfactants may be replaced with biological surfactants. Biosurfactants are green surfactants which possess both amphiphilic and antimicrobial properties, which may be extracted from bacteria, fungi or plants, etc., or through their metabolites. These biosurfactants may also be produced through chemical synthesis using natural extracts [33, 34]. Such biomolecules possess many desired properties which are needed in agro-industrial activities and find suitable applications, such as bioremediation of soil, biodispersants, biostimulants, biofertilizers, and biopesticides. However, research on biosurfactants in agriculture is still in infancy in comparison with research in food, cosmetics, environmental, health, etc. [28, 30, 35–37]. Biosurfactants, which are amphiphilic in nature, are prepared from several microorganisms which are able to dissociate complex biomolecules present in agricultural and industrial wastes. Further biodegradation and fermentation processes help in their transformation into other significant biosurfactants. Biosurfactants may be produced easily with the help

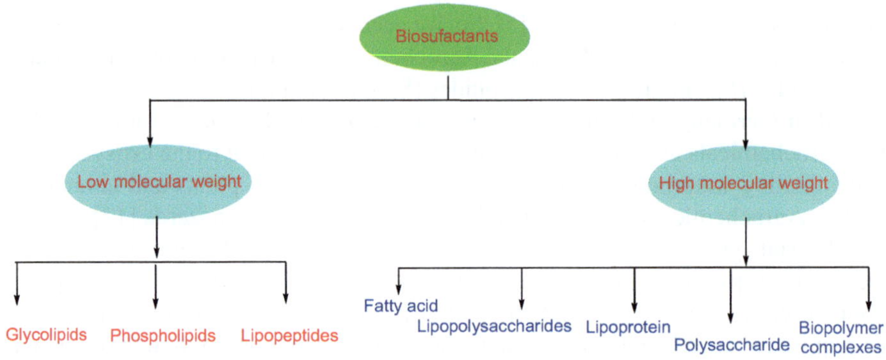

Fig. 2 Broad classification of biosurfactant

of crop remains, byproducts in food processing, industrial effluents, etc. [38]. Optimization of the production process is a little bit challenging and depends upon the selection of microbial strain, conditions of fermentation, and downstream processing. Further research is desired for better understanding of microbial metabolism and optimization of bioprocess with higher efficiency and sustainability in biosurfactants production through industrial and agricultural. This chapter highlights the crucial role of biosurfactants in industrial agriculture and emphasizes the need for further efforts for higher efficiency and promotion of sustainable industrial agriculture safeguarding the environment.

The biosurfactants may be classified as given in Fig. 2.

Biosurfactants, which are fatty acids, are basically spiculisporic acid and phosphatidyl ethanolamine, glycolipids which include mannosyl erythritol, trehaloselipid, sophorolipid, rhamnolipid, while lipopeptides include gramicidin, subtilisin, lichenysin, serrawettin, surfactin, iturin, while polymeric biosurfactants include emulsan, biodispersan, mannan-lipid protein, alasan, and liposan. Rhamnolipids are significantly produced by *Pseudomonas aeruginosa* and emulsify hydrocarbons and possess significant antimicrobial activities [39, 40]. Trehaloselipids, which are hydrogen solubilizer and emulsifying agents, are produced from *Rhodococcus sp.* and *Arthrobacter sp.* and employed in food cosmetics and microbial oil recovery [41]. Sophorolipids are also biosurfactants from the yeast *Candida sp.* and *Torulopsis sp.* employed for stabilizing emulsions of water/oil in therapeutics with immune regulatory, anti-inflammatory and antimicrobial effects [42]. The lipopeptide biosurfactants, iturin and surfactin, are produced from *B. subtilis*; polymyxin from *B. polymyxa;* gramicidin from *B. brevis* and lichenysin by *B. licheniformis* [43]. They exhibit potential thrombolytic, antitumor, antiviral, antimycoplasma, anti-inflammatory and antibacterial properties [44]. Microbial oxidation of alkanes, phospholipids and fatty acids, which are used in gene carrier systems, while fatty acids find applications in food industry [45]. The type of substrate used may also help in the classification of microbial biosurfactants in terms of water insoluble carbon sources which include *Arthrobacterium sp* and *Corynebacterium sp.* while

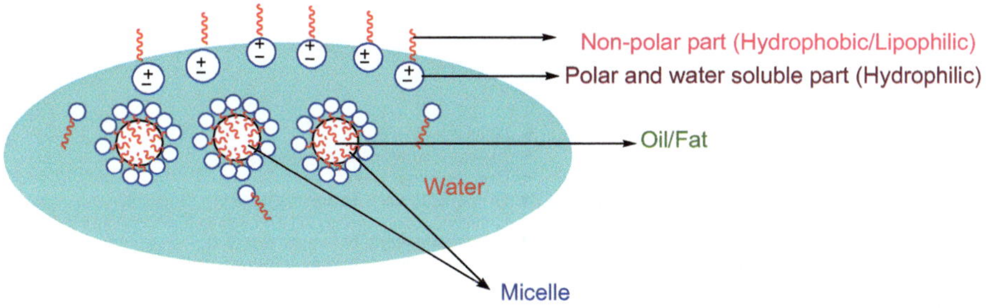

Fig. 3 Emulsification action by the biosurfactants

water-soluble carbon source includes *Bacillus sp.* Biosurfactants are amphiphilic compounds having both hydrophilic and hydrophobic parts and thus lower interfacial and surface tension by binding with different phases (Fig. 3). The hydrophobic part is formed from long-chain fatty acids of amino acids, alcohol, carboxylic acid, phosphate, carbohydrate, cyclic peptides etc. and helps in their concentration in micelles or air-water interface for lowering the surface tension. The hydrophilic part, which is usually a polar chain, enhances the solubility of surfactants in water. Biosurfactants reduce the interfacial surface tension, reduce mobility of the organic part and form micelles. The biosurfactants are characterized by chemical structure, critical micelle concentration (CMC), and hydrophilic-lipophilic balance (HLB). The biosurfactants with high HLB have increased solubility in water [46]. An ideal surfactant reduces the surface tension of water and interfacial tension from 72 to 25 mN/m and 40 to 1 mN/m, respectively [47]. Biosurfactants have higher efficiency as compared to chemical surfactants based on their low CMC of 30–0.15 mg/L [48]. The biosurfactants are more stable and have a better ability to adapt and recover from adverse pH and temperature changes [47].

3 Production of Biosurfactants

The production of the biosurfactants involves the identification of the source, its optimization, extraction, purification, and assessment, which are briefly described in Fig. 4.

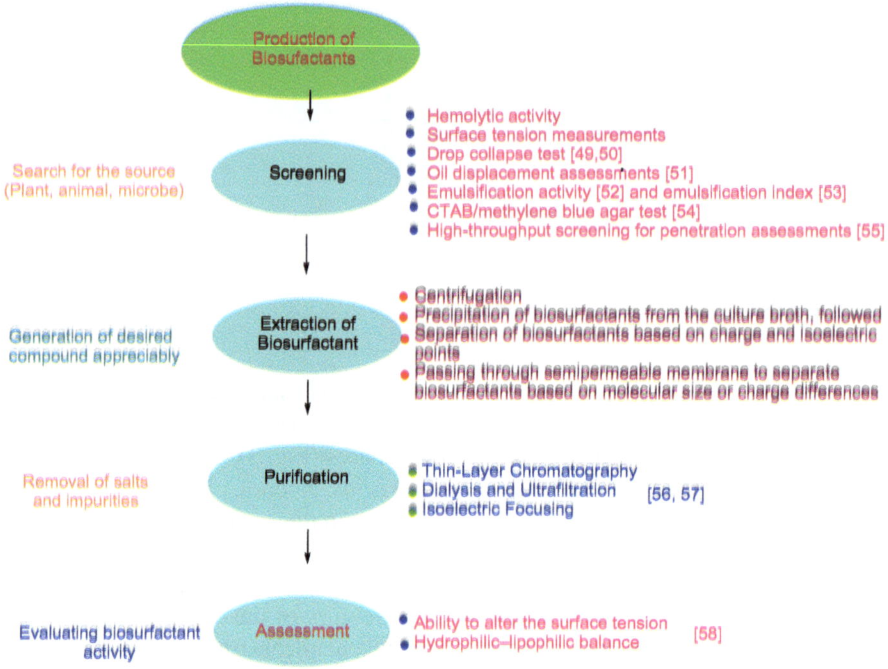

Fig. 4 Steps in the formation of biosurfactant

4 Recent Applications of Biosurfactants in Industrial Agriculture

Biosurfactants possess potential emulsification action and are useful in the industrial agriculture industry for the removal of soil pollutants, and amelioration of the micronutrients in soil control of pathogens. They may play a significant role in replacing chemical surfactants with pesticides, since the microbes in soil use them as carbon sources. The chemical structures of prominent molecules exhibiting the properties of biosurfactants are given in Fig. 5.

4.1 *Improvement of Soil Quality*

Biosurfactants are naturally formed in the soil by some bacteria in the rhizosphere, which facilitate the formation of hydrophobic molecules needful for the growth of plants through easy availability of the nutrients and wetting of soil [49]. Biosurfactants protect plants due to their combination with other chemical compounds

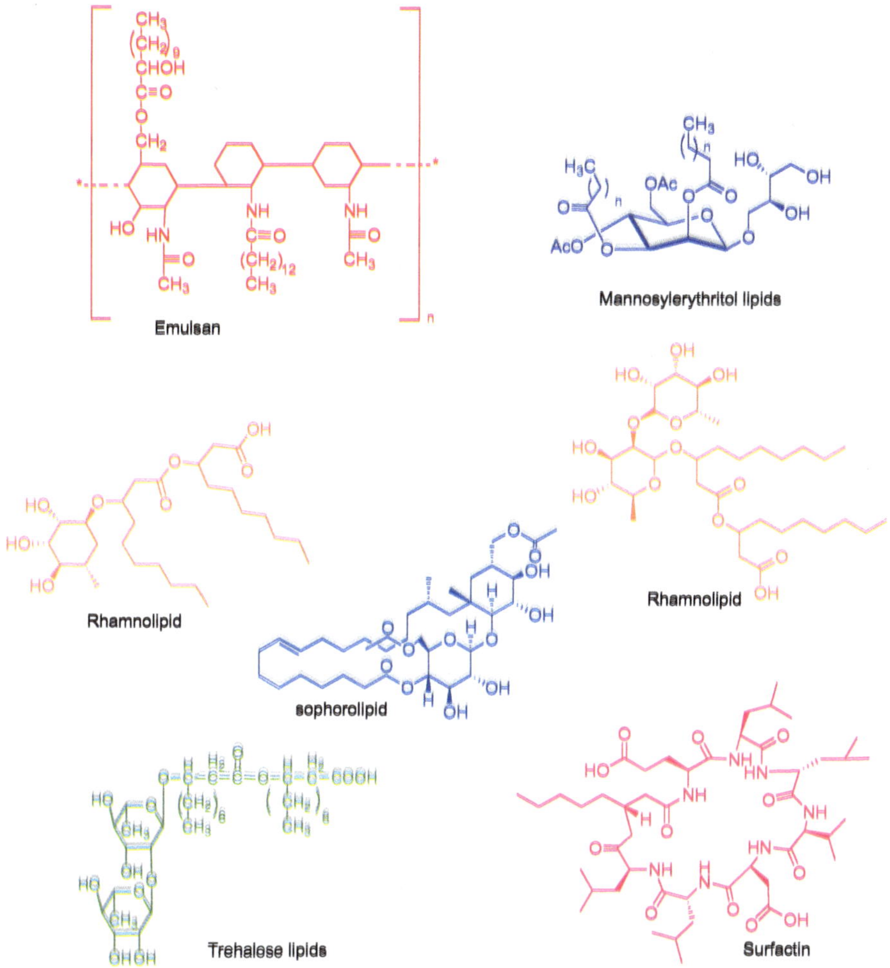

Fig. 5 Chemical structures of promising biosurfactants

provided to the plants. Biosurfactants act as antimicrobials supporting the absorption of biogenic substances and facilitate the germination of seed [50]. Biosurfactants of low-molecular-weight (< 1 kDa) decrease interfacial tension and surface tension of immiscible liquids and thus increase the dispersion or solubility of organic compounds, making their decomposition easier by microorganisms. While biosurfactant with higher molecular weights (>1 kDa) effectively stabilize the emulsions of o/w types by forming a protective layer around oil drops and preventing their coalesce. The stabilization of hydrophobic organic compounds is very important in industrial agriculture using a variety of chemicals in the form of pesticides, insecticides, and fertilizers and will certainly help in increasing the crop yield without degrading the environment [51–53]. Biosurfactants thus help in increasing the nutrients in soil and promote sustainable agriculture.

Biosurfactants are helpful in controlling the pathogens and make the availability of the nutrients to the plant-friendly microorganism which enhances the soil quality. Thus, they may replace the chemical surfactants and help in mitigating the pollution and degrading environment due to the chemicals [49]. The amphiphilicity of biosurfactants helps in its segregation into two immiscible liquids thereby increasing the solubility of hydrophobic compounds. They are significant due to their eco-friendly, non-toxic, and resilient nature to changing pH and high temperatures [54]. Biosurfactants dissociate hydrocarbons and increase the bio-availability of substrates for microorganisms and thus help in dealing with the hydrocarbon contamination in soil [55]. Improved soil conditions will help in dealing with the barren lands or soil degradation [53] and prevention of human migration in search of better agricultural produce.

4.2 Control and Eradication of Plant Pathogen

Biosurfactants of microbial origin have exhibited promising antimicrobial activities [32]. Biosurfactants have shown activities against yeast, Gram-negative and positive bacteria [56]. Biosurfactant from *P. aeruginosa* UCP 0992 was effective in preventing the growth of *Staphylococcus aureus* and *Escherichia coli* with minimum inhibitory concentration ((MIC) of 20 μg/mL). Similarly, biosurfactants obtained from *Candida bombicola* URM 3718 and *P. aeruginosa* UCP 0992 have inhibited the growth of *Streptococcus mutans* with MIC of 20 μg/mL. *C. bombicola* URM 3718, *Bacillus cereus* UCP 1615, and *P. aeruginosa* UCP 0992 exhibited the same effect against *Candida albicans* (MIC: 40 μg/mL) [57]. *Candida sphaerica* UCP 0995 have been used to produce biosurfactant which have exhibited bacteriostatic action to *Klebsiella pneumoniae* and *Klebsiella pneumoniae* [58]. Refinery waste in the presence of *C. lipolytica* UCP 0988 gave a biosurfactant, Rufisan, which exhibited promising antimicrobial activity against *Streptococcus* species above its CMC [59].

The controlling mechanism of the microbes involves hypovirulence, induced resistance, competition, antibiosis, and parasitism for promoting useful microbes and their output [60]. The biosurfactants have shown significant in vitro and in vivo insecticidal activities of surfactants [49]. Fungus *Myrothecium verrucaria*, when combined with a surfactant, helps in the prevention of weeds and helps in preservation of land fertility and safeguard biodiversity [61].

Biosurfactant stops the aflatoxins production by *Aspergillus sp.* to protect maize, cotton, and peanut crops in the storehouse [62]. Phytopathogens have been successfully prevented by isolates of biosurfactant-producing *Bacillus* and *Pseudomonas* [63]. Rhamnolipids inhibited resistant plant pathogens [64] and showed insecticidal activities. Biosurfactant from a *Pseudomonas* strain exhibited insecticidal action against *Myzus persicae* (aphids) [65]. Biosurfactants obtained from *Pseudomonas putida* prevented lysis of zoospores of cucumber [66]. Lipopeptide biosurfactant was formed from Bacillus strains and prevented the proliferation of phytopathogenic fungi from the *Aspergillus* genera and *Fusarium* [67]. The *Brevibacillus brevis* HOB1 strain produced a surfactant produced by *Brevibacillus brevis*, showed promising activities against bacteria and fungi and may be further screened for control of phytopathogen [68]. Biosurfactants obtained from Pseudomonas fluorescens have also yielded biosurfactant with promising antifungal activities [69]. The soil bacterium *Bacillus subtilis* yielded a biosurfactant which effectively controlled the growth of *Colletotrichum gloeosporioides*, which damages the leaves of a papaya plant [70]. Thus, we find that there are numerous microbials stains which give rise to a variety of biosurfactants and possess the potential for inhibiting the growth of pathogens. These biosurfactants have the potential to effectively replace chemical surfactants used as insecticides and pesticides used in farming. Biosurfactants may also enhance the conditions necessary for the growth of microbes in composting [71]. Biosurfactants, owing to their surface tension lowering properties, help in getting rid of organic matter via bioremediation [72]. Biosurfactants inhibiting phytopathogens also influence other plant species, so specific biosurfactant may be obtained by varying the chemical structure by changing the protocol for their production [49]. Glycolipid surfactants derived from *Lactobacillus rhamnosus* inhibited adhesion of bacteria and prevented their antibiofilm activities [73]. Cyclic lipopeptide iturin, which is obtained from *B. subtilis* promotes natural defense of plants against pathogens and boosts their immunity against infections and may replace chemical fungicides [74]. The marine bacteria *B. subtilis* (MC6B-22) yielded lipopeptides which exhibited promising inhibition to 10 phytopathogens of tropical crops and thus act as an effective fungicide in agriculture [75]. *B. Velezensis* PW192 of *Bacillus* species biosurfactants which effectively curbed fungal infections in agriculture [76]. Biosurfactant from corn steep liquor on fermentation by bacteria *L. casei* was active against *E. coli* and *P. aeruginosa* [77]. These examples justify the usage of biosurfactant in reducing the dependency of industrial agriculture on chemical preservatives and pesticides [78].

4.3 Potential of Biosurfactants in Pesticide Industries

Bacillus amyloliquefaciens C11 strain of bacteria was used to produce a Lipopeptide biosurfactant which enhanced the solubility of pesticides, namely chlorpyrifos, iprodione and atrazine, in water by 7.5, 1.3, 1.4 times, respectively. This biosurfactant of environmental concern was investigated. This biosurfactant with CMC of 51.5 mg L^{-1} is highly efficient in decreasing the surface tension of water from 72–28 mN m^{-1}. *B. amyloliquefaciens* C11 enhanced the bioavailability of pesticides to the plants with effective bioremediation processes [78].

Biosurfactant production from *Bacillus amyloliquefaciens* strain AG1. This is a lipopeptide biosurfactant which effectively prevented the growth of *Tuta absoluta* larvae, which are the most damaging pest of the tomato crop. The biosurfactant prevented the growth of larvae with $LC_{50} = 180$ ng/cm^2 in an ecofriendly manner.

Biosurfactant severely damaged midgut tissues, disintegrated and ruptured cellular vacuolization and epithelial layer [79].

Bacillus velezensis PHP1601 culture led to the production of a lipopeptide biosurfactant that inhibited the growth of the agricultural pest blowfly (*Lucilia cuprina*) at larval stage. 20 $\mu g\ g^{-1}$ of the compound killed larvae of second-instar *L. cuprina* due to highly polar surfactin. The biosurfactant produced showed potency of with LC50 value of 9.87 $\mu g\ g^{-}$. Such compounds being insecticidal against blowfly larvae have been first reported [80].

Biosurfactant rhamnolipid (RML) is found to be effective in the prevention of the insect *Bemisia tabaci*. Also, the combination of RML with fungi *Cordyceps javanica* and *Beauveria bassiana* effectively prevented the growth of the pest. It was found that 0.075% and 0.1% (w/v) of RML killed almost 100% of the pests. Conidial suspensions of fungi increased the efficiency of RML to induce nymphal mortalities with a reduction in LC_{50} values [81].

Bacillus sp. gives lipopeptide biosurfactants which are promising amphiphilic with surface-active properties and thus make them effective in disrupting pest membranes and inducing cytotoxic processes in them which cause leakage of the cellular material inhibiting cellular signaling [82].

Pseudomonas aeruginosa LBI 2A1 was used to produce rhamnolipid biosurfactants which were found to exhibit promising insecticidal activities against larvae of *Aedes aegypti* [83]. Glycolipid biosurfactants have also shown a promising role in the biocontrol of several pests, like grasshoppers, boxelder bugs, arachnids, larvae, and eggs. The biosurfactant produced by *Bacillus halotolerans* LYSX1 was employed to produce a biosurfactant which effectively controlled the nematode *Meloidogyne javanica*, a common pest of tomato. *Pseudomonas fluorescens, Pasteuria spp., Panibacillus* have shown potential in controlling plant nematodes [83].

Lipoprotein generated from *Bacillus* species showed promising anti-nematode activity [84]. The activity of the biosurfactant depends upon the composition and the prevailing environmental conditions, e.g., mono-rhamnolipid is highly effective and active in comparison with di-rhamnolipid in the range of 5–20 mg/L [84].

4.4 Environmental Remediation

Biosurfactants play a key role in remediation of the environment as they enhance the surface area of substrates and promote emulsification [85]. Biosurfactants also enhance solubility, thus promoting degradability by distributing pollutants into the aqueous phase, making their bioavailability higher [86]. This method may increase the mobilization of hydrophobic contaminants and facilitate their removal by solubilization and micelle formation and subsequently biodegradation [87]. Heteroatoms present in the biosurfactants form complexes with heavy metal ions and facilitate their removal [86]. These biosurfactants developed a new and efficient technique for the biological removal of heavy metal ions [88]. Biosurfactants may also be employed for medication of cellular properties like permeability and hydrophobicity [86] which may be further used in enhancing the efficiency of bio extraction or biodegradation [89].

4.4.1 Removal of Heavy Metals by Biosurfactants

Biosurfactants are significant in microbe-mediated remediation processes [90]. Extracellular substances produced by microorganisms on biosorption help in immobilizing heavy metals as they bind to the functional groups on cell surfaces [91]. Anionic surfactants form strong ionic bonds with heavy metals in comparison with bonding between metal and soil complexes and thus facilitate the desorption of metal–surfactant complex from the soil matrix, while cationic surfactants, due to ion exchange, compete with charged ions on negatively charged surfaces. Heavy metals in soil may also be removed by surfactant micelles [92, 93]. Various steps involved in heavy metal elimination from soil by employing biosurfactant are given in Fig. 6 [94].

Microorganisms like *Rhizopus Stolonifera*, *Saccharomyces cerevisiae*, *Nostoc linckia*, *Lecythophora sp.*, and *B. subtilishave* potentially modified heavy metals [95]. *Sporosarcina ginsengisoli* reduced arsenic in soil and produced urease in small amounts [96]. In another research, it was found that 76% of Cd2 + was effectively absorbed by *Bacillus subtilis* in contaminated soil and proved its efficiency in heavy metal remediation from soil [97]. *Saccharomyces cerevisiae* showed absorbance efficiency of up to 70% of Cd2 + from soil. Thus, we find biosurfactant are helpful in mitigating heavy metal pollution and promote environmental remediation in appreciable ways.

Soil micronutrients are necessary for the physiological processes in plants and help in controlling metabolism, protein formation and providing immunity to the plants [97]. Chemical fertilizers are supplied to the plants for their growth, but they are not available as they undergo complexation with the soil. Excessive fertilizers disturb normal bioactivities and chemical balance in the soil by altering pH and ionic charges in soil [32, 98].

Biosurfactant reduces the interfacial tension and increase mobility and solubility of electrolytic plant nutrients and thus enhance the nutrient uptake by plants [33, 92]. Anionic biosurfactants form stable complexes with metal and soil. So, they cause desorption of metals from soil and make them accessible to the plants while the

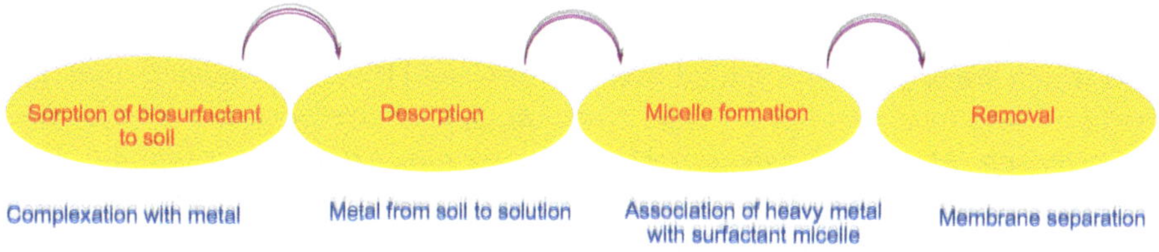

Fig. 6 Steps in the removal of heavy metals using biosurfactant

cationic biosurfactant substitutes the metal ions complexed with soil based on higher affinity (ion exchange mechanism) [94].

Glycolipids like trehalose lipid, rhamnolipid, sophorolipid are prominent glycolipids employed as biosurfactants for metal complexation [56, 94]. Surfactin enhances the emulsification and thus facilitates the nutrient absorption. The biosurfactants promote colony formation by *B. amyloliquefaciens* in Arabidopsis thaliana [99] and wheat by B. subtilis strains [100].

Biosurfactant obtained from the microbes *Torulopsis bombicola, Pseudomonas aeruginosa, Bacillus cereus,* and *Bacillus subtilis,* have shown properties of reclaiming heavy metals like Pb, Cr, Ni, and hydrocarbon pollutants in soil. Thus, biosurfactants may help in the exploration and exploitation of heavy metals and degradation of hydrocarbons, offering new ways for the promotion of agriculture in a sustainable manner [101].

Pseudomonas aeruginosa PTCC1340 has been used to produce rhamnolipid which have been successfully employed for the removal of Ni, Cr, and Cd with efficiencies of 43.05%, 34.73%, and 52.81%, respectively [102].

The rhamnolipids, under optimized conditions, have efficiently removed 72.4% of Cd and 84.4% of phenanthrene in the soil co-contaminated with phenanthrene and cadmium [103].

4.4.2 Removal of Hydrocarbon by Biosurfactants

Biosurfactants also perform emulsification of hydrocarbons in the presence of water, thereby increasing their interaction with water. Various surfactants like rhamnolipids, surfactin, and lichenisins have shown a promising role in mitigating the hydrocarbon pollution in the ecosystem [86, 104]. Marine bacteria have shown remarkable properties of hydrocarbon degradation. Such biosurfactants from marine bacteria have successfully dealt with oil slicks, by forming emulsion with water and promoting their biodegradation. Hydrocarbonoclastic bacteria are well-known marine bacteria which have degraded hydrocarbons and are used in the protection of the marine ecosystem [86]. Surfactin derived from *Bacillus coagulans* 30 formed emulsions with crude oil and helps in oil recovery at higher rates. Surfactin are also stabile in different conditions of salinity, pH, and temperature, and so they are helpful in oil recovery. Various research has proved that microbes like bacteria and fungi use hydrocarbons as a source of carbon for producing energy necessary for their growth and thus help in degradation of the hydrocarbons [105]. Bacteria are primary decomposers of petroleum compounds, which effectively decompose numerous hydrocarbons which are accidentally spilled over soil or water [106]. *Corynebacterium spp., Pseudomonas spp., Alcaligenes spp., Nocardia spp., Flavobacterium spp., Achromobacter spp., Mycobacterium spp., Marinobacter spp., Arthrobacter spp.*, etc., have been isolated successfully from groundwater and soil which exhibited notable properties of decomposing hydrocarbon [33]. Some fungi, *Penicillium simplicissimum, Cyclothyrium spp., Psilocybe spp.,* have potentially oxidized and transformed polycyclic aromatic hydrocarbons (PAHs) into the least toxic byproducts [107, 108].

Lipopeptide biosurfactant derived from *Pseudomonas aeruginosa* strain LBP9 (originated from soil polluted with petroleum), when added to soil at a concentration of 400 mg/L increased solubilization of pyrene, fluoranthene and phenanthrene by 45, 33, 19 times, respectively, in comparison with control [108].

The bacterium *Chelidonium majus L.* has exhibited promising activity in degrading hydrocarbons and simultaneous generation of biosurfactant [109]. Rhamnolipids derived from *Pseudomonas aeruginosa* have explicitly degraded hexadecane, showing the specific action of biosurfactant towards a particular hydrocarbon [86].

5 Conclusions

Biosurfactants have exhibited their potential in the bioremediation of the problems associated with industrial agriculture, where large amounts of chemicals are used in terms of pesticides, insecticides, fertilizers, etc. To conclude, we find that sophorolipid extracted from *Torulopsis bombicola*, rhamnolipid from *Pseudomonas aeruginosa* and surfactin from *Bacillus subtilis* are the promising biosurfactants which have shown promising contribution in the bioremediation of heavy metals. Biosurfactants are promising emulsifiers and wetting solvents which have potential and promising applications in pesticide production use and uptake by plants. Biosurfactants are antagonists to many pathogens and are potential candidates for biopesticides. The main concern of biosurfactant is the high cost of production associated with them due to which they cannot compete with synthetic chemicals and so they are not used on a large scale. Their commercialized production is still challenging in comparison to synthetic surfactants [56, 110]. The substrate to produce biosurfactant, its purification is very costly till date [56, 111]. There is a need for the exploration of cost-effective techniques to produce large scale biosurfactants for application in agriculture by the poor farmers.

In this respect, some research has shown that agricultural waste may be employed as a cheap substrate for the microbial production of biosurfactants [111]. *Bacillus bacteria* are rich in enzyme amylase and may use starch available in agricultural and agroindustry waste [112]. In this regard, bagasse, molasses, potato, banana, and orange peel have been used as substrate by *Bacillus subtilis* to produce biosurfactants [113]. Effluent from breweries is a big source of carbon and has been successfully utilized as a carbon source to produce biosurfactant by *Bacillus subtilis* [114]. Sophorolipids are glycolipid biosurfactants that are produced at lower cost. Methods like continuous removal of biosurfactants through *in-situ* foam fractionation and ultrafiltration may help in cost-cutting in the production of biosurfactants. Genetic engineering may also help in cost-cutting the process of biosurfactant production. Metagenomics is new technique for the characterization of microorganisms which helps in the degradation of pollutants. Metagenomics may help in novel methods for the production of biosurfactant in a cost-effective manner. Clubbing these microbes with biofertilizer strains may enhance nutrient availability, suppression of disease, and insecticidal action will help in improving agricultural produce [115]. Biosurfactant may enhance the efficiency and activity of biopesticides and bioherbicides, which may also be the major advantage of the biosurfactant of synthetic chemicals in the era of industrial agriculture. Myconano technology is an emerging technology which may help in the production of biosurfactants using nanotechnology in detecting and controlling phytopathogens in industrial agriculture. Novel bioformulations with nanocarrier systems need to be developed in the future to improve the nutrient level in soil. Biostimulants may be developed using the biosurfactant for stimulating the growth of root, nutrient uptake, and tolerance to stress. The intricacies of the relation between plant biosurfactants and microbiomes need to be explored at the next level for the agricultural revolution in a sustainable manner. Sustainability needs to be ensured by evaluation of the ecotoxicological effect of biosurfactants in agriculture and their persistence, biodegradability, and ecological risks.

Conflict of Interest The authors declare that there is no conflict of interest regarding the manuscript.

References

1. Mrabet R (2023) Sustainable agriculture for food and nutritional security. In: Sustainable agriculture and the environment, Elsevier: Amsterdam, The Netherlands, pp 25–90
2. Rebolledo-Leiva R, Moreira MT, González-García S (2023) Progress of social assessment in the framework of bioeconomy under a life cycle perspective. Renew Sustain Energy Rev 175:113162
3. Otieno M (2024) Traditional and industrial farming practices. In: Sustainable agroecological practices in Sub-Saharan Africa in the face of climate change. Advances in geographical and environmental sciences. Springer, Cham. https://doi.org/10.1007/978-3-031-70472-7_6
4. https://www.unep.org/news-and-stories/story/10-things-you-should-know-about-industrial-farming
5. Mincytė D, Bartkienė A, Bikauskaitė R et al (2025) What does it mean to care in industrial agriculture? Agric Hum Values. https://doi.org/10.1007/s10460-025-10709-0
6. Dehnert GK, Freitas MB, Sharma PP, Barry TP, Karasov WH (2021) Impacts of subchronic exposure to a commercial 2,4-D herbicide on developmental stages of multiple freshwater fish species. Chemosphere 263:127638
7. Zhang C, Zhou T, Xu Y, Du Z, Li B, Wang J, Wang J, Zhu L (2020) Ecotoxicology of strobilurin fungicides. Sci Total Environ 742:140611
8. Calista N, Haikael MD, Athanasia MO, Neema K, Judith K (2022) Does Pesticide exposure contribute to the growing burden of non—Communicable diseases in Tanzania. Sci Afr 17:e01276
9. Carranca T (2021) Agrotóxico Mais Usado do Brasil Está Associado a 503 Mortes Infantis por ano, Revela Estudo. BBC News Brasil. Available online: https://www.bbc.com/portuguese/brasil-57209799
10. Campanhola C, Bettiol W (eds) (2003) Métodos Alternativos de Controle Fitossanitário; Embrapa Meio Ambiente: Jaguariúna, SP, Brazil. ISBN 85-85771-22-4
11. Ward LT, Hladik ML, Guzman A, Winsemius S, Bautista A, Kremen C, Mills NJ (2022) Pesticide exposure of wild bees and honey bees foraging from field border flowers in intensively managed agriculture areas. Sci Total Environ 831:154697

12. Rigotto RM, Vasconcelos DPE, Rocha MM (2014) Pesticide use in Brazil and problems for public health. Cad Saude Publica 30:1360–1362
13. Jhansi Rani S, Usha R (2013) Transgenic plants: types, benefits, public concerns and future. J Pharm Res 6:879–883
14. Rauchecker M (2019) The territorial and sectoral dimensions of advocacy—The conflicts about pesticide use in Argentina. Polit Geogr 75:102067
15. Srivastava N, Rathore BK, Singh S, Singh S (2022) Thermoacoustical study of intermolecular interactions in binary liquid mixtures of benzaldehyde with methanol and ethanol. Chem Phys Impact 6:100144. https://doi.org/10.1016/j.chphi.2022.100144
16. Srivastava N, Chaturvedi D, Kishore R, Kaur M, Kaur R (2022) Triton-B/CS2 mediated novel synthesis of substituted thioureas: a novel class of anti-cancer agents. Indian J Chem 61:870. https://doi.org/10.56042/ijc.v61i8.65053
17. Srivastava N, Rathore BK, Vishnoi RK, Bajpai S (2020) A simple and convenient method for synthesis of artemisinin dimer through aliphatic nitro compounds. Asian J Chem 33(1):49. https://doi.org/10.14233/ajchem.2021.22918
18. Vishnoi RK, Kishore R, Chaturvedi D, Shukla M, Bajpai S, Srivastava S (2021) Synthesis and antimicrobial activity of cyclic dithiocarbamates employing Triton-B/CS2 system. Asian J Chem 33(5):1133. https://doi.org/10.14233/ajchem.2021.23173
19. Srivastava N, Kishore R, Chaturvedi D (2021) Novel and efficient method for the synthesis of cyclic trithiocarbonates employing Cs_2CO_3 and CS_2 system. Research J Chem Environ 25(12):142–148. https://doi.org/10.25303/2512rjce142148
20. Anisree S, Shanmugavelan P, Vijaylakshmi P, Kishore R, Srivastava N (2024) Synthesis, characterization and anticancer screening of novel phenylbenzylidene thiosemicarbazone derivatives. Phos Sulf Silicon Rel Elem 199(4):267. https://doi.org/10.1080/10426507.2024.2330935
21. Srivastava A, Srivastava N, Srivastava N (2024) Adsorptive removal of Crystal Violet Dye from aqueous solutions utilizing fox nutshell activated carbon (FNAC). South African J Chem 78:227–234. https://hdl.handle.net/10520/ejc-chem-v78-n1-a29
22. Srivastava N, Srivastava A et al (2025) Synthesis, characterization of pine chir based biochar nanocomposites and their application in heavy metal ion detection. Next Mater 8:100805. https://doi.org/10.1016/j.nxmate.2025.100805
23. Srivastava A, Dohare RK, Srivastava N (2025) Efficient removal of Arsenic(III) from aqueous solutions using polypyrrole based magnetic nano-adsorbent. J Incl Phenom Macrocycl Chem. https://doi.org/10.1007/s10847-025-01310-3
24. Goswami MK, Srivastava A (2025) Polypyrrole/$CoFe_2O_4$ nanocomposite for the simultaneous in-situ adsorptive and reductive removal of Chromium(VI): kinetics and mechanism. Colloid Polym Sci. https://doi.org/10.1007/s00396-025-05446-z
25. Srivastava A, Srivastava N, Singh R (2025) Efficient removal of lead(II) from aqueous solutions using polypyrrole based magnetic nano-adsorbent. Prog Reac Kinet Mech 50:e006. https://doi.org/10.48130/prkm-0025-0006
26. Srivastava A, Singh R, Srivastava N (2025) Adsorptive removal of malachite green using fox nutshell activated carbon (FNAC): adsorption isotherms and kinetic study. Indian J Chem Technol 32:230–240. https://doi.org/10.56042/ijct.v32i2.11071
27. Goswami MK, Srivastava A (2025) Polypyrrole/$CoFe_2O_4$ nanocomposite for the removal of basic blue 3 dye from wastewater: kinetic, adsorption isotherm and thermodynamic study. J Appl Polym Sci 142:e56713. https://doi.org/10.1002/app.56713
28. Bailey KL, Boyetchko SM, Langle T (2010) Social and economic drivers shaping the future of biological control: a Canadian perspective on the factors affecting the development and use of microbial biopesticides. Biol Control 52:221
29. Cavalcante BDAM, Scapini T, Camargo AF, Ulrich A, Bonatto C et al (2021) Orange peels and shrimp shell used in a fermentation process to produce an aqueous extract with bioherbicide potential to weed control. Biocatal Agric Biotechnol 32:101947
30. Ulrich A, Lerin LA, Camargo AF, Scapini T, Diering NL et al (2021) Alternative bioherbicide based on trichoderma koningiopsis: enzymatic characterization and its effect on cucumber plants and soil organism. Biocatal Agric Biotechnol 36:102127

31. Mann R, Bidwell J (2001) The acute toxicity of agricultural surfactants to the tadpoles of four Australian and two exotic frogs. Environ Pollut 114:195–205
32. Sangwan S, Kaur H, Sharma P, Sindhu M, Wati L (2022) Routing microbial biosurfactants to agriculture for revitalization of soil and plant growth. In: New and future developments in microbial biotechnology and bioengineering, Elsevier: Amsterdam, The Netherlands, pp 313–338
33. Farias CBB, Almeida FCG, Silva IA, Souza TC et al (2021) Production of green surfactants: market prospects. Electron J Biotechnol 51:28–39
34. Sivapathasekaran C, Sen R (2017) Origin, properties, production and purification of microbial surfactants as molecules with immense commercial potential. Tenside Surfactants Deterg 54:92
35. Maldaner J, Oliveira MN, De Alexandria SD, Simote Silva SY et al (2022) Bioherbicide and anesthetic potential of Aniba canelilla essential oil, a contribution to the demands of the agricultural sector. Biocatal Agric Biotechnol 42:102353
36. Serrano-Carreón L, Aranda-Ocampo S, Balderas-Ruíz KA, Juárez AM, Leyva E et al (2022) A case study of a profitable mid-tech greenhouse for the sustainable production of tomato, using a biofertilizer and a biofungicide. Electron J Biotechnol 59:13
37. Shokouhi D, Seifi A (2020) Organic extracts of seeds of Iranian Moringa peregrina as promising selective biofungicide to control Mycogone perniciosa. Biocatal Agric Biotechnol 30:101848
38. Abid A, Mekhadmi NE, Mlik R, Bireche, K (2025) Role of microorganisms in biosurfactant production from agricultural and industrial wastes. In: Sharma K (ed) Bio-prospecting of novel microbial bioactive compounds for sustainable development. Sustainable landscape planning and natural resources management. Springer, Cham. https://doi.org/10.1007/978-3-031-82178-3_1
39. Li Z, Zhang Y, Lin J, Wang W, Li S (2019) High-yield di-rhamnolipid production by Pseudomonas aeruginosa YM4 and its potential application in MEOR. Molecules 24(7):1433
40. Liu G, Zhong H, Yang X, Liu Y, Shao B, Liu Z (2018) Advances in applications of rhamnolipids biosurfactant in environmental remediation: a review. Biotechnol Bioeng 115(4):796–814
41. Kuyukina MS, Ivshina IB, Baeva TA, Kochina OA et al (2015) Trehalolipid biosurfactants from nonpathogenic Rhodococcus actinobacteria with diverse immunomodulatory activities. N Biotechnol 32(6):559
42. Gaur VK, Regar RK, Dhiman N, Gautam K, Srivastava JK, Patnaik S, Manickam N (2019) Biosynthesis and characterization of sophorolipid biosurfactant by Candida spp: application as food emulsifier and antibacterial agent. Bioresour Technol 285:121314
43. Caldeira AT, Arteiro JS, Coelho AV, Roseiro JC (2011) Combined use of LC–ESI-MS and antifungal tests for rapid identification of bioactive lipopeptides produced by Bacillus amyloliquefaciens CCMI 1051. Process Biochem 46(9):1738–1746
44. Carrillo C, Teruel JA, Aranda FJ, Ortiz A (2003) Molecular mechanism of membrane permeabilization by the peptide antibiotic surfactin. Biochim Biophys Acta Biomembr 1611:91
45. Sharma R, Singh J, Verma N (2018) Production, characterization and environmental applications of biosurfactants from Bacillus amyloliquefaciens and Bacillus subtilis. Biocatal Agric Biotechnol 16:132
46. Van Hamme JD, Singh A, Ward OP (2006) Physiological aspects. Part 1 in a series of papers devoted to surfactants in microbiology and biotechnology. Biotechnol Adv 24:604
47. Vijayakumar S, Saravanan V (2015) Biosurfactants-types, sources and applications. Res J Microbiol 10:181–192
48. Pichot R, Watson RL, Norton IT (2013) Phospholipids at the interface: current trends and challenges. Int J Mol Sc 14:11767–11794
49. Sachdev DP, Cameotra SS (2013) Biosurfactants in agriculture. Appl Microbiol Biotechnol 97:1005–1016
50. Simões LA, Fernandes NAT, dos Santos Junior NA, Dias DR (2023) Biosurfactants and their benefits for seeds. In: Advancements in biosurfactants research, Springer: Cham, Switzerland, pp 309–329

51. Patel P, Patel R, Mukherjee A, Munshi NS (2023) Microbial biosurfactants for green agricultural technology. In: Sustainable agriculture reviews 60: microbial processes in agriculture, Springer: Cham, Switzerland, pp 389–413
52. Daxini A, O'Donoghue C, Ryan M, Buckley C, Barnes AP, Daly K (2018) Which factors influence farmers' intentions to adopt nutrient management planning? J Environ Manag 224:350–360
53. Bretzel F, Caudai C, Tassi E, Rosellini I, Scatena M, Pini R (2018) Culture and horticulture: protecting soil quality in urban gardening. Sci Total Environ 644:45–51
54. Kumar A, Singh SK, Kant C, Verma H, Kumar D, Singh PP, Modi A, Droby S, Kesawat MS, Alavilli H (2021) Microbial biosurfactant: a new frontier for sustainable agriculture and pharmaceutical industries. Antioxidants 10:1472
55. Sah D, Rai J, Ghosh A, Chakraborty M (2022) A review on biosurfactant producing bacteria for remediation of petroleum contaminated soils. 3 Biotech 12:218
56. Santos D, Rufino R, Luna J, Santos V, Sarubbo L (2016) Biosurfactants: multifunctional biomolecules of the 21st century. Int J Mol Sci 17:401
57. Hashemi SZ, Fooladi J, Ebrahimipour G, Khodayari S (2016) Isolation and identification of crude oil degrading and biosurfactant producing bacteria from the oil-contaminated soils of Gachsaran. Appl Food Biotechnol 3:83–89
58. Sobrinho HBS., Luna JM, Rufino RD, Porto ALF, Sarubbo LA (2013) Assessment of toxicity of a biosurfactant from Candida sphaerica UCP 0995 cultivated with industrial residues in a bioreactor. Electron J Biotechnol 16. https://doi.org/10.2225/vol16-issue4-fulltext-4
59. Rufino RD, Luna JM, Sarubbo LA, Rodrigues LRM, Teixeira JAC, Campos-Takaki GM (2011) Antimicrobial and anti-adhesive potential of a biosurfactant Rufisan produced by candida lipolytica UCP 0988. Colloids Surf B Biointerfaces 84:1–5
60. Kim PI, Bai H, Bai D, Chae H, Chung S, Kim Y, Park R, Chi Y-T (2004) Purification and characterization of a lipopeptide produced by Bacillus thuringiensis CMB26. J Appl Microbiol 97:942–949
61. Boyette CD, LynnWalker H, Abbas HK (2010) Biological control of kudzu (Pueraria lobata) with an isolate of Myrothecium verrucaria. Biocontrol Sci Technol 12:75–82
62. Rodriguez SB, Mahoney NE (1994) Inhibition of Aflatoxin production by surfactants. Appl Environ Microbiol 60:106
63. Krzyzanowska DM, Potrykus M, Golanowska M, Polonis KA, Gwizdek-Wisniewska E, Lojkowska E, Jafra S (2012) Rhizosphere bacteria as potential biocontrol agents against soft rot caused by various Pectobacterium and Dickeya spp. strains. J Plant Pathol 94:367
64. Vatsa P, Sanchez L, Clement C, Baillieul F, Dorey S (2010) Rhamnolipid biosurfactants as new players in animal and plant defense against microbes. Int J Mol Sci 11:5095
65. Kim SK, Kim YC, Lee S, Kim JC, Yun MY, Kim IS (2011) Insecticidal activity of Rhamnolipid isolated from pseudomonas sp. EP-3 against green peach aphid (Myzus persicae). J Agric Food Chem 59:934–938
66. Kruijt M, Tran H, Raaijmakers JM (2009) Functional, genetic and chemical characterization of biosurfactants produced by plant growth-promoting pseudomonas putida 267. J Appl Microbiol 107:546–556
67. Velho RV, Medina LFC, Segalin J, Brandelli A (2011) Production of lipopeptides among bacillus strains showing growth inhibition of phytopathogenic fungi. Folia Microbiol 56:297
68. Haddad NIA, Wang J, Mu B (2008) Isolation and characterization of a biosurfactant producing strain, brevibacilis brevis HOB1. J Ind Microbiol Biotechnol 35:1597–1604
69. Nielsen TH, Sørensen J (2003) Production of cyclic lipopeptides by pseudomonas fluorescens strains in bulk soil and in the sugar beet Rhizosphere. Appl Environ Microbiol 69:861
70. Kim PL, Ryu J, Kim YH, Chi Y-T (2010) Production of biosurfactant lipopeptides Iturin A, fengycin and surfactin A from Bacillus subtilis CMB32 for control of Colletotrichum gloeosporioides. J Microbiol Biotechnol 20:138
71. Gayathiri E, Prakash P, Karmegam N, Varjani S, Awasthi MK, Ravindran B (2022) Biosurfactants: potential and eco-friendly material for sustainable agriculture and environmental safety—a review. Agronomy 12:662

72. De Giani A, Zampolli J, Di Gennaro P (2021) Recent trends on biosurfactants with antimicrobial activity produced by bacteria associated with human health: different perspectives on their properties, challenges, and potential applications. Front Microbiol 12:678
73. Patel M, Siddiqui AJ, HamadouWS SM, Awadelkareem AM, Ashraf SA, Alreshidi M, Snoussi M, Rizvi SMD, Bardakci F (2021) Inhibition of bacterial adhesion and antibiofilm activities of a glycolipid biosurfactant from Lactobacillu rhamnosus with its physicochemical and functional properties. Antibiotics 10:1546
74. Yaraguppi DA, Bagewadi ZK, Patil NR, Mantri N (2023) Iturin: a promising cyclic lipopeptide with diverse applications. Biomolecules 13:1515
75. Guillén-Navarro K, López-Gutiérrez T, García-Fajardo V, Gómez-Cornelio S, Zarza E, De la Rosa-García S, Chan-Bacab M (2023) Broad-spectrum antifungal, biosurfactants and bioemulsifier activity of Bacillus subtilis subsp. spizizenii—A potential biocontrol and bioremediation agent in agriculture. Plants 12:1374
76. Jumpathong W, Intra B, Euanorasetr J, Wanapaisan P (2022) Biosurfactant-producing Bacillus velezensis PW192 as an anti-fungal biocontrol agent against Colletotrichum gloeosporioides and Colletotrichum musae. Microorganisms 10:1017
77. López-Prieto A, Xanel Vecino X, Rodríguez-López L, Moldes AB, Cruz JM (2019) A multifunctional biosurfactant extract obtained from corn steep water as bactericide for agrifood industry. Foods 8:410
78. Schalchli H, Lamilla C, Rubilar O, Briceño G, Gallardo F et al (2023) Production and characterization of a biosurfactant produced by Bacillus amyloliquefaciens C11 for enhancing the solubility of pesticides. J Environ Chem Eng 11(6):111572. https://doi.org/10.1016/j.jece.2023.111572
79. Khedher SB, Boukedi H, Kilani-Feki O, Chaib I, Laarif A, Tounsi S et al (2015) Bacillus amyloliquefaciens AG1 biosurfactant: putative receptor diversity and histopathological effects on Tuta absoluta midgut. J Invertebr Pathol 132:42–47. https://doi.org/10.1016/j.jip.2015.08.010
80. Ramesar DR, Hunter CH (2023) Elucidation of lipopeptide biosurfactants responsible for the larvicidal activity of Bacillus velezensis PHP1601 towards Lucilia cuprina larvae. Biocontrol Sci Tech 33(8):772. https://doi.org/10.1080/09583157.2023.2231180
81. Silva JDN, Mascarin GM, de Paula R, de Castro V, Castilho LR, Freire DMG (2019) Novel combination of a biosurfactant with entomopathogenic fungi enhances efficacy against Bemisia whitefly. Pest Manag Sci 75(11):2882–2891. https://doi.org/10.1002/ps.5458
82. Cossus L, Roux-Dalvai F, Kelly I, Nguyen TTA, Antoun H, Droit A, Tweddell RJ (2021) Interactions with plant pathogens influence lipopeptides production and antimicrobial activity of Bacillus subtilis strain PTB185. Biol Control 154(1):104497. https://doi.org/10.1016/j.biocontrol.2020.104497
83. Adetunji CO (2022) Applications of biosurfactant in agriculture. Academic Press: Cambridge, MA, USA
84. Huang S, Zheng X, Luo L, Ni Y, Yao L, Ni W (2021) Biostimulants in bioconversion compost of organic waste: a novel booster in sustainable agriculture. J Clean Prod 319:128704. https://doi.org/10.1016/j.jclepro.2021.128704
85. Karlapudi AP, Venkateswarulu T, Tammineedi J, Kanumuri L, Ravuru BK, Ramu Dirisala V, Kodali VP (2018) Role of biosurfactants in bioremediation of oil pollution-a review. Petroleum 4:241–249
86. Ławniczak Ł, Marecik R, Chrzanowski Ł (2013) Contributions of biosurfactants to natural or induced bioremediation. Appl Microbiol Biotechnol 97:2327
87. Srivastava A, Srivastava N, Singh R (2024) Cu(II) mediated oxidation of L-phenylalanine with chromic acid in micellar medium: a kinetic and mechanistic approach. Main Group Chem 24(1):24–36. https://doi.org/10.1177/10241221241291408
88. Sci Y, Nurbas M, Acıkel YS (2008) A comparative study for the sorption of Cd (II) by soils with different clay contents and mineralogy and the recovery of Cd (II) using rhamnolipid biosurfactant. J Hazard Mater 154:663–673

89. Johnsen A, Karlson U (2004) Evaluation of bacterial strategies to promote the bioavailability of polycyclic aromatic hydrocarbons. Appl Microbiol Biotechnol 63:452–459
90. Mulligan CN (2005) Environmental applications for biosurfactants. Environ Pollut 133:183
91. Sun W, Cheng K, Sun KY, Ma X (2021) Microbially mediated remediation of contaminated sediments by heavy metals: a critical review. Curr Pollut Rep 7:201
92. Sales da Silva IG, Gomes de Almeida FC, Padilha da Rocha e Silva NM, Casazza, AA, Converti A, Asfora Sarubbo L (2020) Soil bioremediation: overview of technologies and trends. Energies 13:4664
93. Pacwa-Płociniczak M, Płaza GA, Piotrowska-Seget Z, Cameotra SS (2011) Environmental applications of biosurfactants: recent advances. Int J Mol Sci 12:633–654
94. Sarubbo LA, Rocha RB, Luna JM, Rufino RD, Santos VA, Banat IM (2015) Some aspects of heavy metals contamination remediation and role of biosurfactants. Chem Ecol 31:707–723
95. Raklami A, Meddich A, Oufdou K, Baslam M (2022) Plants—Microorganisms-based bioremediation for heavy metal cleanup: recent developments, phytoremediation techniques, regulation mechanisms, and molecular responses. Int J Mol Sci 23:5031
96. Achal V, Pan X, Fu Q, Zhang D (2012) Biomineralization based remediation of As (III) contaminated soil by Sporosarcina ginsengisoli. J Hazard Mater 201:178–184
97. Adhikari S, Anuragi H, Chandra K, Tarte SH, Dhaka SR, Jatav HS, Hingonia K (2023) Molecular basis of plant nutrient use efficiency—concepts and challenges for its improvement. Sustain Plant Nutr 107. https://doi.org/10.1016/B978-0-443-18675-2.00001-8
98. Osman KT (2013) Plant nutrients and soil fertility management. In: Soils principles, properties and management. Springer: Dordrecht, The Netherlands, pp 129–159
99. Dietel K, Beator B, Budiharjo A, Fan B, Borriss R (2013) Bacterial traits involved in colonization of arabidopsis thaliana roots by bacillus amyloliquefaciens FZB42. Plant Pathol J 29:59
100. Le Mire G, Siah A, Brisset M-N, Gaucher DM, Jijakli M (2018) Surfactin protects wheat against Zymoseptoria Tritici and activates both salicylic acid- and Jasmonic acid-dependent defense responses. Agriculture 8:11
101. Ayangbenro AS, Babalola OO (2020) Genomic analysis of Bacillus cereus NWUAB01 and its heavy metal removal from polluted soil. Sci Rep 10(1):1–12
102. Kholghi N, Amani H, Malekmahmoodi S, Amiri A (2020) Investigation on heavy metal removal from a crude oil contaminated soil using rhamnolipid biosurfactant as a new eco-friendly method. Tenside Surfact Deterg 57(6):515–520
103. Zhang X, Zhang X, Wang S et al (2022) Improved remediation of co-contaminated soils by heavy metals and PAHs with biosurfactant-enhanced soil washing. Sci Rep 12:3801. https://doi.org/10.1038/s41598-022-07577-7
104. Eze MO, Thiel V, Hose GC, George SC, Daniel R (2022) Bacteria-plant interactions synergistically enhance biodegradation of diesel fuel hydrocarbons. Commun Earth Environ 3(1):1–10
105. Koshlaf E, Bal AS (2017) Soil bioremediation approaches for petroleum hydrocarbon polluted environments. AIMS Microbiol 3:25
106. Brooijmans RJ, Pastink MI, Siezen RJ (2009) Hydrocarbon-degrading bacteria: the oil-spill clean-up crew. Microb Biotechnol 2:587
107. Germida J, Frick C, Farrell R (2002) Phytoremediation of oil-contaminated soils. In: Developments in soil science, Elsevier: Amsterdam, The Netherlands, vol 28. pp 169–186
108. Bezza FA, Chirwa EMN (2017) Biosurfactant-assisted bioremediation of polycyclic aromatic hydrocarbons (PAHs) in liquid culture system and substrate interactions. Polycycl Aroma Compd 37:375–394
109. Marchut-Mikolajczyk O, Drożdżyński P, Pietrzyk D, Antczak T (2018) Biosurfactant production and hydrocarbon degradation activity of endophytic bacteria isolated from Chelidonium majus L. Microb Cell Factor 17(1):1–9
110. Hazra C, Kundu D, Ghosh P, Joshi S, Dandi N, Chaudhari A (2011) Screening and identification of Pseudomonas aeruginosa AB4 for improved production, characterization and application of a glycolipid biosurfactant using low-cost agro-based raw materials. J Chem Technol Biotechnol 86:185–198

111. Freitas BG, Brito JGM, Brasileiro PPF, Rufino RD, Luna JM, Santos VA, Sarubbo LA (2016) Formulation of a commercial biosurfactant for application as a dispersant of petroleum and by-products spilled in oceans. Front Microbiol 7:1–9
112. Lan G, Fan Q, Liu Y, Chen C, Li G, Liu Y, Yin X (2015) Rhamnolipid production from waste cooking oil using Pseudomonas SWP-4. Biochem Eng J 101:44–54
113. Rane AN, Baikar VV, Ravi Kumar V, Deopurkar RL (2017) Agro-industrial wastes for production of biosurfactant by Bacillus subtilis ANR 88 and its application in synthesis of silver and gold nanoparticles. Front Microbiol 8:492. https://doi.org/10.3389/fmicb.2017.00492
114. Moshtagh B, Hawboldt K, Zhang B (2018) Optimization of biosurfactant production by Bacillus subtilis N3–1 P using the brewery waste as the carbon source. Environ Technol. https://doi.org/10.1080/09593330.2018.1473502
115. Nurmayulis, Eris FR, Maida H, Hastuti D, Denny YR (2022) Effect of biofertilizer formulation with addition of consortium of microbes and biosurfactant DEA Palm Olein to Growth of Cacao (Theobroma cacao) Seedling. In: In IOP conference series: earth and environmental science, 978. IOP Publishing, 012013. https://doi.org/10.1088/1755-1315/978/1/012013

Investigation of Synergistic Effects of Mixed Surfactant for Enhanced Oil Recovery

Nazima Sultana, Ajmol Hussain Barbhuiya, Rocktim Ranjan Borah, Umme Humiara Begum, and Damaphihi F. Dkhar

Abstract Enhanced oil recovery (EOR) is considered a crucial strategy for optimizing the extraction of hydrocarbons from established reservoirs, so far traditional techniques frequently encounter difficulties in challenging salinity, temperature, and reservoir heterogeneity conditions. Surfactants are essential for boosting oil mobilization, changing wettability, and lowering interfacial tension. The synergistic utilization of mixed surfactant systems, which perform better physicochemical than the individual surfactants, is highlighted by recent studies. This chapter provides a comprehensive overview of surfactant systems, including their various forms, essential traits, and methods of operation in porous media. In addition to that, thermodynamic models and molecular interactions are utilized to study the fundamentals of synergism, and then experimental and theoretical methods are employed to describe the synergistic effects of the mixed surfactant systems. The potential of surfactant combinations such as anionic-cationic, anionic-nonionic, gemini-conventional, etc., systems to improve oil recovery across sandstone and carbonate reservoirs is demonstrated by different case studies and their practical applications. Besides, important issues are also covered, such as formulation stability, surfactant adsorption, and environmental concerns. Finally, future prospects are drawn with an emphasis on integrating nanotechnology, polymers, green surfactants, and computational modeling to develop economical and sustainable surfactant formulations for the next generation of EOR systems.

Keywords Mixed surfactants · Enhanced oil recovery · Synergistic effects · Interfacial tension · Wettability alteration

N. Sultana (✉) · A. H. Barbhuiya · R. R. Borah · U. H. Begum · D. F. Dkhar
Department of Chemistry, University of Science and Technology, Ri-Bhoi, Meghalaya, India
e-mail: sultana.nazima47@gmail.com

M. Abdul Rub et al. (eds.), *Industrial Applications of Individual and Mixed Surfactant Systems—Volume 2*, Sustainable Materials and Technology,
https://doi.org/10.1007/978-981-95-9195-4_8

1 Introduction

The continuous rise in global demand, connected with reducing access to cost-effective oil reservoir, makes it rising challenging to extract the maximum volumes from existing fields. There are inherent limitations to conventional oil extraction techniques, such as primary recovery that depends on natural reservoir pressure and secondary recovery that uses gas or water injection. Only around 30–40% of the oil that was initially present in the reservoir is usually extracted using these recovery methods, leaving a sizable amount trapped in the pore network of the rock. Enhanced oil recovery (EOR) is the collective term for an advanced tertiary procedure that must be adopted, when reservoirs grown-up due to relying on these conventional methods and outcome becomes insufficient [1, 2]. Surfactant-based flooding has become one of the most promising EOR approaches. In general, surfactants are amphiphilic compounds that have both hydrophilic and hydrophobic areas. They improve the residual oil mobility, change reservoir rock's wettability, and lower oil–water interfacial tension. In particular, they are useful in the case of carbonate and sandstone reservoirs due to the capacity to release oil that is combined into the rock matrix [3]. However, in real-world applications, single surfactant systems frequently encounter various difficulties including low effectiveness in complex reservoir conditions, adsorption losses, and instability under salinity or temperature [4].

However, growing interest has been directed the use of mixed surfactant systems to overcome these issues. In contrast to employing single surfactants, blending of several surfactant classes including anionic-cationic, anionic-nonionic, or combinations of gemini/polymeric surfactant and other conventional surfactants often produces cooperative interactions that improve the overall physicochemical performance of the mixed surfactant systems. These synergistic formulations often exhibit improved structural stability, reduced critical micelle concentrations and better interfacial properties. Consequently, it enhances the effectiveness of oil displacement in reservoir conditions. A promising basis for developing optimal surfactant formulations for enhanced oil recovery is supported by both theoretical and experimental data with a deep understanding of these synergistic mechanisms. This primary objective of this chapter is to investigate how combinations of various surfactants can have synergistic effects that raise enhanced oil recovery's efficiency. Mixed surfactant systems such as anionic-cationic, anionic-nonionic, or gemini/polymeric-conventional combinations, etc., might interact in ways that improve their performance compared to a single surfactant. For instance, a carefully formulation might significantly reduce the interfacial tension of the mixtures, enhancing oil displacement, whereas a single surfactant may show those effect moderately. This chapter also offers a structured discussion that starts with fundamental characteristics of surfactants and the key attributes of synergism. The mechanisms underlying enhanced performance are then examined to assess these mixed surfactant formulations followed by different experimental methodology. Formulation stability and compatibility with reservoir conditions including cost factors are explored to understand the real-world deployment issues and practical industrial applications. Finally, the chapter concludes by outlining new developments such as combination of surfactant blends with polymer, nanoparticles, and other sustainable substances. In addition to that, it emphasizes how machine learning and computational modeling are becoming increasingly important in predicting and designing more efficient formulations. As a result, our overall aim is to demonstrate why surfactant synergism is crucial for formulating surfactant compositions that are more effective in developing sustainable and scalable enhanced oil recovery technologies.

2 Fundamentals of Surfactant Systems

2.1 Definitions and Classifications

Surface-active agents, often known as surfactants, are a special class of amphiphilic molecules that are essential to play fundamental roles in many industrial, chemical, and biological processes. The main attributes of a surfactant comprise of two opposing structural segments [5]:

- A polar hydrophilic head group with a strong propensity for aqueous phases.
- A hydrophobic tail, which is non-polar and repels water, is typically a hydrocarbon chain that interacts more favorably with non-polar surroundings or oils.

As surfactants can preferentially adsorb at interfaces between polar and non-polar phases (such as water/oil or air/water) due to their amphiphilic nature, which lowers the surface tension and interfacial tension of surfactant. Many of the practical uses of surfactants are based on their ability to self-assemble into a wide range of supramolecular structures including micelles, vesicles, bilayer, and liquid crystalline phases [6]. Based on the charge of their hydrophilic head groups, surfactants are classified into four major categories, i.e., anionic, cationic, nonionic, and zwitterionic. Each kind has unique interfacial characteristics that are especially important for applications involving enhanced oil recovery. In particular, anionic surfactants are frequently used in petroleum reservoirs due to their potent capacity to dramatically lower the oil–water interfacial tension, which enhance the mobilization of standard oil. Contrarily, cationic surfactants have positively charged head groups that interact strongly with the negatively charged surface of the reservoir rocks. They are useful for EOR applications owing to this interaction, which often produces significant adsorption as well as effective wettability alteration from oil-wet to water-wet. However, ionic surfactants may become ineffective in critical reservoir conditions such as high salinity and extreme temperature. In addition to that, nonionic surfactants, which contain polar but uncharged head groups, work well in high salinity and high temperature conditions as they rely on hydrogen bonding rather than electrostatic interactions. This molecular interaction prevents precipitation in presence of multi-valent ions and makes

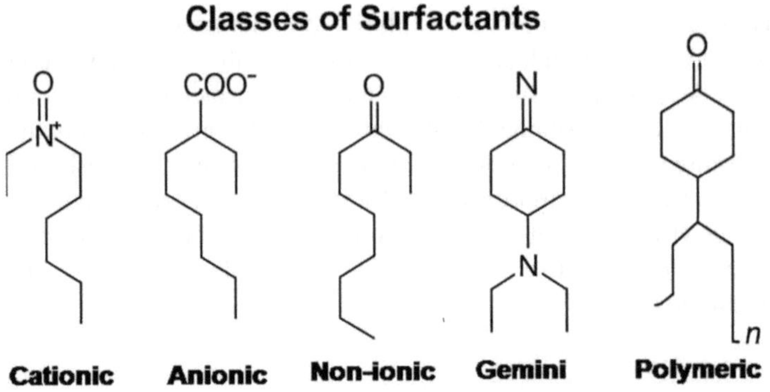

Scheme 1 Structural differences between cationic, anionic, non-ionic, gemini, and polymeric surfactants are illustrated

certain thermal stability for enhanced oil recovery. Similarly, zwitterionic surfactants are another type of surfactants having both positive and negative charges which combine with great stability with high tolerance to electrolyte. Beyond these classifications, gemini surfactant with two head and two tails part connected by a spacer is another significant form of surfactants which are appealing for EOR application due to very critical micelle concentration values as well as higher interfacial tension reduction efficiency. In a similar vein, polymeric surfactants improve mobility control and synergize with other flooding agents. Besides, bio-surfactants are eco-friendly alternatives with a strong thermal stability. The hydrophilic-liophilic balance (HLB), which determines the surfactant capacity to modify the wettability, lower interfacial tension and increase oil displacement efficiency for their effectiveness in enhanced oil recovery. A schematic representation of various classes of surfactants is shown in Scheme 1.

2.2 Physicochemical Characteristics of Surfactants in Controlling Enhanced Oil Recovery

The stability of surfactant-oil-water systems, fluid-rock interactions, and interfacial phenomena are the basic features of surfactants in controlling enhanced oil recovery and they all are determined by the physicochemical parameters of surfactants. In EOR applications, the following characteristics are shown crucial performance determinants:

2.2.1 Critical Micelle Concentration (CMC)

Surfactant molecules start to self-assemble into micelles in a particular concentration beyond a particular threshold, which are ordered colloidal aggregates created to reduce the free energy of the system are known as the critical micelle concentration. At this point, surfactant molecules transition from existing as individual monomers to forming organized clusters, optimizing their amphiphilic properties [7, 8]. Below the CMC, surfactants mainly exist as monomers, while above it, micellization occurs, accompanied by abrupt changes in solution properties such as surface tension, electrical conductivity, viscosity, osmotic pressure, and thermal conductivity [9]. The CMC is influenced by multiple factors including surfactant molecular structure (e.g., head group charge and chain length) [10], solution salinity and ionic strength [11], pH and temperature [12], pressure [13] and the presence of additives such as short-chain alcohols or polymers, etc. Several experimental techniques have been developed to determine the CMC of surfactants, which are broadly classified into **direct methods** where abrupt changes in solution properties such as surface tension, viscosity, electrical conductivity, refractive index, or osmotic pressure are monitored. However, the **indirect methods** rely on the response of an external probe or solubilized species, such as in spectroscopic, dye adsorption, or solubilization approaches. In recent years, advanced techniques including **voltammetry, dynamic light scattering, and calorimetry** have been increasingly employed, offering more accurate and complementary insights into micellization behavior [14–16]. To visualize the structural organization within micelles, Scheme 2 illustrates a spherical ionic micelle depicting the arrangement of surfactant molecules, bound counterions, and the electrical double layer.

In the context of EOR, **CMC** plays a pivotal role in dictating the efficiency of surfactant formulations. At concentrations slightly above the CMC, surfactants are capable of reducing the **interfacial tension (IFT)** which is defined as the force per unit length existing at the interface between two immiscible fluids such as

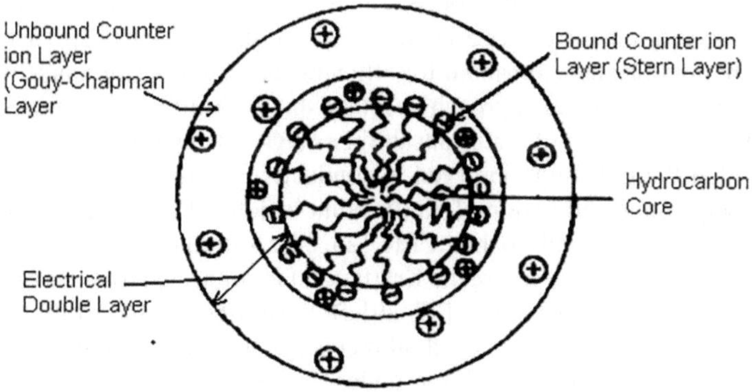

Scheme 2 Schematic representation of a spherical ionic micelle illustrating the arrangement of surfactant molecules, bound counterions, and the surrounding electrical double layer

oil and water to ultralow values (10^{-3}–10^{-2} mN/m), which significantly mobilizes trapped residual oil from pore throats. Similarly, micelles greatly increase solubilization capacity by permitting hydrocarbons to dissolve inside their hydrophobic cores. In the end, this action increases the displacement efficiency under reservoir circumstances by stabilizing the dispersed oil phase. Another important factor that contributes to better oil recovery is wettability alteration, which changes the behavior of rock surface from oil-wet to water-wet paths. For instance, cationic surfactants like CTAB can strongly adsorb onto negatively charged carbonate surfaces and change their path from oil-wet to water-wet carbonate reservoir, but nonionic ethoxylate surfactants retain stability and functioning at high reservoir temperatures. Through improved oil displacement and flow channels inside the reservoir, this transition improves the efficiency of water flooding. While individual surfactants increase oil recovery through processes such as decreased interfacial tension and altered wettability, mixed micelle systems formed at or above the CMC provide additional advantages. These systems can create stable microemulsions that efficiently transport oil, reduce capillary forces, and enhance sweep efficiency by raising the mobility ratio. As a result, the CMC becomes more than simply a basic physicochemical characteristic in chemical EOR operations; it becomes a strategic design parameter for determining the optimal surfactant concentration that balances technical efficiency and economic sustainability.

2.2.2 Krafft Temperature and Krafft Point

The Krafft temperature (T_k), often known as the critical micelle temperature is a crucial factor controlling surfactant micellization [17, 18]. It is commonly described as the hydrated surfactant melting point, where sufficient concentrations dissolve in the aqueous phase to enable the formation of micelles [18]. These are insoluble or only weakly soluble below this temperature, which restrain the formation of micelles and restricts their utility [19–21]. The solubility of surfactant increases dramatically at the Krafft temperature, and when the concentration of the surfactant surpasses the critical micelle concentration, micelles start to form upon further heating [22]. Therefore, this temperature provides as a threshold for efficient surfactant performance in aqueous media. A widely employed approach to establish the Krafft temperature is to assess the electrical conductivity of surfactant above a range of temperature. The Krafft point is identified at the inflection observed in conductivity versus temperature plot [23, 24]. A closely connected notion of Krafft point is defined the temperature at which surfactant's solubility equals to their CMC [25, 26]. And solubility measurements are determined experimentally where the Krafft point is connected to the inflection point of solubility curve with CMC curve [27–29]. Both Krafft point and Krafft temperature are of critical significant in enhanced oil recovery due to the influence of surfactant solubility and micellization condition under the reservoir. Specially, high salinity and increased temperature range are two common parameters during the surfactant flooding which can strongly affect the overall performance of the formulation. For example, if a surfactant's Krafft temperature is higher than the reservoir temperature, it will partially insoluble or unable to form micelles or microemulsions, resulting in inefficient interfacial tension reduction and limited wettability alternation. On the other hand, low Krafft temperature surfactants assurance solubility and micellization in reservoir environment, enhanced oil dislocation efficiency. For instance, anionic surfactants like sodium dodecyl sulfate (SDS) have Krafft temperatures close to room temperature, that makes their uses limited in low-temperature reservoir conditions unless cosurfactant, alcohols, or salts are added to reduce the T_k. In result, it is crucial to formulate surfactant systems with properly calibrated Krafft behavior to control their stability and functioning in EOR applications.

2.2.3 Cloud Point

The cloud point is defined as the temperature at which aqueous solutions of nonionic surfactants exhibit sudden cloudiness or turbidity [30, 31]. The solubility of a broad range of nonionic surfactants, such as those of the polyethoxylated type, depends on forming hydrogen bonds between the head-groups of surfactant and water molecules [32, 33]. The increase in solution temperature, however, weakens these bonds and results in the dehydration of polyoxyethylene chains existing in the head-groups of the surfactant [34, 35]. Therefore, above the cloud point, surfactant solutions tend to undergo phase separation (i.e., they separate into two distinguished phases). One phase contains a high surfactant concentration, while the second phase is a bulk aqueous phase with a comparatively lower surfactant concentration [36, 37]. Puerto et al. state that nonionic surfactants, like alkoxylated glycidyl ether sulfonates (AGES), may perform poorly in high-temperature reservoirs due to their cloud-point behavior [38]. Author reported that the AGES surfactant separates into two different phases when the solution temperature hits the cloud point and in addition of electrolytes like NaCl, this action intensifies to further reduction of cloud-point temperature. In result, it decreases the stability and efficacy of surfactant compositions in reservoir conditions. However, such issue can resolve by mixing AGES with anionic surfactant which leads enhance solubility in aqueous NaCl solution at higher temperature. In this context, therefore cloud-point behavior is played crucial as phase separation reduces the availability of surfactant composition at oil–water surface to make lower interfacial tension and inadequate mobilization of trapped oil. Additionally, sweep efficiency and solubilization capacity may be exposed through the instability of non-ionic surfactant solutions at reservoir relevant salinity and temperature. In order to assure, superior phase stability and increased oil displacement efficiency in challenging reservoir's environment and surfactant formulation ratio can be tailored to exhibit high cloud points or blended with more thermally stable anionic surfactant.

2.2.4 Hydrophilic-Lipophilic Balance (HLB)

The hydrophilic-lipophilic balance is a pragmatic measurement of the relative contributions of hydrophilic and lipophilic groups in a surfactant molecule. It has a significant impact on the performance of surfactants in oil–water systems through the influence of micelle formation, emulsification propensity and micro-emulsion stability. To produce ultra-flow interfacial tension, stabilize oil-in-water emulsions and preserve competence under reservoir salinity and temperature conditions, EOR applications are preferred by choosing surfactant with appropriate HLB values. Researcher have developed a number of equations including Griffin's and Davies approaches which can link to the molecular structure of surfactant in order to estimate the HLB values via hydrophilic-liophilic contributions. Griffin's method for nonionic surfactants is one of the most effective approaches:

$$\text{HLB} = 20 \times \frac{M_{\text{hydrophilic}}}{M_{\text{total}}}$$

where "M_{total}" is the surfactant's total molecular weight and "$M_{\text{hydrophilic}}$" is the molecular weight of hydrophilic part of surfactant. Similarly, Davies' group contributes an approach which is frequently used for a surfactant with a more complicated structure:

$$\text{HLB} = 7 + H_i - K_j$$

where the group numbers "H_i" and "K_j" stand for hydrophilic and lipophilic parts, respectively.

2.2.5 Micellar Solubilization Capacity

Micellar solubilization capacity is the ability of a surfactant to incorporate the hydrophobic oil parts interested in the hydrophobic hub of micelles, once surfactant concentrations surpass the CMC. This property is essential to improved oil recovery as it directly persuades the mobilization of enduring oil trapped in pore spaces subsequent to conventional water-flooding. Surfactants having high solubilization ability can encapsulate superior amounts of hydrocarbons, raising the oil's clear solubility in the aqueous phase. This process reduces the oil–water interfacial tension to ultra-low levels and facilitates the formation of microemulsions, both of which are essential for displacing capillary-trapped oil. Moreover, effective micellar solubilization improves sweep efficiency by enabling uniform distribution of surfactant solutions throughout the reservoir. For example, SDS micelles effectively solubilize hydrophobic hydrocarbons, and this capacity is further enhanced in mixed micellar systems particularly with nonionic co-surfactant like Triton X-100 and due to the synergistic interactions of these mixed surfactants. It lowers the critical micelle concentration and increase solubilization efficiency [39].

2.2.6 Microemulsions and Winsor's *R* Ratio

Microemulsions are thermodynamically stable, transparent, and isotropic mixtures composed of oil, water, surfactant, and often a co-surfactant. Depending on phase behavior, they can form oil-in-water (Winsor I), water-in-oil (Winsor II), or bi-continuous middle-phase systems (Winsor III). In EOR, the formation of Winsor III microemulsions is particularly beneficial, because they enable ultralow interfacial tension (~ 10^{-3} to 10^{-4} mN/m), high solubilization of both oil and water, and enhanced oil mobilization. For instance, a castor oil-based novel surfactant combined with propan-2-ol formed stable Winsor III microemulsions that significantly reduced IFT under optimal salinity, making them promising candidates for surfactant flooding in EOR applications [40]. Winsor's *R*-ratio represents the balance between a surfactant's affinities for the oil phase versus the aqueous phase. It is defined as:

$$R = \frac{A_s - o}{A_s - w}$$

- when $R < 1$, an oil-in-water (Winsor I) micro-emulsion is favored,
- when $R = 1$, a middle-phase bi-continuous system (Winsor III) forms which is ideal for EOR due to ultralow interfacial tension and maximum solubilization capacity,
- when $R > 1$, a water-in-oil (Winsor II) system forms.

Phase behavior studies demonstrate that optimized surfactant formulations designed to maintain $\mathbf{R \approx 1}$ can generate Winsor III microemulsions, which have been shown to dramatically improve oil recovery even in challenging heavy oil systems by significantly reducing viscosity and enhancing displacement efficiency. For instance, phase behavior studies were used to develop an alkali–surfactant–polymer (ASP) formulation to create stable Winsor III microemulsions with viscous heavy oil and author reported that this system recovered over 80% of the material from Berea sandstone cores [41].

2.2.7 Role of Surfactant and Molecular Packing Parameter

A key feature that gearshift how surfactants self-assemble into a diverse aggregates morphology is governed by the molecular packing parameter, commonly referred to as the critical packing parameter (CPP). It is described as:

$$\mathrm{CPP} = \frac{v}{a_0 \cdot l_c}$$

where "l_c" is the hydrophobic chain length, "a_0" is the effective headgroup area, and "v" is the hydrophobic tail volume [42]. The preferred surfactant morphology is determined by the value of CPP, which is displayed in Table 1.

Table 1 Molecular packing parameter (CPP) and aggregate morphologies

Critical packing parameter (CPP)	Preferred aggregate structure	Shape of aggregates
CPP < 1/3	Spherical micelles	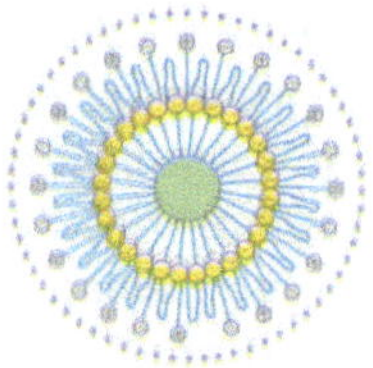
1/3 < CPP < 1/2	Cylindrical (rod-like) micelles	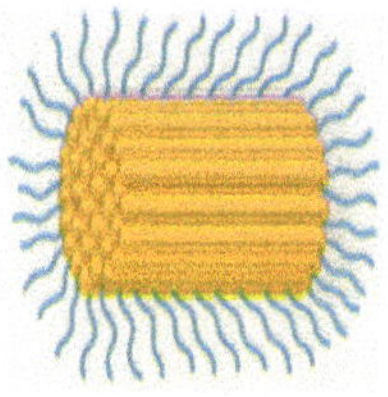
1/2 < CPP < 1	Bilayers, vesicles	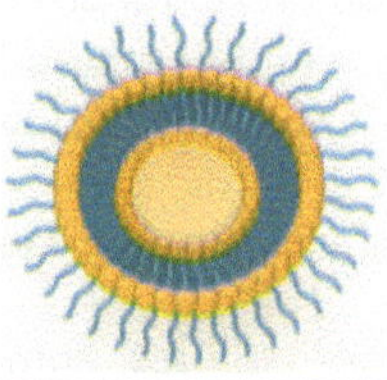
CPP ~ 1	Lamellar phases	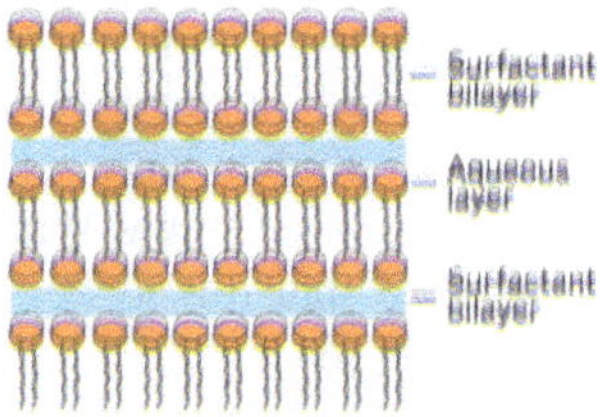
CPP > 1	Reverse micelles	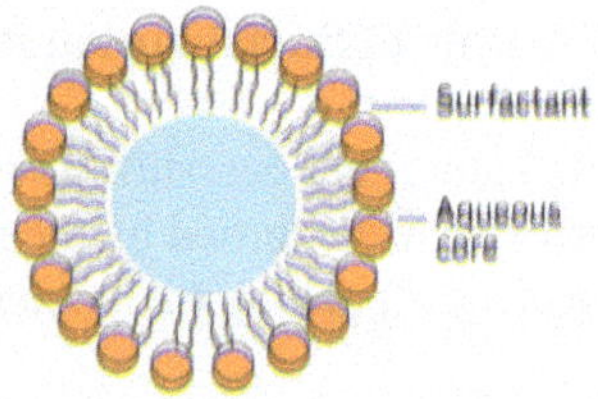

2.2.8 Other Significant Properties

Surfactants must function efficiently under difficult physicochemical conditions present in petroleum reservoirs when employed in enhanced oil recovery. Among these, thermal stability is an essential because, surfactant formulations continue to effectively reduce oil=water interfacial tension that must retain their structural integrity and interfacial action at high reservoir temperature condition [43–45]. Similarly, salt tolerance is an important property as it regulates the ability of surfactants to maintain ultralow interfacial tension and produce stable microemulsions in saline brine environments, particularly in the formations with high concentrations of divalent ions [46–49]. Besides, advanced EOR methods prioritize biodegradability and environmental compatibility in addition to these performances attributes, making sure that surfactant formulation not only achieve high oil recovery but also adhere to legal and environmental safety requirements [50–52]. Beyond inherent characteristics, hydrophilic–lipophilic deviation (HLD) concept also provides a significant framework for predicting the surfactant phase behavior under various reservoir conditions. HLD quantitatively illustrates the steadiness between hydrophilic and lipophilic forces in the formulation of a surfactant system via accounting various variables like salinity, oil type, temperature, and surfactant structure. In this case, system promotes the development of bi-continuous Winsor III microemulsions when HLD ≈ 0. These microemulsions are perfect for EOR because they maximize oil mobilization from a porous media and attain ultralow interfacial tension. Hydrophilic systems (oil-in-water, Winsor I) are indicated by negative HLD values, whereas lipophilic systems (water-in-oil, Winsor II) are indicated by positive values. Therefore, by facilitating the methodical optimization of surfactant formulations for a particular reservoir circumstance, HLD framework directly supplements the previously described physicochemical features, guaranteeing both technical efficiency and field adaptability [53, 54]. As was previously discussed, HLB offers helpful molecular-level guidance for choosing surfactants in emulsification systems; nonetheless, it is still a fixed empirical parameter and does not account for various specific conditions that are unique to reservoirs. The HLD, on the other hand, is a dynamic formulation parameter that takes into account various outside factors including temperature, salinity, oil type, and surfactant structure and is particularly useful for customizing surfactant formulations in enhanced oil recovery because it provides a prediction framework for micro-emulsion phase behavior, whereas HLB helps with initial surfactant screening.

2.3 Mechanisms of Surfactant-Based Enhanced Oil Recovery

The unique ability of surfactants to change the reservoir's surface and interfacial properties is crucial to their successful application in enhanced oil recovery. Due to strong capillary forces, poor wettability conditions, and ineffective displacement at the pore scale, conventional water flooding frequently leaves behind a sizeable portion of residual oil. Surfactants overcome these limitations by simultaneously mobilizing the stored oil through a variety of physicochemical mechanistic pathways when they are added to the reservoir. For instance, interfacial tension reduction, wettability modification, micellar solubilization, emulsification, and stability of microemulsions are important mechanistic considerations that enhance various physiochemical property including sweep efficiency and displacement efficiency. Designing effective surfactant formulations that continue to function under the various environmental reservoir conditions, particularly temperature, salinity, and pressure which require a deep understanding of these processes.

2.3.1 Reduction of Interfacial Tension (IFT)

One of the primary ways for surfactants enhance oil recovery is by reducing the interfacial tension between the composition of oil and water. Capillary forces, which are openly related to the quantity of oil–water IFT, can cause crude oil to remain trapped inside the pore spaces in petroleum reservoirs. And this is due to the low capillary number; therefore, the residual oil cannot be eliminated by traditional water flooding [55]. Surfactants reduce the interfacial free energy by orienting their hydrophilic head groups toward the aqueous phase and their hydrophobic tails toward the oil phase, when they adsorb at the oil–water contact. This reduction can dramatically increase the capillary number and facilitate the mobilization of trapped oil by lowering IFT from several tens of mN/m to ultralow rate in the range of 10^{-3}–10^{-2} mN/m. For example, researcher reveals the study on CTAB with smart seawater (SW) in carbonate reservoirs, where they highlight the reduction mechanism of IFT as a crucial factor in enhanced oil recovery. In this study, author reported the ultra-low IFT values (< 0.087 mN/m) which were attained by the optimizing surfactant formulation by reducing capillary forces and released trapped oil. Additionally, a transition to a substantially water-wet state increased the displacement efficiency. Even though author improved the adsorption of CTAB onto the rock surface by smart water, however, the adsorption levels stayed below 1 mg/g rock, which is considered as an acceptable condition for EOR applications [56]. Similarly, another study reported that another significant physiochemical features, i.e., HLB determines whether the surfactant prefers oil-in-water or water-in-oil systems, and HLD concept assimilate reservoir-specific factors like salinity, temperature, and oil type into consideration for controlling the effectiveness of IFT reduction. Authors stressed that in order to forecast the phase behavior; the HLD approach should incorporate all factors that affecting the surfactant-oil-water systems formulation. As we mentioned earlier, when $HLD = 0$, the system is in Winsor III conditions and that allow for effective oil mobilization due to ultra-low IFT. Therefore, with immediate implications for IFT reduction as a crucial mechanism in EOR, mapping emulsion performance through HLD offers a methodical framework to analyze the transitions between emulsion types to optimize the system for industrial processes [57]. As we understand now that the negative or positive HLD values suggest O/W or W/O emulsions, respectively, which are less beneficial for EOR. Therefore, in addition to being a crucial mechanism of surfactant flooding, the lowering of IFT is a property-driven process where vigilant molecule blueprint ensures the optimal performance under various reservoir circumstances.

2.3.2 Wettability Alteration

Another significant way that surfactants improve oil recovery is through the wettability alteration methods. The majority of carbonate reservoirs are mostly oil-wet or mixed-wet systems, which limits the flow of injected water and keeps a sizable proportion of oil stranded can be trapped [58, 59]. Studies show that the reservoir rock becomes more water-wet system when surfactants are added and this is due to the adsorption of surfactant formulated composition onto the rock surface to alter their surface free energy [60]. This modification enhances the water imbibitions into the pore spaces, also raises water's relative permeability to facilitate the more efficient displacement of previously trapped oil. As a result, wettability modification and interfacial tension reduction can work together to create a dual mechanism that greatly increases the displacement efficiency and total oil recovery. If we consider a carbonate reservoir, there are two basic ways that surfactants might change their wettability. Anionic surfactants work via forming a monolayer of surfactant molecules on the rock, exposing hydrophilic head groups and increasing the surface's moisture content. In context to that, Jia et al., investigated the combined effects of SDS and 1-dodecyl-3-methylimidazolium chloride (M12) on the efficiency of formulated solution properties and their applications in EOR. They reported that under high salinity and temperature conditions, the combined SDS–M12 system reduced IFT more significantly than individual surfactants [61]. While all systems altered rock wettability, the mixed surfactants showed stronger effects due to the formation of a pseudo two-tailed co-surfactant. Importantly, the SDS–M12 system improved oil recovery by over 10% under harsh reservoir conditions. However, with cationic surfactants, a cleaning mechanism dominates, where ion-pair interactions strip the adsorbed oil layer and restore the rock's original water-wet nature. Hammond and Unsal demonstrated that water-soluble cationic surfactants restore wettability in oil-wet capillaries through surfactant-contaminant complexation, effectively removing the oil film from rock surfaces [62]. These findings support the cleaning mechanism as a key pathway for wettability alteration in carbonates. Overall, cationic surfactants generally show better performance in carbonate reservoirs due to stronger ion-pair interactions compared to the hydrophobic interactions of anionic surfactants, while similar mechanisms may operate in sandstones depending on the surfactant type.

2.3.3 Micellar Solubilization Mechanism

In addition, mechanism of surfactant for enhance oil recovery can also be understand through the micellar solubilization, where oil molecules are incorporated into the hydrophobic cores of surfactant micelles. Both single and mixed surfactant systems have the ability to solubilize the residual oil above the critical micelle concentration, which actually increases its mobility and facilitate the transportation with the aqueous phase. For example, Salager et al. noted the effectiveness of HLD framework to evaluate the strong influence of surfactant formulation with different variables including oil type on emulsion inversion [63]. Emulsion inversion is a crucial surfactant mechanism for improving oil mobilization in EOR since ultra-low IFT and microemulsions production occur at circumstances close to HLD = 0. In this regard, micelle formation above this threshold promotes the solubilization process and make it necessary for phase transitions and the process is intimately related to the micellar concentration. Similarly, in difficult reservoir conditions of high salinity and temperature, mixed surfactant systems frequently show synergistic effects via creating more stable form of surfactant aggregates with enhanced solubilization capacity in comparison with singular surfactants. This process is complimentary to ultra-low IFT and wettability modification in tertiary recovery. Acosta et al. demonstrated how the HLD-NAC "equation of state" can estimate the microemulsions phase transitions, solubilization capacity, and interfacial tensions in a system containing nonionic surfactants like alcohol-ethoxylates, nonylphenol ethoxylates, etc., by expanding the studies of net-average curvature (NAC) model to those surfactant mixtures [64]. In their studies, the highlighted key features of surfactant formulations such as characteristic curvature (C_c), characteristic length (L), and molecular area, and reported that the extended nonionic surfactants with ethoxylate or other hydrophilic chains have extremely low micellar concentration. And this lower CMC value promotes significantly micelle formation, enhances solubilization capability, and facilitates the generation of microemulsions under comparatively mild reservoir conditions. Furthermore, their findings show that the integration of CMC data with HLD-NAC modeling is played crucial role for the development of ideal surfactant formulations that maximize oil solubilization and generate ultralow interfacial tension, hence improving performance in EOR applications.

3 Principles of Synergism in Surfactant Mixtures

Synergistic interactions in surfactant mixers are caused by a number of chemical mechanisms, including hydrogen bonding between polar groups, enhanced hydrophobic packing of alkyl chains, and electrostatic attraction between oppositely charged head groups. These interactions have an impact on important micellization features such as the CMC, micellar size and shape, aggregation number, and degree of counterion binding constant. As a result, mixed surfactant systems often perform better than individual surfactants due to superior structural organization [65–71]. Mixed surfactant solutions typically lower CMC values and increase aggregation number, producing a larger and more stable surfactant aggregates with superior solubilization capabilities. In order to improve oil mobilization and interfacial coverage, surfactant morphology may also shift from spherical to rod-like or worm-like structures. Additionally, by screening the electrostatic repulsions between the surfactant mixers or additives, counterion binding stabilizes the micelles and encourages closer molecular packing parameter at oil–water interface. A mathematical framework for describing such non-ideal mixing processes and predicting the formation of efficient micelle is provided by considering different thermodynamic models such as Rubingh, Clint, and Motomura's models [72–74]. Importantly, during the micellization process, scientist identified a cooperative mechanism that propels a quick transition from monomeric surfactant to aggregate micelles of sufficiently large size [75, 76]. And synergistic interactions in mixed surfactant systems amplify these cooperative effects, resulting in optimal micellization behavior. Furthermore, different environmental variables can easily regulate the physiochemical properties of surfactants by altering their molecular packing and interfacial structural formation for an effective enhanced oil recovery. Therefore, the following sections offer a comprehensive description of these features.

3.1 *Thermodynamic Basis of Synergism*

In mixed surfactant systems, the idea of synergism is essentially thermodynamic and occurs from non-ideal interaction processes between various surfactant species. By assessing the equilibrium between enthalpic and entropic contributions of the system during the micellization process, thermodynamic models aid in the quantification of these effects. Notably, the models put forth by different thermodynamic models such as Clint, Rubingh, and Motomura models have been extensively used to evaluate this micellization behavior of binary and ternary surfactant mixtures. For instance, Rubingh's model takes into consideration of interaction parameters that quantify the deviations from ideality and reflecting their antagonistic or synergistic effects. While Clint's ideal mixing model offers a baseline for forecasting the CMC of mixed systems. This approach was expanded by Motomura models by integrating the comprehensive thermodynamic calculations of mixed micelles, including the information on micellar stability, activity coefficients, and compositions. In addition, the characteristics of mixed micelles also provide a logical foundation for designing the surfactant formulations. Such thermodynamic knowledge is essential in the context of EOR because it directs the design of mixed surfactant systems to produce improved mobilization of trapped oil, complimentary wettability alteration, and shrinking interfacial tension.

3.1.1 Clint Model

Mixed micelles can display ideal or non-ideal performance based on how surfactant component interact. Clint's model offers the simplest thermodynamic explanation of ideal mixing, assuming that there are no specific interactions occur between the surfactant monomers within the micelle [72]. Under such ideal environment, critical micelle concentration of the mixed micelle is calculated using the following relationship:

$$\frac{1}{C_{0,m}} = \sum_i \frac{\alpha_i}{C_{0,i}} \quad (1)$$

where α_i represents bulk mole fraction of the component "*i*" and $C_{0,i}$ represents critical micelle concentration of the pure component "*i*."

For *non-ideal systems*, deviations occur due to attractive or repulsive interactions between different surfactant molecules. This relationship is determined by establishing the activity coefficient (f_i) into the model system and it follows by:

$$\frac{1}{C_{0,m}} = \sum_i \frac{\propto_i}{f_i C_{0,i}} \quad (2)$$

Here, (f_i) represents the non-ideality of component "*i*" in the mixed micelle. The chemical potentials of surfactants in the bulk micellar phases at equilibrium are considered to obtain those expressions. Despite the simplicity, Clint model is a useful tool to the initial estimate of micellar formation of mixed surfactant which is usually relevant to potential application in enhanced oil recovery due to the synergistic reduction in micellar concentration that improves immediately the interfacial activity and oil displacement effectiveness [77]. For example, a mixture containing nonionic surfactant like polyethoxylated fatty alcohols ($C_{12}E_6$) and anionic surfactant, i.e., sodium dodecyl benzenesulfonate (SDBS), deviated from ideal mixing and showed synergistic behavior with improved microemulsions stability and lower micelle concentration, as predicted by the Clint equation. Besides, more effective mobilization of residual oil is made possible by optimizing solubilization capacity, higher thermodynamic stability, and improved interfacial tension reduction. Therefore, it demonstrates how Clint's thermodynamic model offers a theoretical basis for rational formulation of mixed surfactant in oil recovery applications.

3.1.2 Rubingh Model

In order to account for non-ideal interactions in mixed surfactant micelles, the Rubingh model goes beyond ideal mixing [73]. It computes how mixed surfactant component deviate from their ideal behavior, by introducing an interaction parameters "β_m" in relation to this approximation. According to this approximation, activity coefficient of each surfactant components is directly related to "β_m" which controls the degree of non-ideality in mixed micelle and that can be explained by following relation:

$$\beta_m = \ln \frac{f_1}{(1 - x_1)^2} = \ln \frac{f_2}{x_1^2} \quad (3)$$

$$f_i = \frac{\infty_i C_{0,m}}{x_i, C_{0,m}} \quad (4)$$

At the mixed critical micelle concentration $\left(C_{0,m}\right)$, substitution of the activity coefficients into Eq. (3) defers the following equation:

$$x_1^2 \ln \frac{\infty_1 C_{0,m}}{x_1, C_{0,m}} = (1 - x_1)^2 \ln \frac{(1 - \infty_1)C_{0,m}}{(1 - x_1)C_{0,2}} \quad (5)$$

The mole fraction in the mixed micelle can be calculated from Eq. (5) using an iterative approach, while the interaction parameter "β_m" is determined from Eq. (3). A negative "β_m" denotes the synergistic interactions, whereas a positive "β_m" signifies the repulsive interactions and a value of zero indicates to the ideal micellization process between the mixed surfactants system. For example, researcher reported that mixed surfactant system of anionic SDS and zwitterionic OAB-30 surfactants demonstrates notable non-ideal synergistic relations, resulting in lower critical micelle concentrations with enhanced micelle formulation capacity and improved viscoelastic and interfacial characteristics [78]. Therefore, the Rubingh model helps to quantify these synergistic effects of surfactant-surfactant interactions which also include improving micellar solubilization, wettability alteration, and oil displacement in heterogeneous reservoirs. Using this approach EOR efficiency may be greatly increased by adjusting surfactant composition based on thermodynamic non-ideality in comparison with traditional chemical flooding.

3.1.3 Rodenas, Valiente, and Villafruela Model (RVV)

Rodenas, Valiente, and Villafruela (RVV) [79] established a thermodynamic approach to calculate the micellar mole fraction x_1, without using the standard approximations by using the Gibbs-Duhem relation. The expression for x_1 in this model is provided as:

$$x_1 = \alpha_1 - \alpha_1(1 - \alpha_1)\left[\mathrm{dlnC}_{0,m}/\mathrm{d}\alpha_1\right] \quad (6)$$

where the bulk mole-fraction of "surfactant 1" is represented by "α_1" and the mixed critical micelle concentration is represented by and "$C_{0,m}$." The slope [dln$C_{0,m}$/ dα], which is obtained by least–squares filling of "ln$C_{0,m}$" versus "α_1" that allows direct calculation of micellar composition from experimental data. The RVV model provides a precise thermodynamic structure for describing the non-ideal mixed surfactant system via unambiguously accounting the activity co-efficient of surfactant in the micellar media. In comparison with the ideal hypothesis of the Clint equation and the deviation of Rubingh's model, RVV approach connects the excess Gibbs free energy of micellization directly to the micellar activity coefficients; facilitate more accurate calculation of synergistic or antagonistic behavior in mixed micelle systems. Using Murgules type expression of excess free energy, this model offers a thermodynamically reliable framework for calculating micellar compositions, activity coefficients, and mixed critical micelle concentrations [80]. For instance, Pei et al. reported the significance of tailoring surfactant formulations to obtain an ultralow

IFT for enhanced oil recovery [81]. In these studies, author found combinations that illustrate the strong synergistic effects, improved oil displacement efficiency via methodically assessing the surfactant compositions. Here, a favorable surfactant mixer interactions are indicated by negative value of excess free energy and activity coefficient found below unity under the RVV framework. Therefore, this integrated strategy provides a potent path toward the development of highly effective surfactant formulations in the future through the combinations predictive thermodynamic models with systematic experimental screening. Additionally, a number of models to assess micellar composition and micellization energy within the range of mixed surfactant have been presented in various literatures. As a result of anticipating and optimizing, such interactions, micellar structure, and stability of mixed surfactants, these thermodynamic approaches are essential to improved oil recovery via facilitating the more potent formulations.

3.2 *Types of Synergistic Interactions in Surfactant Mixtures*

Three important synergistic interactions are the main drivers of synergistic effects in surfactant mixtures and are explained briefly in the following section:

3.2.1 Electrostatic Interactions

These interactions arise from the attraction of surfactants with oppositely charges, which improves micellar stability and lowers CMC. Anionic and cationic surfactants, for example, can significantly reduce the CMC, increasing the efficiency of oil recovery. Researcher studies and demonstrates how important the electrostatic interactions of surfactant in determining the nature of micellization and adsorption at the rock-fluid interface. In contrast to monovalent cations like Na^+, divalent cations (e.g., Mg^{2+}) effectively compress the electrical double layer with lowering the CMC and increased adsorption in presence of zwitterionic ions surfactant adsorption onto the calcite surfaces [82]. Author observed the synergistic effects in the mixed surfactant cationic-anionic surfactant systems, where electrostatic attractions stabilize micelle by reducing the repulsions between the head groups and encourage the cooperative adsorptions. In consequence, such electrostatic-driven synergy is helpful for the effective surfactant formulations for enhanced oil recovery as it has not only increased the interfacial tension reduction but also enhances the wettability alteration.

3.2.2 Hydrophobic Interactions

The solubilization of oil components can be improved via the development of highly stable and compact micelles' structure due to the hydrophobic tail group interactions between surfactants. It helps to improve the oil displacement and interfacial tension reduction through this kind of interaction. The significance of hydrophobic interaction in oil recovery applications is demonstrated by a recent study where author mentioned the synergistic effects of two nonionic surfactants, Tergitol 15-S-12 and PEG 600 that leads to EOR applications [83]. Here, author described the cooperative interactions of surfactants that facilitated the formation of stable mixed micelle, leading to significant level of IFT reduction, i.e., 0.672 mN/m at optimal salinity concentration. Such type of hydrophobic-driven micellar assembles were improved oil solubilization with promoted wettability alteration of sandstone from oil-wet to water-wet reservoir conditions.

3.2.3 Hydrogen Bonding

In addition, stabilized micellar structures and phase behavior of mixed surfactant can also be influenced significantly via hydrogen bonding interactions between surfactant head groups and surrounding water molecules. In particular, non-ionic surfactant with ethoxylate head group generates hydrogen bonding networks with water. These interactions increase the solubilization of hydrocarbons, reduced the free energy of micellization and persuade the formation of compact and stable micelles. Hydrogen bonding assists more effective oil displacement in context of EOR via developing compatibility toward wettability alteration and interfacial tension reduction. For instance, a study on nonionic surfactant demonstrated that variations in the number of ethylene oxide groups affected the micellization, surface activity, and foaming characteristics of the system, highlighting the function of hydrogen bonding in regulating interfacial phenomena and oil recovery efficiency [84].

3.3 Factors Influencing Synergy: Salinity, Temperature, Oil Composition, pH

The competence of surfactant mixtures in EOR is significantly influenced by reservoir parameters as temperature, salinity, pH, and crude oil content. Although adsorption onto the reservoir rock surfaces can restrict the efficiency and increasing operating costs, but surfactant chemical flooding successfully lowers interfacial tension and releases stored oil. Salinity with divalent ions plays crucial role due to compressing the electrical double layer, improving micellar stability and effective electrostatic screening as well as surfactant packing parameter at oil–water interface. Temperature is another significant factor that has two conflicting effects, i.e., (i) while excessively high temperature, it breaks hydrogen bonding networks and destabilize aggregates and (ii) in moderate rises of temperature, it encourages hydrophobic interactions and micellization. Similarly, pH affects the adsorption, micelle formation and wettability alteration via regulating the surfactant ionization and reservoir rock surface charge. Surfatant oil interactions are also greatly influenced the compositions of crude oil, particularly polar and aromatic constituents like resins and asphaltenes which regulate emulsion formation, interfacial tension reduction and overall displacement efficiency [85]. A recent study demonstrated how nonionic surfactant mixtures, like Tergitol 15-S-12 + PEG-600, can form compact micelles with varying salinity and temperature by significantly lowering IFT and changing wettability, leading to an extra 17.73% of oil recovery in a core-flood test [83]. Similarly, Zhao et al. demonstrated the significance of electrostatic screening and nanoparticle-assisted stabilization by demonstrating that a zwitterionic surfactant combined with silica nanoparticles maintains ultralow interfacial tension and stable wettability under high-salinity and elevated-temperature conditions [46]. Besides, oil composition is further defined the synergistic performances, especially in heavy crude oils. Ionic interactions between surfactant head groups and aromatic components like asphaltenes surfaces affect the adsorption and micellar stability, whereas polar group interactions between surfactants and oil compositions encourage emulsion formulation and enhanced oil mobilization. Therefore, the balance of these interactions governs the overall effectiveness of surfactant flooding, highlighting the importance of understanding crude oil compositions to design mixed surfactant formulations that exploit synergistic mechanisms for enhanced micellar aggregates formation, reduction of interfacial tension and wettability alteration [86].

4 Experimental Approaches and Characterizations

A brief understanding of mixed surfactant behavior and performances in enhanced oil recovery, it is essential to integrate the interfacial, micellar, and core-level experimental approaches. Such multi-scale measurements enable the precise assessment of surfactant effectiveness, identification of surfactant interactions as well as optimization of their accurate formulations for specific reservoir conditions.

4.1 Surface and Interfacial Tension Measurements

Determination of surface tension and interfacial tension are fundamental analytical methods in evaluating the performances of mixed surfactant in EOR. The ability of surfactant to alter wettability, releases trapped oil, and to reduce capillary forces is clarified using this analysis. Drop shape analysis, pendant drop method, and spinning drop tensiometry are the common techniques that used to quantify IFT between oil–water phases. Furthermore, contact angle measurements on rock surfaces help to assess the wettability alteration, a crucial factor in enhanced oil displacement efficiency. In a comprehensive study, researcher reported the higher interfacial properties of a green surfactant derived from cashew and phenol in oil–water phase with a considerable reduction up to approximately 10^{-1} mN/m [87]. This reduction is significant for improving oil recovery, as it facilitates to separate out of oil droplets from the rock surfaces and then mobilize subsequently. It also demonstrated the ability of surfactant to endure with the changes in salinity and temperature, makes it viable candidate for EOR applications in case of challenging reservoir conditions. Similarly, the role of silica nanoparticles in conjunction with surfactant like CTAB and SDBS for increased oil recovery [88]. Standard tensiometric techniques were used to carefully quantify the surface tension and interfacial tension and zeta potential analysis to reveal the interactions between surfactants and nanoparticles. These findings showed that nanoparticles interact with surfactants and adsorb on both oil–water and air–water interfaces, that reduces the interfacial tension more effectively and alter wettability via stabilizing the formulated emulsions and improves micellar solubilization.

4.2 Advanced Methodologies to Probe Micellization in EOR Applications

Experimental techniques are required to clarify the physiochemical characteristics of surfactant that can regulate the effectiveness of enhanced oil recovery. The tendency of a surfactant to lower oil–water interfacial tension to extremely lower levels and thereby releasing trapped oil which is directly demonstrated by measuring the surface tension and interfacial tension using a variety of techniques like Wilhelmy plate, Du Nouy ring, spinning drop tensiometer, etc. Besides, wettability alteration is another important factor which is evaluated by contact angle measurements and spontaneous imbibitions analysis which shows how surfactant adsorption properties and micellar structure alter the rock-fluid interactions. A variety of methods are used to investigate the micellization behavior including conductivity measurements to ascertain critical micelle concentrations and the interactions between surfactant-surfactant or ion-surfactant; fluorescence spectroscopy using different probes like pyrene or Nile red, etc., offers information on micellar polarity, aggregation number and microstructural changes; nuclear magnetic resonance (NMR) specifically diffusion-ordered spectroscopy (DOSY) provides quantitative data on micellar size, molecular arrangement and dynamics. To understand the micellization of mixed surfactant is further enhanced using different complementary techniques like differential scanning calorimetry (DSC) or isothermal titration calorimetry (ITC) for thermodynamic parameters like enthalpy and entropy, UV–Visible absorption study provides information for aggregation monitoring and dynamic light scattering (DLS) gives information for particle size distribution. Therefore, in result these multi-scale analytical approaches enable the logical design of surfactant formulations that are appropriate for different reservoir conditions considering factors like salinity, temperature, pH, and oil compositions. It helps to establish the mechanistic connections between the micellization behavior of surfactant, reduced interfacial tension and wettability alteration for establishing an effective enhanced oil recovery system.

4.3 Core Flooding, Micro-fluidics, and Reservoir Simulations

The above discussed molecular-scale characterization analytical methods accomplish by linking basic surfactant behavior with reservoir scale performance via core-flooding, micro-fluidic imaging, and reservoir simulations. In order to replicate reservoir conditions, surfactant solutions are injected into the cylindrical rock cores under carefully regulated the pressure drop and surfactant adsorption. These experiments establish a clear connection between the residual oil displacement and micellization characteristics. For instance, a study looks into the impact of surfactant by observing how oil recovery is affected in complex reservoir conditions. Author reported the significant improvements in oil efficiency by altering wettability at the pore scale to reduce their interfacial tension. Their results demonstrated the useful application of surfactant characteristics in improving the oil recovery process and are consistent with the molecular scale insights, obtained from micellization investigations [89]. Similarly, micro-fluidics visualization observes fluid and surfactant dynamics in real-time via using micro-scale channels that resembles reservoir pore networks. The mechanisms of pore-scale displacement, emulsion formulations and wettability alteration, are considered as challenging to capture in bulk experiments. In order to investigate pore-scale mechanisms controlling increased oil recovery, Yang et al. evaluated micro-model experiments utilizing anionic, cationic, and nonionic surfactant systems. According to their research, surfactants increase the efficacy of microscopic sweeps by reducing interfacial tension, changing wettability, and releasing trapped oil structures. The study uses micro-fluidic visualization to show real-time behaviors in mixed surfactant systems, including cooperative interactions, emulsification, and film thinning [90]. In addition, those laboratory and micro-model result of surfactant formulations are further extended to field-based predictions via reservoir simulations. Computations models can replicate fluid flow, chemical transport, and oil displacement more extensively than the vast spatial and temporal dimensions using experimental inputs such as micellization, core-flood recovery factors, and IFT measurements. For example, Al-Shalabi and Sepehrnoori provided a thorough analysis of low-salinity and engineered-water injection processes in carbonate and sandstone reservoirs, highlighting their impact on fines migration, multi-ion exchange, and wettability alteration. The study gathers computational and experimental research demonstrating how altered salinity lowers residual oil saturation and improves displacement efficiency. Crucially, authors show how reservoir models that take these physicochemical variables into account can forecast field-scale performance, optimize injection plans, and direct the creation of engineered-water formulations. Thus, their work establishes a crucial connection between large-scale recovery forecasts and pore-scale dynamics [91].

4.4 *Synergistic Effects of Advanced Mixed Surfactants Case Studies in EOR*

The importance of synergistic interactions in mixed surfactant systems for improved oil recovery has been demonstrated by number of experimental investigations. For example, Kumari et al. showed a significant decrease in oil–water interfacial tension, improving oil displacement efficiency between oil and water via analysis of mixed SDS-CTAB surfactant combinations [92]. In comparison with either surfactant alone, their research demonstrated that optimum SDS–CTAB ratios provided ultralow IFT due to improved electrostatic attraction and closed interfacial packing and I results author demonstrated the synergistic effects that greatly increase oil recovery, highlighting their significance in creating effective EOR formulations. Similarly, Maleki et al. used homogeneous glass micromodels to study the effectiveness of polyacrylamide nanocomposites based on silica and alumina, namely NCSP, NCAP, and NCSAP in improving recovery of light and heavy oils under different salinities [93]. The study assessed mobility control, wettability modification, and interfacial tension as important factors. According to the results, NCSAP had the highest light oil recovery, i.e., 95.83% at 250,000 ppm salinity, mainly due to the ability of the combinations of nanocomposites with high surface area to reduce IFT. The addition of CTAB surfactant improved wettability, and a mixture of polymeric surfactants, such as polyacrylamides (PAM) via improving the mobility control. In this case, mobility control dominated the recovery for heavy oil with PAM, performing around 17% better than combination with those nanocomposites. Therefore, author emphasizes their work, under the relevant reservoir conditions, integrating processes such as IFT reduction, wettability modification, and mobility control through nanocomposites-based surfactant-polymer systems offers synergistic enhancement in oil displacement. Another study by Pablo et al. analyzed numerical simulations of a two-phase, five-component systems, i.e., aqueous and organic phases containing water, surfactant, petroleum, polymer, and salt in porous media to examine the combined use of surfactant and polymer (SP) flooding for EOR [94]. The study focused on the interaction between surfactant and polymer, showing that injecting polymer first with a slightly overlapping surfactant slug, reduces surfactant adsorption, increasing oil recovery and sweep efficiency. The significance of solution chemistry in SP flooding was highlighted by the fact that the presence of salt affected the adsorption rates but had no effect on IFT or viscosity. According to the theoretical calculations of simulation studies, carefully planned injection schemes that take advantage of SP interaction which can improve recovery factors and possibly lower chemical costs. Therefore, under this study author offers a framework for developing SP formulations under reservoir-relevant conditions and highlights the synergistic advantages of coupled chemical EOR. Furthermore, other studies reported that ZrO_2 nanoparticles impact on the interfacial characteristics of SDS and CTAB mixed surfactant system at air–water and n-heptanes-water interfaces [95]. A synergistic effect was highlighted due to the strong connections between surfactant mixture and nanoparticles and it was confirmed by zetapotential studies, improving the adsorption at the interfaces. It enhanced the formations of stabile micelle for potential use of EOR via encouraging the effectiveness of oil displacement and wettability alteration. Therefore, metal oxide incorporated nanoparticle surfactant mixture highlights the significance in attaining ultralow interfacial tension and optimizing the chemical flooding techniques. Similarly, Yekeen et al. examined how different salinity, temperature, and SiO_2-nanoparticle conditions affected the adsorption behavior of anionic (like SDS), cationic (like CTAB) and nonionic (like Tween-80) surfactants onto Malaysian shale formations [96]. According to their findings, anionic surfactants had the lowest absorption whereas cationic surfactants had the highest adsorption because of their strong electrostatic attraction to negatively charged shale surfaces. However, nonionic surfactants showed intermediate adsorption that was primarily controlled by surface polarity and hydrogen bonding. While adding SiO_2 nanoparticles decreased the surfactant retention by offering competitive binding sites and improving surfactant stability in solution, increasing salinity, and temperature usually increased adsorption. Besides, shale mineralogy and clay concentration also had an impact on adsorption behavior of this study, where significance reservoir conditions and surfactant-nanoparticles synergy help in developing the successful EOR techniques. In another study, Saien et al. examined the synergistic effect of combining anionic surfactants like SDS or SDBS with a benzimidazolium-based gemini ionic liquid, [C_4benzim-C_6-benzimC_4][Br_2] [97]. The investigation showed that this type of surfactant mixture produced exceptional synergistic gains in interfacial properties, achieving an ultra-low IFT value of 0.04 mN/m at ideal combined surfactant mixture's (GSAIL) mole fraction, i.e., approximately 0.2, decreasing the oil–water interfacial tension by up to 99.6% range. They reported that the min interactions between oppositely charged surfactants was highlighted significant result than either surfactant alone, attributed to cooperative interfacial packing and enhanced micellization. The mixed formulations also achieved more effective wettability alteration and significantly improved oil recovery in core-flood experiments. The Frumkin isotherm and RVV model were used to further evaluate the adsorption behavior of individual surfactant and their mixtures at the interface.

5 Industrial Applications and Challenges

5.1 Industrial Application of Mixed Surfactants in EOR

Chemical enhanced oil recovery has focused largely on combined formulations containing alkali-surfactant polymer (ASP) and surfactant-polymer (SP) floods due to blend mechanism like IFT reduction, wettability alteration, and mobility control that translate better to large scale recovery of mixed surfactants than single component injections. Unlike laboratory-scale studies, field-scale applications show that surfactants are mainly deployed in the presence of other additives. The majority of these additives are alkalis and polymers. Alkalis are usually incorporated to increase the pH, thereby reducing surfactant adsorption on reservoir rock. They also contribute to additional IFT reduction by the in-situ generation of surfactants. On the other hand, polymers are added to provide further mobility control, ensuring stable displacement fronts and improved volumetric sweep efficiency. For instance, the Daqing and Shengli oilfields in China provide the best-documented industrial experiences like commercial ASP/SP pilots there demonstrated substantial incremental recovery (typical reported ranges ~ 10–25% original oil in place (OOIP) in successful blocks) and highlighted critical screening parameters such as water quality, surfactant stability in high salinity/brine composition, polymer viscosity retention, and surfactant adsorption control. Reviews of these field deployments also stress that careful chemical selection and injection sequencing (for example, polymer pre-flush or overlapped polymer-surfactant slugs) reduce surfactant losses and improve sweep efficiency [98]. Beyond ASP/SP, pilot and laboratory-to-field translational work has explored mixer of (surfactant-nanoparticle) systems and novel surfactants mixer (e.g., gemini/ionic-liquid surfactants) to improve interfacial stability and salt/temperature tolerance. Nanoparticles (silica, ZrO_2, etc.) can reduce surfactant adsorption on rock, stabilize emulsions, and help maintain ultralow IFT under harsh conditions, improving the robustness of surfactant formulations in high-salinity/high-temperature reservoirs. Field-scale economics remain the main constraint including chemical cost, required dosage, produced-water handling and potential environmental impacts drive the need for lower-dosage, higher-efficiency mixtures (including gemini surfactants or polymer–nanocomposites) and for optimizing injection strategies by coupling lab screening, core floods, and reservoir simulation. In result, the industrial record shows that mixed-surfactant and hybrid chemical formulations can work at scale but only when formulation chemistry, rock/fluid screening, adsorption mitigation, and injection design are addressed holistically.

5.2 Synergism of Anionic-Cationic, Anionic-Nonionic, and Gemini-Other Conventional Surfactants in Field Studies

Field and near-field studies have increasingly demonstrated that mixed surfactant systems outperform single surfactants in terms of interfacial activity, adsorption control, and oil displacement efficiency. The first field-scale application of ASP flooding using novel anionic/cationic surfactant mixtures was conducted in a high-temperature, high-water-cut mature sandstone reservoir at the Daqing Oilfield, China. Laboratory studies had shown that these surfactant blends exhibit extremely low critical micelle concentrations (3.50×10^{-4} wt%), ultralow interfacial tension (~10^{-4} mN/m), high oil solubilization (23), and minimal adsorption (0.27 mg/g sand), making them suitable for negatively charged sandstone formations. Core-flooding experiments indicated that ASP flooding with a pre-slug of cross-linked polymer increased oil recovery by 24.2% of OOIP over conventional water-flooding under the optimized chemical formulation. During field deployment, alternating injections of ASP slugs and cross-linked polymer slugs for conformance control reduced the water cut from 97.9 to 90.2%, with an estimated ultimate oil recovery enhancement of 14.2% OOIP, reaching a cumulative recovery of 67.5% OOIP. These results highlight the ability of anionic/cationic surfactant mixtures to maintain interfacial stability, minimize adsorption losses, and deliver substantial incremental recovery under harsh reservoir conditions [99]. Moreover, anionic-nonionic mixed surfactant systems have been tailored for high-temperature and hyper-saline reservoirs, where two different classes of synergistic interactions allow for ultralow interfacial tension, enhanced salt tolerance and improved thermal stability. These combined formulations decrease the surfactant adsorption onto mineral surfaces, greatly increase the volumetric sweep efficiency and remain stable in different reservoir conditions in assessments of both field and large-scale studies. Additionally, in difficult hyper-saline reservoir conditions, alternating or co-injection with polymers offers better mobility control, enabling more effective residual oil displacement and eventually increasing overall oil recovery [100, 101]. In addition to these binary systems, gemini surfactants have been tested in carbonate reservoirs subjected to high salinity and temperature as co-surfactants with traditional surfactants. Stronger interfacial adsorption, lower CMC, and increased surface activity are all facilitated by their distinct molecular structure, which makes it possible to maintain ultralow IFT and stabilize emulsions in heterogeneous carbonate matrices. Large-scale research shows that gemini surfactants offer a viable and reliable alternative for field-level EOR operations in complicated mature carbonate reservoirs [102]. This helps reduce adsorption-related losses and improves sweep efficiency.

5.3 Challenges, Limitations, and Future Outlook

There are still a number of challenges remain to overcome despite tremendous advancements in the formulation and application of mixed surfactant solutions for EOR. Inconsistent adsorption behavior and decreased surfactant mixture's effectiveness are frequently caused by complex reservoir heterogeneity, strong mineral surface reactivity, and unpredictable changes in different salt composition. For instance, surfactant efficiency is still being lowered by adsorption losses, especially in carbonate and shale formations [103]. Sometimes, micellization and interfacial performances can be adversely affected by harsh reservoir conditions such as high temperature (> 100 °C), high salinity, (particularly when divalent ions like Ca^{2+} and Mg^{2+} are present) and high pressure [101]. Besides, in large scale chemical use also raise the economic difficulties due to cost-performance trade-offs, which is required to be analyzed carefully.

Additionally, a successful laboratory scale does not always interpret in terms of reliability to the field; a widespread application is further limited by various environmental factors such as surfactant toxicity, persistence in reservoirs, and regulatory compliance. A higher cost of non-toxic surfactant like gemini and nonionic surfactants are also restricted the use of effective formulation in large scale analysis.

Therefore, integrating the advanced molecular modeling, machine learning, and real-time reservoir monitoring can significantly help to design and prediction of synergistic interactions. Moreover, development of environmental-friendly, biodegradable surfactant mixtures and smart formulation with self-adaptive behavior beneath vibrant reservoir conditions will further enhance the viability of surfactant-based enhanced oil recovery system. In continuation, large scale testing and improved cost optimization will enable to depend on sustainable field with synergistic surfactant mixers. Future, EOR formulations will need to comply with more stringent safety and environmental regulations, particularly in relation to biodegradability, toxicity, and groundwater preservation. Open life-cycle assessments and the development of eco-friendly surfactants will be necessary to obtain regulatory approvals. Harmonizing international regulations for the use of chemical EOR could also promote more consistent, secure, and eco-friendly field practices in a range of reservoir settings.

6 Future Perspectives and Conclusions

A significant progress has been made in enhanced oil recovery via employing the surfactant-based techniques, especially with mixed surfactant systems such anionic/cationic, anionic/nonionic, and gemini/conventional mixtures. Even in difficult reservoir conditions with high salinity, high temperatures, and heterogeneous moiety, these formulations achieve ultralow interfacial tension, effective wettability alteration and increased oil displacement efficiency [99–102]. Various laboratory formulations of mixed surfactant or advanced surfactant mixtures and their findings have been confirmed via implementations of field-scale level, particularly in context of ASP and SP flooding in solid sandstone and carbonate reservoir conditions, which provide significant additional recovery while lowering surfactant adsorption and preserving formulation stability. There are still difficulties in spite of these achievements and broader applicability is still hampered by various factors including adsorption losses in reactive formations, instability in high-salinity compositions and extremely high temperatures, etc., the cost of large-scale chemical usage, environmental issues, and scale-up constraints also considered another essential point for the recovery of challenges in enhanced oil [103]. To solve the aforementioned issues, surfactant carries crucial experimental techniques for better oil recovery in the design of hybrid systems. In order to transport surfactants through a porous media, various advanced formulation minimize the adsorption losses, improve interfacial characteristics and improved oil recovery factors. Polymeric surfactant or supramolecular systems can perform as effective carriers with lowering mobility ratio concern in different viscous oil media. In addition to that, inorganic nanoparticles or carbon resources-based carriers can easily alter wettability, increasing flow pathways and block isolated pores of the reservoir for improving oil recovery. Therefore, future work should focus on combining improved carriers with synergistic mixed surfactant formulation with data-driven modeling for accurate field scale prediction. As a result, a thorough analysis of realistic reservoir conditions with high salinity, pressure and temperature will continue to be crucial. It follows the path toward ecologically responsible enhanced oil recovery which is further strengthened via the emergence of sustainable conditions.

References

1. Sheng J (2011) Modern chemical enhanced oil recovery: theory and practice. ISBN: 978-1-85617-745-0. https://doi.org/10.1016/C2009-0-20241-8
2. Gbadamosi AO, Junin R, Manan MA, Agi A, Yusuff AS (2019) An overview of chemical enhanced oil recovery: recent advances and prospects. Int Nano Lett 9:171–202. https://doi.org/10.1007/s40089-019-0272-8
3. Olajire AA (2014) Review of ASP EOR (alkaline surfactant polymer enhanced oil recovery) technology in the petroleum industry: prospects and challenges. Energy 77:963–982. https://doi.org/10.1016/j.energy.2014.09.005
4. Qian C, Telmadarreie A, Dong M, Bryant S (2020) Synergistic effect between surfactant and nanoparticles on the stability of foam in EOR processes. SPE J 25:883–894. https://doi.org/10.2118/195310-PA
5. Cullum DC (1994) Introduction to surfactant analysis, 1st edn. Springer Netherlands, Dordrecht, pp 17–41. ISBN: 978-94-010-4574-2. https://doi.org/10.1007/978-94-011-1316-8
6. Iglauer S, Wu Y, Shuler P, Tang Y, Goddard WA (2010) New surfactant classes for enhanced oil recovery and their tertiary oil recovery potential. J Petrol Sci Eng 71:23–29. https://doi.org/10.1016/j.petrol.2009.12.009
7. Bhosle MR, Joshi SA, Bondle GM (2020) An efficient contemporary multicomponent synthesis for the facile access to coumarin-fused new thiazolyl chromeno[4,3-b]quinolones in aqueous micellar medium. J Heterocycl Chem 57:456–468. https://doi.org/10.1002/jhet.3802
8. Naseri N, Ajorlou E, Asghari F, Pilehvar-Soltanahmadi Y (2018) An update on nanoparticle-based contrast agents in medical imaging artificial cells. Nanomed Biotechnol 46:1111–1121. https://doi.org/10.1080/21691401.2017.1379014
9. Sabahi N, Razfar MR, Hajian M (2017) Experimental investigation of surfactant-mixed electrolyte into electrochemical discharge machining (ECDM) process. J Mater Process Technol 250:190–202. https://doi.org/10.1016/j.jmatprotec.2017.07.017
10. Glennie AR, Mohareb MM, Palepu RM (2006) Thermodynamic and related properties of alkyl cationic surfactants based on dimethyl and diethyl ethanol amines. J Dispers Sci Technol 27:731–738. https://doi.org/10.1080/01932690600660608
11. Bratovcic A, Nazdrajic S (2020) Viscoelastic behavior of synthesized liquid soaps and surface activity properties of surfactants. J Surfactants Deterg 6:3150–3178. https://doi.org/10.1002/jsde.12444
12. Harutyunyan LR, Harutyunyan RS (2019) Effect of amino acids on micellization and micellar parameters of anionic surfactant alpha olefin sulfonate C_{14}–C_{16} in aqueous solutions: surface tension, conductometric, volumetric, and fluorescence studies. J Chem Eng Data 64:640–650. https://doi.org/10.1021/acs.jced.8b00886
13. Thiruvengadam S, Murphy M, Tan JS, Miller K (2020) A generalized theoretical model for the relationship between critical micelle concentrations, pressure and temperature for surfactants. J Surfactants Deterg 23:273–303. https://doi.org/10.1002/jsde.12360

14. Nesmerak K, Nemcova I (2006) Determination of critical micelle concentration by electrochemical means. Anal Lett 39:1023–1040. https://doi.org/10.1080/00032710600620302
15. Haq B, Liu J, Liu K, Al Shehri D (2020) The role of biodegradable surfactant in microbial enhanced oil recovery. J Petrol Sci Eng 189:106688. https://doi.org/10.1016/j.petrol.2019.106688
16. Liu J, Liu Z, Yuan T, Wang C, Gao R, Hu G, Xu J, Zhao J (2020) Synthesis and properties of zwitterionic gemini surfactants for enhancing oil recovery. J Mol Liq 311:113179. https://doi.org/10.1016/j.molliq.2020.113179
17. Nazrul Islam M, Sharker KK, Sarker KC (2015) Salt-induced modulation of the Krafft temperature and critical micelle concentration of benzyldimethylhexadecylammonium chloride. J Surfactants Deterg 18:651–659. https://doi.org/10.1007/s11743-015-1696-4
18. Roy JC, Islam MN, Aktaruzzaman G (2014) The effect of NaCl on the Krafft temperature and related behavior of cetyltrimethylammonium bromide in aqueous solution. J Surfactants Deterg 17:231–242. https://doi.org/10.1007/s11743-013-1510-0
19. Dicharry C, Diaz J, Torré JP, Ricaurte M (2016) Influence of the carbon chain length of a sulfate-based surfactant on the formation of CO_2, CH_4 and CO_2–CH_4 gas hydrates. Chem Eng Sci 152:736–745. https://doi.org/10.1016/j.ces.2016.06.034
20. Dölle S, Lechner BD, Park JH, Schymura S, Lagerwall JPF, Scalia G (2012) Utilizing the Krafft phenomenon to generate ideal micelle-free surfactant stabilized nanoparticle suspensions. Angew Chem Int Ed 51:3254–3257. https://doi.org/10.1002/anie.201106793
21. Moon TL, Blanzat M, Labadie L, Perez E, Rico-Lattes I (2001) Synthesis and physicochemical study of new surfactants derived from carboxylic acid sugars. J Dispers Sci Technol 22:167–176. https://doi.org/10.1081/DIS-100105202
22. Malik NA, Ali A (2016) Krafft temperature and thermodynamic study of interaction of glycine, diglycine, and triglycine with hexadecylpyridinium chloride and hexadecylpyridinium bromide: A conductometric approach. J Mol Liq 213:213–220. https://doi.org/10.1016/j.molliq.2015.11.017
23. Bales BL, Benrraou M, Zana R (2002) Krafft temperature and micelle ionization of aqueous solutions of cesium dodecyl sulfate. J Phys Chem B 106:9033–9035. https://doi.org/10.1021/jp021297l
24. Song B, Zhao J, Wang B, Jiang R (2009) Synthesis and self-assembly of new light-sensitive Gemini surfactants containing an azobenzene group. Colloids Surf A 352:24–30. https://doi.org/10.1016/j.colsurfa.2009.09.044
25. Peris-Vicente J, Albiol-Chiva J, Roca-Genovés P, Esteve-Romero J (2016) Advances on melamine determination by micellar liquid chromatography: a review. J Liq Chromatogr Relat Technol 39:325–338. https://doi.org/10.1080/10826076.2016.1152482
26. Zapf A, Beck R, Platz G, Hoffmann H (2003) Calcium surfactants: a review. Adv Colloid Interface Sci 100–102:349–380. https://doi.org/10.1016/S0001-8686(02)00065-9
27. Inada T, Koyama T, Tomita H, Fuse T, Kuwabara C, Arakawa K, Fujikawa S (2017) Anti-ice nucleating activity of surfactants against silver iodide in water-in-oil emulsions. J Phys Chem B 121:6580–6587. https://doi.org/10.1021/acs.jpcb.7b02644
28. Miyake M, Oyama N (2009) Effect of amidoalkyl group as spacer on aggregation properties of guanidine-type surfactants. J Colloid Interface Sci 330:180–185. https://doi.org/10.1016/j.jcis.2008.10.047
29. Shinoda K, Hutchinson E (1962) Pseudo-phase separation model for thermodynamic calculations on micellar solutions. J Phys Chem 66:577–582. https://doi.org/10.1021/j100810a001
30. Aktar S, Saha M, Mahbub S, Halim MA, Rub MA, Hoque MA, Islam DMS, Kumar D, Alghamdi YG, Asiri AM (2020) Influence of polyethylene glycol on the aggregation/clouding phenomena of cationic and non-ionic surfactants in attendance of electrolytes (NaCl & Na_2SO_4): An experimental and theoretical analysis. J Mol Liq 306:112880. https://doi.org/10.1016/j.molliq.2020.112880
31. Khattab A, Mohamed M, Basalious EB (2020) Design of self-nanoemulsifying system to enhance absorption and bioavailability of poorly permeable Aliskiren hemi-fumarate. J Drug Deliv Sci Technol 57:101646. https://doi.org/10.1016/j.jddst.2020.101646

32. Alvarez MS, Longo MA, Deive FJ, Rodríguez A (2019) Non-ionic surfactants and combination for aqueous two-phase systems. Fluid Phase Equilib 502:112302. https://doi.org/10.1016/j.fluid.2019.112302
33. Illous E, Ontiveros JF, Lemahieu G, Lebeuf R, Aubry JM (2020) Amphiphilicity and salt-tolerance of ethoxylated and propoxylated anionic surfactants. Colloids Surf A 601:124786. https://doi.org/10.1016/j.colsurfa.2020.124786
34. Mukhopadhyay C, Suba M, Sivakumar D, Dhamodharan K, Rao RVS (2019) Cloud point extractive spectrophotometric method for determination of uranium in raffinate streams during spent nuclear fuel reprocessing. J Radioanal Nucl Chem 322:743–750. https://doi.org/10.1007/s10967-019-06704-5
35. Turan K, Kaur P, Manhas D, Sharma J, Verma G (2020) Novel insights into the dispersed and acid-mediated surface modification of the carbon nanofibers. Mater Chem Phys 239:121978. https://doi.org/10.1016/j.matchemphys.2019.121978
36. Kachangoon R, Vichapong J, Santaladchaiyakit Y, Srijaranai S (2020) Cloud point extraction coupled to in-situ metathesis reaction of deep eutectic solvents for preconcentration and liquid chromatographic analysis of neonicotinoid insecticide residues in water, soil and urine samples. Microchem J 152:104377. https://doi.org/10.1016/j.microc.2019.104377
37. Pan T, Wang R, Xiao K, Ye W, Dong W, Xu M (2019) Continuous degradation of phenanthrene in a cloud point system by reuse of sphingomonas polyaromaticivorans cells. AMB Express 9:8. https://doi.org/10.1186/s13568-019-0736-2
38. Puerto M, Hirasaki GJ, Miller CA, Barnes JR (2012) Surfactant systems for EOR in high-temperature, high-salinity environments. SPE J 17:11–19. https://doi.org/10.2118/129675-PA
39. Shi Z, Chen J, Yin X (2013) Effect of anionic–nonionic-mixed surfactant micelles on solubilization of PAHs. J Air Waste Manage Assoc 63:694–701. https://doi.org/10.1080/10962247.2013.778918
40. Pal N, Saxena N, Mandal A (2017) Phase behavior, solubilization, and phase transition of a microemulsion system stabilized by a novel surfactant synthesized from castor oil. J Chem Eng Data 62:1278–1291. https://doi.org/10.1021/acs.jced.6b00806
41. Sim SS, Wassmuth FR, Bai J (2014) Identification of Winsor Type III micro-emulsion for chemical EOR of heavy oil. SPE Conf Proc: SPE-170018-MS. ISBN: 978-1-61399-334-7. https://doi.org/10.2118/170018-MS
42. Yamada S (2015) Surfactant assemblies (micelles, vesicles, emulsions, films, etc.): an overview. Encycl Polym Nanomater 2431–2436. https://doi.org/10.1007/978-3-642-36199-9_158-1
43. Hou J, Huang T, Alotaibi M, Sofi A (2024) Long-term thermal stability of ionic surfactants for improving oil production at high-salinity high-temperature conditions. ACS Omega 9:11976–11986. https://doi.org/10.1021/acsomega.3c09734
44. Memon MK, Ansari U, Memon HUZ (2020) Thermal stability and compatibility of surfactants in presence of formation water salinity under reservoir conditions. Mehran Univ Res J Eng Technol 39:826–830. https://doi.org/10.22581/muet1982.2004.15
45. Zulkifli NN, Mahmood SM, Akbari S, Manap AA, Kechut NI, Elrais KA (2020) Evaluation of new surfactants for enhanced oil recovery applications in high-temperature reservoirs. J Pet Explor Prod Technol 10:283–296. https://doi.org/10.1007/s13202-019-0713-y
46. Zhong X, Li C, Li Y, Pu H, Zhou Y, Zhao JX (2020) Enhanced oil recovery in high salinity and elevated temperature conditions with a zwitterionic surfactant and silica nanoparticles acting in synergy. Energy Fuels 34:2893–2902. https://doi.org/10.1021/acs.energyfuels.9b04067
47. Ghosh B, Obasi D (2013) Eco-friendly surfactant for EOR in high temperature, high salinity carbonate reservoir. SPE Conf Proc. https://doi.org/10.2118/165219-MS.ISBN:978-1-61399-258-6
48. Maia KCB, Francisco AD, Dos S, Moreira MP, Nascimento RSV, Grasseschi D (2024) Advancements in surfactant carriers for enhanced oil recovery: Mechanisms, challenges, and opportunities. ACS Omega 9:36874–36903. https://doi.org/10.1021/acsomega.4c04058

49. Nyah F, Ridzuan N, Nwaichi PI, Umunnawuike C, Agi A (2024) Comprehensive review on the role of salinity on oil recovery mechanisms during chemical flooding. J Mol Liq 415:126308. https://doi.org/10.1016/j.molliq.2024.126308
50. Wang J, Gu F, Han W, Fu L, Dong S, Zhang Z, Ren Z, Liao K (2023) Green surfactant made from cashew phenol for enhanced oil recovery. ACS Omega 8:2057–2064. https://doi.org/10.1021/acsomega.2c05647
51. Shaikhah D, Loise V, Angelico R, Porto M, Calandra P, Abe AA, Testa F, Bartucca C, Rossi CO, Caputo P (2024) New trends in biosurfactant: from renewable origin to green enhanced oil recovery applications. Molecules 29:301. https://doi.org/10.3390/molecules29020301
52. Abdurrahman M, Kamal MS, Ramadhan R, Daniati A, Arsad A, Abdul Rahman AF, Rita N (2023) Ecofriendly natural surfactants in the oil and gas industry: a comprehensive review. ACS Omega 8:41004–41021. https://doi.org/10.1021/acsomega.3c04450
53. Leal E, Torres-Mancera P, Ancheyta J (2022) Catalyst stacking technology as a viable solution to ultralow sulfur diesel production. Energy Fuels 36:3201–3218. https://doi.org/10.1021/acs.energyfuels.1c04038
54. He D, Jiang P, Xu R (2019) Pore-scale experimental investigation of the effect of supercritical CO_2 injection rate and surface wettability on salt precipitation. Environ Sci Technol 53:14744–14751. https://doi.org/10.1021/acs.est.9b03323
55. Cong Y, Zhang W, Liu C (2020) Composition and oil-water interfacial tension studies in different vegetable oils. Food Biophys 15:1–11. https://doi.org/10.1007/s11483-019-09617-8
56. Massarweh O, Abushaikha AS (2023) The synergistic effects of cationic surfactant and smart seawater on the recovery of medium-viscosity crude oil from low-permeability carbonates. J Mol Liq 389:122866. https://doi.org/10.1016/j.molliq.2023.122866
57. Salager JL, Anton RE, Briceno MI, Choplin L, Marquez L, Pizzino A, Rodriguez MP (2003) The emergence of formulation engineering in emulsion making: transferring know-how from research laboratory to plant. Polym Int 52:471–478. https://doi.org/10.1002/pi.1112
58. Jing W, Huiqing L, Genbao Q, Yongcan P, Yang G (2019) Investigations on spontaneous imbibitions and the influencing factors in tight oil reservoirs. Fuel 236:755–768. https://doi.org/10.1016/j.fuel.2018.09.053
59. Yu H, Lu X, Fu W, Wang Y, Xu H, Xie Q, Qu X, Lu J (2020) Determination of minimum near miscible pressure region during CO_2 and associated gas injection for tight oil reservoir in Ordos Basin, China. Fuel 263:116737. https://doi.org/10.1016/j.fuel.2019.116737
60. Standnes DC, Austad T (2000) Wettability alteration in chalk: II. Mechanism for wettability alteration from oil-wet to water-wet using surfactants. J Pet Sci Eng 28:123–143. https://doi.org/10.1016/S0920-4105(00)00084-X
61. Jia H, Lian P, Leng X, Han Y, Wang Q, Jia K, Niu X, Guo M, Yan H, Luv K (2019) Mechanism studies on the application of mixed cationic/anionic surfactant systems to enhance oil recovery. Fuel 258:116156. https://doi.org/10.1016/j.fuel.2019.116156
62. Hammond PS, Unsal E (2011) Spontaneous imbibition of surfactant solution into an oil-wet capillary: wettability restoration by surfactant–contaminant complexation. Langmuir 27:4412–4429. https://doi.org/10.1021/la1048503
63. Salager JL, Forgiarini A, Marquez R, Pena AA, Pizzino A, Rodriguez MP, Rondon-González M (2004) Using emulsion inversion in industrial processes. Adv Colloid Interface Sci 108–109:259–272. https://doi.org/10.1016/j.cis.2003.10.008
64. Acosta E (2008) The HLD–NAC equation of state for microemulsions formulated with nonionic alcohol ethoxylate and alkylphenol ethoxylate surfactants. Colloids Surf A 320:193–204
65. Dey J, Sultana N, Ismail K (2015) Counter ion binding in solutions of sodium dodecylsulfate in water–ethylene glycol mixtures and the Corrin-Harkins equation. J Mol Liq 207:107–111. https://doi.org/10.1016/j.molliq.2015.03.030
66. Thapa U, Ray D, Dey J, Sultana N, Aswal VK, Ismail K (2015) Influence of hydrotropic coions on the shape transition of sodium dioctylsulfosuccinate aggregates in an aqueous medium. RSC Adv 5:45956–45964. https://doi.org/10.1039/C5RA04151A

67. Dey J, Sultana N, Kumar S, Aswal VK, Chowdhury S, Ismail K (2015) Controlling aggregation of sodium dodecylsulfate in aqueous poly(ethylene glycol) solutions: the role of added sodium chloride. RSC Adv 5:74744–74752. https://doi.org/10.1039/C5RA13096D
68. Sultana N, Ismail K (2016) Specific ion effects of chloride vis-à-vis acetate, propionate, and butyrate counterions on the headgroup of cetylpyridinium chloride at the micelle–solution and air–solution interfaces. J Mol Liq 213:145–152. https://doi.org/10.1016/j.molliq.2015.11.016
69. Sultana N (2017) Role of ammonium ion on the aggregation and adsorption properties of sodium dodecylsulfate. J Disp Sci Technol 38:1–8. https://doi.org/10.1080/01932691.2017.1297720
70. Sultana N, Dey J, Ismail K (2019) Effect of added counterions on the micellization of cetylpyridinium chloride in the presence of acetonitrile–water mixture. J Chem Eng Data 64:4193–4205. https://doi.org/10.1021/acs.jced.9b00260
71. Sultana N, Deb DK, Ismail K (2023) A comprehensive assessment of interfacial and bulk properties of sodium dodecylsulfate via addition of competitive Hofmeister ions in presence of cosolvent: a combined experimental and theoretical study. J Mol Liq 390:122978. https://doi.org/10.1016/j.molliq.2023.122978
72. Clint JH (1975) Micellization of mixed nonionic surface active agents. J Chem Soc Faraday Trans 1(71):1327–1334. https://doi.org/10.1039/F19757101327
73. Rubingh DN (1979) Mixed micelle solutions. In: Mittal KL (ed) Solution chemistry of surfactants, vol 1. Plenum Press, pp 337–354. ISBN: 978-1-4899-0702-8. https://doi.org/10.1007/978-1-4615-7880-2_15
74. Motomura K, Aratono M, Matuura R (1984) Thermodynamic consideration of the mixed micelle of surfactants. Colloid Polym Sci 262:948–955. https://doi.org/10.1007/BF01490027
75. Mukerjee P (1967) The nature of the association equilibria and hydrophobic bonding in aqueous solutions of association colloids. Adv Colloid Interface Sci 1:242–275. https://doi.org/10.1016/0001-8686(67)80005-8
76. Tanford C (1980) The hydrophobic effect: formation of micelles and biological membranes. Wiley Intersci 233. ISBN 978-0471048930. https://doi.org/10.1002/pol.1980.130181008
77. Zhang B, Guan B, Tao Y, Liu W, Peng B, Lv K (2024) Surfactant synergistic effect and interfacial properties of microemulsions compounded with anionic and nonionic surfactants using dissipative particle dynamics. ACS Omega 9:23903–23916. https://doi.org/10.1021/acsomega.4c01933
78. Ling C, Liu Y, Yang X, Ye Q, Zhou D (2025) Flow characteristics and enhanced oil recovery performance of anionic and zwitterionic viscoelastic surfactant system. Gels 11:627. https://doi.org/10.3390/gels11080627
79. Rodenas E, Valiente M, Villafruela MS (1999) Different theoretical approaches for the study of the mixed tetraethylene glycol mono-n-dodecyl ether/hexadecyltrimethylammonium bromide micelles. J Phys Chem B 103:4549–4554. https://doi.org/10.1021/jp981871m
80. Bakshi MS, Kaur I, Sood R, Singh J, Singh K, Sachar S, Singh KJ, Kaur G (2004) Mixed micelles of benzyldimethyltetradecylammonium chloride with tetradecyltrimethylammonium and tetradecyltriphenylphosphonium bromides: a head group contributions. J Colloid Interface Sci 271:227–231. https://doi.org/10.1016/j.jcis.2003.10.019
81. Pei H, Shan J, Zhang G, Zheng J, Zhao J (2021) Selection of optimum surfactant formulations with ultralow interfacial tension for improving the oil washing efficiency. ACS Omega 6:23952–23959. https://doi.org/10.1021/acsomega.1c02930
82. Bazyari A, Hashemi A, Soulgani BS (2025) Experimental and theoretical investigation of zwitterionic surfactant adsorption on calcite for enhanced oil recovery. Sci Rep 15:22730. https://doi.org/10.1038/s41598-025-05707-5
83. Saw RK, Sinojiya D, Pillai P, Prakash S, Mandal A (2023) Experimental investigation of the synergistic effect of two nonionic surfactants on interfacial properties and their application in enhanced oil recovery. ACS Omega 13:12445–12455. https://doi.org/10.1021/acsomega.3c00401

84. Bera A, Mandal A, Belhaj H (2017) Enhanced oil recovery by nonionic surfactants considering micellization, surface, and foaming properties. Pet Sci 14:362–371. https://doi.org/10.1007/s12182-017-0156-3
85. Belhaj AF, Elraies KA, Mahmood SM, Zulkifli NN, Akbari S, Hussien OSE (2020) The effect of surfactant concentration, salinity, temperature and pH on surfactant adsorption for chemical enhanced oil recovery: a review. J Pet Explor Prod Technol 10:125–137. https://doi.org/10.1007/s13202-019-0685-y
86. Yeh-Seng L, Al-Shaikh M, Hascakir B (2020) Intermolecular interaction between heavy crude oils and surfactants during surfactant-steam flooding process. ACS Omega 42:27383–27392. https://doi.org/10.1021/acsomega.0c00193
87. Jiang R, Li K, Horne R (2017) A mechanism study of wettability and interfacial tension for EOR using silica nanoparticles. In: SPE annual technical conference and exhibition. SPE-187096-MS. ISBN: 978-1-61399-542-6. https://doi.org/10.2118/187096-MS
88. Ma H, Luo M, Dai LL (2008) Influences of surfactant and nanoparticle assembly on effective interfacial tensions. Phys Chem Chem Phys 10:2207–2213. https://doi.org/10.1039/B718427C
89. Yarveicy H, Habibi A, Pegov S (2018) Enhancing oil recovery by adding surfactants in fracturing water: a Montney case study. SPE Can Unconv Resour Conf. https://doi.org/10.2118/189829-MS.ISBN:97861399-582-2
90. Yang W, Lu J, Wei B, Yu H, Liang T (2021) Micromodel studies of surfactant flooding for enhanced oil recovery: a review. ACS Omega 9:6064–6069. https://doi.org/10.1021/acsomega.0c05750
91. Al-Shalabi EW, Sepehrnoori K (2016) A comprehensive review of low-salinity/engineered water injections and their applications in sandstone and carbonate rocks. J Pet Sci Eng 139:137–161. https://doi.org/10.1016/j.petrol.2015.11.027
92. Kumari R, Kakati A, Nagarajan R, Sangwai J (2018) Synergistic effect of mixed anionic and cationic surfactant systems on the interfacial tension of crude oil–water and enhanced oil recovery. J Disp Sci Technol 40:1–13. https://doi.org/10.1080/01932691.2018.1489280
93. Maleki A, Sedaee B, Bahramian A, Gharechelou S, Sarlak N, Mehdizad A, Rasaei MR, Dehghan A (2024) Effect of alumina and silica nanocomposite based on polyacrylamide on light and heavy oil recovery in presence of formation water using micro model. Petroleum 10:338–353. https://doi.org/10.1016/j.petlm.2023.03.001
94. Druetta P, Picchioni F (2020) Surfactant–polymer interactions in a combined enhanced oil recovery flooding. Energies 13:6520. https://doi.org/10.3390/en13246520
95. Esmaeilzadeh P, Hosseinpour N, Bahramian A, Fakhroueian Z, Arya S (2014) Effect of ZrO_2 nanoparticles on the interfacial behavior of surfactant solutions at air–water and n-heptane-water interfaces. Fluid Phase Equilib 361:289–295. https://doi.org/10.1016/j.fluid.2013.11.014
96. Yekeen NP, Padmanabhan E, Idris AK, Ibad SM (2019) Surfactant adsorption behaviors onto shale from Malaysian formations: influence of SiO_2 nanoparticles, surfactant type, temperature, salinity, and shale lithology. J Pet Sci Eng 179:841–854. https://doi.org/10.1016/j.petrol.2019.04.096
97. Saien J, Shokri B, Kharazi M (2024) Effective synergistic action of benzimidazolium nano-Gemini ionic liquid and conventional surfactant for chemical enhanced oil recovery. ACS Omega 20:22336–22344. https://doi.org/10.1021/acsomega.4c01768
98. Sun C, Guo H, Li Y, Song K (2020) Recent advances of surfactant-polymer flooding enhanced oil recovery field tests in China. Geofluids 2020:8286706. https://doi.org/10.1155/2020/8286706
99. Li Y, Kong B, Zhang W, Bao X, Jin J, Wu X, Guo Y, Liu Y, Wang Y, He X, Zhang H, Shen Z, Sha O (2020) Field application of alkali/surfactant/polymer flood with novel mixtures of anionic/cationic surfactants for high-temperature and high-water-cut mature sandstone reservoir. SPE Reserv Eval Eng 23:1093–1104. https://doi.org/10.2118/195056-PA
100. Ma L, Xu P, Wang L, Xia K, Du H, Gao R, Chen Z (2024) Study on anionic–nonionic mixed surfactant for enhanced oil recovery in a hypersaline reservoir. RSC Adv 14:552–559. https://doi.org/10.1039/D3RA07116B

101. Li Y, Jin X, Dong Y, Dai C (2024) Compound anionic–nonionic surfactant system with excellent temperature and salt resistance for enhanced oil recovery. Energy Fuels 38:1047–1055. https://doi.org/10.1021/acs.energyfuels.3c04523
102. Al-Azani K, Abu-Khamsin SA, Kamal MS, Patil S, Hussain SMS, Norrman K (2025) Enhanced oil recovery in heterogeneous carbonate rocks using a Gemini surfactant under harsh conditions. Energy Fuels 39:3433–3445. https://doi.org/10.1021/acs.energyfuels.4c05284
103. Maia KCB, Francisco ADS, Moreira MP, Nascimento RSV, Grasseschi D (2024) Advancements in surfactant carriers for enhanced oil recovery: mechanisms, challenges, and opportunities. ACS Omega 35:36874–36903. https://doi.org/10.1021/acsomega.4c04058

Clouding Behaviour of Surfactant Systems and Their Applications

Pintu Sar

Abstract The understanding of unique phase separation nature of aqueous surfactant solution is an interesting topic of research. The phenomenon of phase separation, as manifested by temperature-induced cloudiness of the surfactant solutions, is highly sensitive to the presence of different additives. This chapter describes the clouding phenomena of surfactants and their numerous applications in different fields of science. We start with a brief introduction of the clouding-assisted phase separation of surfactants. Thereafter, we address the basic principles of clouding along with the detailing about the various experimental approaches of cloud point measurement. The differences in clouding behaviour of several ionic and nonionic surfactants are highlighted. Further, we describe the impact of numerous additives on the shifting of cloud point (CP) of surfactant. The description about the effects of surfactant structures on the CP is noted. We also highlight the clouding-induced morphological switching of surfactant aggregates in solution. The temperature-dependent energetic parameters of clouding are reported from the thermodynamic viewpoint. With a view of its exercise in separation and preconcentration of compounds, we recapitulate its various applications. Finally, we make a remark on the future scope of the clouding mediated extractive approach.

Keywords Cloud point · Nonionic surfactant · CMC · Phase separation · Turbidity · Cloud point extraction

1 Introduction

Solution properties of amphiphilic compounds (surfactants, polymers, drugs, etc.) are one of the widely investigated topics of research in surface and colloid science. In aqueous media, the solution phase properties of ionic surfactants is distinctly different

P. Sar (✉)
Department of Chemistry, Bhagalpur National College, Tilka Manjhi Bhagalpur University, Bhagalpur 812007, Bihar, India
e-mail: pintu.chem@bncollegebgp.ac.in; pintusar1@gmail.com

M. Abdul Rub et al. (eds.), *Industrial Applications of Individual and Mixed Surfactant Systems—Volume 2*, Sustainable Materials and Technology,
https://doi.org/10.1007/978-981-95-9195-4_9

from that exhibited by nonionic types. Among the various properties, clouding behaviour of surfactant systems plays an important role in industrial applications for cleaning, foaming, and wetting [1]. Clouding is a process of phase separation which is quite familiar in particular to the nonionic surfactants. Nevertheless, the clouding in ionic surfactant solutions is rarely perceived under special conditions. Usually, the surfactant solution becomes turbid upon heating and produces two different phases above a particular temperature, called the cloud point (CP) temperature. The surfactant is almost insoluble in aqueous media and is considered as the solubility limit at or above this threshold temperature [2]. As the surfactant solutions heated to T_{CP}, two distinct phases (one is surfactant-rich and another is surfactant-lean) appeared. Basically, this clouding behaviour is reversible since two-phase clouded solutions transform into a single homogeneous phase upon cooling the solution. The phase separation behaviour (clouding) emerges from a critical thermodynamic balance between the hydrophobic (surfactant tail aggregation) and hydrophilic (polar head group hydration) interactions of the aggregates with the aqueous environment. Clouding behaviour of surfactants is not limited to academic interest only but is also useful in certain extraction processes and soil remediation approaches [3–5].

Prior to unveiling the clouding behaviour, we need to understand the self-organization or micellization of surfactant in aqueous solution. Typically, two compartments of surfactant, the hydrophobic tail and the hydrophilic head tend to drive the micellization after reaching a particular concentration, recognized as critical micellization concentration (CMC) [6]. Surfactants of the ionic type have a high CMC value in comparison with nonionic surfactants since the induced repulsion among the electrically polar heads of comparable charge prevents the ionic surfactant monomer to construct the micelle-like aggregate in the solution phase [7]. The micelle formation begins at a point where the hydrophobic interaction predominates over the repellent force. In general, heating the surfactant solution induces the phase separation. The upper dilute phase, basically the major aqueous phase, settled by the surfactant concentration proximal to CMC, separated by the surfactant-rich compact phase.

The clouding of surfactants, typically reversible in nature, is governed by the interplay of enthalpic and entropic factors [2]. The energetically controlled clouding process is related to the changes in hydration of micellar head group region, which ultimately leads to formation of large micellar colloidal dispersion. The process often involves the detachment of water molecules from the micellar surfaces and therefore increases the entropy (disorder) of the system. The micelles tend to coalesce, i.e. the desolvation occurs after reaching the clouding temperature, which is an indication of heat release (exothermic in nature) [5].

This chapter emphasizes the clouding behaviour of surfactant systems in detail, including basic principles of clouding, determination of CP temperature, additive influence on clouding of surfactant, and utility of cloud point phenomena in various fields of applications.

2 Fundamental Aspects of Surfactant-Clouding in Aqueous Media

Clouding phenomenon is known to reveal the inequality of surfactant concentration in aqueous media; hence, acquiring valuable insights about the phase separation at a particular temperature is quite essential. Noticeably, the commencement of phase separation is associated with an intensification of scattered light, as witnessed by the cloudiness of the surfactant solution. Usually, the appearance of turbidity at CP is considered to be the development of bulky aggregates in the solution phase, while the solubility of surfactants is presumed to be extreme at CP [8]. The bulky aggregates, or clusters, are shaped by the combination of micelles since the attractive force operates a crucial role. Another fundamental basis of phase separation is the substantial intermicellar repulsions that are governed by the extensive dehydration of polar segments of surfactant molecules [9]. So, the solvation and desolvation of surfactant micelles at high temperature are the prime factors responsible for the clouding behaviour. Of course, this temperature-regulated solvation $\rightleftharpoons$ desolvation equilibrium could be controlled by a number of approaches, for instance, varying the nature of surfactants, adding the salt, altering the pH, etc.

3 Methods of Cloud Point Determination

The amphiphilic nature of surfactant is familiar with the formation of self-assembly in aqueous solution, as accounted earlier. The cloud point of surfactant is associated with their state of aggregation at a definite temperature. The appearance or disappearance of solution turbidity is mostly visible to the naked eye [10]; while several experimental approaches are available in order to measure the CP precisely. One of the most universal methods for the quantification of cloud point temperature is the UV–Vis spectrophotometric technique. Measuring the viscosity of surfactant solution at variable temperatures offers the cloud point as perceived from the abrupt drop in viscosity (η). Kabir-ud-Din et al. [11] carried out the small-angle neutron scattering (SANS) study to investigate the clouding phenomena of sodium dodecylsulfate (SDS) in association with tetra-*n*-butylammonium bromide (Bu_4NBr). Dynamic light scattering (DLS) and small angle X-ray scattering (SAXS) techniques have also been adopted to point out the boundaries of phase transition of various surfactant systems. Yao et al. [12] examined the thermally induced phase separation phenomena of various nonionic surfactants following the quantitative structure–property relationship (QSPR) study, which is linked with structural, spatial, and thermodynamic properties.

4 Clouding Behaviour of Different Surfactants

Clouding behaviour of anionic surfactants is not as much investigated as that of the nonionic type. The surfactants of ionic type having bulky hydrophobic counter ions are noticeably prone to clouding; for example, tetrabutylammonium dodecylsulfate (TBADS) exhibits the temperature-dependent phase separation, while pure SDS does not show clouding [13]. The clouding phenomena of 11.37 mM TBADS was initiated at 50.6 °C, which reduced gradually to 25.5 °C on increasing the (TBADS) to 469.0 mM. Higher concentrations of TBADS in solution indicating a large content of tetrabutylammonium ion (TBA^+), which helps to crosslink two or more micelles; as a result, CP diminishes. The homologous tetrabutylammonium tetradecylsulfate (TBATS) becomes opalescent on heating the solution above 40 °C, and a similar decrease in T_{cp} was noticed as the concentration of TBATS in solution increased [14]. The cloud curves for those two (TBADS and TBATS) surfactants look like U-shapes signifying a minimum T_{cp} at higher (surfactant), and the clouding temperature increases prior to this minimum concentration of TBADS or TBATS. The molecular composition of the head group influences the CP of three anionic surfactants: sodium dodecylbenzenesulfonate (SDBS), sodium dodecylsulfonate (SDSo), and SDS, as accounted by Yadav et al. [15–17]. As usual, the CP decreases on increasing the concentration of such counter ions. Another anionic surfactant, tetrabutylammonium dodecylbenzenesulfonate (TBADBS), is reputed for its reversible phase separation under the heating–cooling interchange process. Using the temperature-regulated water bath, the cloud points of TBADBS at various dilutions were recorded, which range from 54.5 to 40.5 °C. The observed CP value of TBADBS is lower with respect to TBADS and TBATS on account of the presence of the benzene ring in the dodecylbenzenesulphonate (DBS^-) ion in addition to the stronger hydrophobic interactions operating between TBA^+ and DBS^- [18]. Kumar et al. [19] reported a Hofmeister-like series, which offers a clear understanding about the cloud point of numerous anionic surfactants in the presence of TBA^+ ions. TBA^+ is reported to form a close ion-pair with alkyl sulfonate (and/or alkyl sulphate) through a stronger attractive force than that of alkyl carboxylate. Also reported that the presence of TBA^+ (35.0 mM TBAB) ion could be effective to reveal the clouding phenomena of α-sulphonato palmitic acid methyl ester (PES ~ 75.0 °C) and α-sulphonato myristic acid methyl ester (MES ~ 76.8 °C) at their 5 mM concentration. Along with TBADS, Moulik and workers have successfully prepared another anionic surfactant, tetrapropylammonium dodecylsulfate (TPADS), by the combination of tetrapropylammonium bromide (TPAB) and SDS [20]. As usual, TBADS witnessed its phase separation point around 60.3 °C at 20 mM isolated surfactant concentration, while for TPADS surfactant, no phase separation was noticed. While talking about the cationic surfactant, the clouding phenomenon is very unusual and rarely reported. Needless to mention that the surfactant solutions alone do not qualify for phase separation; the presence of salt having a hydrophobic part causes the clouding of cationic surfactants too, like the anionic type. A few examples of cationic surfactants that are reputed for the temperature-dependent demixing in solution are listed in Table 1. Not very recently, the clouding features of erucyl bis(hydroxyethyl)methylammonium chloride (EHAC) surfactant were presented rationally [21]. Indeed, the EHAC solution achieves phase separation even at very low salt concentrations. In order to make the system temperature sensitive, the salts, e.g. sodium tosylate (NaTos) and sodium salicylate (NaSal) were utilized. Noticeably, the strong binding affinity of the tosylate counter ion (90 mM NaTos) to the positively charged head group of EHAC micelles (40 mM) produces a neutral complex at the micellar surface. Accounted that, the extent of salt adsorption to the micellar surface is the prime factor to produce the lowest CP following the creation of pseudo-nonionic micelles [21]. Raghavan et al. [21] reveal that, for the EHAC/salicylate system, the microstructural alterations from linear micelle to branched micelle with saturated network eventually cause the phase separation at higher salt contents.

The temperature sensitive macroscopic phase separation of the nonionic type is commonly observed, as reported by numerous researchers [25]. Nonionic surfactants contain polyoxyethyelene (POE) chains mostly in their molecular structure. The solubility of nonionic surfactant containing polyoxyethyelene (POE) chains becomes lower at the cloud temperature since the oxyethylene units get dehydrated upon reaching the CP. Since the degree of oxyethylene group increases in the molecular structure of surfactant, the clouding occurs at elevated temperature (Table 2). In addition, the surfactant belongs to alkylglucoside family; for example, n-octyl-β-D-thioglucoside (OTG) could not phase separate alone; however, the binary combination of OTG/TX-100 mixed systems shows clouding at 87.2 °C when TX-100 is present in the mixture as $\alpha_{TX\text{-}100} = 0.2$ [26].

Table 1 Impact of electrolytes on cloud points of certain cationic surfactants

Surfactant	Salt	CP	Ref
Dodecyltributylammonium bromide (46 wt%)	–	58 °C (in water) 48 °C (in D_2O)	[22]
Butyltetradecyldimethylammonium bromide (0.951 wt%)	NaBr (4 M)	36 °C (in water)	[23]
Dodecylarnmonium chloride (1.5446 wt%)	KCl (0.28 M)	23 °C (in water)	[24]

Table 2 Clouding temperature of various nonionic surfactants

Surfactant	CP	Ref
TX-100 (1.0 wt%)	66 °C	[27, 28]
TX-114 (1.0 wt%)	24 °C	[29]
TX-405 (1.0 wt%)	116 °C	[30]
Brij-35 (1.0 wt%)	118.5 ± 0.5 °C	[31]

5 Impact of Various Additives on the Clouding of Surfactant

Tuning the solubility of surfactant through the cloud point approach plays a crucial role in view of its applications in several fields. A number of additives could modify the phase separation point of surfactants to the desired range. Whatever the class, the various functional additive induces the clouding of surfactant (either decreases or increases) even at a very low concentration. Usually, additives impart several modifications, like fluctuations in CMC, deviations in micellar size, alterations in water-surfactant interactions, and ultimately alterations in the phase separation point. Parikh et al. [29] investigated the temperature-dependent phase behaviour of two nonionic surfactants, Triton X-100 and Triton X-114 in the presence of aliphatic alcohols and inorganic salts. All the electrolytes NaCl, Na_2SO_4, Na_2CO_3, KCl, K_2SO_4, K_2CO_3, $MgCl_2$, and $MgSO_4$ promote the clouding of TX-100 surfactant below the temperature at which TX-100 exhibits phase separation alone. Anions have more impact on reducing the clouding temperature than that of cations. Except Cl^- all the cations and anions impart a salting-out effect which elucidates the reduction of the clouding temperature of nonionic Triton X-100. Noted, the sulphate ion decreases the CP more effectively in comparison with carbonate and chloride ions. From the thermodynamic point of view, a negative value of ΔH°_{CP} was accounted which implies the clouding process is associated with an exothermic energy change in the presence of those electrolytes. TX-114 exhibits the reduction of CP also in the presence of such electrolytes. In addition to, the phase separation temperature of both nonionic surfactants (TX-114 and TX-100) reduces with the incremental addition of electrolytes into the solution (Fig. 1).

The effect of short chain alcohols on the clouding behaviour of TX-100 and TX-114 was elucidated explicitly. Alcohols, e.g. methanol, ethanol, propanol, and butanol produced an opposite impact on the clouding of Triton X-100 and Triton X-114. The alcohols, especially CH_3OH, CH_3CH_2OH, and $CH_3CH_2CH_2OH$, facilitate to break

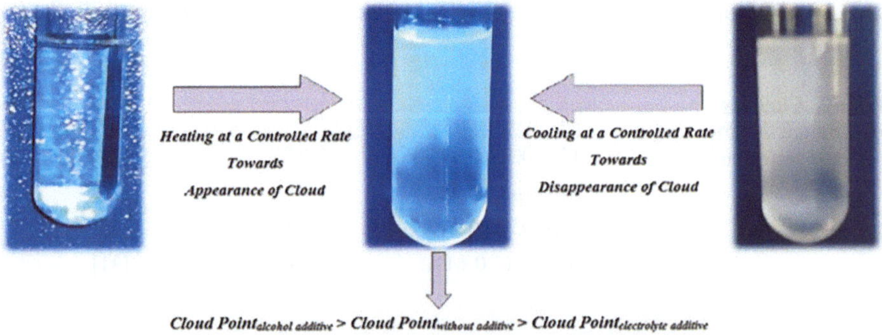

Fig. 1 Temperature-regulated appearance of clouding behaviour of nonionic TX-100 surfactants along with the order of CP in presence of various additives [29]. Adopted from Ref.

the water structure, while comparatively hydrophobic butanol molecules are water structure maker. The ability of short chain alcohols to disrupt or to make the structure of water results in a decrease or increase in aggregate size in solution and thereby alters the CP. Noticeably, the elevation of CP of Triton X-100 and Triton X-114 was reported as the homologous series of alcohols stretched from methanol to propanol, while butanol does not influence the CP much more. In another study Batıgöç and Akbaş addressed the impact of inorganic salts such as KNO_3, K_2SO_4, K_3PO_4, Li_2SO_4, Na_2SO_4, and $(NH_4)_2SO_4$ in regulating the CP of two nonionic surfactants Brij-35 and Triton X-405 [32]. The addition of electrolytes encourages the disruption of surfactant-H_2O hydrogen bonding interactions, and ultimately surfactant aggregates get desolvated. The PO_4^{3-}, SO_4^{2-}, and NO_3^- ions lower the aqueous solubility of Brij-35 (118 °C at pure state) and TX-405 (116 °C at pure state) effectively with the decreasing order of phase separation point: KNO_3 (112.05 °C) > K_2SO_4 (70.35 °C) > K_3PO_4 (61.35 °C) at 0.4 M salt concentration for 1.0 wt% Brij-35 solution. The influence of cations on the cloud point of TX-405 and Brij-35 was accounted in the order $Na^+ \approx K^+ > NH_4^+ > Li^+$. In comparison with NH_4^+ ion the smaller size K^+ and Na^+ ions are mostly hydrated, resulting in less hydration of the POE chain of TX-405 and Brij-35. Li^+ ion, in contrast, not reduces the cloud point of Brij-35 and Triton X-405 more efficiently as like Na^+ and K^+ do. This is possibly caused by the coordination of Li^+ ions to the ether oxygen (–O–) of POE chains, which hydrated the surfactants eventually; as a consequence, the reduction of CP value was observed in the order 64.25 °C (Na_2SO_4) < 68.05 °C (K_2SO_4) < 77.45 °C ($(NH_4)_2SO_4$) < 77.85 °C (Li_2SO_4) for 1 wt% TX-405 surfactant.

In the case of poly(oxyethylene)$_n$oleyl ether surfactants, the clouding temperature increases with the increment of n initially, and after attaining maximum, reduces to some extent [33]. Another set of nonionic surfactants, Brij 56 and Brij 76, exhibit the reduction of CP in a low concentration range and then rise after passing through the lowest value [34]. Most of the nonionic surfactants exhibit concentration dependent phase separation, like a decrease in CP and thereafter a slight increase. Growing the number density of micellar aggregates in solution is attributed to the decrease in CP of nonionic surfactant at a lower concentration range while in addition to micellar growth, the attractive interactions between aggregates play a crucial role at a higher range of concentration.

The introduction of a drug into the surfactant solution offers various molecular interactions like hydrophobic interaction, H-bonding, electrostatic forces, π–π interaction, van der Waals attraction, etc. Drug molecules are sometimes hydrophobic in nature, while the amphiphilic nature of the drugs is worth mentioning also. Many drugs have a heterocyclic ring structure. The amphiphilic drug by itself shows phase separation under special conditions, like at optimum pH, definite temperature, suitable additive, etc. Mahajan et al. [35] investigated the phase separation studies in detail together with the interaction between phenothiazine-based drugs and surfactants. The impact of neutral (Triton X-114 and Triton X-100), positively charged (CPC, CPB, TTAB, and HTAB), negatively charged (SDS and SDBS), and dicationic (16-6-16, 16-5-16 and 16-4-16) surfactants on the clouding behaviour of promethazine hydrochloride (PMT) was explored by Kabir-ud-Din et al. [36].

They have investigated the cloud temperature variation of another important drug, imipramine hydrochloride (IMP), in the presence of several conventional surfactants like SDS, SDBS, CPC, CPB, TTAB, Tween 80, Tween 60, Tween 40, Tween 20, Brij-35, Brij 56, Brij-57, TX-100, TX-114, and gemini (16-6-16, 16-5-16, 16-4-16, 14-6-14, 14-5-14, 14-4-14) [37, 38]. The temperature-regulated phase separation of clomipramine hydrochloride (CLP) was observed in attendance of various additives like sodium salicylate, sodium benzoate, sodium tosylate, pyrogallol, resorcinol, *O*-toluidine HCl, aniline HCl, *p*-toluidine HCl, sodium cholate, sodium deoxycholate, sodium palmitate, sodium decanoate, mannitol, sorbitol, and cyclodextrin [39]. Hoque et al. have explored the thermodynamic aspects of phase separation of various drug molecules [e.g. metformin hydrochloride (MNH), ciprofloxacin hydrochloride (CFH), and ceftriaxone sodium trihydrate (CFT)] in association with TX-100 in the presence of aliphatic alcohols [40]. Amitriptyline hydrochloride (AMT), another tertiary amine containing amphiphilic drug, remains protonated at pH 6.7 and shows variable clouding temperature at ethylene glycol-H_2O (EG-WR) and glycerol-H_2O (GL-WR) media. The temperature-induced clouding of AMT is fluctuated under the influence of various electrolytes: potassium salts (KF, KCl, KBr) and chloride salts (LiCl, NaCl, NH_4Cl) [41]. The clouding behaviour of various amphiphilic drug-additive systems, along with their physicochemical aspects, has been analysed in detail by a number of published reviews and recent works (Table 3) [34, 42–46].

6 Impact of Molecular Architecture of Nonionic Surfactants on Clouding

Molecular structures of nonionic surfactants have a great impact on their clouding behaviour [50]. The surfactant's head groups are hydrophilic in nature, and the assembly of constituent head groups decorates the surface of micelle [51]. In this contextualization, a series of nonionic surfactants, as synthesized by varying polyethoxylate chains, reveal the temperature-regulated phase separation at different temperatures. As the surfactant chain length increases with 7, 11, and 16 units of ethylene oxide (EO) monomer, the clouding temperature is heightened by 16 °C, 43 °C, and 64 °C, respectively [52]. Scamehorn et al. [53] have accounted for the similar increase in clouding temperature as a function of EO number and alkyl chain length for nonionic ethoxylated surfactants. This fact is explained on the basis of falling in inter-micellar attractive forces, which in fact enhances the micellar excluded volume; as a result, the clouding temperature increases following the upgradation of surfactant structure with the number of ethylene oxide units. The clouding behaviour was observed at lower temperatures when the alkyl carbon number of ethoxylated surfactant increases (and/or hydrophobic branching of nonionic surfactant increases). A Bisphenol-based series of nonionic surfactants showed variable temperature-dependent clouding behaviours [54]. These surfactants were synthesized following a multistep process:

Table 3 Clouding behaviour of few important drug molecules in presence of different additives

	Drugs/Mixture	Additives	CP	Remarks	Refs.
1	MNH (2 mmol kg^{-1}) + TX-100 (92.7 mmol kg^{-1})	Water	70 °C	Addition of more alcohols to the mixture increases the CP; lower alcohols are hydrophilic in nature and resist micelle formation, which elevates the CP of mixture	[40]
		MeOH	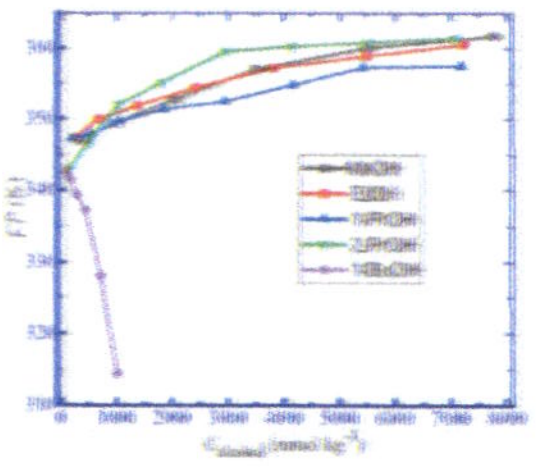 (Adapted from Ref. [40])		
		EtOH			
		1-PrOH			
		1-BuOH		Higher alcohol penetrates into the micellar palisade layer and favours the micellization resulting in a reduction in CP	
2	CFH + TX-100	Water	64.7 °C[a]	CP decreases with an increase of glucose/fructose concentration. Glucose or fructose molecules are highly solubilized, and therefore, TX-100 becomes less hydrated, which facilitates the interaction among the nanomicelles, thus lowering CP	[28]
		Glucose (10.04 mmol kg^{-1})	60.53 °C[b]		
		Fructose (10.07 mmol kg^{-1})	60.23 °C[c]		
3	Tween 80 (3.05 × 10^{-2} mol kg^{-1}) + neomycin sulphate	Water	68.58 °C[d]	CP of mixture of Tween 80 and NS decreases in presence of electrolyte as: $CP_{Na_2SO_4} > CP_{NaNO_3} > CP_{NaCl}$	[47]
		NaCl	62.80 °C[e]		
		$NaNO_3$	63.09 °C[f]		
		Na_2SO_4	60.49 °C[g]		

(continued)

Table 3 (continued)

	Drugs/Mixture	Additives	CP	Remarks	Refs.
4	AMT (50 mmol kg^{-1})	Water	25.2 °C	Both organic solvents media reduces the dielectric constant of media which repel the polar head group of AMT to come close to each other in order to form giant aggregates. As a result of amplified electrostatic repulsion, the CP of AMT increases Salt effect: $CP_{KF} < CP_{KCl} < CP_{KBr}$ Cation effect: $CP_{NH^{4+}} > CP_{Li^+} > CP_{Na^+} > CP_{K^+}$	[41]
		EG-WR GL-WR KF KCl KBr LiCl NaCl NH_4Cl	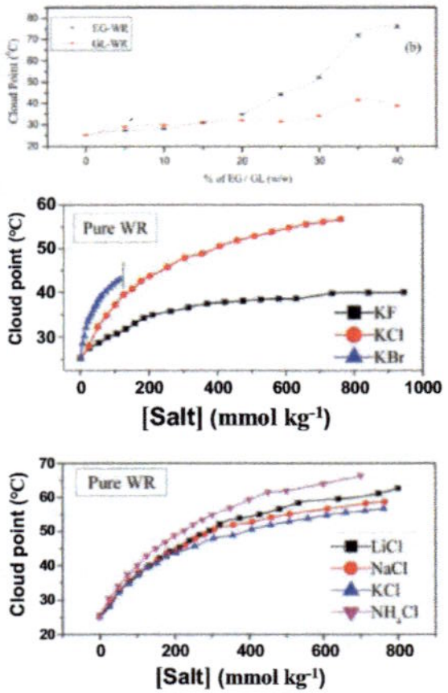 (Adapted from Ref. [41])		

(continued)

Table 3 (continued)

	Drugs/Mixture	Additives	CP	Remarks	Refs.
5	IMP	Water	49.15 °C[i]	IMP is cationic and promotes the inter-micellar repulsion with the polar amino acids. Acidic amino acids are negatively charged and are prone to associate with cationic IMP; thus, hydration of micelles occurs efficiently, resulting in the increase of CP Sugars help to make the water structure, and this property declines the CP of IMP sharply	[48]
		Proline[i]	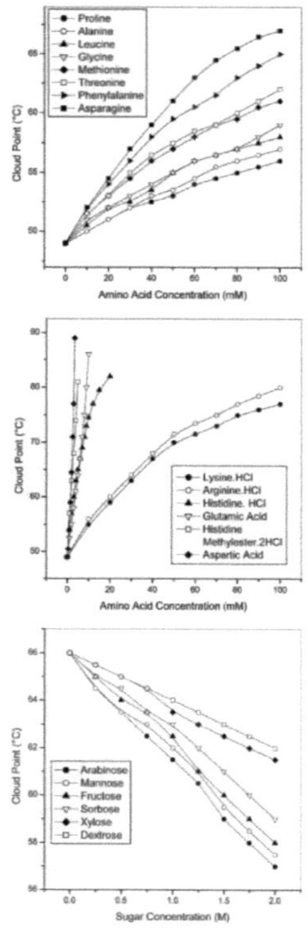 (Adapted from Ref. [48])		
		Alanine[i]			
		Leucine[i]			
		Glycine[i]			
		Methionine[i]			
		Threonine[i]			
		Phenylalanine[i]			
		Asparagine[i]			
		Arabinose[j]			
		Mannose[j]			
		Fructose[j]			
		Sorbose[j]			
		Xylose[j]			
		Dextrose[j]			

(continued)

Table 3 (continued)

	Drugs/Mixture	Additives	CP	Remarks	Refs.
6	CFT + TX-100	Water	64.95 °C[k]	The presence of more ethanol in the mixture alters the shape of micelle, as a result hydration of the micelle increases; consequently, the CP value is augmented	[49]
		5% EtOH	72.95 °C[k]		
		10%EtOH	82.73 °C[k]		

[a][CFH] = 1.03 mmol kg^{-1}, [Triton X-100] = 25.12 mmol kg^{-1}
[b][CFH] = 1.05 mmol kg^{-1}, [Triton X-100] = 24.91 mmol kg^{-1}
[c][CFH] = 0.97 mmol kg^{-1}, [Triton X-100] = 24.97 mmol kg^{-1}
[d][NS] = 2.05 mmol kg^{-1}, pressure $P = 0.1$ MPa
[e][NS] = 2.02 mmol kg^{-1}, [NaCl] = 2.03 mmol kg^{-1}, pressure $P = 0.1$ MPa
[f][NS] = 2.01 mmol kg^{-1}, [$NaNO_3$] = 1.99 mmol kg^{-1}, pressure $P = 0.1$ MPa
[g][NS] = 2.03 mmol kg^{-1}, [Na_2SO_4] = 2.01 mmol kg^{-1}, pressure $P = 0.1$ MPa
[i][IMP] = 100 mM, Sodium phosphate buffer = 10 mM, pH of IMP = 6.7
[j][IMP] = 150 mM, Sodium phosphate buffer = 10 mM, pH of IMP = 6.7
[k][CFT] = 1.0 mmol kg^{-1}, [TX-100] = 5.0% (w/w)

- attachment of bisphenol with acetone and/or acetophenon and/or cyclohexanone.
- bisphenol frame with acetone (BA) is ethoxylated with different degree of ethylene oxide unit (27, 35 and 43).
- ethoxylated bisphenols were esterified by one molecule of fatty acids separately. Used fatty acids are: oleic, linolenic, linoleic, lauric, myristic, palmitic, decanoic, and stearic.

Bisphenol-derived surfactants with the rise of ethylene oxide content exhibit clouding at higher temperatures. The surfactant structure having bisphenol of acetophenone (BAC) ethoxylated with 43 ethylene oxide units and monoesterified with fatty acids showed an augmented phase separation temperature as the alkyl chain of fatty acids increases (C10–C18, Fig. 2).

7 The Morphological Study of Clouding

The phase separation with precipitation of a concentrated phase occurs as a result of the effective dehydration of the micelle or ionic head group of the surfactant; hence, during the cloud formation of surfactant, the morphological changes in the solution were detected as the witness of CP. The clouding behaviour of surfactant could be realized by a number of experimental parameters like DLS, ^{1}H-NMR, SANS, etc. It is worthwhile to provide an in-depth knowledge of the morphologies resulting from a variety of surfactants in aqueous media. The structural change of TX-100 during phase transition was assessed with the help of light scattering technique [55]. The ionic surfactant, TBADS when heated in a range of temperature including its cloud point, exhibits two morphologies: individual micelle of 5 nm size and giant aggregates of 500 nm size, as evidenced by DLS measurement. The cross-linking of individual small micelles may be the cause of giant aggregate formation near and beyond CP. In addition, Kumar et al. [56] have performed the SANS experiment with 50 mM TBADS (Fig. 3) in order to provide the information regarding individual micellar fractions before and after phase separation. The chemical environment of two marked protons of TBADS, –O–CH_2 of dodecylsulphate anion (DS^-) and –N–CH_2 of tetrabutylammonium cation (TBA^+), are different as indicated by different δ (3.95 ppm and 3.2 ppm, respectively) values in ^{1}H-NMR spectra. As noticed from the ^{1}H-NMR spectra of TBADS (Fig. 3), the downfield shift with broadening of peak under the examining temperature indicates the very low degree of solvation of TBA^+ ion and micellar head group portion as the aqueous solution of TBADS is heated from 36 °C to 45 °C. Again, the splitting of peak describes the aforementioned protons responsible for the generation of two morphological structures.

The similar splitting of peak and gradual peak broadening was accounted in the temperature variable ^{1}H-NMR spectra for the synthesized anionic surfactant, tetra-*n*-butylphosphonium dodecylsulphate (TBPDS) [57]. The morphological switching below and above its CP (48.5 °C) was observed with the 0.0124 M TBADS, as indicated by the SANS spectral curve (Fig. 4). The deterioration of scattering intensity

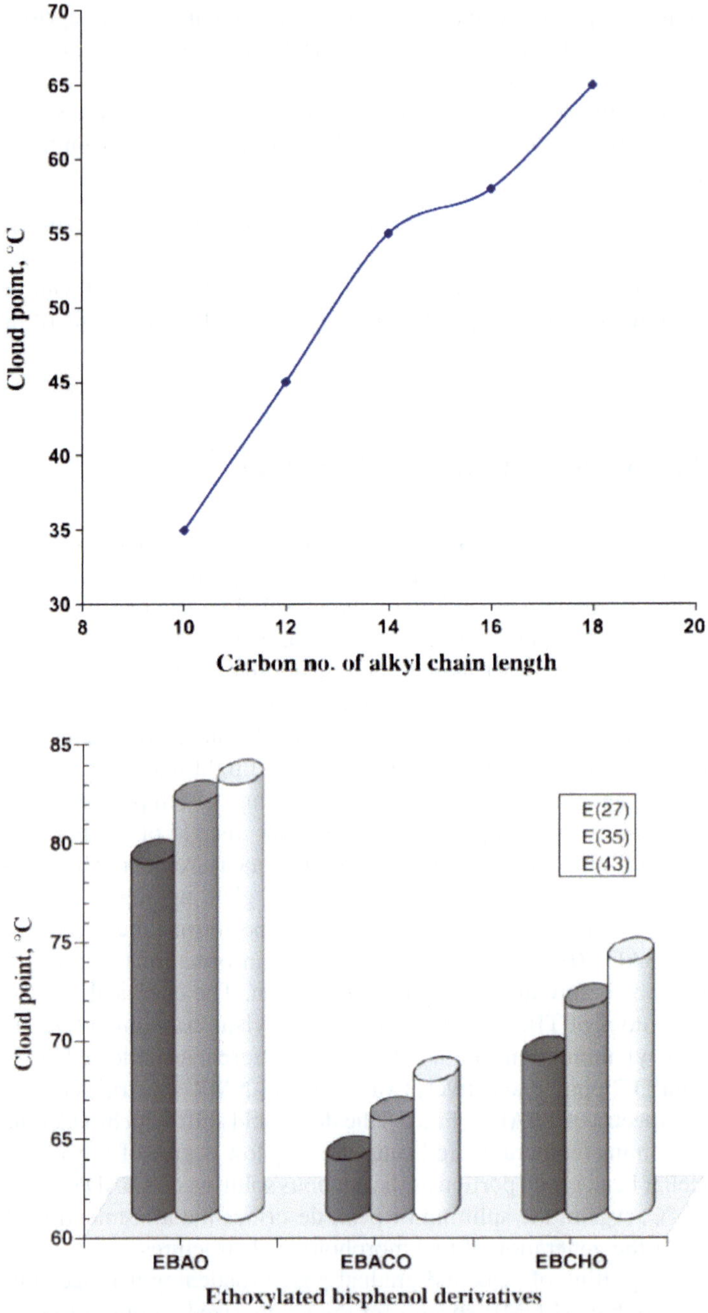

Fig. 2 Impact of alkyl chain length (top) and ethylene oxide content (bottom) on the cloud point of some ethoxylated bisphenol monoester surfactants (Adapted from Ref. [54])

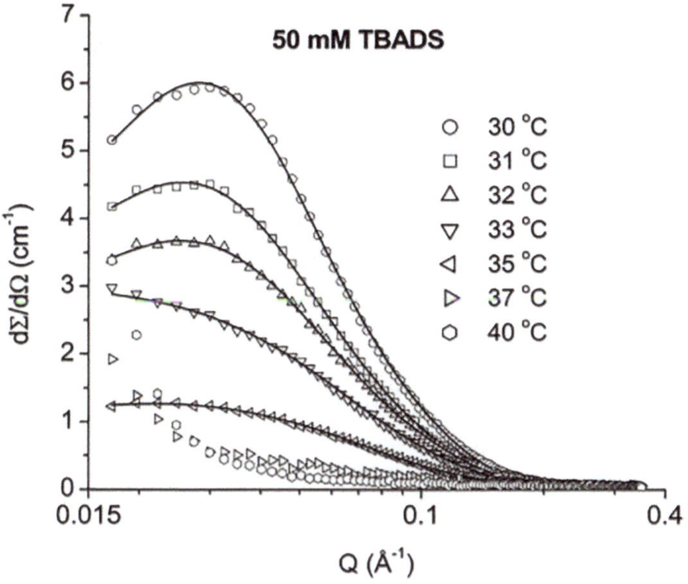

a)

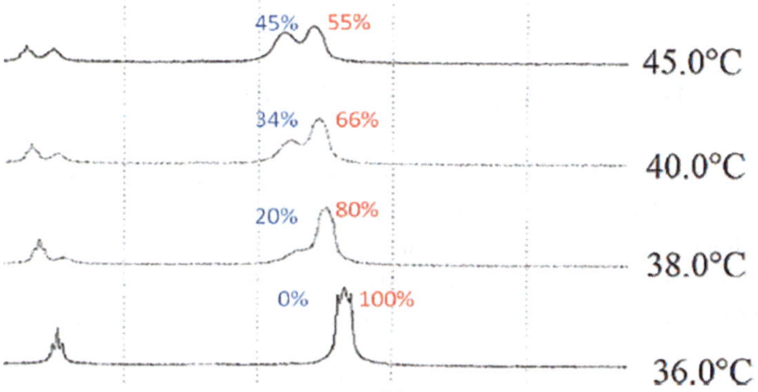

b)

Fig. 3 SANS data of aqueous solution of 50 mM TBADS at variable temperature (top). ^{1}H-NMR spectra of TBADS solution at several temperatures with focusing on the –O–CH_2 (3.95 ppm) and –N–CH_2 (3.2 ppm) protons, particularly (bottom) (Adapted from Ref. [56])

plot in the lower Q zone at the temperature higher than CP signifies the reduction of individual micelles with the generation of larger aggregates by fusion. The polarized optical microscopic (POM) images of 0.1 M TBADS were inspected at room temperature, at which the surfactant reach phase (holds the larger aggregates) is assessed in POM. The clouding phenomena is explained, herein, on the basis of large-sized aggregates formation by the combination of individual micelles.

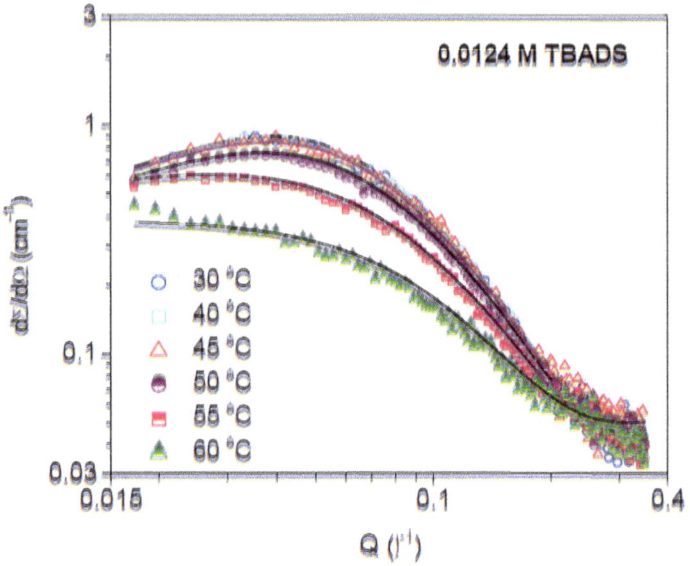

Fig. 4 Temperature-dependent SANS spectral profile of TBADS (0.0124 M) (Adapted from Ref. [57])

8 Thermodynamic Properties of Clouding

Clouding-induced phase segregation is an energy-driven process; thus, the assessment of the energetic parameters could offer further understandings about the clouding phenomena. The phase separation model suggests to consider the CP as the critical solubility concentration of the surfactant; hence, the standard Gibbs energy ($\Delta_s G°$) for the dissolution of surfactant is evaluated [58] using the formula

$$\Delta_s G° = RT \ln x_s$$

where x_s representing the mole fraction of the surfactant at T_{CP}; T is the clouding temperature in Kelvin and R is the gas constant.

The standard enthalpy of clouding ($\Delta_s H°$) is evaluated from the plot of $\Delta_s G°/T$ versus $1/T$ (see below equation); additionally, one can figure out the standard entropy of clouding ($\Delta_s S°$) following the Gibbs–Helmholtz equation:

$$\Delta_s H° = \frac{\partial(\Delta_s G°/T)}{\partial(1/T)}$$

$$\Delta_s S° = (\Delta_s H° - \Delta_s G)/T$$

The overall heat change for the clouding-induced phase separation involves different steps which need to be accounted. Two important contributions: *dehydration of surfactant head groups* which is endothermic in nature involving heat adsorption and the subsequent *clustering of less hydrated nanomicelles* involving the liberation of heat which is exothermic in nature [58]. However, other approaches such as disassembly of nanomicelles, interfacial adsorption, and nanomicellar growths affect the energetics of clouding. The relative contribution of all these factors controls the $\Delta_s H^\circ$ and $\Delta_s S^\circ$ [59].

Several reports about the thermodynamics of the clouding process of different surfactants reveal that the process involves discharge of heat, demonstrating that clustering of less hydrated nanomicelles by reducing inter-micellar interaction is the leading factor [60]. Sometimes the process of clouding is endothermic, as observed from the positive $\Delta_s H^\circ$ value.

The negative $\Delta_s G^\circ$ value indicates the spontaneity of the clouding process and the Gibbs free energy is measured by the balance between enthalpy and entropy involvement. From a thermodynamic viewpoint, since H_2O molecules are absolutely separated from the micellar aggregates at the cloud point, the net entropy change becomes extremely raised, and the change in free energy becomes fairly more negative, thereby stimulating the initiation of phase separation.

9 Exploitation of Surfactant-Clouding in Several Area

The temperature-regulated phase separation behaviour of aqueous surfactant solution has potential applicability in the field of extraction through the entrapment of compounds within the aggregates. The cloud point-based approach in separation science is crucial in order to advancement of purification, extraction, and preconcentration schemes for the desired analyte. Clouding phenomena is essential also in the field of contaminant soil remediation since it appeared as the alternative and green approach for the extraction of microchemicals from soil. The extraction of proteins and enzymatic transformations could be achieved also by the temperature-controlled phase separation approach. Cloud point extraction (CPE) or micelle-mediated extraction is a sustainable approach that provides multiple benefits compared to other traditional extraction processes [3].

- **Isolations of Targeted Inorganic Compounds, Pesticide Residues, and Proteins**

The diversity of the clouding process is that it makes the system heterogeneous and useful to recover the water-insoluble compounds from the site of contamination. Tuning various factors such as surfactant nature, temperature, pH, additive systems, time, and reagents facilitates the biphasic separation, and subsequently, the extraction of chemical and biological samples is achieved proficiently. As an environmentally friendly technique, the CPE method acknowledged considerable attention in extracting and preconcentrating the numerous organic substances like pesticides, phenolic derivatives, polycyclic aromatic hydrocarbons (PAH), polychlorinated compounds, polybrominated biphenyl ethers, aryl amines, and some inorganic compounds, including a few metal ions, parabens, phthalates, etc. [52, 61–65]. To get a clear understanding of the mechanism of extraction through cloud point methods, it must be realized clearly that the aggregation nature of amphiphilic compounds or surfactants is a fundamental phenomenon in addition to their temperature responsive solution behaviour. Noticeably, nonionic surfactants play a crucial role compared to cationic, anionic, and/or their mixtures to pursue the clouding-initiated extraction process. Typically, an aqueous solution of surfactant is introduced to the solution holding organic/inorganic substances. In this contextualization, a higher amount of surfactant should be employed compared to the CMC of that surfactant so that the targeted substances solubilize within the formed micelle-like aggregates. Afterwards, heating the mixture of solution to that particular temperature, known as CP, results in separation of two phases, and the encapsulated organic/inorganic substances tend to stay at the surfactant-dense phase. Collection of the surfactant-rich phase through decantation and further instrumental analysis of surfactant-rich phase concluded the CPE protocol. The graphical illustration of the cloud point extraction technique is highlighted in Fig. 5.

While illustrating the nature of introducing surfactant, three categories are mostly utilized in cloud point extraction method, as depicted below:

- Nonionic: TX-100 (64 °C), TX-114 (24 °C), Tween 20 (76 °C), Tween 80 (65 °C)
- Anionic: SDS (> 100 °C), SDBS, ammonium lauryl sulphate, sodium laureth sulphate
- Cationic: CTAB (24.5 °C), CPC

In CPE method, the concentration factor ($C_F = V_{aq}/V_s$), quantified as the ratio of analyte quantity located in surfactant-dense layer with respect to the original sample matrix, has a pivotal role on the extraction efficiency. C_F depends on concentration of employed surfactant as well as nature of nonionic surfactant (alkyl chain length). In most studies, it has been accounted that extraction efficiency stimulated with the employed surfactant concentration up to a maximum value.

The medium pH also influences the extraction procedure involving temperature responsive phase separation of surfactant for ionizable analytes since the protonation

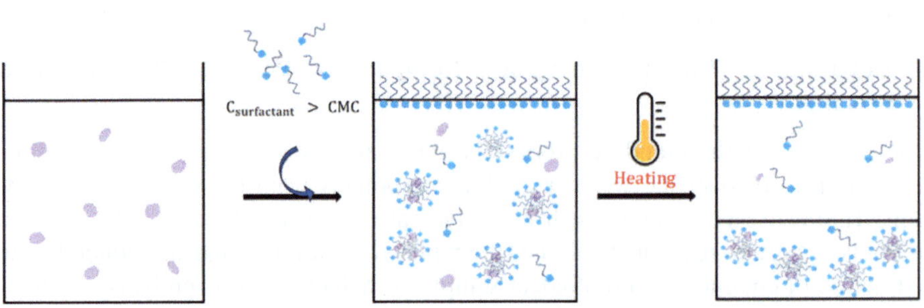

Fig. 5 Presentation of cloud point extraction (CPE) technique in aqueous solution

$\rightleftharpoons$ deprotonation equilibrium of extracting compounds has strong impact on the interaction between substrates (analyte) and micellar aggregates.

The incorporation of electrolytes or salts in the CPE procedure is significant since electrolytes greatly impact on the micellization of surfactant, i.e. CMC of surfactant as well as the extraction efficiency of polar substrates. The electrolytes also control the dehydration of micellar aggregates which ultimately affect the hydrophobic interactions between aggregates and substrates (analytes).

Centrifugation is a significant step involve in CPE approach; however, it has no impact on the generation of micellar aggregates, interaction between analytes and network micelle, but the time of centrifugation is important in phase separation following the cloud formation.

The analysis of collected surfactant-rich phase occurs by several methods like, capillary electrophoresis, liquid chromatography, capillary electrochromatography, fluorography, autoradiography, sodium dodecylsulfate/polyacrylamide gel electrophoresis (SDS-PAGE), etc. [66].

Few important points need to be considered for CPE techniques:

- Temperature must be kept constant during centrifugation in CPE method because the extraction efficiency might be affected sometimes upon fluctuating the temperature.
- Addition of small amount of appropriate solvent (H_2O, MeOH, EtOH, CH_3CN, aqueous solution of another surfactant) facilitates the dilution of surfactant-enrich layer by decreasing the viscosity of surfactant-rich phase [67].

Inorganic metal ions, form chelated complex with the suitable ligands, can be extracted using cloud point approach in which the addition of surfactant followed by the phase separation upon heating the solution ensues the extraction [68, 69]. Suitable pH also impacts on the extraction of trace amount of metal ions from the systems. The extraction of metal ions is basically regulated by the introduction of suitable complexing agents; thus, choice of the complexing ligands is fundamental step [70]. Among the employed ligands, azo dyes and dithiocarbamates are the most versatile complex forming agents that are used frequently in the extraction of metal ions through CPE method. Those agents are poorly soluble in aqueous media and offers a large number of metal ions to bind with coordinating modes [71]. The produced metallic complex becomes hydrophobic, and thereafter following the CPE approach, the extracted metal ions can be determined qualitatively and quantitatively using various analytical techniques (Table 4).

Cloud point approach has been adapted also to preconcentrate and separation of numerous organic contaminants in a range of sample matrices on account of its significant recovery efficiencies and concentration factors. The analytical approaches, such as, high performance liquid chromatography (HPLC), gas chromatography (GC), and capillary electrophoresis, combined with CPE make the micelle-mediated extraction simple, safe, and practical. A number of research work developed the successful and cost-effective CPE technique to extract organic contaminant under mild conditions in absence of organic solvents. Table 5 highlights the analytical methodologies associated with CPE for estimating the various organic compounds from different type

Table 4 Extraction of various metal ions using CPE technique

Metal ions/chelating agents	Type of sample	Analytical method	Conditions	Extraction parameters	Ref
Al^{3+}/PAN	In human albumin	GFAAS	Triton X-114, pH = 8.0, Methanol containing HNO_3	$D_L = 0.06$ ng mL^{-1}	[72]
Mn(II)/PAN	In milk and water	AAS	Triton X-114, pH = 9.2, 0.1 mol L^{-1} HNO_3	$D_L = 0.39$ mg L^{-1}	[73]
Pb^{2+}/ 1-phenyl-3-methyl-4-benzoyl-5-pyrazolone (PMBP)	Tap water, rain water, waste water,	FAAS	Triton X-114, pH = 5.5, 0.1 M HNO_3	$D_L = 1.49$ µg L^{-1}	[74]
Co/PAN	In urine	FAAS	Triton X-114, pH = 3.7, 200 µL MeOH	$D_L = 0.38$ mg L^{-1}	[75]
Ag(I)/2-Mercapto benzothiazole (MBT)	Samples in aqueous phase	AAS	Triton X-114, pH = 2.0, 300 µL MeOH with 0.1 mol L^{-1} HNO_3	$D_L = 2.2$ µg L^{-1}	[76]
Fe(III), Co(II), Ni(II)/Ammonium pyrrolidine dithiocarbamate	In aqueous media	AAS	Triton X-114, pH = 5.0, MeOH with 70 g L^{-1} HCl	19 mg L^{-1} 5 mg L^{-1} 11 mg L^{-1}	[77]
Au nanoparticles	In water or in agricultural soil	Asymmetric flow field-flow fractionation in combination with ICP-MS	TX-114 and TX-100	LLOQ (calibration curve) = 0.05 µg kg^{-1}	[78]
Sm^{3+}/ammonium salt of aurintricarboxylic acid (aluminon)	Rock samples, waste water	ICP-OES	Triton X-114, pH = 2.5, 1.0 mol L^{-1} HNO_3	$D_L = 0.06$ µg L^{-1}	[79]

(continued)

Table 4 (continued)

Metal ions/chelating agents	Type of sample	Analytical method	Conditions	Extraction parameters	Ref
Ag, Au/*O,O*-diethyldithiophosphosric acid (DDTP)	Geological	AAS	TX-114, pH = 1.0, MeOH with 0.1 mol L^{-1} HNO_3	D_L: 460 μg L^{-1} (Ag) 530 μg L^{-1} (Au)	[80]
Te/ammonium pyrrolidine dithiocarbamate (APDC)	Soil, water, sediment samples	ETAAS	TX-114, pH = 4.0	D_L: Te(IV) = 1.1 ng L^{-1} Te(VI) = 1.7 ng L^{-1}	[81]

D_L = limit of detection, C_F = concentration factor

PAN 1-(2-Pyridylazo)-2-naphtol, *AAS* Atomic Absorption Spectrometry, *FAAS* Flame Atomic Absorption Spectrometry, *GFAAS* Graphite Furnace Atomic Absorption Spectrometry, *ETAAS* Electrothermal Atomic Absorption Spectrometry, *ICP-OES* Inductively Coupled Plasma-Optical Emission Spectrometry of samples. The extraction efficiency of utilized surfactants varied from analyte to analyte at variable conditions. Sometimes mixed surfactant systems have recently been noticed to beat single surfactants in extracting organic pollutants.

The application of cloud point extraction method is not limited to the isolation of organic contaminants and metal ions; however, it has long been utilized in the extraction of bioactive compounds from food processing by-products. The details of cloud point technology involving the isolation of lycopene from the tomato peel using SPAN-20 surfactant were accounted by Papaioannou et al. [91]. In a similar approach, Katsoyannos et al. [92] investigated the extraction of carotenoids and phenols from red-flesh orange juice using a number of surfactants: Tween 80, Tween 20, PEG 400, and SPAN-20. The extraction efficiencies of TX-100 and TX-114 are greatly acknowledged in a number of reports where efficient extraction of phenolic compounds from wastewater was emphasized in the light of CPE. Certain ionic surfactants intensely modify the clouding phenomena of nonionic surfactants, as mentioned earlier. The mixture of nonionic polyoxyethylene(7.5)nonylphenylether (PONPE 7.5) and anionic SDS was selected for the extraction of fluoroquinolone-based antimicrobial compounds: ofloxacin and gatifloxacin from the aqueous solution [93]. The concentration of NaCl greatly impacts the CP of surfactant mixture, which offers to entrap ofloxacin (and/or gatifloxacin) in the surfactant-dense zone; in that way, the highest extraction proficiency was achieved by reducing the phase volume ratio.

In this context, Arya et al. [94] highlighted the significant role of CPE in improving the extraction of bioactive compounds from food processing by-products (Table 6).

Wang et al. [99] have investigated the conversion of penicillin G to 6-aminopenicillanic acid (APA) through the enzymatic hydrolysis in low pH using cloud point techniques. The nonionic surfactant, Tergitol TMN-3, with the CP of 3–5 °C, facilitates to partitioning the hydrolysis product APA in the coacervate phase. The cloud point-mediated enzymatic hydrolysis as well as separation of APA from phenylacetic acid (PAA) was achieved in the absence of pH regulation, as accounted by Qi et al. [100]. In another study, Ritter and Smirnova have utilized Tergitol NP-7 to generate the two-phase system for continuous biocatalytic extraction of PAA from the mixture of product, both of which are produced upon the hydrolysis of penicillin G [101]. The potential of cloud point system as the sustainable two-phase partitioning medium for biotransformation has been emphasized comprehensively by Wang [102]. The application of cloud point system in extractive microbial fermentation of organic compounds is really worth mentioning since conventional microbial processes undergo various limitations like insoluble nature of the substrate, substrate-product inhibition, and degradation of products, etc. [103].

The cloud point preconcentration approach is utilized in the separation of numerous dyes like chrysoidine dye, direct yellow, sudan dyes (I, II, III, IV), coralene red, blue BG, blue 19, allura red, etc. [104–108]. The CPE method offers the efficient extraction of crystal violet (CV), ethyl violet (EV), malachite green (MG), and brilliant green (BG) dyes, which were partitioned in the coacervate phase as produced by the aqueous TX-114 solution. The efficiency of extraction of those four dyes increases with the increase in temperature, [TX-114], [NaCl], and [$CaCl_2$] as

Table 5 Extraction of several organic compounds using CPE technique

Compounds	Type of sample	Analytical method	Conditions	Extraction parameters	References
Paracetamol	Human urine	Spectrophotometry	TX-100	LOQ = 0.13 μg mL^{-1}	[82]
Norfloxacin, ciprofloxacin, sarafloxacin	Water (lake, river, waste water)	HPLC–DAD Fluorescence analysis	SDS, HCl (mol L^{-1})	D_L (μg mL^{-1}): 0.007 (norfloxacin), 0.010 (ciprofloxacin) 0.009 (sarafloxacin), 0.013 (gatifloxacin)	[83]
Bisoprolol	Human plasma	LC-ESI-MS/MS	TX-114, EtOH	LLOQ = 0.3 ng mL^{-1}	[84]
Vitamin E	In water	HPLC-DAD	Sodium dodecanesulfonic acid (SDSA), 12 M HCl	LLOQ = 0.1 μg mL^{-1}	[85]
violacein	Biomass	UV–Visible spectrophotometry	SDS with tetrabutylammonium chloride $[N_{4444}]$Cl–SDS or TX-114 EtOH	D_L: 98.1 μg mL^{-1} 109.9 μg mL^{-1}	[86]
PAH	Water	HPLC	Genapol X80, 2.0% NaCl	$D_L = 2.6–6.8$ ng L^{-1}	[87]
PCBs	In water	HPLC	Brij 30 or Brij 97	1.5–16.3 ng mL^{-1}	[88]
Organophosphorus	In water	HPLC	TX-114	D_L = 0.03–0.08 ng mL^{-1}	[89]
Anthracene	Aqueous solution	HPLC analysis with fluorescence detection	TX-114	$C_F = 40$	[90]

D_L = limit of detection, C_F = concentration factor, *LOQ* limit of quantification

Table 6 CPE techniques for estimation of bioactive compounds

Bioactive compounds isolated	Sources of crude material	Technique employed	Extracting phase	Comments	References
Lycopene	Tomato peel biomass	Enzyme-facilitated CPE	SPAN-20	Yield increased by a factor of four compared to EAE and by a factor of ten relative to SLE	[91]
Phenols and carotenoids	Juice of red-fleshed citrus and olive oil mill	CPE	Tween 80	Extracted as much as 94.4% retaining radical quenching ability	[92]
Phenolic derivatives	Aqueous waste from olive processing	Ultrafiltration enhanced CPE	TX-100	The maximum yield was obtained at 10% Triton X-100 at 90 °C	[95]
Polyphenolic compounds	Olive plant leaf	CPE	Surfactant mixture comprising Tween 80 and Genapol X-080 with polyethylene glycol 8000	Isolated while preserving antioxidant activity	[96]
Xanthohumol	Beer	CPE	TX-114	Enhanced the thermal stability while maintaining the AOX function	[97]
Thymol	Ajwain seeds	CPE	SPAN 80	Order of optimal thymol extraction: SPAN 80 > TX-100 > Tween 80	[98]

reported by Guo et al. [109]. The effective isolation of remazol turquoise blue G-133 (RTB G-133) dye has been accomplished successfully by the utilization of cloud point behaviour of TX-114 [110]. The preconcentration of trace amounts allura red in food samples was effectively determined with the help of CPE technique using a mixture of surfactants: CTAB, TX-100, TX-114. The LOD of allura red is 7.8 $\mu g\ L^{-1}$ as determined by Pourreza et al. [111]. The surfactant, nonylphenol with 9.5 ethoxylation degree (NP9.5EO), has been employed in the extraction of toxic dye, Reactive blue 19 following cloud point approach. The extraction efficiency is strongly influenced by the concentration of NP9.5EO, temperature of the CPE system, and initial dye concentration, as investigated by Melo et al. [112].

10 Conclusion and Future Outlooks

Clouding of surfactant solution is an excellent topic of research, getting most priority to the researchers working in the advancement of wastewater treatment, drug delivery, accumulation of proteins and biomolecules, biotransformation, etc. The particular temperature at which separation of two phases into comparatively intense turbid phase and dilute aqueous phase occur, can be tuned by altering numerous variable parameters, for example: pH of the media, concentration of surfactants, buffer media, structural features of surfactant, ionic strength of the media, presence of salts, and other additives, etc. Certain compounds elevated the cloud point and known as cloud point boosters. The UV–visible spectrophotometer is mostly used to measure the cloud point of amphiphiles. In view of the temperature-induced phase separation characteristic, micelle-mediated separation technique has been developed for several application purposes. The mechanism of CPE is driven by the solubilization of compound (analyte) within the surfactant-micelle and the uneven partitioning of surfactant between two immiscible phases. The hydrophobic-lyophilic balance of surfactant, polarity of analytes, pH and ionic strength of the system, and operating temperature, etc. have significantly impact on the phase volume ratio, partitioning coefficient, and extraction efficiency. Entropy-enthalpy compensation was accounted in all studies. The advantages of CPE are not limited to

- Rapid and simple with lower extraction time
- Use of benign surfactant and lessen the toxicity
- Offers high preconcentration factors even with low surfactant concentration
- Affordable technique
- Insignificant environmental impact

Recently, several new approaches have been introduced to improve the conventional CPE procedures, like ultrasound supported cloud point extraction, microwave-enabled cloud point extraction, flow-controlled cloud point extraction, rapid synergistic cloud point extraction, etc. [5, 113–115].

In addition to several advantages, CP may pose a limitation in pharmaceutical applications due to the propensity of certain amphiphilic drugs to exhibit clouding behaviour. The recovery and repetitive use of nonionic surfactant have a crucial impact on its economic and environmentally friendly process.

Bibliography

1. Mukherjee P, Padhan SK, Dash S, Patel S, Mishra BK (2011) Adv Coll Interface Sci 162:59–79
2. Khan MAR, Amin MR, Rahman M, Rub MA, Hoque MA, Khan MA, Asiri AM (2019) J Chem Eng Data 64:668–675
3. Samaddar P, Sen K (2014) J Ind Eng Chem 20:1209–1219
4. Noorashikin MS, Sohaimi NM, Suda N, Aziz HZ, Zaini SRM, Kandasamy S, Suresh K (2017) J Sustain Sci Manage 12:79–95
5. Mortada WI (2020) Microchem J 157:105055

6. Sar P, Ghosh A, Scarso A, Saha B (2019) Res Chem Intermed 45:6021–6041
7. Sar P, Saha B (2020) Adv Coll Interface Sci 284:102241
8. Kabir-ud-Din, Al-Ahmadi MDA, Naqvi AZ, Akram M (2007) J Surfactant Deterg 10:231–236
9. Inoue T, Ohmura H, Murata D (2003) J Colloid Interface Sci 258:374–382
10. Rahman M, Rub MA, Hoque MA, Khan MA, Asiri AM (2020) J Mol Liq 312:113366
11. Kabir-Ud-Din D, Sharma ZA, Khan VK, Aswal S (2006) Kumar. J Colloid Interface Sci 302:315–321
12. Yao HL, Shi YC, Yuan SL, Li GZ (2009) J Dispersion Sci Technol 30:1223–1230
13. Bales BL, Zana R (2004) Langmuir 20:1579–1581
14. Zana R, Benrraou M, Bales BL (2004) J Phys Chem B 108:18195–18203
15. Yadav SK, Yadav RB, Kumar S (2016) J Surfactant Deterg 19:1063–1069
16. Kumar S, Patel H, Patil SR (2013) Colloid Polym Sci 291:2069–2071
17. Kumar S, Sharma D (2002) Kabir-ud-din J Surf Sci Technol 18:25–33
18. Kumar S, Bhadoria A (2012) J Chem Eng Data 57:521–525
19. Kumar S, Patel H, Patil SR (2013) Colloid Polym Sci 291:2069–2077
20. Mitra D, Chakraborty I, Bhattacharya SC, Moulik SP (2007) Langmuir 23:3049–3061
21. Raghavan SR, Edlund H, Kaler EW (2002) Langmuir 18:1056–1064
22. Warr GG, Zemb TN, Drifford M (1990) J Phys Chem 94:3086–3092
23. Kubota K, Kuwahara N, Sato H (1994) J Chem Phys 100:4543–4547
24. Martin A, Lopez I, Monroy F, Casielles AG, Ortega F, Rubio RG (1994) J Chem Phys 101:6874–6879
25. Lindman B, Medronho B, Karlström G (2016) Curr Opin Colloid Interface Sci 22:23–29
26. Molina-Bolívar JA, Ruiz CC (2012) Fluid Phase Equilib 327:58–64
27. Batıgöç Ç, Akbas H, Boz M (2011) J Chem Thermodyn 43:1800–1803
28. Mahbub S, Rub MA, Hoque MdA (2019) J Chem Eng Data 64:4181–4188
29. Parikh J, Rathore J, Bhatt D, Desai M (2013) J Dispersion Sci Technol 34:1392–1398
30. Akbaş H, Batıgöç Ç (2009) Fluid Phase Equilib 279:115–119
31. Batıgöç Ç, Akbaş H (2011) Fluid Phase Equilib 303:91–95
32. Batıgöç Ç, Akbaş H (2017) J Mol Liq 231:509–513
33. Shigeta K, Olsson U, Kunieda H (2001) Langmuir 17:4717–4723
34. Ghosh S, Moulik SP (1999) Indian J Chem Sect A 38:201–208
35. Mahajan S, Mahajan RK (2013) Adv Coll Interface Sci 199–200:1–14
36. Kabir-ud-Din, Rub MA, Naqvi AZ (2012) J Dispers Sci Technol 33:1667–1673
37. Naqvi AZ, Khatoon S, Kabir-ud-Din (2011) Colloids and Surfaces B 88:568–573
38. Kabir-ud-Din, Rub MA, Naqvi AZ (2012) J Mol Liquids 167:103–109
39. Rub MA, Asiri AM, Azum N, Khan A, Khan AAP, Kabir-ud-Din (2013) Tenside Surfactant Detergent 50:376–384
40. Maya SA, Jahan I, Khan JM, Ahsan SkMdA, Rana S, Rahman MM, Hoque MdA, Goni MdA, Khan MA (2024) J Sol Chem 53:1527–1543
41. Pulikkal AK, Gurung J (2018) New J Chem 42:4402–4411
42. Gurung J, Anjudikkal J, Pulikkal AK (2020) J Mol Liq 318:114221
43. Naqvi AZ, Kabir-ud-Din, Phase separation phenomena in drug systems. In: Encyclopedia of biocolloid and biointerface science 2V Set, Chap 69, vol I, II
44. Alam MdS, Kabir-ud-Din, Mandal AB (2013) Colloids Surf B Biointerfaces 105:236–245
45. Hasan T, Rub MA, Md TR, Joy S, Rana F, Khan MdA, Hoque MK (2022) J Mol Liq 354:118853
46. Rahim MdA, Mahbub S, Ahsan SkMdA, Alam M, Saha M, Shahriar I, Rana S, Halim MA, Hoque MdA, Kumar D, Khan JM (2021) J Mol Liquids 322:114683
47. Khan MAR, Amin MdR, Patoary MOF, Rub MA, Hoque MdA, Khan MA, Kumar D, Asiri AM, Khan F, Alfaifi SYM (2021) Phys Chem Liquid 59:781–794
48. Alam MdS, Naqvi AZ, Kabir-ud-Din (2010) J Dispers Sci Technol 31:449–455
49. Rahman M, Anwar SkJ, Molla MR, Rana S, Hoque MdA, Rub MA, Khan MA, Kumar D (2019) J Mol Liquids 292:111322
50. Huibers PDT, Shah DO, Katritzky AR (1997) J Colloid Interface Sci 193:132–136

51. Gu T, Sjoblom J (1992) Colloids Surf 64:39–46
52. Mushtaq M, Tan IM, Ismail L, Lee SYC, Nadeem M, Sagir M (2014) J Dispersion Sci Technol 35:322–328
53. Taechangam P, Scamehorn JF, Osuwan S, Rirksomboon T (2009) Colloids Surf A 347:200–209
54. Al-Sabagha AM, Nasser NM, Migahed MA, Kandil NG (2011) Egypt J Pet 20:59–66
55. Streletzky K, Phillies GDJ (1995) Langmuir 11:42–47
56. Kumar S, Bhadoria A, Patel H, Aswal VK (2012) J Phys Chem B 116:3699–3703
57. Bhadoria A, Kumar S, Aswal VK, Kumar S (2015) RSC Adv 5:23778–23786
58. Khan ZA, Albishri HM, Ghazali A, Ahmad T (2013) J Surfactants Deterg 16:25–31
59. Alam MdS, Kabir-ud-Din, Mandal AB (2010) J Chem Eng Data 55:1693–1699
60. Kabir-ud-Din, Rub MA, Alam MdS (2010) J Dispersion Sci Technol 31:1182–1187
61. Sznek B, Kupczyk O, Czyrski A (2025) Processes 13:430
62. Quina FH, Hinze WL (1999) Indus Eng Chem Res 38:4150–4168
63. Martınez RC, Gonzalo ER, Cordero BM, Pavon JLP, Pinto CG, Laespada EF (2000) J Chromatogr A 902:251–265
64. Ojeda CB, Rojas FS (2009) Anal Bioanal Chem 394:759–782
65. Hunga KC, Chena BH, Yu LE (2007) Sep Purif Technol 57:1–10
66. Halko R, Hagarová I, Andruch V (2023) J Chromatogr A 1701:464053
67. Semysim FA, Shabaa GJ, Azooz EA, Snigur D (2025) Trends Environ Anal Chem 45:e00250
68. Ojeda CB, Rojas FS (2012) Microchim Acta 177:1–21
69. Mandal S, Lahiri S (2022) Microchem J 175:107150
70. Pytlakowska K, Kozik V, Dabioch M (2013) Talanta 110:202–228
71. Bezerra MA, Arruda MAZ, Ferreira SLC (2005) Appl Spectrosc Rev 40:269–299
72. Sun M, Wu Q (2010) J Hazard Mater 176:901–905
73. Rod AR, Borhani S, Shemirani F (2006) Eur Food Res Technol 223:649–653
74. Manzoori J, Abdolmohammad-Zadeh H (2007) Acta Chim Slov 54:378–384
75. Manzoori JL, Karim-Nezhad G (2003) Anal Sci 19:579–583
76. Shemirani F, Kozani RR, Assadi Y (2007) Microchim Acta 157:81–85
77. Giokas DL, Paleologos EK, Tzouwara-Karayanni SM, Karayannis MI (2001) J Anal Atom Spectrom 16:521–526
78. Hadri HE, Hackley VA (2017) Environ Sci Nano 4:105–116
79. Mortada WI, Awad AA, El-Defrawy MM, Khalifa ME (2022) Anal Sci 38:1489–1495
80. Mesquita da Silva MA, Frescura VLA, Aguilera FJN, Curtius AJ (1998) J Anal Atom Spectrom 13:1369–1373
81. Mauricio L, Wuilloud RG (2020) Talanta 212:120802
82. Filik H, Sener I, Cekic SD, Kilic E, Apak R (2006) Chem Pharm Bull 54:891–896
83. Xia Q, Yang Y, Liu M (2014) Spectrosc Lett 47:206–213
84. Giebułtowicz J, Kojro G, Bús-Kwásnik K, Rudzki PJ, Marszałek R, Lés A, Wroczýnski P (2015) J Chromatogr A 1423:39–46
85. Casero I, Sicilia D, Rubio S, Perez-Bendito D (1999) Anal Chem 71:4519–4526
86. Schaeffer N, Kholany M, Veloso TLM, Pereira JL, Ventura SPM, Nicaud J-M, Coutinho JAP (2019) Chem Commun 55:8643–8646
87. Boeckelen A, Niessner R, Fresenius J (1993) Anal Chem 346:435
88. Fernandez AE, Ferrera ZS, Rodriguez JJS (1998) Anal Chim Acta 358:145–155
89. Garcıa-Pinto C, Perez-Pavon JL, Moreno-Cordero B (1995) Anal Chem 67:2696
90. Cordero BM, Pavon JLP, Pinto CG, Laespada MEF (1993) Talanta 40:1703
91. Papaioannou EH, Karabelas AJ (2012) Acta Biochim Pol 59:71–74
92. Katsoyannos E, Gortzi O, Chatzilazarou A, Athanasiadis V, Tsaknis J, Lalas S (2012) J Sep Sci 35:2665–2670
93. Wu H, Zhao G, Du L (2010) Spectrochimica Acta Part A 75:1624–1628
94. Arya SS, Kaimal AM, Chib M, Sonawane SK, Show PL (2019) J Food Sci Technol 56:524–534
95. El-Abbassi A, Kiai H, Raiti J, Hafidi A (2014) J Environ Chem Eng 2:1480–1486
96. Stamatopoulos K, Katsoyannos E, Chatzilazarou A (2014) Antioxidants 3:229–244

97. Chen L, Zhao Q, Jin H, Zhang X, Xu Y, Yu A, Zhang H, Ding L (2010) Talanta 81:692–697
98. Chatterjee S, Jain A, De S (2017) J Food Sci Technol 54:4353–4361
99. Wang Z, Guo Y, Bao D, Qi H (2006) J Chem Technol Biotechnol 81:560–565
100. Wang L, Wang Z, Xu JH, Bao D, Qi H (2006) Bioprocess Biosyst Eng 29:157–162
101. Ritter E, Smirnova I (2018) Chem Ing Tec 90:348–357
102. Wang Z (2007) Appl Microbiol Biotechnol 75:1–10
103. Wang Z, Dai Z (2010) Enzyme Microb Technol 46:407–418
104. Tatara E, Materna K, Schaadt A, Bart HJ, Szymanowski J (2005) Environ Sci Technol 39(9):3110–3115
105. Shi J, Chen L (2014) Anal Methods 6:8129–8135
106. Bhatt DR, Maheria KC, Parikh JK (2016) Int J Environ Sci Technol 13:589–598
107. Bhatt DR, Maheria KC, Parikh JK (2015) J Environ Chem Eng 3:1365–1371
108. Purkait MK, DasGupta S, De S (2006) J Hazard Mater 137:827–835
109. Pan T, Xu M, Chen X, Sun G, Guo J (2013) Sep Sci Technol 48:1040–1048
110. Saravanan DD, Murugesan T, Arunagiri A (2017) Desalin Water Treat 60:228–241
111. Pourreza N, Rastegarzadeh S, Lark A (2011) Food Chem 126:1465–1469
112. Melo RPF, Neto ELB, Moura MCPA, Dantas TNC, Neto AAD, Oliveira HNM (2014) Sep Purif Technol 138:71–76
113. Hagarová I, Urík M (2016) Curr Anal Chem 12:87–93
114. Snigur D, Azooz EA, Zhukovetska O, Guzenko O, Mortada W (2023) Trends Anal Chem 164:117113
115. Mandal S, Poi R, Hazra DK, Ansary I, Bhattacharyya S, Karmakar R (2023) J Chromatogr B 1215:123587

Physicochemical Behavior and Applications of Biosurfactants

Nikil Satpute, Abhinav Sati, and Amit P. Pratap

Abstract Biosurfactants are a varied class of amphiphilic compounds primarily synthesized by microorganisms which exhibit remarkable surface-active properties. Biosurfactants are environmentally friendly, biodegradable agents that remain efficient even under extreme temperature, pH, and salinity conditions, making them promising sustainable alternatives to conventional synthetic surfactants. This review highlights the physicochemical behavior of biosurfactants, focusing on their ability to reduce surface and interfacial tensions, form micelles and emulsions, and increase the accessibility and solubility of hydrophobic compounds. Particular attention is given to their impact on microbial cell surfaces, membrane interactions, and their role in processes such as biofilm disruption and pollutant degradation. Additionally, the paper discusses the expanding industrial and environmental applications of biosurfactants, including bioremediation, wastewater treatment, enhanced oil recovery, food processing, agriculture, and medical therapeutics. They are promising sustainable substitutes for synthetic surfactants, highlighted by the rising market demand, technological advancements in fermentation, and the utilization of renewable substrates. Overall, biosurfactants represent a promising class of biomolecules that bridge environmental responsibility with industrial innovation.

Keywords Biosurfactants · Applications · Physiochemical behaviour · Industrial · Environmental · Medical · Agricultural applications

N. Satpute · A. Sati · A. P. Pratap (✉)
Department of Oils, Oleochemicals and Surfactants Technology, Institute of Chemical Technology, Mumbai, India
e-mail: ap.pratap@ictmumbai.edu.in
URL: https://www.ictmumbai.edu.in

M. Abdul Rub et al. (eds.), *Industrial Applications of Individual and Mixed Surfactant Systems—Volume 2*, Sustainable Materials and Technology,
https://doi.org/10.1007/978-981-95-9195-4_10

1 Introduction

Surfactants are chemical substances composed of amphipathic molecules containing both hydrophilic and hydrophobic counterparts, enabling them to accumulate and function at physical interfaces. The hydrophobic portion typically consists of long hydrocarbon chains, while the hydrophilic part can be cationic, anionic, nonionic, or amphoteric. This unique structure enables surfactants to lower surface and interfacial tensions facilitating the formation of microemulsions, where hydrocarbons can be dispersed in water or vice versa [1].

Most synthetic surfactants originate from the petrochemical industry, making them cost-effective and easy to produce in large quantities. However, their production is often considered unsustainable and conflicts with the United Nations' sustainable development goals, various national initiatives aimed at fostering a green economy and growing consumer preference for environmentally friendly products. Additionally, synthetic surfactants can pose toxicity and biocompatibility concerns, leading to harmful effects on ecosystems, which further restricts their use [2–4]. In contrast, biosurfactants, particularly microbial biosurfactants, are metabolic by-products of bacterial or fungal fermentation. These natural compounds are produced through sustainable methods and have minimal environmental impact. Compared to synthetic surfactants, biosurfactants generally exhibit lower toxicity. As a result, they are increasingly regarded as a viable alternative, leading to a steady rise in their application and market share over the past 15 years. Like synthetic surfactants, biosurfactants exhibit a wide variety of chemical structures and are produced by microorganisms that can utilize either insoluble substrates such as oils, residues, and hydrocarbons or soluble substrates like sugars and other carbohydrates [5–7]. Substituting synthetic surfactants with biosurfactants has the potential to reduce lifetime CO_2 emissions by about 8%, preventing roughly 1.5 million tons of CO_2 from being released into the atmosphere. Currently, biosurfactants make up nearly 10% of global surfactant production, which translates to around ten million tons annually. These naturally derived surfactants find applications across a wide range of industries, including petroleum, food (as emulsifiers), pharmaceuticals (in creams, moisturizers, and drugs), medicine (as antimicrobial agents), agriculture (in fertilizers), and civil engineering (for wastewater and sewage treatment) [8–10]. Research into microbial biosurfactant production began in the 1960s, and over the past six decades, it has advanced considerably, resulting in the development and commercialization of numerous biosurfactant-based products. In the last ten years, with sustainability becoming a global priority, interest in biosurfactant production has surged. These molecules are highly valued for their biodegradability, low toxicity, specificity, and stability under extreme conditions, such as low temperatures, harsh pH, and high salinity. Moreover, they can be produced from renewable substrates, offering both economic advantages and environmentally friendly biodegradability [11, 12].

2 Physiological Behavior of Biosurfactants

Biosurfactants exhibit amphiphilic properties, reducing surface tension and forming stable emulsions, micelles, and biofilms. They enhance the bioavailability of hydrophobic substances, aiding in bioremediation by facilitating pollutant degradation. Their stability under extreme conditions, lesser toxicity as well as biodegradability elevates their significance for applications in medicine, agriculture, pharmaceuticals, and environmental remediation [13]. Above critical micelle concentration (CMC), biosurfactant molecules self-assemble to form micelles as shown in the diagram. Here, hydrophobic pollutants such as hydrocarbons and polycyclic aromatic compounds get into the core of the structure, increasing the solubility in water. This process not only enhances pollutant dispersion but also improves their bioavailability to microorganisms, facilitating enzymatic degradation, and accelerating overall bioremediation efficiency (Fig. 1).

2.1 Impact on Microbial Cell Surface Properties

Biosurfactants impact cell surface properties, leading to notable changes in hydrophobicity of surfaces, electrokinetic potential, cell morphology, and surface chemical functional groups. As discussed earlier, these compounds can alter cell surface characteristics, facilitating microbial adhesion to pollutants or improving pollutant partitioning and bioavailability for microbial uptake. One of the most assessed parameters in this context is cell surface hydrophobicity (CSH), primarily due to the ease of

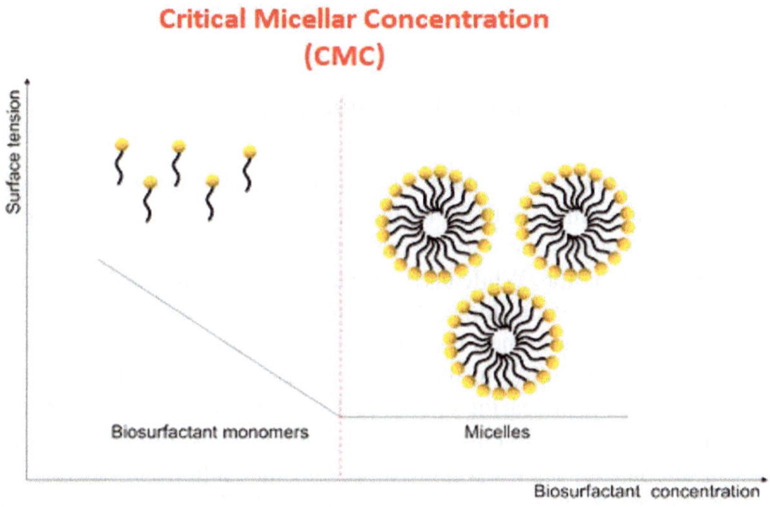

Fig. 1 Diagram showing micelle assembly by biosurfactants after reaching the CMC

measurement. The microbial adhesion to hydrocarbons (MATH) assay is often used to evaluate the propensity of cells to attach to hydrophobic surfaces, such as emulsion droplets [14, 15]. In a study by Patel et al., *Lactobacillus rhamnosus* biosurfactant production was quantified at 4.32 ± 0.19 g/L, with a critical micelle concentration (CMC) of 3.0 g/L. It significantly reduced surface tension from 71.12 ± 0.73 mN/m to 41.76 ± 0.60 mN/m, enhancing its surface-active properties. The biosurfactant demonstrated strong anti-adhesion and antibiofilm activity against *Bacillus subtilis, Pseudomonas aeruginosa, Staphylococcus aureus,* and *Escherichia coli.* It inhibited bacterial adhesion to surfaces, resulting in biofilm disruption through the reduction of exopolysaccharide (EPS) production and altering cell viability. The structural analysis via GC–MS revealed the presence of glycolipid analogues contributing to its activity.

These findings confirm that biosurfactants from *L. rhamnosus* effectively modify microbial cell surface properties, reducing adhesion, and destabilizing biofilms [16]. In another study by Silva et al., derived biosurfactants were synthesized and tested for stability under various pH, salinity, and temperature conditions. Glyceryl laurate and hydroxystearic acid were synthesized. The surfactants exhibited excellent stability, emulsifying properties, and nearly 100% inhibition of bacterial adhesion for *Pseudomonas aeruginosa* and *Bacillus cereus.* Quantitative comparison demonstrates that bio-based surfactants such as glyceryl laurate and hydroxystearic acid inhibited bacterial adhesion by nearly 100%, marginally surpassing the inhibition of surface colonization achieved with SDS (≈ 70–80%). Furthermore, they have reduced extracellular polymeric substances (EPS) production—protein content decreased by approximately 70% and carbohydrate levels by 60%, compared to reductions of 45% and 35% with SDS, respectively. These results suggest that lupin-derived bio-based surfactants could be promising non-biocide alternatives for controlling microbial fouling and biofilm formation on various surfaces [17].

2.2 *Effects on Microbial Cell Membranes*

Microbial biosurfactants, particularly glycolipids and lipopeptides, have garnered significant attention due to their unique characteristics and versatility. Their amphiphilic nature allows them to interact with phospholipid membranes, altering their structure and function. While early research primarily focused on model phospholipid membranes, recent studies have shifted toward exploring their interactions with biological membranes, including those of mammals and microorganisms [18]. For instance, biofouling in membrane bioreactors (MBRs) causes clogging and reduced efficiency, increasing operational costs. Chemical surfactants are toxic and damaging, while biosurfactants, such as rhamnolipids, offer a biodegradable, low-toxicity alternative. They disrupt hydrophobic interactions, effectively cleaning membranes, improving flux recovery, and reducing fouling, offering a sustainable solution for MBR maintenance [19]. A study by Essghaier et al. focused on a rapid method for biosurfactant monitoring using hemolysis, oil spreading, and emulsification tests. Biosurfactants produced by *Pantoea alhagi* showed an 82% emulsification index and identified as glycolipid with predominant fatty acids, oleic acid, and methyl 2-hyroxydodecanoate. The biosurfactants exhibited antimicrobial and antibiofilm properties, with strong antibacterial activity against *E. coli* and *S. aureus* and antifungal effects against *Fusarium sp.*, along with notable antioxidant activity. These characteristics suggest potential industrial applications for the BS, particularly in biofilm control and pathogen inhibition [20]. In another study by Mandal et al., the biosurfactant-coated CuO nanoparticles exhibited superior antibacterial activity against *Micrococcus luteus* and *Escherichia coli* and attributed to enhance reactive oxygen species (ROS) generation. ROS disrupt bacterial cell membranes, causing oxidative stress and cell death.

Nanoparticles coated with biosurfactants exhibited better dispersion and reduced particle sizes, which enhanced their antimicrobial effectiveness. This approach also provides a more environmentally friendly method for nanoparticle synthesis [21]. Al-Shamir et al. investigated the antibiofilm and antimicrobial effects of *Lactobacillus rhamnosus* and its biosurfactants, lactic acid (LA) and acetic acid (AA), against *Acinetobacter baumannii*, a multidrug-resistant biofilm-forming pathogen. LA and AA exhibited strong bactericidal effects, disrupting the bacterial cell membrane, causing cell envelope discontinuity, leakage of cell contents, and protein lysis. They also damaged bacterial DNA and reduced the expression of biofilm and DNA repair genes. These findings highlight the potential of probiotics and their biosurfactants as novel, eco-friendly alternatives for preventing and treating biofilm-related infections in clinical settings [22]. Another study by McKenna explored the production and effectiveness of biosurfactants from *Lactobacillus rhamnosus*, isolated from human breast milk, in inhibiting bacterial adhesion and biofilm formation. The biosurfactants exhibited strong anti-adhesion and antibiofilm properties, reducing surface tension and altering bacterial cell integrity. The biosurfactants caused damage to bacterial cell membranes, leading to cell viability loss and reduced exopolysaccharide (EPS) production, suggesting their potential as eco-friendly alternatives to chemical antibiofilm agents in controlling biofilm-related infections [16]. Amirinejad et al. investigated the antimicrobial and antibiofilm potential of a glycolipid biosurfactant (GBB12) produced by *Shewanella algae* B12. GBB12 enhanced the effectiveness of antibiotics against resistant bacteria such as *MRSA* and *Acinetobacter baumannii*, showing increased inhibition of biofilm formation and disruption. Scanning electron microscopy confirmed that GBB12 disrupted bacterial cell membranes. The study highlights the promising role of biosurfactants in enhancing antibiotic efficacy and controlling biofilm-related infections [23]. Ren et al. investigated that fatty acyl glutamic acid (FA-glu) biosurfactant inhibited the growth of foodborne pathogens *L. monocytogenes* and *E. coli O157:H7* by disrupting their cell envelopes. FA-glu caused rupture and perforation of the bacterial cell membranes through selective interactions with phospholipids, altering the lipid bilayer's fluidity. This led to cell envelope damage, inhibiting bacterial growth [24].

2.3 Enhancement of Biodegradation Processes

Biosurfactants boost the breakdown of hydrophobic pollutants by making them more accessible to microbes, lowering surface tension, and facilitating better microbial interaction with the contaminants. They stimulate microbial activity, promote emulsification, and accelerate pollutant breakdowns. Applications include oil spill cleanup, soil and water bioremediation, and wastewater treatment. Optimizing conditions like pH, temperature, and microbial strains improve efficiency. Biosurfactants enhance biodegradation by increasing petroleum hydrocarbon solubility and microbial access. Oleophilic microorganisms producing biosurfactants accelerate bioremediation. The Potential Microorganisms Immobilized on Biochar (PMIBC) system offers a cost-effective solution for aquatic pollution cleanup. This eco-friendly approach requires further research to optimize large-scale applications for sustainable environmental remediation [25]. Bacteria in polycyclic aromatic hydrocarbons (PAH)-contaminated environments form biofilms that produce biosurfactants and exopolymeric substances (EPS), enhancing PAH biodegradation. EPS facilitates PAH adsorption and emulsification, while biosurfactants improve solubilization by reducing interfacial tension. Key genes for exopolysaccharide synthesis and biosurfactant production aid PAH uptake. Regulatory mechanisms like quorum sensing and cyclic-di-GMP control these processes. Understanding these pathways enhances bioremediation efficiency, optimizing PAH degradation for environmental cleanup [26].

A study by Parthipan et al. demonstrated the enhanced degradation of mixed PAHs (anthracene, pyrene, and benzo(a)pyrene) using biosurfactant-producing bacterial strains and iron nanoparticles. The biosurfactant, produced by *Bacillus subtilis* A1, increased PAH bioavailability, while iron nanoparticles boosted bacterial growth, biomass, and PAH adsorption. This integrated approach achieved 85% degradation efficiency and shows potential for effectively removing complex PAH pollutants from soil and aquatic environments [27]. Another study by Shin et al. evaluated the solubilization–biodegradation process for phenanthrene removal using a soil-packed column. The injection of a 150 mg/L biosurfactant solution enhanced solubility, with the highest removal observed at pH 5. After 10 weeks of biodegradation, phenanthrene levels significantly decreased, except at pH 4. These findings suggest that biosurfactant flushing can effectively enhance biodegradation without negatively impacting subsurface environments [28]. In a separate study, Thirumuruga et al. isolated bacterial strains from soil polluted with used motor oil to explore their potential for producing biosurfactants. Two strains, DTS1 and DTS2, were selected based on primary screening methods and showed significant crude oil and anthracene degradation. Biosurfactant production enhanced the bioavailability of hydrophobic hydrocarbons, aiding degradation. The strains were identified through 16S rRNA gene sequencing, and their potential for hydrocarbon bioremediation makes them suitable for environmental cleanup applications [29]. Taghavi et al. examined the biodegradation of UV-pretreated and un-pretreated plastics (PE, PS, PET) by a microbial consortium and biosurfactant (rhamnolipid). The combination of UV pretreatment, bio-treatment, and biosurfactant was most effective for PS, showing significant surface degradation and biofilm formation. However, biosurfactant negatively impacted PE and PET degradation. Chemical and thermal analyses revealed transformation in PS, while PE and PET showed minimal changes, highlighting the varying effects of biosurfactants on different plastics [30]. In another study, Das et al. isolated *Bacillus* spp. from crude oil-contaminated soil and evaluated their hydrocarbon degradation efficiency. *Bacillus thuringiensis* SSL1 and *Bacillus cereus* SSL3 showed high asphaltene degradation (76.4% and 67.4%, respectively), with biosurfactants enhancing hydrocarbon availability. *B. thuringiensis* SSL1 was particularly effective for asphaltene, petroleum, and polyaromatic hydrocarbon degradation, offering potential for improved crude oil remediation strategies using biosurfactants for efficient cleanup [31]. Xia et al. identified a biosurfactant-producing strain, Pseudomonas mosselii MP-6, to improve the breakdown of high-molecular-weight polycyclic aromatic hydrocarbons (HMW-PAHs) in wastewater. The rhamnolipid biosurfactant reduced surface tension, improving PAH solubilization, and degradation. Environmental stress tests confirmed the robustness of the MP-6 biosurfactant, highlighting its potential for wastewater treatment, particularly in industries like coking, where HMW-PAHs are prevalent [32]. Zhou et al. investigated the effect of biosurfactant and nutrient amendment on heavy oily sludge bioremediation. Biosurfactants produced by *Pseudomonas sp. CH1* enhanced petroleum hydrocarbon degradation, particularly n-alkanes, and enriched microbial communities capable of hydrocarbon degradation. However, excessive biosurfactant addition inhibited biodegradation. This suggests the importance of biosurfactants in promoting biodegradation of petroleum-based hydrocarbons present in oily sludge, though balanced application is crucial for optimal bioremediation [33].

2.4 Applications in Bioremediation

Biosurfactants, for example rhamnolipids and surfactin, are used in bioremediation for oil spill cleanup, soil and water decontamination, and wastewater treatment [34]. Biosurfactants improve the bioavailability of hydrophobic pollutants by lowering surface tension, thereby enhancing microbial degradation. Common producers include Pseudomonas aeruginosa, Bacillus subtilis, and Candida species. For example, Ng et al. studied a biosurfactant from Acinetobacter venetianus AMO1502, isolated from an oil spill. Produced in a NaCl-supplemented medium, it displayed strong emulsifying activity, reduced surface tension by 41.7%, and remained stable across varying temperatures, pH levels, and salinity. This biosurfactant successfully removed about 25% of engine oil from sand, 12% from water, and decreased oil stains on contaminated stone [35]. In another study by Pandey et al., a novel *Aeromonas hydrophila* RP1 strain was isolated for biosurfactant production, which was characterized as a glycolipopeptide. The biosurfactant decreased surface tension and remained stable over a broad range of temperatures, pH levels, and salinity conditions. It enhanced oil recovery by 28% and degraded hydrocarbons like diesel and n-hexadecane by 77.33% and 55.98%, respectively, demonstrating significant potential for bioremediation and enhanced oil recovery (EOR) applications [36]. D'Almeida et al. studied the biosurfactant produced by *Acinetobacter venetianus* AMO1502, isolated from an oil spill off the Brazilian coast, for bioremediation. It showed high emulsifying activity, reducing water surface tension by 41.7%. The biosurfactant demonstrated stability across temperature, pH, and NaCl variations, aiding in the removal of engine oil from sand (25%) and aqueous medium (12%) and reducing oil stains on contaminated stones [37]. Manikkasundaram et al. investigated the biosurfactant produced by *Streptomyces enissocaesilis* HRB1, isolated from petroleum-contaminated soil, for its biomedical and bioremediation potential. The glycolipid biosurfactant demonstrated antimicrobial, antioxidant, and anticancer properties, inhibiting biofilm formation, quorum sensing, and leukemia cell proliferation. It also showed effective decolorization of malachite green and minimal toxicity. The strain holds promise for low-cost biosurfactant production with diverse environmental and biomedical applications [38]. Faccioli et al., focused on biosurfactant production from *Pseudomonas cepacia* CCT6659 in bioreactors, evaluating its stability over 120 days. The biosurfactant showed low toxicity in Artemia salina larvae and enhanced oil solubilization in seawater compared to chemical surfactants. It effectively remediated petroleum-contaminated sand, removing 94.5% of the oil, and inhibited corrosion and microbial biofilm formation, highlighting its potential for industrial applications in bioremediation [39]. Another study by Gidudu and Chirwa evaluated a hybrid system combining bioremediation, biosurfactants, and electrokinetic remediation to enhance petrochemical hydrocarbon removal from contaminated soil. The rhamnolipid biosurfactant showed no significant toxicity to microbial growth but affected plant root elongation. The bio-electrokinetic system, optimized with 30 V, 185 mm electrode spacing, and 84 g/L biosurfactant, achieved 92% hydrocarbon removal and enhanced microbial growth, demonstrating an effective and sustainable remediation approach [40]. Purnomo et al. explored the use of biosurfactant-producing bacteria (Pseudomonas aeruginosa, Bacillus subtilis, and Ralstonia pickettii) immobilized in a PVA/SA/bentonite matrix to remediate oil-contaminated soil. FTIR and SEM analyses confirmed the structural integrity of the immobilized beads, while GC–MS results demonstrated complete degradation of various hydrocarbons. This study highlights the promise of eco-friendly, immobilized bacterial systems for effective soil bioremediation [41] (Fig. 2).

3 Applications of Biosurfactants

1. Environmental Applications

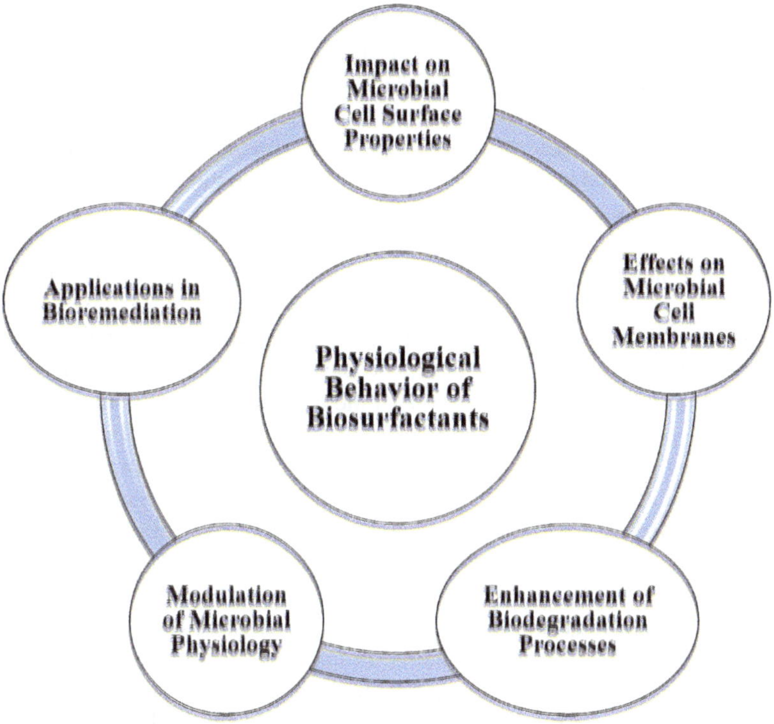

Fig. 2 Physiological behavior of biosurfactants

3.1 Bioremediation

The large-scale production and use of hydrocarbons directed significant environmental contamination. Because of their toxicity, persistence, and harmful effects on biological entities, it is crucial to remediate polluted sites. Being hydrophobic compounds, hydrocarbons tend to adhere to the soil matrix and are not very soluble in groundwater, often accounting for 90–95% or more of the total contaminant mass. This partitioning results in hydrocarbons being difficult to recover through physicochemical methods that lowers their bioavailability for microorganisms and limits their accessibility for oxidative as well as reductive chemicals in both in-situ and ex-situ remediation efforts [42]. The application of biosurfactants is a promising strategy to improve the efficiency of bioremediation in environments contaminated by hydrocarbons. Improvement of hydrocarbon bioremediation is done through two primary mechanisms: firstly, by raising bioavailability of substrates for microorganisms, and another, by enhancing the hydrophobicity of the cell surface through interaction, making it easier for hydrophobic substrates to attach to bacterial cells [43]. Through the reduction of surface and interfacial tensions, biosurfactants facilitate the dispersion of hydrophobic compounds, thereby increasing their effective surface area and consequently enhancing the mobility and bioavailability of hydrocarbons for microbial degradation. Consequently, this promotes the biodegradation and elimination of hydrocarbons. Hydrocarbon biodegradation is improved by the addition of biosurfactants via processes like emulsification, solubilization, and mobilization [44–46]. The mechanism of mobilization occurs when the concentration of surfactant is lower than critical micelle concentration. At these concentrations, biosurfactants decrease the surface and interfacial tension at the air–water and soil–water interfaces. When biosurfactants are present below their critical micelle concentration (CMC), they reduce the capillary forces that hold oil tightly to soil particles. This change increases the contact angle in soil–oil systems, making it easier for the oil to detach. Once the concentration rises above the CMC, a different process occurs—biosurfactant molecules begin clustering together to form micelles. In these micelles, the water-repelling ends gather inside, while the water-attracting ends remain exposed to the surrounding solution. The inner core of the micelle creates a favorable space for trapping hydrophobic organic compounds, allowing them to dissolve water through a process called solubilization [47]. Emulsification refers to the process of forming an emulsion, where fine droplets of oil or fat are dispersed within a liquid, usually water. High molecular weight biosurfactants serve as powerful emulsifiers and are often added to boost bioremediation by aiding the removal of oil-based pollutants from the environment. Recent studies emphasize their importance in bridging interactions between hydrocarbons and microorganisms, making them key players in the breakdown and cleanup of contaminants. Franzetti et al. [48] highlighted the role of biosurfactants in regulating cell surface hydrophobicity, a key factor influencing how microorganisms interact with hydrocarbons. When cells exhibit high hydrophobicity, they can directly attach to oil droplets or solid hydrocarbons, while lower hydrophobicity favors adhesion to micelles or emulsified oils. The study described three primary mechanisms of interaction: uptake of hydrocarbons dissolved in water, direct contact with large oil droplets, and association with emulsified or pseudosolubilized oil. The findings suggest that biosurfactants can alter these access routes depending on the microbial growth stage. Research on Gordonia sp. strain BS29 revealed the production of both cell-bound glycolipid biosurfactants and extracellular bioemulsifiers, with noticeable shifts in surface hydrophobicity during growth on hexadecane [48, 49]. *Cameotra and Singh* et al. [50] shed light on how Pseudomonas aeruginosa takes up *n*-alkanes and the role of rhamnolipids in this process. Their study revealed a novel mechanism in which hydrocarbons are internalized by bacterial cells for degradation. Rhamnolipids were shown to disperse hexadecane into tiny microdroplets, making the hydrocarbons more accessible to bacteria. Electron microscopy indicated that these biosurfactant-coated droplets were absorbed through a process of resembling active pinocytosis—an uptake mechanism not previously documented in bacteria. Although much is known about biosurfactants facilitating the breakdown of water-insoluble substrates, several details of these processes remain to be fully understood.

3.2 Soil Enhancement

Soil washing technology makes use of the chemical and physical properties of biosurfactants, rather than focusing on their effects on microbial metabolism or cell surfaces. Even so, this method can indirectly improve contaminant bioavailability, supporting further bioremediation. In practice, aqueous biosurfactant solutions are applied to remove low-solubility compounds from soils and similar materials through the washing process. Kang et al. [51] compared sophorolipids with nonionic surfactants such as Tween 80, Tween 60, Tween 20, and Span 20, 80, and 85 for their ability to wash 2-methylnaphthalene from contaminated soil. The results showed that sophorolipids outperformed all of the tested surfactants, except for Tween 80. The stronger performance of Tween 80 was linked to its high hydrophilic-lipophilic balance (HLB), since surfactants with higher HLB values tend to increase the solubility of 2-methylnaphthalene more effectively [51]. Lai et al. evaluated rhamnolipid, surfactin, Tween 80, and Triton X-100 for removing total petroleum hydrocarbons (TPH) from soils with low (3000 mg kg^{-1}) and high (9000 mg kg^{-1}) contamination. At 0.2% concentration, rhamnolipid and surfactin achieved higher removal efficiency than the synthetic surfactants, particularly in highly contaminated soils (63 and 62% vs. 40 and 35%). The results highlight the stronger potential of biosurfactants for remediating oil-polluted soils [51]. Franzetti et al. [5] investigated the surface-active compounds produced by *Gordonia* sp. strain BS29 for their potential application in soil remediation. In microcosm experiments, these biosurfactants facilitated the bioremediation of soils contaminated with both aliphatic and aromatic hydrocarbons. Complementary batch experiments demonstrated their efficiency in the removal of crude oil, polycyclic aromatic hydrocarbons (PAHs), and heavy metals. This study was the first to highlight the potential of these compounds in soil decontamination, expanding upon previous research that had characterized their physicochemical properties and role in enhancing hydrocarbon bioavailability [52]. Gordonia sp. BS29 grew on aliphatic hydrocarbons and produced extracellular bioemulsans and cell-bound biosurfactants. While bioemulsans slightly improved biodegradation of branched hydrocarbons, they were more effective in soil washing, removing about 33% of crude oil. Franzetti et al. [55] further demonstrated that the bioemulsions produced by *Gordonia* sp. BS29 function as effective washing agents for soils contaminated with hydrocarbons. In addition, these compounds enhanced the removal of heavy metals, including Cu, Cd, Pb, Zn, and Ni, although their removal efficiency was comparatively lower than that observed with rhamnolipids.

3.3 Wastewater Treatment

The rapid growth of industrialization has had a profound impact on aquatic ecosystems, introducing a range of harmful contaminants into our water sources. In addition, the unchecked extraction of water for industrial purposes has greatly reduced the availability of freshwater. This problem is further exacerbated by the growing global population, which puts even more strain on limited water resources, leading to severe water shortages. Insufficient wastewater treatment and improper management of sludge and solid waste have led to the discharge of large quantities of pollutants into freshwater ecosystems. As a result, water bodies are increasingly becoming dumping grounds for industrial chemicals. The rising demand for new materials has also led to the emergence of "emerging contaminants," which pose significant threats to both aquatic life and human health. The United Nations World Water Development Report (2017) estimates that more than 80% of wastewater is discharged into aquatic environments without adequate treatment. Major contributors to this pollution include industries such as textiles, tanneries, food processing, petroleum, petrochemicals, and fertilizers. The ever-growing need for chemicals in our industrialized world continues to exert immense pressure on the natural water bodies we rely on [53–55].

3.4 Contaminant Degradation

Rhamnolipid biosurfactants effectively enhance Bisphenol A (BPA) removal from wastewater via laccase. BPA, a common industrial wastewater pollutant with harmful effects on human health, is degraded more efficiently in the presence of rhamnolipids, achieving a 65% removal rate. Unlike synthetic surfactants, rhamnolipids maintain enzymatic activity in the wastewater medium. Molecular docking and dynamics simulations by Liu et al. revealed that hydrogen bonding and van der Waals interactions bind phenols to laccase, while surfactants alter enzyme conformation to promote degradation. Rhamnolipids outperform Triton X in facilitating enzymatic breakdown of phenolic compounds. Total petroleum hydrocarbons (TPH), phenolic compounds, and polycyclic aromatic hydrocarbons (PAHs) in petroleum wastewater can all be reduced by 70–90% using rhamnolipids made from Stenotrophomonas sp. that was isolated from contaminated river water. This degradation is attributed to enhanced activities of phenol hydroxylase, catechol 1,2-dioxygenase, and catechol 2,3-dioxygenase, alongside reduced surface tension and improved hydrocarbon bioavailability [56–58]. Due to its antibacterial and antifungal properties, triclosan is widely used in personal care formulations, but it is challenging to remove from wastewater due to low bioavailability. A lipopeptide biosurfactant from *Bacillus licheniformis* achieved 47% triclosan removal within 2 h of treatment. Alternatively, aerobic degradation with di-rhamnolipids in water–sediment systems achieved 93.87% efficiency after 56 h, highlighting their effectiveness in triclosan bioremediation [56, 57]. The performance of vineyard pruning waste with hydrogel is improved by biosurfactants made from corn steep liquor, which result in a 62% improvement in sulfate removal from wastewater and a 10% increase in dye removal. These eco-friendly biomolecules show significant potential for developing green industrial adsorbents. Industrial dyes, often toxic and carcinogenic, pose serious health risks when discharged untreated. Functionalized iron oxide nanoparticles with rhamnolipid biosurfactants from *Pseudomonas aeruginosa* demonstrate superior photocatalytic biodegradation of methylene blue, achieving a 93% removal efficiency compared to bare nanoparticles [59, 60]. Biosurfactants have shown great efficacy in treating heavily contaminated poultry wastewater containing high levels of soluble chemical oxygen demand (SCOD) and oil and grease (OG). Surfactin, produced by *Bacillus subtilis* LB5a, achieved over 80% SCOD and 71% OG removal under optimal anaerobic conditions. Similarly, rhamnolipids from *Pseudomonas aeruginosa* enabled enzymatic anaerobic treatment with 95.8% OG removal. Additionally, bovine bile, a biosurfactant with high critical micelle concentration (CMC) values, effectively degraded oils, fats, and grease in poultry waste, allowing 100% bile recovery [61–63]. Ammonia from wastewater of China was effectively treated using biosurfactants derived from microorganisms cultivated on the same wastewater. The use of multiple soil layer bioreactors supplemented with these biosurfactants enhanced ammonia removal efficiency, achieving a 93% reduction in the effluent [64].

3.5 Oil–Water Separation

Oil spills, fuel transport accidents, and effluent discharge from the petroleum, oil, and food industries pose serious risks to aquatic ecosystems, mainly due to the toxic and hazardous nature of hydrocarbons. These substances, which include alkanes, phenols, cycloalkanes, aromatic hydrocarbons, and BTEX compounds, are highly damaging to the environment. Additionally, carcinogenic polycyclic aromatic hydrocarbons (PAHs) are released from partial combustion of fossil fuel, petrochemical processes, oil spills, and biomass burning. Research on biosurfactants produced by microbial communities has shown promise in breaking down PAHs by improving their solubility and bioavailability in water. However, the ability to degrade these harmful compounds diminishes as the number of carbon rings in the PAH structure increases, making it more challenging to treat high-molecular-weight PAHs effectively [65, 66].

Rhamnolipids, a type of biosurfactant, have become a focal point in bioremediation research for both water and soil environments. Their effectiveness in breaking down contaminants is influenced by factors like pH and ionic strength. One notable example is Pseudomonas aeruginosa, a bacterium that produces rhamnolipids and has been used to tackle oil pollution in marine waters off the coast of Brazil. This bacterium showed its best performance at pH 8, achieving a 100% dispersion index. The oil decontamination process operates through the solubilization and mobilization of hydrophobic compounds. A recent study reported that the combined application of sugarcane biochar, rhamnolipids, and nitrogen supplementation achieved an 80% degradation of total petroleum hydrocarbons (TPH) in crude oil-contaminated coastal soil within only 20 day [67, 68]. Fats, oils, and greases (FOGs) in wastewater can form sticky, soap-like deposits when they combine with other substances, which can block pipes and interfere with oxygen transfer. The food industry is a major contributor to

FOGs, with animal fats and vegetable oils being the primary culprits. Research on the biodegradation of FOGs, including substances like butter, vegetable oils, animal fats, and margarine, using biosurfactants has shown promising results in breaking down most of these compounds, except for butter. The biodegradation process was monitored by measuring oxygen consumption and carbon dioxide release. Adjusting the pH was found to improve the efficiency of degradation, with animal fat showing 80% removal at pH 8.5 and pig lard achieving 72% removal at pH 8 [69, 70].

3.6 *Removal of Heavy Metals*

The widespread reliance on heavy metals in various industrial processes creates serious environmental challenges, as these substances are persistent and resistant to natural degradation. Major sources include mining, metallurgical processes, electronics manufacturing, electroplating, and tanneries. Although heavy metals cannot be eliminated entirely, microorganisms can modify their chemical forms, reducing their toxicity. Research by Junior et al. showed that biosurfactants derived from the yeast *Candida tropicalis*, cultivated on molasses, used frying oil, and corn steep liquor, were able to extract 30–80% of zinc and copper from polluted soil within just 30 min using repeated column washing. In another example, biosurfactants obtained from *Ziziphus spina-christi*—a plant with saponin-rich leaves traditionally applied in detergents—proved effective in removing lead and copper ions from water, underscoring their promise for heavy metal cleanup [71–73]. Combining electrokinetic methods with biosurfactants has proven effective in improving heavy metal removal from sludge. The electrokinetic process reduces the mobility of heavy metals by shifting them into more stable forms, while also enhancing electroosmotic flow and electrical conductivity of the medium. In another strategy, researchers treated grass surfaces with rhamnolipids, achieving up to 85% removal of cadmium (Cd^{2+}) from aqueous solutions at neutral pH. Similarly, Wang et al. modified two-dimensional Ti_2CTX MXene with biosurfactants, which expanded the nanosheet spacing and significantly improved ion exchange and adsorption capacity, leading to more efficient Pb (II) removal [74–76].

3.7 *Flotation of Effluents*

Flotation is a widely used separation method for eliminating metals, oils, and other pollutants from industrial wastewater. The process works on the principles of gravity and colloidal aggregation in aqueous systems, where rising air bubbles within the column induce the aggregation of contaminants, causing them to float to the surface—a process known as dissolved air flotation. Conventionally, synthetic surfactants are added to help collect these aggregates at the liquid surface, but concerns over their toxicity question the long-term sustainability of this approach. Recent studies suggest that biosurfactants such as tea saponin and rhamnolipids can serve as greener alternatives to chemical agents like Triton X-100, improving bubble behavior, and enhancing the overall eco-friendliness of the flotation process [77, 78]. Even in small quantities, tea saponin can strongly influence bubble dynamics in flotation systems. It has been shown to lower the aspect ratio of individual bubbles by 33%, reduce their terminal velocity by 35%, and decrease trajectory amplitude by 27%, effects largely attributed to the Marangoni phenomenon. Biosurfactants also promote higher bubble swarm density and generate smaller bubbles, thereby increasing the overall surface area available for interaction. In one study, Zhao et al. demonstrated the use of tea saponin in flotation to selectively separate polyethylene from a ternary plastic mixture containing polyethylene, acrylonitrile–butadiene–styrene (ABS), and thermoplastic rubber (TPR), highlighting its potential in efficient material recovery [79]. In a recent study, Silva et al. extracted a biosurfactant from Pseudomonas aeruginosa and applied it in dissolved air flotation to treat oily wastewater. By improving the adhesion of oil particles, the biosurfactant boosted the process efficiency, achieving oil removal rates ranging from 65 to 95% [80].

2. Industrial Applications

3.8 Oil and Gas Industry

Biosurfactants have emerged as a pivotal class of bio-based surface-active compounds with extensive applicability across diverse industrial sectors, including food processing, cosmetics, household cleaning formulations, pharmaceuticals, and the oil and gas industry. The global biosurfactant market demonstrated substantial economic significance, generating revenues exceeding USD 1.5 billion in 2019. Furthermore, market projections indicate a compound annual growth rate (CAGR) surpassing 5.5% between 2020 and 2026, reflecting the expanding industrial adoption of these environmentally benign and functionally versatile biomolecules [81]. Notably, the predominant share of this market is driven by applications in the household care, cosmetic, personal care, and food manufacturing sectors, where biosurfactants are increasingly favored as sustainable alternatives to synthetic surfactants [81].

3.9 Soil Bioremediation

Soil contamination with hydrocarbons is a widespread environmental concern, often arising from drilling operations, leakages of pipelines, storage tank failures, and transportation accidents. Hydrocarbons, along with heavy metals, tend to bind strongly to soil particles, making them hydrophobic and difficult to remove. Conventional remediation strategies—such as excavation, incineration, soil washing, and solvent extraction—can be effective but are often costly, disruptive to soil structure, and harmful to biodiversity. In contrast, bioremediation provides a more sustainable and cost-effective approach by utilizing native soil microorganisms to degrade petroleum hydrocarbons into carbon dioxide, water, and microbial biomass. The success of this process depends on multiple factors, including soil moisture, aeration, pH, temperature, microbial activity, nutrient supply, and the specific properties of hydrocarbon pollutants. Careful optimization of these conditions is crucial for achieving efficient bioremediation outcomes [82, 83]. Bioremediation of contaminated soils can be performed either in-situ, where treatment occurs directly at the site, or ex-situ, where excavated soil is treated elsewhere. In-situ approaches generally target the top 30 cm of soil and often involve the addition of fertilizers to stimulate native microorganisms that degrade hydrocarbons. Although this method is widely preferred, it carries the risk of leaching dissolved hydrocarbons into underlying aquifers. To improve its efficiency, researchers have applied partially purified biosurfactants, which increase the solubility and bioavailability of hydrocarbons and other hydrophobic pollutants by expanding their accessible surface are [84, 85]. A field trial on LaTouche Island, Alaska, demonstrated the potential of PES-51, a biologically derived surfactant, which removed about 30% of semi-volatile petroleum hydrocarbons from subsurface beach sediments. Despite this success, most biosurfactant-based bioremediation studies have been limited to laboratory settings. For example, in Argentina, surfactin produced by *Bacillus subtilis* strain O9 significantly enhanced crude oil removal from sandy loam soil over a 300-day period. Similarly, the addition of rhamnolipids produced by *Pseudomonas aeruginosa* strain SSC2 to crude oil-contaminated soil sludge resulted in a 98% degradation rate within four weeks, in contrast to only 67% observed in untreated controls. The efficiency of this process was further improved by the addition of nutrients. In another case, an in-situ bioremediation trial conducted near an oil production facility in Pakistan achieved up to 77% crude oil degradation through the combined use of a specialized bacterial consortium, rhamnolipids, and nutrient amendments [86–89]. Baek et al., evaluated the effectiveness of mannosylerythritol lipids (MELs) produced by Candida antarctica SY16 in promoting crude oil degradation in soil. The findings revealed that the combined application of biostimulation, bioaugmentation with the hydrocarbon-degrading strain *Nocardia* sp. H17-1, and the addition of mannosylerythritol lipids (MELs) resulted in the most rapid degradation of total petroleum hydrocarbons (TPHs) during the initial four weeks of treatment. Nevertheless, by the conclusion of the 100-day incubation period, the residual hydrocarbon concentrations were observed to be comparable among all experimental treatments, indicating a convergence in degradation efficiency over time [90].

3.10 Microbial Enhanced Oil Recovery (MEOR)

Microbial enhanced oil recovery (MEOR) entails the introduction of microorganisms or their metabolic by-products into mature oil reservoirs to extract residual crude oil left after primary and secondary recovery. The underlying principle is that, under suitable reservoir conditions, the injected microbes multiply and produce metabolites that help mobilize the remaining oil [91]. Although MEOR presents several advantages, it also has certain limitations. Multiple aspects of its applications have been thoroughly examined in previous studies, with a recent comprehensive review presented by Nikolova and Gutierrez (2020) [92]. MEOR operates on two key principles. The first mechanism entails improving oil mobility within porous rock formations by altering the interfacial characteristics of the oil–water–mineral system, thus enhancing displacement efficiency. This can be achieved by reducing interfacial tension (IFT) to increase permeability, boosting reservoir pressure, enhancing oil fluidity through miscible flooding or viscosity reduction, and optimizing sweep efficiency via selective plugging and mobility control. Another principle emphasizes microbial degradation as well as elimination of sulfur and heavy metal contaminants from heavy crude oils.

In most field trials, indigenous or pre-cultured bacteria—or bacterial consortia—are injected along with nutrients such as oxygen and nitrogen, allowing microbes to produce biosurfactants in-situ, which lowers operational costs. Ex-situ produced biosurfactants can also support microbial growth in reservoirs. In cases where less recovery of oil is caused by low rock permeability or high oil viscosity, biosurfactants enhance recovery by decreasing the interfacial tension between water and residual oil, which decreases the capillary forces that trap oil in rock pores. For effective mobilization, IFT reduction must be at least two orders of magnitude. Since typical hydrocarbon–water IFT values range from 30 to 40 mN/m, biosurfactants need to lower this to approximately 10^{-3} mN/m to significantly impact MEOR performance [93–96]. To date, no known biosurfactants have achieved the IFT reductions required for optimal MEOR performance, suggesting that their practical effectiveness may be limited. Additionally, the type of reservoir—such as sandstone or carbonate—together with residual oil saturation and the desired incremental recovery can determine the amount of biosurfactant required. Achieving a 30–60% increase in oil recovery may therefore require large amounts, potentially making the approach uneconomical. Beyond lowering IFT, biosurfactants can also alter rock wettability, emulsify crude oil, and enhance microbial metabolism of viscous oil, contributing to improved recovery [97, 98]. Despite these challenges, several studies have reported encouraging outcomes for the use of biosurfactants in MEOR. Lipopeptides have been the most widely studied in laboratory MEOR research due to their capacity to reduce IFT to below 0.1 mN/m. Both laboratory-scale studies and in-situ lipopeptide production by *Bacillus* strains have shown effectiveness in enhancing oil recovery, even in wells approaching the end of their productive life [99–101].

Surfactins have shown remarkable stability and activity across a broad spectrum of temperatures, pH levels, and salinities, making them particularly effective for recovering oil trapped in sand. For example, Bacillus subtilis produced surfactin at elevated temperatures, achieving 90% emulsification of diesel, and recovering over 60% of oil from sand core samples. More recently, surfactin has also been found to alter the wettability of CO_2 in subsurface rock formations, demonstrating its potential for carbon capture and storage applications [102, 103]. Lichenysin has been shown to lower interfacial tension (IFT) to below 10^{-2} mN/m even at low concentrations

(10–60 mg/L). It exhibits remarkable stability under extreme conditions, including temperatures up to 140 °C, pH values from 6 to 10, salinities up to 10% NaCl, and calcium levels up to 340 mg/L (as $CaCl_2$). In core flooding experiments, partially purified lichenysin was able to recover up to 40% of residual oil from sandstone cores, significantly outperforming chemical surfactants, which achieved only around 10% recovery [103].

Incorporating biosurfactants like rhamnolipids during chemical surfactant flooding can significantly enhance oil recovery. Rhamnolipids help reduce the adsorption of chemical surfactants onto sandstones, making them more available to displace oil. They also behave as sacrificial agents by binding to the oil-bearing sand and can modify the wettability of the porous media, further improving oil mobilization [104]. Macromolecular biopolymers, such as emulsan, have demonstrated high efficiency in removing pre-adsorbed crude oil from limestone core samples, achieving up to 98% removal even at low concentrations of 0.5 mg/mL. Additionally, a biopolymer produced by *Rhizobium viscosum* CECT908 has shown better performance than xanthan gum in recovering heavy oil, highlighting the strong potential of these biopolymers in enhancing oil recovery processes [105, 106].

3.11 Oil Spill Response in Marine

Crude oil is strongly hydrophobic and consists of a complex mixture of hydrocarbons, non-hydrocarbons, and metals, each with different levels of water solubility. When released into water, oil generally floats because it is less dense than water. It is spreading across the water surface, and dispersion into subsurface layers is influenced by properties such as viscosity and surface tension. Lowering the interfacial tension between surfaces improves its ability to spread and disperse, which is a critical factor in applications like oil recovery and bioremediation [107]. To improve the solubility of oil in water and enhance its biodegradation, chemicals are often applied to oil slicks. These agents lower the interfacial tension between oil and water, enabling the formation of fine, neutrally buoyant droplets, typically ranging from 1 to 70 μm in diameter. The smaller droplet size increases the surface area-to-volume ratio, making the oil more accessible to hydrocarbon-degrading bacteria. This promotes biodegradation, as microorganisms biologically oxidize hydrocarbons into simpler, less harmful compounds. In marine environments, the fate of crude oil results from a complex interplay of physical dispersion, chemical weathering, and microbial activity, which has been extensively studied over the years [108–113]. Studies on marine bioremediation have mainly explored the use of fertilizers and the introduction of seed cultures of highly efficient oil-degrading microorganisms, though the results have often been inconsistent. A major limitation of relying solely on indigenous microbes is that their population in oil-contaminated open water is generally too low to effectively emulsify the oil and promote biodegradation [112, 113]. Both biosurfactants and synthetic surfactants help break up or emulsify oil, making it easier for microorganisms to degrade it more quickly. While biosurfactants have shown promise in laboratory studies for enhancing crude oil dispersion and biodegradation, large-scale field applications are often constrained by logistical, financial, and regulatory challenges. In one lab experiment, researchers used a bacterial consortium made up of oil-degrading *Ochrobactrum* and *Brevibacillus* strains to treat crude oil in one-liter water tanks, both with and without added rhamnolipids. The results showed that the consortium alone removed slightly less oil than when combined with the biosurfactant. Interestingly, rhamnolipids improved the biodegradation of longer-chain alkanes (nC15+) but reduced the solubility and increased the stability of shorter-chain alkanes (nC13–C15). Even so, the consortium paired with rhamnolipids removed more n-alkanes than the control, and the same improvement was seen for PAHs and other biomarker compounds [114]. The findings of this study align with earlier work that evaluated rhamnolipids used alongside a previously acclimated bacterial consortium (Nikolopoulou et al., 2013a). The effects became especially notable under nutrient-supplemented conditions, with the degradation rate of various n-alkanes being substantially higher than the control. Rhamnolipids also promoted the rapid proliferation of hydrocarbon-degrading microorganisms within a few days, enabling them to break down nearly half of the crude oil-rich fraction. Moreover, low-molecular-weight PAHs, including biomarkers like pristane and phytane, were markedly degraded in the presence of rhamnolipids [115].

The type of nutrients added alongside rhamnolipids—whether organic, lipophilic, or water-soluble—can further enhance the breakdown of crude oil in marine water as well as sediment systems [115–117]. Bioemulsifiers have also shown considerable promise for oil spill response. For instance, the exopolysaccharide bioemulsifier EPS2003, produced by Acinetobacter calcoaceticus, was found to improve microbial breakdown of crude oil in natural seawater test systems. The addition of EPS2003 promoted the growth of hydrocarbon-degrading bacteria and accelerated total oil biodegradation, making it significantly faster than in microcosms without EPS. However, most studies have focused on a limited set of bacteria, without assessing the response of the broader indigenous marine microbial community. Gaining insight into how native microbial communities react to crude oil, whether biosurfactants are present or not, is essential for evaluating their effectiveness in oil spill remediation. Practical application is further limited by the high cost of biosurfactants and logistical challenges in field studies [118] surfactin produced by *Bacillus sp.* strain H2O-1 was evaluated against the synthetic dispersant Ultrasperse II. While surfactin-enriched hydrocarbon-degrading bacteria more effectively than the synthetic dispersant, overall oil biodegradation was similar for both treatments. In a more recent study, rhamnolipid from *Pseudomonas aeruginosa* was compared with a chemical dispersant regarding their effects on the natural bacterial community exposed to crude oil. Rhamnolipid supported a greater diversity of oil-degrading bacteria, whereas overall crude oil biodegradation was higher with the chemical dispersant, except for the aromatic fraction. Significantly, the synthetic dispersant inhibited Cycloclasticus, a genus of bacteria specialized in degrading aromatic hydrocarbons [119–121].

3.12 Food Industry

The food industry uses rigorous standards to ensure safety to ensure that microorganisms do not compromise product safety, thereby protecting consumer health. Equally important is maintaining the quality and appeal of the final product, particularly as consumer preference shifts toward foods with fewer synthetic ingredients and a higher content of natural components [122, 123] To meet this demand, new product formulations are increasingly replacing chemical additives with natural, plant-derived options such as lecithin and gum Arabic, which have gained strong acceptance in the marketplace.. However, these natural ingredients may experience a reduction in their functional performance during processes like microwave cooking or irradiation. Consequently, researchers have developed and identified novel additives that provide thickening, stabilizing, and emulsifying functions, along with antioxidant, antibiofilm, and antimicrobial activities, to enhance modern food formulations [124, 125].

3.13 Biosurfactants as Emulsifying and Stabilizing Agents

Emulsifiers are crucial for producing stable emulsions, as they facilitate the formation of lipid droplets during homogenization and help maintain their stability over time (Fig. 3). By regulating the aggregation of these droplets, emulsifiers help prevent phase separation, extend product shelf life and maintain the stability of heterogeneous systems [126].

In the food industry, which relies on proteins, carbohydrates, fats, and lipids, thickeners, stabilizers, and emulsifiers are among the most used additives. These substances are essential for ensuring the proper functionality of the final product, as emulsions naturally form from various combinations of these ingredients. An emulsion, consisting of lipid droplets dispersed within an aqueous phase, is often challenging to stabilize because of its complex microstructural characteristic [127]. With increasing consumer demand for natural ingredients, biosurfactants have emerged as a promising alternative to emulsifiers derived from genetically modified sources such as soy, which may affect the flavor and texture of food. They have the potential to replace synthetic emulsifiers while enhancing product quality. Although their direct application as emulsifying agents is not yet extensively documented, existing studies indicate that biosurfactants offer multiple benefits, including improved food characteristics and greater stability under the harsh conditions of food processing [128–130]. Biosurfactants with strong emulsifying abilities are often referred to as bioemulsifiers, particularly in multiphase systems. It is important to recognize that emulsifying agents do not always function by lowering the surface tension between water and hydrocarbons. Instead, they can reduce the surface energy between phases and create electrostatic or steric barriers that prevent particle coalescence. Like other

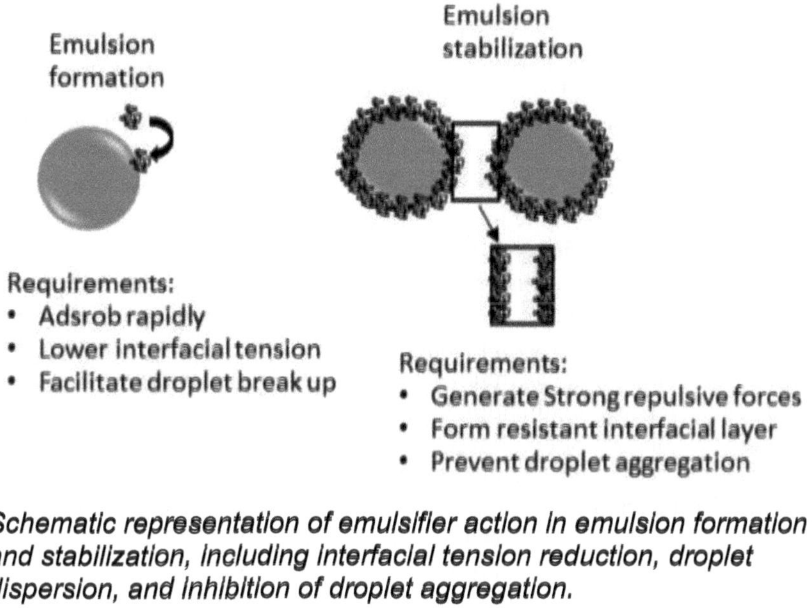

Fig. 3 Schematic representation of emulsifier action in emulsion formation and stabilization, including interfacial tension reduction, droplet dispersion, and inhibition of droplet aggregation

biosurfactants, bioemulsifiers are also capable of lowering interfacial tension within immiscible interface [131, 132].

3.14 *Salad Dressings*

Microbial surfactants are being used by researchers as a substitute for emulsifiers in food items such as mayonnaise and salad dressings, for enhancing texture and viscosity. For example, a biosurfactant from Candida utilis, composed of a carbohydrate-protein-lipid complex, was added to salad dressings to enhance stability. When used at a concentration of 0.7% (p/v) together with guar gum and carboxymethylcellulose, the dressing maintained its consistency even after a month of storage. While thermal treatment can degrade some additives—affecting their rheological behavior and the maintenance of granular structure—biosurfactants retain excellent emulsifying and solubilizing abilities under extreme temperatures and pH levels. Research has also demonstrated that yeast-derived biosurfactants surpass conventional stabilizers such as gum arabic, guar gum, and carboxymethylcellulose in terms of both stabilization efficiency and thermal stability [131, 132]

3.15 Baked Products

In recent years, bread, cakes, and cookies based on flour are becoming popular, driven by consumer demand for innovative and healthier options that meet specific dietary needs. Researchers have explored incorporating ingredients with enhanced nutritional value or reduced calorie content—such as soy protein concentrate in gluten-free cookies and soluble fiber (β-glucan) from barley in chocolate cookies—to enhance both health benefits and functional properties [133–135]. Biosurfactants have also shown promise as ingredients in baked goods, offering the potential to reduce dependence on commercial additives while enhancing the products' rheological properties. Although this application remains relatively underexplored, studies indicate that biosurfactants can improve dough texture and reduce the energy content of muffins and cookies, highlighting their potential for industrial use [136].

Over the past decade, studies have shown that incorporating rhamnolipid biosurfactants can markedly enhance dough consistency, texture, volume, and the shelf life of baked products [137]. Supplementing bread with a bioemulsifier produced by *Bacillus subtilis* has been shown to improve its texture and reduce sensitivity to microbial growth. Likewise, incorporation of this bioemulsifier into a wheat flour and sesame seed cookie formulation improved the texture, yielding softer cookies with superior overall quality compared to those prepared with the commercial emulsifier glycerol monostearate [138]. A lipopeptide biosurfactant from *Nesterenkonia sp.* reduced hardness and chewiness in muffins while improving cohesion and elasticity, yielding a softer, high-quality product. Similarly, replacing egg yolk with a glycolipid biosurfactant from *S. cerevisiae* URM 6670 in cookies-maintained texture and physical properties, demonstrating its effectiveness as an emulsifier and its added nutritional benefits [139]. In a recent study, a biosurfactant from *C. bombicola* URM 3718 was added to muffin formulations, showing thermal stability and antioxidant activity. It maintained the muffins' physical and textural qualities while enhancing nutritional value by lowering the trans fatty acid content typically found in vegetable fats [140, 141].

3.16 Ice Cream

Although no specific studies have reported the use of biosurfactants in frozen desserts, studies indicate their potential as emulsifiers in ice cream. Their inclusion could improve texture, creaminess, and overall quality, likely due to their ability to control consistency, preserve physical properties, and solubilize aromatic oils that enhance flavor. In ice cream production, partial coalescence is essential. Biosurfactants should form a thin, permeable layer around fat droplets, allowing fat crystals to penetrate the liquid phase without fully blocking the process. Furthermore, some studies emphasize the potential of surface-active molecules like hydrophobins-globular proteins made by fungi-as stabilizers in aerated foods such as ice cream. These proteins function as adsorption and adhesion promoters, offering strong resistance to denaturation. Low-molecular weight biosurfactants also show great potential for stabilizing complex emulsions such as ice cream, as they rapidly lower surface tension and achieve equilibrium more efficiently than proteins. A blend of proteins and low-molecular weight biosurfactants can also be utilized, provided that the proportion fosters a synergistic interaction instead of competition for adsorption at the phase interface [142, 143].

3.17 Biosurfactants as Antimicrobial and Antibiofilm Agents in Food Systems

Beyond their role in emulsifying and stabilizing food products, biosurfactants can also contribute to extending shelf life through their antimicrobial and anti-adhesive activities. These natural compounds prevent microbial growth by inhibiting the attachment of microorganisms to food surfaces and processing equipment, ensuring the production of safe, high-quality products. For instance, biosurfactants have been proposed as effective anti-adhesive agents in dairy processing, where microbial adhesion to heat exchangers can cause encrustation [144, 145]. A biosurfactant from *Lactobacillus plantarum* showed strong anti-adhesive and antimicrobial activity against common foodborne pathogens, including *E. coli, Salmonella typhi, Yersinia enterocolitica,* and *Staphylococcus aureus*, highlighting its potential as a biocontrol agent. Similarly, a glycolipid biosurfactant from *Wickerhamomyces anomalus* demonstrated antimicrobial effects against pathogens like *Candida albicans* and *Staphylococcus epidermidis* in microdilution tests [146]. A biosurfactant from the marine yeast *Cyberlindnera saturnus* SBPN-27 showed strong antimicrobial activity, significantly inhibiting pathogens like *Salmonella paratyphi, Klebsiella oxytoca, Vibrio cholerae, and S. aureus*, while completely suppressing the growth of *E. coli, Bacillus cereus,* and *Bacillus subtilis* [147].

3.18 Biosurfactants as Antioxidant Agents in Foods

Antioxidant activity is vital in the food industry because antioxidant-rich foods can lower the risk of heart disease, protect against degradative conditions, and extend shelf life by slowing oxidation. This has driven consumer demand for healthier, antioxidant-enriched food products [147, 148]. The antioxidant activity of biosurfactant functionality is a critical factor in assessing their suitability for food applications. Different tests measure varying aspects of this activity, such as the ability to neutralize free radicals or reduce oxidized compounds, and results can differ depending on the method used. The chemical structure of biosurfactants, especially the presence of unsaturated fatty acids, plays a central role in their effectiveness, as these molecules can donate electrons or hydrogen atoms to stabilize reactive species. Studies on a biosurfactant from *Candida utilis* showed it had a higher capacity for reducing complexes than for radical scavenging, with activity exceeding 70% at moderate concentrations, and a clear positive relationship between concentration and antioxidant effect.

Biosurfactants from *Lactobacillus* species have also demonstrated strong radical-sequestering activity in DPPH assays, achieving around 75% inhibition. This indicates their ability to neutralize free radicals, which are responsible for oxidation in foods, thereby protecting nutritional quality and extending shelf life. Such antioxidant properties signify the ability of biosurfactants as functional ingredients in food formulations, not only for preserving product stability but also for contributing to consumer health benefits [149].

4 Medical Applications

The widespread use of antimicrobial agents has contributed to the rapid rise of drug-resistant microbial strains, creating a serious challenge for global healthcare systems. According to the World Health Organization, antibiotic resistance is linked to around 700,000 deaths each year, including roughly 25,000 in Europe, which also imposes an economic burden of about EUR 1.5 billion. In the United States, resistant infections cause an estimated 23,000 deaths each year, with associated costs estimated between USD 55 and 70 billion [150]. Amid these challenges, microbial metabolites have become an important source of bioactive compounds. Among them, biosurfactants have attracted considerable interest for their potent antibacterial and antifungal properties. Some biosurfactants, such as daptomycin, and echinocandin-derived antifungals like caspofungin, micafungin, and anidulafungin, have already been successfully developed into commercially available antibiotics [151–154].

4.1 Lipopeptides and Glycolipids as Antimicrobial Agents

Lipopeptides and glycolipids are diversely studied biosurfactants with antimicrobial activity. Key lipopeptides include fengycin, iturin, surfactin, and lichenysin, while widely recognized glycolipids include rhamnolipids, sophorolipids, and mannosylerythritol lipids [155–159] Yang et al. [160] a new cationic lipopeptide produced by an environmental isolate of *Brevibacillus laterosporus*. This compound exhibited strong antibacterial activity against several Gram-positive pathogens—such as MRSA, vancomycin-resistant *Lactobacillus plantarum*, and *Enterococcus faecalis*—showing a level of effectiveness on par with vancomycin. Additionally, a lipopeptide from *B. subtilis* SPB1, known for broad inhibitory range targeting bacteria and phytopathogenic fungi, was incorporated into a toothpaste formulation, showing strong inhibition of pathogens such as *E. faecalis*, and *Micrococcus luteus* [161, 162].

Cordeiro et al. [162] identified lipopeptide mixture TIM96 effectively eliminates *Trichosporon inkin* and *Trichosporon asahii* within 48 h by lowering cellular ergosterol, modifying surface hydrophobicity, and increasing membrane permeability. Similarly, Basit et al. [163] investigated the ability of native *Bacillus cereus* strains from garden soil to produce biosurfactants and assessed their antimicrobial and antioxidant properties. The strains, labeled MMIC 1, MMIC 2, and MMIC 3, produced gray-white biosurfactant precipitates most efficiently at 37 °C on the seventh day of incubation, with 0.5% NaCl, pH 7.0, and corn steep liquor as the preferred carbon source. Characterization using biochemical tests, thin-layer chromatography, agar double diffusion, and FTIR confirmed that these biosurfactants were cationic lipopeptides. They displayed strong antibacterial and antifungal activity against pathogens including *S. aureus, E. coli, P. aeruginosa, K. pneumoniae, A. niger,* and *C. albicans* at 30 mg/ml, and antiviral activity against NDV at 10 mg/ml. Safety was supported by cytotoxicity testing, which showed 63% cell survival in BHK-21 cells at 10 mg/ml. Additionally, the biosurfactants exhibited notable antioxidant effects in ferric-reducing and DPPH scavenging assays at 2 mg/ml. These findings indicate that lipopeptide biosurfactants produced by native *B. cereus* strains are safe, multifunctional agents with potential applications as antimicrobials, emulsifiers, and antioxidants.

More recently, Medeot et al. [164] studied Fengycin from *Bacillus amyloliquefaciens* MEP218 altered the surface structure of *P. aeruginosa*, reducing cell height and causing intracellular content loss. Mixtures of surfactin and rhamnolipids showed strong antimicrobial activity against pathogens, including antibiotic-resistant *S. aureus, E. coli,* and *C. albicans* [165]. Glycolipids exhibit potent antimicrobial properties. A biosurfactant from *Staphylococcus saprophyticus* SBPS 15 fully inhibited several major pathogens, and sophorolipid-coated gold surfaces were likewise effective, destroying both Gram-positive and Gram-negative bacteria by disrupting their cell membranes [166, 167]. Elshikh et al. [168, 169] performed the inhibitory potential of rhamnolipid and lactonic sophorolipid mixtures against oral pathogens with MIC values between 0.1 and 0.4 mg/mL. Additionally, Sen et al. [170] studied rhamnolipid from P. aeruginosa SS14 showed strong antifungal activity against *Trichophyton rubrum*. At 0.5 mg/mL, it disrupted cell membranes, inhibited spore germination and hyphal growth, and effectively treated induced cutaneous dermatophytosis in mice within 21 days. The antimicrobial effects of biosurfactants are mainly due to their amphiphilic nature, which disrupts microbial membranes. Additionally, biosurfactants can influence microbial interactions through quorum sensing, as seen in *P. aeruginosa*, where mixed microbial communities enhance rhamnolipid production via AI-2 signaling, highlighting their ecological and industrial potential [170–172] (Fig. 4).

Nanocarriers have emerged as powerful tools for improving drug delivery and boosting bioavailability. Biosurfactants, thanks to their unique surface-active and functional properties, are increasingly being incorporated into these delivery systems. In the fight against multidrug-resistant infections and cancer, researchers are exploring how biosurfactants can enhance the effectiveness and precision of drug transport. When used alongside conventional drugs, they can work through multiple

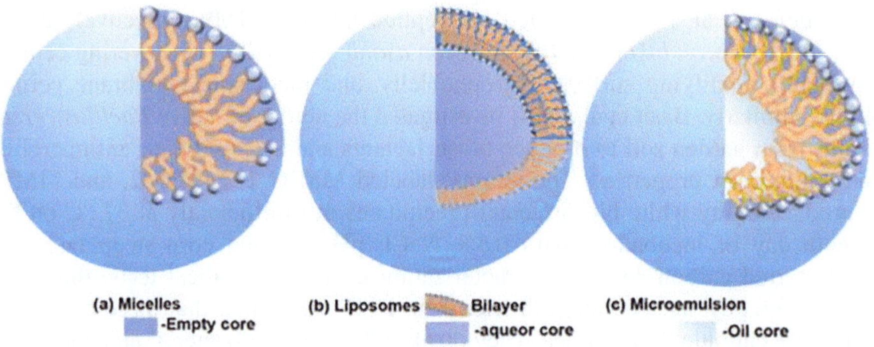

Fig. 4 Self-assembled structure of biosurfactants: **a** micelles, **b** liposomes, **c** microemulsion

mechanisms to improve therapeutic outcomes. Their ability to interact with cell membranes aids targeted delivery, while their multifunctional nature may synergize with drugs to amplify treatment effects. In nanocarriers, biosurfactants are typically employed in the form of micelles, liposomes, or microemulsions. Understanding their interactions within these systems is key to maximizing their benefits in drug delivery [173]. In micelle-based drug delivery systems, surfactin A has been utilized as a carrier for Itraconazole (ITZ), a widely used antifungal agent. ITZ is known to cause hemolytic side effects, but when encapsulated in surfactin A micelles—which exhibit lower hemolytic activity—this risk is significantly reduced. The micellar formulation limits direct interaction between ITZ and red blood cells, resulting in a notable reduction in hemolytic activity by over 50% compared to conventional solutions. Additionally, this approach enhances antimicrobial effectiveness against pathogens such as Aspergillus fumigatus, Aspergillus niger, and Candida albicans. Overall, biosurfactant-based carriers for hydrophobic drugs like ITZ not only improve bioavailability and safety but also contribute to superior therapeutic outcomes [174].

Liposomes are valuable tools for treating genetic disorders because they can offer foreign genes directly to target cells. Their stability is further enhanced when biosurfactants are incorporated. For example, combining di-rhamnolipids with dioleoyl phosphatidylethanolamine (DOPE) produces pH-sensitive liposomes that remain stable under neutral conditions but release their payload in acidic environments. This property allows the liposomes to fuse with endosomal membranes, ensuring efficient delivery of therapeutic agents into the cytoplasm. Research by Niaz et al. [175] demonstrated that rhamnolipids improve liposomal functionality by improvising the encapsulation efficiency, improving stability, and modulating the controlled release of nisin, thereby enhancing its antimicrobial effects [176] (Fig. 5).

Mannosylerythritol lipid-A (MEL-A) demonstrates a strong ability to bind human immunoglobulin G (HIgG). When used in nanocarriers, MEL-A interacts with lectins on target cell surfaces, promoting membrane fusion and facilitating effective DNA

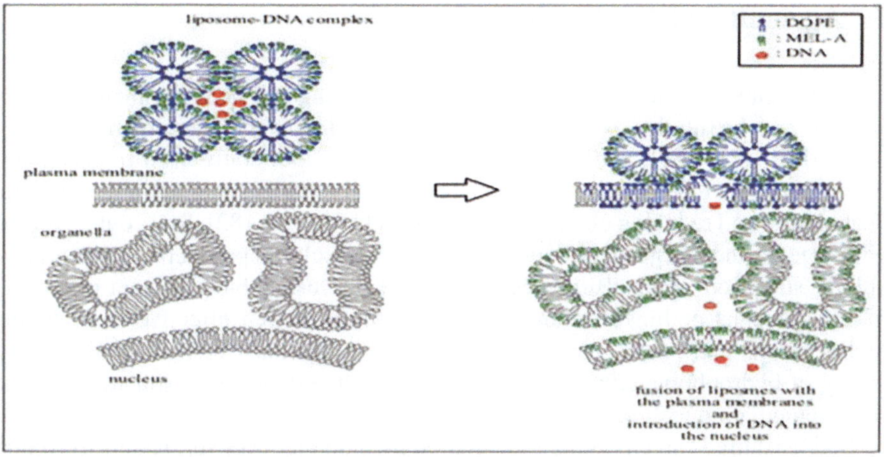

Fig. 5 Schematic showing the initial gene delivery step, driven by fusion of MEL-A nanovectors with the target cell membrane

delivery to the nucleus. The binding efficiency of MEL-A is influenced by the unsaturation level in its fatty acid chains, which also affects DNA encapsulation, membrane fusion, and overall gene transfer performance. Studies indicate that an unsaturated fatty acid content around 21.5% optimizes gene delivery compared to higher or lower levels. By selectively targeting specific cells, biosurfactants like MEL-A improve the precision and effectiveness of therapeutic delivery [177]. In cancer treatment, liposomes carrying betulinic acid show significantly improved anticancer activity when combined with MEL-A. This combination targets cancer cells by compromising mitochondrial integrity, inhibiting RNA and protein production, blocking progression through the S phase of the cell cycle, and promoting programmed cell death. These findings highlight the valuable role of biosurfactants in nanocarrier systems, offering enhanced precision and effectiveness in delivering anticancer therapies [178–180].

In microemulsion-based drug delivery, incorporating etoposide into a submicron microemulsion using acidic sophorolipids as the carrier greatly improved drug loading compared to conventional formulations. This sophorolipid–etoposide microemulsion also induced significantly higher apoptosis in A2780 cancer cells than standard etoposide treatments. Moreover, MEL-A has shown strong potential for forming water/oil microemulsions without additional surfactants, accommodating a substantial oil phase while maintaining encapsulation efficiency comparable to soybean lecithin [180, 181]. In summary, biosurfactants serve as essential components in drug carrier systems, enhancing drug solubility, promoting targeted adhesion, improving bioavailability, and mitigating multidrug resistance commonly observed in conventional treatments. Their inclusion paves the way for developing advanced and more effective drug delivery strategies, particularly in cases where traditional approaches prove inadequate.

5 Agricultural Applications

Biosurfactants are crucial in agriculture for improving soil health, enhancing plant growth, and facilitating pollutant biodegradation due to their antimicrobial properties [177, 178, 182]. Unlike synthetic surfactants commonly used in pesticides, biosurfactants are potential carbon sources for soil microbes, aiding in their natural degradation and reducing environmental impact [177–180, 182]. Soil contamination with organic and inorganic pollutants can lower agricultural productivity and impose abiotic stress on crops. Bioremediation strategies using biosurfactant-producing microorganisms or biosurfactants themselves have proven effective in removing hydrocarbons and heavy metals. These natural compounds increase the bioavailability of hydrophobic contaminants, improving the efficiency of soil washing and other remediation techniques [182].

Soil washing is an emerging remediation method that uses surfactants to solubilize hydrophobic pollutants bound to soil particles. Different types of surfactants—anionic, cationic, zwitterionic, and nonionic—have been employed to improve contaminant extraction. Effective surfactants should have strong desorption ability, perform efficiently at low concentrations, and be economically feasible for large-scale use [182]. Surfactant-based soil washing can be performed either ex situ or in situ. In ex-situ methods, the contaminated soil is dug up, mixed with surfactant solutions, and mechanically agitated to detach and remove pollutants. The treated soil is then placed back at the site, and the wash solutions can be recovered and recycled. In contrast, in-situ techniques involve delivering surfactant solutions directly into the polluted subsurface, where they generate micelles that solubilize contaminants, allowing them to be extracted, treated, or reinjected as part of the cleanup process [182]. Surfactant adsorption onto soil particles influences their effectiveness in removing pollutants. At low concentrations, surfactants accumulate at solid–liquid or liquid–liquid interfaces, displacing water and lowering surface tension. As the concentration rises, micelles form, greatly increasing contaminant solubility. These micelles feature hydrophilic exteriors and hydrophobic interiors that promote the dispersion of pollutants, support their release from soil particles, and increase their bioavailability for microbial breakdown or removal by physical mechanisms. Effective remediation of surfactant-contaminated soils requires consideration of several key factors, including how surfactants absorb soil, their ability to solubilize and remove contaminants, and their toxicity and biodegradability. Economic factors, such as surfactant cost and the scale of contamination, are also important. Ideally, a surfactant should offer high efficiency, strong desorption capability, and a low critical micelle concentration (CMC) to reduce the amount needed in washing solutions, thereby lowering overall remediation costs [183] Successful remediation of surfactant-contaminated soils depends on multiple factors, including the extent of surfactant adsorption to soil, its capacity to solubilize and extract pollutants, and its toxicity and biodegradability. Economic considerations—such as the price of the surfactant and the size of the contaminated site—also play a significant role. An ideal surfactant should provide high efficiency, strong contaminant desorption, and a low critical micelle concentration (CMC), minimizing the amount required in washing solutions and thereby reducing overall remediation costs [184, 185].

When surfactants interact with soil–water systems, some surfactant molecules are absorbed onto soil particles, which can reduce their ability to solubilize pollutants. As surfactant concentrations increase, soil hydrophobicity can rise, sometimes causing dissolved organic contaminants to reattach to soil surfaces. At low concentrations, surfactants accumulate solid–liquid or liquid–liquid interfaces as individual molecules, displacing interfacial solvents like water, lowering surface tension, and enhancing contaminant dissolution. At higher concentrations, micelles form, with hydrophilic exteriors and lipophilic cores, improving the solubility and dispersion of non-aqueous contaminants. Once solubilized in the aqueous phase, these pollutants become more mobile and can be removed through biological processes, such as microbial degradation or plant uptake, or via physical–chemical methods like soil washing and separation. Biosurfactants remove hydrophobic organic pollutants from soil through two main mechanisms: displacement and solubilization. Displacement takes place when surfactant levels are below the critical micelle concentration (CMC), while solubilization begins once the concentration surpasses the CMC (Fig. 6). In the displacement mechanism, surfactant monomers accumulate at the interface between soil contaminants and water, altering wettability by raising the contact angle between soil particles and hydrophobic contaminants. This leads to the adsorption of biosurfactant molecules on contaminant surfaces, creating repulsion from soil particles and effectively detaching the pollutants. Above the critical micelle concentration (CMC), biosurfactants solubilize hydrophobic pollutants by incorporating them into micelles, greatly increasing their partitioning into the aqueous phase. Pollutants encapsulated within micelles during soil washing can subsequently be separated and treated using techniques such as activated carbon adsorption, electrochemical processes, or demulsification. For cost efficiency, the surfactant mixture can be recycled for reuse or safely disposed of after treatment.

Biosurfactants promote the breakdown of chemical insecticides in agricultural soils, supporting soil remediation and the breakdown of pesticides and hydrocarbons. Certain *Burkholderia* species generate biosurfactants that can contribute to the remediation of pesticide-contaminated environments. Heavy metals from fungicides, sewage, and sludge can be toxic at high levels. Surfactants remove metals via complexation and ion exchange: anionic surfactants form strong complexes with metals for desorption, while cationic surfactants displace metal ions from negatively charged soil surfaces, enabling their removal through micelles [184, 186–190] The removal of heavy metals through ionic surfactants follows three main steps: (1) Biosurfactants bind to metal ions by adsorbing onto soil surfaces, (2) Desorption of metal ions into the aqueous phase, and (3) Heavy metals are captured within surfactant micelles through electrostatic interactions, allowing recovery using membrane separation techniques (Fig. 7).

Biosurfactants from *Bacillus, Pseudomonas,* and *Acinetobacter* species effectively remove heavy metals from soil and accelerate pesticide biodegradation. Rhamnolipids and surfactin can extract metals such as Mg, Ca, Cd, Ni, Mn, Ba, Cu, Li, and Zn. Unlike synthetic surfactants, which often need high concentrations

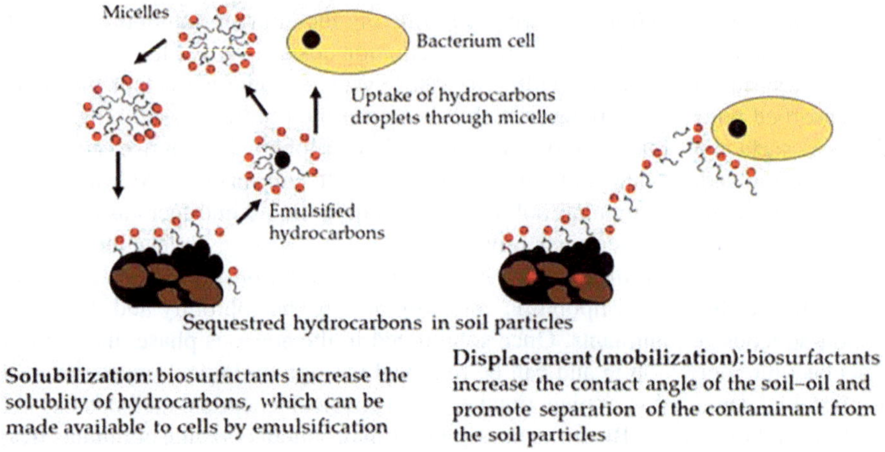

Fig. 6 Biosurfactant-mediated soil washing mechanisms—displacement below the CMC and solubilization above the CMC—in hydrocarbon-contaminated soils

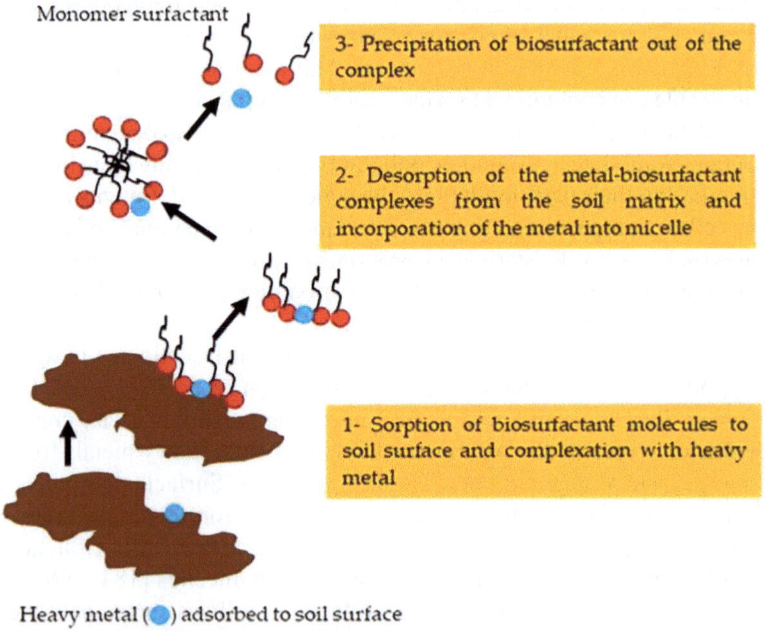

Fig. 7 Mechanistic pathways through which biosurfactants facilitate the removal of heavy metals

and may hinder microbial biodegradation, biosurfactants are more efficient and environmentally friendly [186]. Soil micronutrients are vital for plant physiological processes, supporting functions such as hormone regulation, protein formation, defense responses, and the upkeep of cellular membranes. Although chemical fertilizers supply these nutrients, their availability is often limited due to complexation with soil particles. Moreover, excessive use can disrupt soil structure, chemical balance, and biological activity, with their effectiveness further influenced by soil pH and ionic composition [191] Biosurfactants provide a sustainable way to enhance metal bioavailability for plants by lowering interfacial tension and enhancing the solubility and mobility of ionic nutrients, improving root uptake. Anionic biosurfactants form strong complexes with metals, surpassing their bonds with soil particles, which promotes desorption and better interaction at the soil–water interface. Cationic biosurfactants work via ion exchange, replacing bound metal ions in soil. Glycolipid biosurfactants, such as sophorolipids, rhamnolipids, trehalose lipids, and mannosylerythritol lipids (MELs), are widely studied for metal complexation. Specifically, surfactin promotes nutrient uptake by facilitating emulsification and contributes to biofilm development, enhancing root colonization by beneficial bacteria like *Bacillus amyloliquefaciens* on *Arabidopsis thaliana* and *Bacillus subtilis* on wheat [192].

6 Conclusion

Biosurfactants have become intriguing substitutes for traditional surfactants, merging surface-active effectiveness with environmental suitability. Their amphiphilic structure makes it possible for them to lower surface and interfacial tension, make stable emulsions, and make hydrophobic chemicals more soluble. Because of these qualities, they are useful in a wide range of fields, including environmental cleanup, pharmaceuticals, food technology, and agriculture. Biosurfactants, in contrast to those derived from petrochemicals, are naturally biodegradable, exhibit minimal toxicity, and maintain stability in challenging environments, making them a good fit for international sustainability programs.

Despite their advantages, large-scale adoption is still constrained by challenges such as high production costs, low yields, and the need for optimized fermentation processes. Dependence on expensive substrates and downstream processing further limits economic feasibility. Recent advances in metabolic engineering, strain improvement, and the use of agro-industrial residues as substrates show great promise in reducing costs and enhancing production efficiency. Integrating biotechnological innovations with sustainable feedstocks will be essential for scaling biosurfactant production.

Looking ahead, biosurfactants hold immense potential to reshape industries by replacing synthetic counterparts in critical applications. Their role in addressing pressing issues such as pollution, antimicrobial resistance, and the demand for eco-friendly consumer products highlights their relevance in a rapidly changing world. Continued interdisciplinary research, policy support, and industrial partnerships will be vital to unlock their full potential. Ultimately, biosurfactants represent not only a scientific achievement but also a step toward harmonizing industrial progress with environmental responsibility.

References

1. Farias CBB et al (2021) Production of green surfactants: market prospects. Electron J Biotechnol 51:28–39. https://doi.org/10.1016/j.ejbt.2021.02.002
2. Nagtode VS et al (2023) Green surfactants (biosurfactants): a petroleum-free substitute for sustainability–comparison, applications, market, and future prospects. Am Chem Soc. https://doi.org/10.1021/acsomega.3c00591
3. Silva RD, Almeida DG, Rufino RD, Luna JM, Santos VA, Sarubbo LA (2014) Applications of biosurfactants in the petroleum industry and the remediation of oil spills. MDPI AG. https://doi.org/10.3390/ijms150712523
4. Rebello S, Asok AK, Mundayoor S, Jisha MS (2014) Surfactants: toxicity, remediation and green surfactants. Springer. https://doi.org/10.1007/s10311-014-0466-2
5. Banat IM et al (2010) Microbial biosurfactants production, applications and future potential. Appl Microbiol Biotechnol 87(2):427–444. https://doi.org/10.1007/s00253-010-2589-0
6. Naughton PJ, Marchant R, Naughton V, Banat IM (2019) Microbial biosurfactants: current trends and applications in agricultural and biomedical industries. J Appl Microbiol 127(1):12–28. https://doi.org/10.1111/jam.14243
7. Bezerra KGO, Rufino RD, Luna JM, Sarubbo LA (2018) Saponins and microbial biosurfactants: potential raw materials for the formulation of cosmetics. Biotechnol Prog 34(6):1482–1493. https://doi.org/10.1002/btpr.2682
8. Ribeiro BG, Guerra JMC, Sarubbo LA (2020) Biosurfactants: production and application prospects in the food industry. Biotechnol Prog 36(5):e3030. https://doi.org/10.1002/btpr.3030
9. Banat IM, Carboué Q, Saucedo-Castañeda G, de Jesús Cázares-Marinero J (2021) Biosurfactants: the green generation of speciality chemicals and potential production using Solid-State fermentation (SSF) technology. Bioresour Technol 320:124222. https://doi.org/10.1016/j.biortech.2020.124222
10. Rocha e Silva NMP et al (2019) Natural surfactants and their applications for heavy oil removal in industry. Sep Purif Rev 48(4):267–281. https://doi.org/10.1080/15422119.2018.1474477
11. da Silva IGS, de Almeida FCG, da Rocha e Silva NMP, Casazza AA, Converti A, Sarubbo LA (2020) Soil bioremediation: overview of technologies and trends MDPI AG. https://doi.org/10.3390/en13184664
12. De Almeida DG et al (2016) Biosurfactants: promising molecules for petroleum biotechnology advances. Frontiers Media S.A. https://doi.org/10.3389/fmicb.2016.01718
13. Eras-Muñoz E, Farré A, Sánchez A, Font X, Gea T (2022) Microbial biosurfactants: a review of recent environmental applications. Taylor and Francis Ltd. https://doi.org/10.1080/21655979.2022.2074621
14. Silva-Dias A, Miranda IM, Branco J, Monteiro-Soares M, Pina-Vaz C, Rodrigues AG (2015) Adhesion, biofilm formation, cell surface hydrophobicity, and antifungal planktonic susceptibility: relationship among Candida spp. Front Microbiol 6. https://doi.org/10.3389/fmicb.2015.00205
15. Liu G, Zhong H, Yang X, Liu Y, Shao B, Liu Z (2018) Advances in applications of rhamnolipids biosurfactant in environmental remediation: a review. Biotechnol Bioeng 115(4):796–814. https://doi.org/10.1002/bit.26517

16. Patel M et al (2021) Inhibition of bacterial adhesion and antibiofilm activities of a glycolipid biosurfactant from lactobacillus rhamnosus with its physicochemical and functional properties. Antibiotics 10(12). https://doi.org/10.3390/antibiotics10121546
17. da Silva MD, da Silva ME, de Medeiros AO, Meira HM, Sarubbo LA (2022) Surface properties and biological activities on bacteria cells by biobased surfactants for antifouling applications. Surfaces 5(3):383–394. https://doi.org/10.3390/surfaces5030028
18. Aranda FJ, Teruel JA, Ortiz A (2023) Recent advances on the interaction of glycolipid and lipopeptide biosurfactants with model and biological membranes. Elsevier Ltd. https://doi.org/10.1016/j.cocis.2023.101748
19. Singh D et al (2023) Biofouling in membrane bioreactors: mechanism, interactions and possible mitigation using biosurfactants. Springer. https://doi.org/10.1007/s12010-022-04261-4
20. Essghaier B, Mallat N, Khwaldia K, Mottola F, Rocco L, Hannachi H (2023) Production and characterization of new biosurfactants/bioemulsifiers from Pantoea alhagi and their antioxidant, antimicrobial and anti-biofilm potentiality evaluations. Molecules 28(4). https://doi.org/10.3390/molecules28041912
21. Mandal S et al (2025) Comparative insights of bio and chemical surfactants in CuO nanoparticle synthesis: unveiling the superior antibacterial efficacy of biosurfactants. J Ind Eng Chem. https://doi.org/10.1016/j.jiec.2024.12.046
22. Al-Shamiri MM et al (2023) Probiotic lactobacillus species and their biosurfactants eliminate acinetobacter baumannii biofilm in various manners. Microbiol Spectr 11(2). https://doi.org/10.1128/spectrum.04614-22
23. Amirinejad N, Shahriary P, Hassanshahian M (2023) Investigation of the synergistic effect of glycolipid biosurfactant produced by Shewanella algae with some antibiotics against planktonic and biofilm forms of MRSA and antibiotic resistant Acinetobacter baumannii. World J Microbiol Biotechnol 39(2). https://doi.org/10.1007/s11274-022-03492-1
24. Ren K, Patra P, Lamsal BP, Mendonca A (2021) Antimicrobial mechanism of 'fatty acyl glutamic acid' biosurfactant against Escherichia coli O157:H7 and Listeria monocytogenes involves disruption of the cytoplasmic membrane. ACS Food Sci Technol 1(10):1792–1804. https://doi.org/10.1021/acsfoodscitech.1c00180
25. Zahed MA, Matinvafa MA, Azari A, Mohajeri L (2022) Biosurfactant, a green and effective solution for bioremediation of petroleum hydrocarbons in the aquatic environment. Discover Water 2(1). https://doi.org/10.1007/s43832-022-00013-x
26. Yesankar PJ, Pal M, Patil A, Qureshi A (2023) Microbial exopolymeric substances and biosurfactants as 'bioavailability enhancers' for polycyclic aromatic hydrocarbons biodegradation. Inst Ion. https://doi.org/10.1007/s13762-022-04068-0
27. Parthipan P, Cheng L, Dhandapani P, Elumalai P, Huang M, Rajasekar A (2022) Impact of biosurfactant and iron nanoparticles on biodegradation of polyaromatic hydrocarbons (PAHs). Environ Pollut 306:119384. https://doi.org/10.1016/j.envpol.2022.119384
28. Shin KH, Kim KW, Ahn Y (2006) Use of biosurfactant to remediate phenanthrene-contaminated soil by the combined solubilization-biodegradation process. J Hazard Mater 137(3):1831–1837. https://doi.org/10.1016/j.jhazmat.2006.05.025
29. Thirumurugan D et al (2023) Impact of biosurfactant produced by Bacillus spp. on biodegradation efficiency of crude oil and anthracene. Chemosphere 344:140340. https://doi.org/10.1016/j.chemosphere.2023.140340
30. Taghavi N, Zhuang WQ, Baroutian S (2022) Effect of rhamnolipid biosurfactant on biodegradation of untreated and UV-pretreated non-degradable thermoplastics: Part 2. J Environ Chem Eng 10(1). https://doi.org/10.1016/j.jece.2021.107033
31. Das S, Das N, Choure K, Pandey P (2023) Biodegradation of asphaltene by lipopeptide-biosurfactant producing hydrocarbonoclastic, crude oil degrading Bacillus spp. Bioresour Technol 382. https://doi.org/10.1016/j.biortech.2023.129198
32. Xia M, Wang S, Chen B, Qiu R, Fan G (2023) Enhanced solubilization and biodegradation of HMW-PAHs in water with a pseudomonas mosselii-released biosurfactant. Polymers (Basel) 15(23). https://doi.org/10.3390/polym15234571

33. Zhou H et al (2022) Effects of regular addition and discontinuation of biosurfactant on hydrocarbon biodegradation and microorganisms in heavy oily sludge. J Soils Sediments 22(9):2475–2484. https://doi.org/10.1007/s11368-022-03239-3
34. Markam SS, Raj A, Kumar A, Khan ML (2024) Microbial biosurfactants: green alternatives and sustainable solution for augmenting pesticide remediation and management of organic waste. Curr Res Microb Sci 7:100266. https://doi.org/10.1016/j.crmicr.2024.100266
35. Ng YJ et al (2022) Recent advances of biosurfactant for waste and pollution bioremediation: substitutions of petroleum-based surfactants. Environ Res 212:113126. https://doi.org/10.1016/j.envres.2022.113126
36. Pandey R, Krishnamurthy B, Singh HP, Batish DR (2022) Evaluation of a glycolipopepetide biosurfactant from Aeromonas hydrophila RP1 for bioremediation and enhanced oil recovery. J Clean Prod 345:131098. https://doi.org/10.1016/j.jclepro.2022.131098
37. Portal D'Almeida A, de Albuquerque TL, Melo VMM, Silveira Vieira R, Ponte Rocha MV (2023) Biosurfactant production by Acinetobacter venetianus and its application in bioremediation. Chem Eng Technol 46(6):1106–1114. https://doi.org/10.1002/ceat.202200540
38. Manikkasundaram V, Baskaran A, Kaari M, Angamuthu V, Venugopal G, Manikkam R (2023) Production and characterization of glycolipid biosurfactant from HRB1 and its evaluation for biomedical and bioremediation applications. J Surfactants Deterg 26(4):491–503. https://doi.org/10.1002/jsde.12643
39. da Silva Faccioli YE, da Silva GO, Sarubbo LA (2022) Application of a biosurfactant from Pseudomonas cepacia CCT 6659 in bioremediation and metallic corrosion inhibition processes. J Biotechnol 351:109–121. https://doi.org/10.1016/j.jbiotec.2022.04.009
40. Gidudu B, Chirwa EM (2022) Evaluation of the toxicity of a rhamnolipid biosurfactant for its application in the optimization of the bio-electrokinetic remediation of petrochemical contaminated soil. Clean Eng Technol 9. https://doi.org/10.1016/j.clet.2022.100521
41. Purnomo AS et al (2023) The application of biosurfactant-producing bacteria immobilized in PVA/SA/bentonite bio-composite for hydrocarbon-contaminated soil bioremediation. RSC Adv 13(31):21163–21170. https://doi.org/10.1039/d3ra02249h
42. Pacwa-Płociniczak M, Płaza GA, Piotrowska-Seget Z, Cameotra SS (2011) Environmental applications of biosurfactants: recent advances. https://doi.org/10.3390/ijms12010633
43. Mulligan C, Gibbs BF (2004) Types, production and applications of biosurfactants. Proc Indian Nat Sci Acad 70:31–55
44. Nievas ML, Commendatore MG, Esteves JL, Bucalá V (2008) Biodegradation pattern of hydrocarbons from a fuel oil-type complex residue by an emulsifier-producing microbial consortium. J Hazard Mater 154(1):96–104. https://doi.org/10.1016/j.jhazmat.2007.09.112
45. Rahman KSM, Rahman TJ, Lakshmanaperumalsamy P, Marchant R, Banat IM (2003) The potential of bacterial isolates for emulsification with a range of hydrocarbons. Acta Biotechnol 23(4):335–345. https://doi.org/10.1002/abio.200390043
46. Bai G, Brusseau ML, Miller RM (1997) Biosurfactant-enhanced removal of residual hydrocarbon from soil. J Contam Hydrol 25(1):157–170. https://doi.org/10.1016/S0169-7722(96)00034-4
47. Urum K, Pekdemir T (2004) Evaluation of biosurfactants for crude oil contaminated soil washing. Chemosphere 57(9):1139–1150. https://doi.org/10.1016/j.chemosphere.2004.07.048
48. Franzetti A, Gandolfi I, Bestetti G, Smyth TJP, Banat IM (2010) Production and applications of trehalose lipid biosurfactants. Eur J Lipid Sci Technol 112(6):617–627. https://doi.org/10.1002/ejlt.200900162
49. Franzetti A et al (2009) Potential applications of surface active compounds by Gordonia sp. strain BS29 in soil remediation technologies. Chemosphere 75(6):801–807. https://doi.org/10.1016/j.chemosphere.2008.12.052
50. Cameotra SS, Singh P (2009) Synthesis of rhamnolipid biosurfactant and mode of hexadecane uptake by Pseudomonas species. Microb Cell Fact 8(1):16. https://doi.org/10.1186/1475-2859-8-16

51. Lai C-C, Huang Y-C, Wei Y-H, Chang J-S (2009) Biosurfactant-enhanced removal of total petroleum hydrocarbons from contaminated soil. J Hazard Mater 167(1):609–614. https://doi.org/10.1016/j.jhazmat.2009.01.017
52. Franzetti A, Bestetti G, Caredda P, La Colla P, Tamburini E (2008) Surface-active compounds and their role in the access to hydrocarbons in Gordonia strains. FEMS Microbiol Ecol 63(2):238–248. https://doi.org/10.1111/j.1574-6941.2007.00406.x
53. Rout PR, Zhang TC, Bhunia P, Surampalli RY (2021) Treatment technologies for emerging contaminants in wastewater treatment plants: a review. Sci Total Environ 753:141990. https://doi.org/10.1016/j.scitotenv.2020.141990
54. Kwon G, Bhatnagar A, Wang H, Kwon EE, Song H (2020) A review of recent advancements in utilization of biomass and industrial wastes into engineered biochar. J Hazard Mater 400:123242. https://doi.org/10.1016/j.jhazmat.2020.123242
55. Kasonga TK, Coetzee MA, Kamika I, Ngole-Jeme VM, Momba MN (2021) Endocrine-disruptive chemicals as contaminants of emerging concern in wastewater and surface water: a review. J Environ Manage 277:111485. https://doi.org/10.1016/j.jenvman.2020.111485
56. Alshabib M, Onaizi SA (2020) Enzymatic remediation of bisphenol a from wastewaters: effects of biosurfactant, anionic, cationic, nonionic, and polymeric additives. Water Air Soil Pollut 231(8):428. https://doi.org/10.1007/s11270-020-04806-5
57. Liu Y et al (2018) Effect of surfactants on the interaction of phenol with laccase: Molecular docking and molecular dynamics simulation studies. J Hazard Mater 357:10–18. https://doi.org/10.1016/j.jhazmat.2018.05.042
58. Patel K, Patel M (2020) Improving bioremediation process of petroleum wastewater using biosurfactants producing Stenotrophomonas sp. S1VKR-26 and assessment of phytotoxicity. Bioresour Technol 315:123861. https://doi.org/10.1016/j.biortech.2020.123861
59. Perez-Ameneiro M, Vecino X, Cruz JM, Moldes AB (2015) Wastewater treatment enhancement by applying a lipopeptide biosurfactant to a lignocellulosic biocomposite. Carbohydr Polym 131:186–196. https://doi.org/10.1016/j.carbpol.2015.05.075
60. Bhosale SS, Rohiwal SS, Chaudhary LS, Pawar KD, Patil PS, Tiwari AP (2019) Photocatalytic decolorization of methyl violet dye using Rhamnolipid biosurfactant modified iron oxide nanoparticles for wastewater treatment. J Mater Sci Mater Electron 30(5):4590–4598. https://doi.org/10.1007/s10854-019-00751-0
61. Damasceno FRC, Cavalcanti-Oliveira ED, Kookos IK, Koutinas AA, Cammarota MC, Freire DMG (2018) Treatment of wastewater with high fat content employing an enzyme pool and biosurfactant: technical and economic feasibility. Braz J Chem Eng 35(2):531–542. https://doi.org/10.1590/0104-6632.20180352s20160711
62. Natassia JC, Benedito MG, Simone DG, Ana PRS, Glaucia MP (2017) Use of biosurfactant surfactin produced from cassava wastewater for anaerobic treatment of effluent from a poultry slaughterhouse. Afr J Biotechnol 16(5):224–231. https://doi.org/10.5897/ajb2016.15668
63. Harris PW, Schmidt T, McCabe BK (2018) Bovine bile as a bio-surfactant pre-treatment option for anaerobic digestion of high-fat cattle slaughterhouse waste. J Environ Chem Eng 6(1):444–450. https://doi.org/10.1016/j.jece.2017.12.034
64. Guo J, Jia X, Yang L, Zhang G, Zhang L (2020) Effect of biosurfactant on ammonia removal from anaerobically digested swine wastewater by multi soil layering treatment bioreactors. Environ Technol 41(19):2510–2517. https://doi.org/10.1080/09593330.2019.1574314
65. Souza EC, Vessoni-Penna TC, de Souza Oliveira RP (2014) Biosurfactant-enhanced hydrocarbon bioremediation: an overview. Int Biodeterior Biodegradation 89:88–94. https://doi.org/10.1016/j.ibiod.2014.01.007
66. Bezza FA, Chirwa EM (2016) Biosurfactant-enhanced bioremediation of aged polycyclic aromatic hydrocarbons (PAHs) in creosote contaminated soil. Chemosphere 144:635–644. https://doi.org/10.1016/j.chemosphere.2015.08.027
67. Silva EJ, Correa PF, Almeida DG, Luna JM, Rufino RD, Sarubbo LA (2018) Recovery of contaminated marine environments by biosurfactant-enhanced bioremediation. Colloids Surf B Biointerfaces 172:127–135. https://doi.org/10.1016/j.colsurfb.2018.08.034

68. Wei Z et al (2020) Remediation of crude oil-contaminated coastal marsh soil: Integrated effect of biochar, rhamnolipid biosurfactant and nitrogen application. J Hazard Mater 396:122595. https://doi.org/10.1016/j.jhazmat.2020.122595
69. He X, Leming ML, Dean LO, Lappi SE, Ducoste JJ (2013) Mechanisms of fat, oil and grease (FOG) deposit formation in sewer lines. Water Res 47(13):4451–4459. https://doi.org/10.1016/j.watres.2013.05.002
70. Kis Á, Laczi K, Zsíros S, Rákhely G, Perei K (2015) Biodegradation of animal fats and vegetable oils by Rhodococcus erythropolis PR4. Int Biodeterior Biodegradation 105:114–119. https://doi.org/10.1016/j.ibiod.2015.08.015
71. Saleem H, Pal P, Haija MA, Banat F (2019) Regeneration and reuse of bio-surfactant to produce colloidal gas aphrons for heavy metal ions removal using single and multistage cascade flotation. J Clean Prod 217:493–502. https://doi.org/10.1016/j.jclepro.2019.01.216
72. George S, Jayachandran K (2013) Production and characterization of rhamnolipid biosurfactant from waste frying coconut oil using a novel Pseudomonas aeruginosa D. J Appl Microbiol 114(2):373–383. https://doi.org/10.1111/jam.12069
73. Mohaddes-Kamranshahi M, Jafarizadeh-Malmiri H, Simjoo M, Jafarizad A (2019) Evaluation of the saponin green extraction from Ziziphus spina-christi leaves using hydrothermal, microwave and Bain-Marie water bath heating methods. Green Process Synth 8(1):62–67. https://doi.org/10.1515/gps-2017-0185
74. Tang J et al (2020) Heavy metal removal effectiveness, flow direction and speciation variations in the sludge during the biosurfactant-enhanced electrokinetic remediation. Sep Purif Technol 246:116918. https://doi.org/10.1016/j.seppur.2020.116918
75. Ghaith ES, Rizvi S, Namasivayam C, Rahman PK (2019) Removal of Cd++ from contaminated water using bio-surfactant modified ground grass as a bio-sorbent. In: 2019 Advances in science and engineering technology international conferences (ASET), pp 1–7. https://doi.org/10.1109/ICASET.2019.8714467
76. Wang S, Liu Y, Lü Q-F, Zhuang H (2020) Facile preparation of biosurfactant-functionalized Ti_2CTX MXene nanosheets with an enhanced adsorption performance for Pb(II) ions. J Mol Liq 297:111810. https://doi.org/10.1016/j.molliq.2019.111810
77. Rocha e Silva FC, Rocha e Silva NM, Luna JM, Rufino RD, Santos VA, Sarubbo LA (2018) Dissolved air flotation combined to biosurfactants: a clean and efficient alternative to treat industrial oily water. Rev Environ Sci Biotechnol 17(4):591–602. https://doi.org/10.1007/s11157-018-9477-y
78. Wang H, Yang J, Lei S, Wang X (2013) Comparing the effect of biosurfactant and chemical surfactant on bubble hydrodynamics in a flotation column. Water Sci Technol 68(4):783–790. https://doi.org/10.2166/wst.2013.295
79. Zhao Y, Han F, Abdelaziz IIM, Liu X, Ghazali KH, Mishra P (2019) Application of biosurfactant tea saponin in flotation separation for ternary plastic mixtures: Statistical optimization and mechanism analysis. J Clean Prod 232:499–507. https://doi.org/10.1016/j.jclepro.2019.06.002
80. Silva EJ et al (2019) Treatment of oily effluent using a low-cost biosurfactant in a flotation system. Biodegradation 30(4):335–350. https://doi.org/10.1007/s10532-019-09881-y
81. Desai JD, Banat IM (1997) Microbial production of surfactants and their commercial potential. Microbiol Mol Biol Rev 61(1):47–64. https://doi.org/10.1128/MMBR.61.1.47-64.1997
82. Venosa AD, Zhu X (2003) Biodegradation of crude oil contaminating marine shorelines and freshwater wetlands. Spill Sci Technol Bull 8(2):163–178. https://doi.org/10.1016/S1353-2561(03)00019-7
83. Huesemann MH (2004) Biodegradation and bioremediation of petroleum pollutants in soil. In: Singh A, Ward OP (eds) Applied bioremediation and phytoremediation. Springer, Berlin, pp 13–34. https://doi.org/10.1007/978-3-662-05794-0_2
84. Atlas RM, Hazen TC (2011) Oil biodegradation and bioremediation: a tale of the two worst spills in U.S. history. Environ Sci Technol 45(16):6709–6715. https://doi.org/10.1021/es2013227

85. Ron EZ, Rosenberg E (2002) Biosurfactants and oil bioremediation. Curr Opin Biotechnol 13(3):249–252. https://doi.org/10.1016/S0958-1669(02)00316-6
86. Tumeo M, Braddock J, Venator T, Rog S, Owens D (1994) Effectiveness of a biosurfactant in removing weathered crude oil from subsurface beach material. Spill Sci Technol Bull 1(1):53–59. https://doi.org/10.1016/1353-2561(94)90007-8
87. Cubitto MA, Morán AC, Commendatore M, Chiarello MN, Baldini MD, Siñeriz F (2004) Effects of Bacillus subtilis O9 biosurfactant on the bioremediation of crude oil-polluted soils. Biodegradation 15(5):281–287. https://doi.org/10.1023/B:BIOD.0000042186.58956.8f
88. Tahseen R et al (2016) Rhamnolipids and nutrients boost remediation of crude oil-contaminated soil by enhancing bacterial colonization and metabolic activities. Int Biodeterior Biodegradation 115:192–198. https://doi.org/10.1016/j.ibiod.2016.08.010
89. Cameotra SS, Singh P (2008) Bioremediation of oil sludge using crude biosurfactants. Int Biodeterior Biodegradation 62(3):274–280. https://doi.org/10.1016/j.ibiod.2007.11.009
90. Baek KH, Yoon BD, Kim BH, Cho DH, Lee IS, Oh HM, Kim HS (2007) Monitoring of microbial diversity and activity during bioremediation of crude oil-contaminated soil with different treatments. J Microbiol Biotechnol 17(1):67–73
91. Gao CH, Zekri A (2011) Applications of microbial-enhanced oil recovery technology in the past decade. Energy Sources Part A: Recover Util Environ Eff 33(10):972–989. https://doi.org/10.1080/15567030903330793
92. Nikolova C, Gutierrez T (2020) Use of microorganisms in the recovery of oil from recalcitrant oil reservoirs: current state of knowledge, technological advances and future perspectives. Frontiers Media S.A. https://doi.org/10.3389/fmicb.2019.02996
93. Shibulal B, Al-Bahry SN, Al-Wahaibi YM, Elshafie AE, Al-Bemani AS, Joshi SJ (2014) Microbial enhanced heavy oil recovery by the aid of inhabitant spore-forming bacteria: an insight review. Sci World J 2014(1):309159. https://doi.org/10.1155/2014/309159
94. Brown LR (2010) Microbial enhanced oil recovery (MEOR). Curr Opin Microbiol 13(3):316–320. https://doi.org/10.1016/j.mib.2010.01.011
95. Lazar I, Petrisor IG, Yen TF (2007) Microbial enhanced oil recovery (MEOR). Pet Sci Technol 25(11):1353–1366. https://doi.org/10.1080/10916460701287714
96. Geetha SJ, Banat IM, Joshi SJ (2018) Biosurfactants: production and potential applications in microbial enhanced oil recovery (MEOR). Biocatal Agric Biotechnol 14:23–32. https://doi.org/10.1016/j.bcab.2018.01.010
97. Kowalewski E, Rueslåtten I, Steen KH, Bødtker G, Torsæter O (2006) Microbial improved oil recovery—bacterial induced wettability and interfacial tension effects on oil production. J Pet Sci Eng 52(1):275–286. https://doi.org/10.1016/j.petrol.2006.03.011
98. Shabani Afrapoli M, Alipour S, Torsaeter O (2011) Fundamental study of pore scale mechanisms in microbial improved oil recovery processes. Transp Porous Media 90(3):949–964. https://doi.org/10.1007/s11242-011-9825-7
99. Al-Sayegh A, Al-Wahaibi Y, Al-Bahry S, Elshafie A, Al-Bemani A, Joshi S (2015) Microbial enhanced heavy crude oil recovery through biodegradation using bacterial isolates from an Omani oil field. Microb Cell Fact 14(1):141. https://doi.org/10.1186/s12934-015-0330-5
100. Pereira JFB et al (2013) Optimization and characterization of biosurfactant production by Bacillus subtilis isolates towards microbial enhanced oil recovery applications. Fuel 111:259–268. https://doi.org/10.1016/j.fuel.2013.04.040
101. Gudiña EJ, Pereira JFB, Rodrigues LR, Coutinho JAP, Teixeira JA (2012) Isolation and study of microorganisms from oil samples for application in Microbial Enhanced Oil Recovery. Int Biodeterior Biodegradation 68:56–64. https://doi.org/10.1016/j.ibiod.2012.01.001
102. Park T, Joo HW, Kim GY, Kim S, Yoon S, Kwon TH (2017) Biosurfactant as an enhancer of geologic carbon storage: microbial modification of interfacial tension and contact angle in carbon dioxide/water/quartz systems. Front Microbiol 8(7). https://doi.org/10.3389/fmicb.2017.01285
103. McInerney MJ, Javaheri M, Nagle DP Jr (1990) Properties of the biosurfactant produced by Bacillus licheniformis strain JF-2. J Ind Microbiol 5(2–3):95–101. https://doi.org/10.1007/BF01573858

104. Olivera NL, Nievas ML, Lozada M, del Prado G, Dionisi HM, Siñeriz F (2009) Isolation and characterization of biosurfactant-producing Alcanivorax strains: hydrocarbon accession strategies and alkane hydroxylase gene analysis. Res Microbiol 160(1):19–26. https://doi.org/10.1016/j.resmic.2008.09.011
105. Couto MR, Gudiña EJ, Ferreira D, Teixeira JA, Rodrigues LR (2019) The biopolymer produced by Rhizobium viscosum CECT 908 is a promising agent for application in microbial enhanced oil recovery. N Biotechnol 49:144–150. https://doi.org/10.1016/j.nbt.2018.11.002
106. Perfumo A, Rancich I, Banat IM (2010) Possibilities and challenges for biosurfactants use in petroleum industry. In: Sen R (ed) Biosurfactants. Springer, New York, pp 135–145. https://doi.org/10.1007/978-1-4419-5979-9_10
107. Brakstad OG, Nordtug T, Throne-Holst M (2015) Biodegradation of dispersed Macondo oil in seawater at low temperature and different oil droplet sizes. Mar Pollut Bull 93(1):144–152. https://doi.org/10.1016/j.marpolbul.2015.02.006
108. Wade TL et al (2013) Analyses of water samples from the deepwater horizon oil spill: documentation of the subsurface plume. In: Monitoring and modeling the deepwater horizon oil spill: a record breaking enterprise. Wiley, pp 77–82. https://doi.org/10.1029/2011GM001103
109. Iyer A, Mody K, Jha B (2006) Emulsifying properties of a marine bacterial exopolysaccharide. Enzyme Microb Technol 38(1):220–222. https://doi.org/10.1016/j.enzmictec.2005.06.007
110. Campo P, Venosa AD, Suidan MT (2013) Biodegradability of Corexit 9500 and Dispersed South Louisiana crude oil at 5 and 25 °C. Environ Sci Technol 47(4):1960–1967. https://doi.org/10.1021/es303881h
111. Prince RC, McFarlin KM, Butler JD, Febbo EJ, Wang FCY, Nedwed TJ (2013) The primary biodegradation of dispersed crude oil in the sea. Chemosphere 90(2):521–526. https://doi.org/10.1016/j.chemosphere.2012.08.020
112. Seidel M, Kleindienst S, Dittmar T, Joye SB, Medeiros PM (2016) Biodegradation of crude oil and dispersants in deep seawater from the Gulf of Mexico: insights from ultra-high resolution mass spectrometry. Deep Sea Res Part II 129:108–118. https://doi.org/10.1016/j.dsr2.2015.05.012
113. Hazen TC, Prince RC, Mahmoudi N (2016) Marine oil biodegradation. Environ Sci Technol 50(5):2121–2129. https://doi.org/10.1021/acs.est.5b03333
114. Chen Q, Bao M, Fan X, Liang S, Sun P (2013) Rhamnolipids enhance marine oil spill bioremediation in laboratory system. Mar Pollut Bull 71(1):269–275. https://doi.org/10.1016/j.marpolbul.2013.01.037
115. Nikolopoulou M, Kalogerakis N (2008) Enhanced bioremediation of crude oil utilizing lipophilic fertilizers combined with biosurfactants and molasses. Mar Pollut Bull 56(11):1855–1861. https://doi.org/10.1016/J.MARPOLBUL.2008.07.021
116. Nikolopoulou M, Pasadakis N, Kalogerakis N (2013) Evaluation of autochthonous bioaugmentation and biostimulation during microcosm-simulated oil spills. Mar Pollut Bull 72(1):165–173. https://doi.org/10.1016/j.marpolbul.2013.04.007
117. McKew BA et al (2007) Efficacy of intervention strategies for bioremediation of crude oil in marine systems and effects on indigenous hydrocarbonoclastic bacteria. Environ Microbiol 9(6):1562–1571. https://doi.org/10.1111/j.1462-2920.2007.01277.x
118. Nikolova CN, Ijaz UZ, Magill C, Kleindienst S, Joye SB, Gutierrez T (2021) Response and oil degradation activities of a northeast Atlantic bacterial community to biogenic and synthetic surfactants. Microbiome 9(1):191. https://doi.org/10.1186/S40168-021-01143-5/FIGURES/6
119. Garcia SN, Osburn BI, Jay-Russell MT (2020) One health for food safety, food security, and sustainable food production. Front Sustain Food Syst 4:481782. https://doi.org/10.3389/FSUFS.2020.00001/BIBTEX
120. Head IM, Jones DM, Röling WFM (2006) Marine microorganisms make a meal of oil. Nat Rev Microbiol 4(3):173–182. https://doi.org/10.1038/nrmicro1348
121. Nwoba EG, Ogbonna CN, Ishika T, Vadiveloo A (2020) Microalgal pigments: a source of natural food colors. In: Alam MA, Xu JL, Wang Z (eds) Microalgae biotechnology for food, health and high value products. Springer, Singapore, pp 81–123. https://doi.org/10.1007/978-981-15-0169-2_3

122. Garcia SN, Osburn BI, Jay-Russell MT (2020) One health for food safety, food security, and sustainable food production. Frontiers Media S.A. https://doi.org/10.3389/fsufs.2020.00001
123. Nwoba EG, Ogbonna CN, Ishika T, Vadiveloo A (2020) Microalgal pigments: a source of natural food colors. In: Alam MdA, Xu J-L, Wang Z (eds) Microalgae biotechnology for food, health and high value products. Springer, Singapore, pp 81–123. https://doi.org/10.1007/978-981-15-0169-2_3
124. Nitschke M, Costa SGVAO (2007) Biosurfactants in food industry. Trends Food Sci Technol 18(5):252–259. https://doi.org/10.1016/J.TIFS.2007.01.002
125. Faustino M, Veiga M, Sousa P, Costa EM, Silva S, Pintado M (2019) Agro-food byproducts as a new source of natural food additives. Molecules 24(6):1056. https://doi.org/10.3390/MOLECULES24061056
126. McClements DJ, Gumus CE (2016) Natural emulsifiers—biosurfactants, phospholipids, biopolymers, and colloidal particles: molecular and physicochemical basis of functional performance. Adv Colloid Interface Sci 234:3–26. https://doi.org/10.1016/j.cis.2016.03.002
127. Pascual-Villalobos MJ, Guirao P, Díaz-Baños FG, Cantó-Tejero M, Villora G (2019) Chapter 9—Oil in water nanoemulsion formulations of botanical active substances. In: Koul O (ed) Nano-biopesticides today and future perspectives. Academic Press, pp 223–247. https://doi.org/10.1016/B978-0-12-815829-6.00009-7
128. Nitschke M, Silva SS (2018) Recent food applications of microbial surfactants. Crit Rev Food Sci Nutr 58(4):631–638. https://doi.org/10.1080/10408398.2016.1208635
129. Egolf A, Hartmann C, Siegrist M (2019) When evolution works against the future: disgust's contributions to the acceptance of new food technologies. Risk Anal 39(7):1546–1559. https://doi.org/10.1111/risa.13279
130. Gallo SB et al (2019) Effect of biosurfactant added in two different oil source diets on lamb performance and ruminal and blood parameters. Livest Sci 226:66–72. https://doi.org/10.1016/j.livsci.2019.06.006
131. Campos JM, Stamford TLM, Sarubbo LA (2019) Characterization and application of a biosurfactant isolated from Candida utilis in salad dressings. Biodegradation 30(4):313–324. https://doi.org/10.1007/s10532-019-09877-8
132. Heyman B, De Vos WH, Depypere F, Van der Meeren P, Dewettinck K (2014) Guar and xanthan gum differentially affect shear induced breakdown of native waxy maize starch. Food Hydrocoll 35:546–556. https://doi.org/10.1016/j.foodhyd.2013.07.011
133. Antoniewska A, Rutkowska J, Pineda MM, Adamska A (2018) Antioxidative, nutritional and sensory properties of muffins with buckwheat flakes and amaranth flour blend partially substituting for wheat flour. LWT 89:217–223. https://doi.org/10.1016/j.lwt.2017.10.039
134. Sarabhai S, Indrani D, Vijaykrishnaraj M, Milind, Arun Kumar V, Prabhasankar P (2015) Effect of protein concentrates, emulsifiers on textural and sensory characteristics of gluten free cookies and its immunochemical validation. J Food Sci Technol 52(6):3763–3772. https://doi.org/10.1007/s13197-014-1432-5
135. Frost DJ, Adhikari K, Lewis DS (2011) Effect of barley flour on the physical and sensory characteristics of chocolate chip cookies. J Food Sci Technol 48(5):569–576. https://doi.org/10.1007/s13197-010-0179-x
136. Bandyopadhyay K, Chakraborty C, Bhattacharyya S (2014) Fortification of mango peel and kernel powder in cookies formulation. J Acad Ind Res 2:661
137. Giri SS, Sen SS, Jun JW, Sukumaran V, Park SC (2017) Role of Bacillus licheniformis VS16-derived biosurfactant in mediating immune responses in carp rohu and its application to the food industry. Front Microbiol 8(3). https://doi.org/10.3389/fmicb.2017.00514
138. Mnif I, Besbes S, Ellouze R, Ellouze-Chaabouni S, Ghribi D (2012) Improvement of bread quality and bread shelf-life by Bacillus subtilis biosurfactant addition. Food Sci Biotechnol 21(4):1105–1112. https://doi.org/10.1007/s10068-012-0144-8
139. Kiran GS, Priyadharsini S, Sajayan A, Priyadharsini GB, Poulose N, Selvin J (2017) Production of lipopeptide biosurfactant by a marine Nesterenkonia sp. and its application in food industry. Front Microbiol 8. https://doi.org/10.3389/fmicb.2017.01138

140. Ribeiro BG, Guerra JM, Sarubbo LA (2020) Potential food application of a biosurfactant produced by saccharomyces cerevisiae URM 6670. Front Bioeng Biotechnol 8. https://doi.org/10.3389/fbioe.2020.00434
141. Silva IA et al (2020) Production of cupcake-like dessert containing microbial biosurfactant as an emulsifier. PeerJ 4:2020. https://doi.org/10.7717/peerj.9064
142. Kieliszek M, Kot AM, Bzducha-Wróbel A, Błażejak S, Gientka I, Kurcz A (2017) Biotechnological use of Candida yeasts in the food industry: a review. Fungal Biol Rev 31(4):185–198. https://doi.org/10.1016/J.FBR.2017.06.001
143. Marcelino PR, Gonçalves F, Jimenez IM, Carneiro BC, Santos BB, da Silva SS (2020) Sustainable production of biosurfactants and their applications, pp 159–183. https://doi.org/10.1002/9781119568858.ch8
144. Marcelino PR, Gonçalves F, Jimenez IM, Carneiro BC, Santos BB, da Silva SS (2020) Sustainable production of biosurfactants and their applications. In: Lignocellulosic biorefining technologies, pp 159–183. https://doi.org/10.1002/9781119568858.ch8
145. Campos JM, Montenegro Stamford TL, Sarubbo LA, de Luna JM, Rufino RD, Banat IM (2013) Microbial biosurfactants as additives for food industries. Biotechnol Prog 29(5):1097–1108. https://doi.org/10.1002/btpr.1796
146. Madhu AN, Prapulla SG (2014) Evaluation and functional characterization of a biosurfactant produced by Lactobacillus plantarum CFR 2194. Appl Biochem Biotechnol 172(4):1777–1789. https://doi.org/10.1007/s12010-013-0649-5
147. Amiri A, Mousakhani-Ganjeh A, Amiri Z, Guo YG, Singh AP, Kenari RE (2020) Fabrication of cumin loaded-chitosan particles: characterized by molecular, morphological, thermal, antioxidant and anticancer properties as well as its utilization in food system. Food Chem 310:125821. https://doi.org/10.1016/j.foodchem.2019.125821
148. Hameed A et al (2020) Antioxidant activity of polyphenolic extracts of filamentous fungus Mucor circinelloides (WJ11): extraction, characterization and storage stability of food emulsions. Food Biosci 34:100525. https://doi.org/10.1016/j.fbio.2019.100525
149. Merghni A et al (2017) Antioxidant and antiproliferative potential of biosurfactants isolated from Lactobacillus casei and their anti-biofilm effect in oral Staphylococcus aureus strains. Microb Pathog 104:84–89. https://doi.org/10.1016/j.micpath.2017.01.017
150. La Fauci V, Alessi V (2018) Antibiotic resistance: where are we going? Ann Ig 30(4):52–57. https://doi.org/10.7416/ai.2018.2235
151. Robbel L, Marahiel MA (2010) Daptomycin, a bacterial lipopeptide synthesized by a nonribosomal machinery *. J Biol Chem 285(36):27501–27508. https://doi.org/10.1074/jbc.R110.128181
152. George J, Reboli AC (2012) Anidulafungin: when and how? The clinician's view. Mycoses 55(1):36–44. https://doi.org/10.1111/j.1439-0507.2011.02052.x
153. Ngai AL, Bourque MR, Lupinacci RJ, Strohmaier KM, Kartsonis NA (2011) Overview of safety experience with caspofungin in clinical trials conducted over the first 15 years: a brief report. Int J Antimicrob Agents 38(6):540–544. https://doi.org/10.1016/j.ijantimicag.2011.07.008
154. Emiroglu M (2011) Micafungin use in children. Expert Rev Anti Infect Ther 9(9):821–834. https://doi.org/10.1586/eri.11.91
155. Cochrane SA, Vederas JC (2016) Lipopeptides from Bacillus and Paenibacillus spp.: a gold mine of antibiotic candidates. Med Res Rev 36(1):4–31. https://doi.org/10.1002/med.21321
156. Kitamoto D, Yanagishita H, Shinbo T, Nakane T, Kamisawa C, Nakahara T (1993) Surface active properties and antimicrobial activities of mannosylerythritol lipids as biosurfactants produced by Candida antarctica. J Biotechnol 29(1):91–96. https://doi.org/10.1016/0168-1656(93)90042-L
157. Saini HS et al (2008) Efficient purification of the biosurfactant viscosin from Pseudomonas libanensis strain M9-3 and its physicochemical and biological properties. J Nat Prod 71(6):1011–1015. https://doi.org/10.1021/np800069u
158. Benincasa M, Abalos A, Oliveira I, Manresa A (2004) Chemical structure, surface properties and biological activities of the biosurfactant produced by Pseudomonas aeruginosa LBI from

soapstock. Antonie Van Leeuwenhoek 85(1):1–8. https://doi.org/10.1023/B:ANTO.0000020148.45523.41
159. Grangemard I, Wallach J, Maget-Dana R, Peypoux F (2001) Lichenysin. Appl Biochem Biotechnol 90(3):199–210. https://doi.org/10.1385/ABAB:90:3:199
160. Ghribi D et al (2012) Investigation of antimicrobial activity and statistical optimization of Bacillus subtilis SPB1 biosurfactant production in solid-state fermentation. Biomed Res Int 2012(1):373682. https://doi.org/10.1155/2012/373682
161. Mnif I et al (2016) Purification and identification of Bacillus subtilis SPB1 lipopeptide biosurfactant exhibiting antifungal activity against Rhizoctonia bataticola and Rhizoctonia solani. Environ Sci Pollut Res 23(7):6690–6699. https://doi.org/10.1007/s11356-015-5826-3
162. de Cordeiro RA et al (2018) Inhibitory effect of a lipopeptide biosurfactant produced by Bacillus subtilis on planktonic and sessile cells of Trichosporon spp. Biofouling 34(3):309–319. https://doi.org/10.1080/08927014.2018.1437617
163. Basit M, Rasool MH, Naqvi SA, Waseem M, Aslam B (2018) Biosurfactants production potential of native strains of Bacillus cereus and their antimicrobial, cytotoxic and antioxidant activities The user has requested enhancement of the downloaded file. Biosurfactants production potential of native strains of Bacillus cereus and their antimicrobial, cytotoxic and antioxidant activities [Online]. Available: https://www.researchgate.net/publication/322592670
164. Medeot DB, Fernandez M, Morales GM, Jofré E (2020) Fengycins from Bacillus amyloliquefaciens MEP218 exhibit antibacterial activity by producing alterations on the Cell Surface of the pathogens Xanthomonas axonopodis pv. vesicatoria and Pseudomonas aeruginosa PA01. Front Microbiol 10. https://doi.org/10.3389/fmicb.2019.03107
165. Ndlovu T, Rautenbach M, Vosloo JA, Khan S, Khan W (2017) Characterisation and antimicrobial activity of biosurfactant extracts produced by Bacillus amyloliquefaciens and Pseudomonas aeruginosa isolated from a wastewater treatment plant. AMB Express 7(1):108. https://doi.org/10.1186/s13568-017-0363-8
166. Mani P, Dineshkumar G, Jayaseelan T, Deepalakshmi K, Ganesh Kumar C, Senthil Balan S (2016) Antimicrobial activities of a promising glycolipid biosurfactant from a novel marine Staphylococcus saprophyticus SBPS 15. 3 Biotech 6(2):163. https://doi.org/10.1007/s13205-016-0478-7
167. Valotteau C et al (2017) Antibacterial properties of sophorolipid-modified gold surfaces against Gram positive and Gram negative pathogens. Colloids Surf B Biointerfaces 157:325–334. https://doi.org/10.1016/j.colsurfb.2017.05.072
168. Elshikh M, Funston S, Chebbi A, Ahmed S, Marchant R, Banat IM (2017) Rhamnolipids from non-pathogenic Burkholderia thailandensis E264: Physicochemical characterization, antimicrobial and antibiofilm efficacy against oral hygiene related pathogens. N Biotechnol 36:26–36. https://doi.org/10.1016/j.nbt.2016.12.009
169. Elshikh M et al (2017) Rhamnolipids and lactonic sophorolipids: natural antimicrobial surfactants for oral hygiene. J Appl Microbiol 123(5):1111–1123. https://doi.org/10.1111/jam.13550
170. Sen S, Borah SN, Kandimalla R, Bora A, Deka S (2019) Efficacy of a rhamnolipid biosurfactant to inhibit Trichophyton rubrum in vitro and in a mice model of dermatophytosis. Exp Dermatol 28(5):601–608. https://doi.org/10.1111/exd.13921
171. Yan X et al (2019) Antimicrobial, anti-adhesive and anti-biofilm potential of biosurfactants isolated from Pediococcus acidilactici and Lactobacillus plantarum against Staphylococcus aureus CMCC26003. Microb Pathog 127:12–20. https://doi.org/10.1016/j.micpath.2018.11.039
172. Woźniak-Karczewska M et al (2017) Isolation of rhamnolipids-producing cultures from faeces: influence of interspecies communication on the yield of rhamnolipid congeners. N Biotechnol 36:17–25. https://doi.org/10.1016/j.nbt.2016.12.008
173. Naz S, Banerjee T, Totsingan F, Woody K, Gross RA, Santra S (2021) Therapeutic efficacy of lactonic sophorolipids: nanoceria-assisted combination therapy of nsclc using hdac and hsp90 inhibitors. Nanotheranostics 5(4):391–404. https://doi.org/10.7150/ntno.57675

174. Usman F et al (2023) Itraconazole loaded biosurfactin micelles with enhanced antifungal activity: fabrication, evaluation and molecular simulation. Antibiotics 12(10). https://doi.org/10.3390/antibiotics12101550
175. Niaz T, Shabbir S, Noor T, Imran M (2019) Antimicrobial and antibiofilm potential of bacteriocin loaded nano-vesicles functionalized with rhamnolipids against foodborne pathogens. LWT 116:108583. https://doi.org/10.1016/j.lwt.2019.108583
176. Nakanishi M, Inoh Y, Kitamoto D, Furuno T (2009) Nano vectors with a biosurfactant for gene transfection and drug delivery. J Drug Deliv Sci Technol 19(3):165–169. https://doi.org/10.1016/S1773-2247(09)50031-7
177. Sánchez M, Aranda FJ, Teruel JA, Ortiz A (2011) New pH-sensitive liposomes containing phosphatidylethanolamine and a bacterial dirhamnolipid. Chem Phys Lipids 164(1):16–23. https://doi.org/10.1016/j.chemphyslip.2010.09.008
178. Shu Q, Wu J, Chen Q (2019) Synthesis, characterization of liposomes modified with biosurfactant MEL-A loading betulinic acid and its anticancer effect in HepG2 cell. Molecules 24(21). https://doi.org/10.3390/molecules24213939
179. Inoh Y, Furuno T, Hirashima N, Kitamoto D, Nakanishi M (2010) The ratio of unsaturated fatty acids in biosurfactants affects the efficiency of gene transfection. Int J Pharm 398(1):225–230. https://doi.org/10.1016/j.ijpharm.2010.07.042
180. Worakitkanchanakul W et al (2008) Aqueous-phase behavior and vesicle formation of natural glycolipid biosurfactant, mannosylerythritol lipid-B. Colloids Surf B Biointerfaces 65(1):106–112. https://doi.org/10.1016/j.colsurfb.2008.03.009
181. Ma X et al (2022) Formulation and physicochemical and biological characterization of etoposide-loaded submicron emulsions with biosurfactant of sophorolipids. AAPS PharmSciTech 23(6):181. https://doi.org/10.1208/s12249-022-02329-2
182. Sarubbo LA et al (2022) Biosurfactants: production, properties, applications, trends, and general perspectives. Biochem Eng J 181:108377. https://doi.org/10.1016/j.bej.2022.108377
183. Mao X, Jiang R, Xiao W, Yu J (2015) Use of surfactants for the remediation of contaminated soils: a review. J Hazard Mater 285:419–435. https://doi.org/10.1016/j.jhazmat.2014.12.009
184. Sarubbo LA, Rocha RB Jr, Luna JM, Rufino RD, Santos VA, Banat IM (2015) Some aspects of heavy metals contamination remediation and role of biosurfactants. Chem Ecol 31(8):707–723. https://doi.org/10.1080/02757540.2015.1095293
185. Sales da Silva IG, Gomes de Almeida FC, Padilha da Rocha e Silva NM, Casazza AA, Converti A, Asfora Sarubbo L (2020) Soil bioremediation: overview of technologies and trends. Energies (Basel) 13(18). https://doi.org/10.3390/en13184664
186. Mulligan CN, Wang S (2006) Remediation of a heavy metal-contaminated soil by a rhamnolipid foam. Eng Geol 85(1):75–81. https://doi.org/10.1016/j.enggeo.2005.09.029
187. Adhikari S et al (2023) Chapter six—molecular basis of plant nutrient use efficiency—concepts and challenges for its improvement. In: Aftab T, Hakeem KR (eds) Sustainable plant nutrition. Academic Press, pp 107–151. https://doi.org/10.1016/B978-0-443-18675-2.00001-8
188. Odukkathil G, Vasudevan N (2013) Enhanced biodegradation of endosulfan and its major metabolite endosulfate by a biosurfactant producing bacterium. J Environ Sci Health B 48(6):462–469. https://doi.org/10.1080/03601234.2013.761873
189. Mata-Sandoval JC, Karns J, Torrents A (2001) Influence of rhamnolipids and Triton X-100 on the biodegradation of three pesticides in aqueous phase and soil slurries. J Agric Food Chem 49(7):3296–3303. https://doi.org/10.1021/jf001432w
190. Wattanaphon HT, Kerdsin A, Thammacharoen C, Sangvanich P, Vangnai AS (2008) A biosurfactant from Burkholderia cenocepacia BSP3 and its enhancement of pesticide solubilization. J Appl Microbiol 105(2):416–423. https://doi.org/10.1111/j.1365-2672.2008.03755.x
191. Osman KT (2013) Plant nutrients and soil fertility management. In: Osman KT (ed) Soils: principles, properties and management. Springer, Dordrecht, pp 129–159. https://doi.org/10.1007/978-94-007-5663-2_10

192. Le Mire G, Siah A, Brisset MN, Gaucher M, Deleu M, Jijakli MH (2018) Surfactin protects wheat against zymoseptoria tritici and activates both salicylic acid- and jasmonic acid-dependent defense responses. Agriculture 8:1–14 [Online]. Available: https://api.semanticscholar.org/CorpusID:90483745

Utilization of Mixed Surfactant Systems in Personal Care

Dharmesh Varade and Sandeep Patil

Abstract Surface-active agents, often known as surfactants, are substances that reduce the surface tension between two liquids, a liquid and a solid, or a gas and a liquid. Because their molecules have both a water-loving (hydrophilic) head and a water-fearing (hydrophobic) tail, they are crucial in many products as they enable water to interact with oils and grease. In personal care, surfactants are surface-active substances that lower the surface tension between water and oil, enabling them to combine and carry out essential tasks including foaming, emulsifying, and cleaning. They contribute to the texture, froth, and stability of products and are used in shampoos, body washes, soaps, and conditioners to efficiently remove dirt and oil. Different kinds of surfactants are employed for different purposes; some are better for conditioning or making stable emulsions, while others are better for cleaning. Due to their safety, environmental advantages, and similar or better functional qualities, microbial biosurfactants and plant-based alternatives are increasingly seen as the best alternatives to synthetic surfactants in cosmetics. In this chapter, we review the basic properties of surfactants with specific focus on their use as cleansers in household and personal care applications.

Keywords Mixed surfactants · Cosmetics · Personal care products · Emulsification · Foaming agents · Cleansing agents

1 Introduction

Surfactants are compounds that adsorb to the interface between a solution and a separate phase (gases/solids) and form self-assembled molecular clusters termed micelles in a solution (water or oil phase) as depicted in Fig. 1. Surfactants have a major role in the effectiveness of personal care products, including formulae for skincare and haircare [1–7]. The production of foam, emulsions, and detergency

D. Varade (✉) · S. Patil
School of Science, Navrachana University, Vadodara, Gujarat, India
e-mail: dharmesh.varade@nuv.ac.in

M. Abdul Rub et al. (eds.), *Industrial Applications of Individual and Mixed Surfactant Systems—Volume 2*, Sustainable Materials and Technology,
https://doi.org/10.1007/978-981-95-9195-4_11

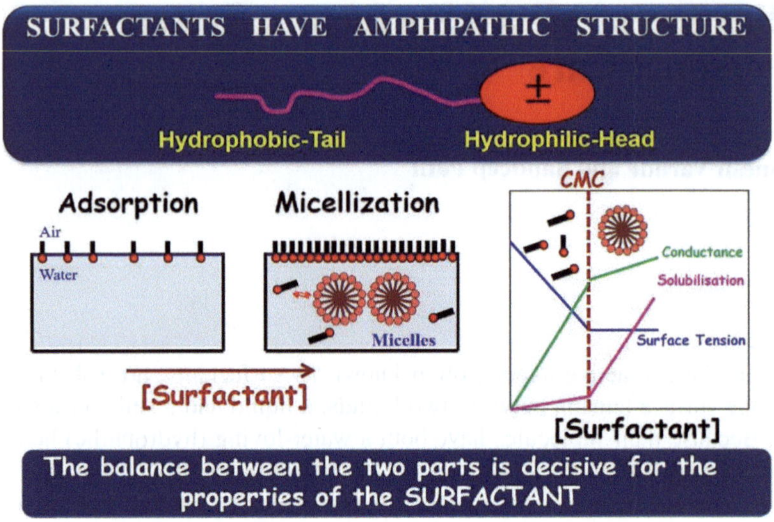

Fig. 1 Structure of surfactants and their interfacial properties

depend on these amphiphilic molecules' ability to lower surface tension and promote the mixing of water and oil phases. The demand for gentle, eco-friendly products is rising due to increased consumer awareness and regulatory concerns.

The critical micelle concentration (CMC) is the surfactant concentration above which micelles can be formed. At the CMC, several physicochemical properties of the solution vary abruptly. Surface tension, electrical conductivity, viscosity, and other changes in the solution's characteristics can be measured directly or indirectly to establish the CMC of surfactants. Other popular approaches include light scattering, sound velocity, electrochemical procedures, and spectroscopic techniques including UV–Vis and fluorescence. The CMC of a particular surfactant depends on factors such surfactant molecular structure (e.g., hydrophobic chain length), pressure conditions, solution salinity, ionic composition, pH, temperature, and others [1–7].

They are classified into ionic surfactants and nonionic surfactants depending upon the charge present on head groups. Surfactants typically have a hydrophobic chain (*R*) with an end group. Anionics have a negatively charged polar group, while the charge on a cationic is positive, and zwitterionics have a positive charge at one end and a negative charge at the other end. Nonionics have neutral end groups (Table 1).

It is very important to know the Krafft point and cloud point of any surfactants before we put them in applications. The Krafft point is the temperature at which the solubility of ionic surfactants in water increases drastically. Hence, the working temperature for ionic surfactants should be above Krafft point. In order to maintain the stability of surfactant solutions at low temperatures, it is necessary to lower their Krafft point by mixing different surfactants. The solubility of ethoxylate (nonionic surfactants) is unusual because it decreases with increasing temperature. This is a result of the ambivalent nature of ethoxylate chains which tend to curl into hydrophilic

Table 1 Classification of surfactants

Surfactants	Functional groups
Anionics	The most crucial functional groups are phosphate, sulfate, carboxylate and sulfonate, and acyl-amino acids and salts
Cationics	Alkylamines, alkylimidazolines, quaternary ammonium compounds, ethoxylated alkylamines, and esterified quaternaries
Nonionics	Fatty alcohols, ethylene oxide/propylene oxide-block polymers, alkyl polygycosides (APGs), ethoxylated fatty acids, sorbitan and sorbitol esters and ethoxylated derivatives, alkanolamides
Zwitterionics	Alkylamido alkyl amines (NMEA) and alkyl substituted amino acids

coils at lower temperatures and when randomized by heat tend to present a more hydrophobic nature on average. In the case of nonionic surfactants, the working temperature should be below the cloud point.

A surfactant's degree of hydrophilicity or lipophilicity is measured by its hydrophilic–lipophilic balance (HLB), which is calculated by dividing the molecular weights of its hydrophilic and lipophilic components by their percentages. A more oil-soluble, lipophilic surfactant is indicated by low HLB values, whereas a more water-soluble, hydrophilic surfactant is indicated by high HLB values as shown in Fig. 2. This approach assists formulators in choosing the appropriate surfactant for particular uses [1–5].

Compared to single-surfactant systems, mixed surfactant systems are significant because they harness synergistic effects to provide improved performance, lower costs, and a less environmental impact [8–20]. Finding appropriate surfactant combinations with a synergistic interaction between the head or tail groups was typically the main goal of research on mixed surfactant systems. By mixing surfactants with various characteristics, these systems enable the manipulation of characteristics like solubility, foamability, and interfacial tension, resulting in a wider variety of self-assembly behaviors and enhanced overall performance in applications as shown in Fig. 3.

In personal care products, mixed surfactants which combine anionic, amphoteric, and nonionic types are used to balance cleansing, mildness, and conditioning properties [10–20]. Nonionic surfactants improve skin feel and solubilize oils, while amphoteric surfactants minimize irritation caused by anionic ones. It is difficult to create personal cleaning solutions that can afford a combination of desired qualities including improved foaming, decent washing, enhanced mildness and skin feel, and appealing product look without imposing any restrictions. Amphoterics and mild anionics have been widely used in skin-cleansing product formulations in recent years. Anionic surfactants with non-sulfated head groups are thought to be gentle on skin, although they are challenging to thicken. High dosage levels of additional thickening agents such as synthetic or natural polymers are needed for such formulations. From offering hygiene and cleansing benefits to the latest technologies that offer sophisticated moisturizing effects in the shower, the cleansing sector has come a long way. For thousands of years, soap which is linked to dry skin has been used

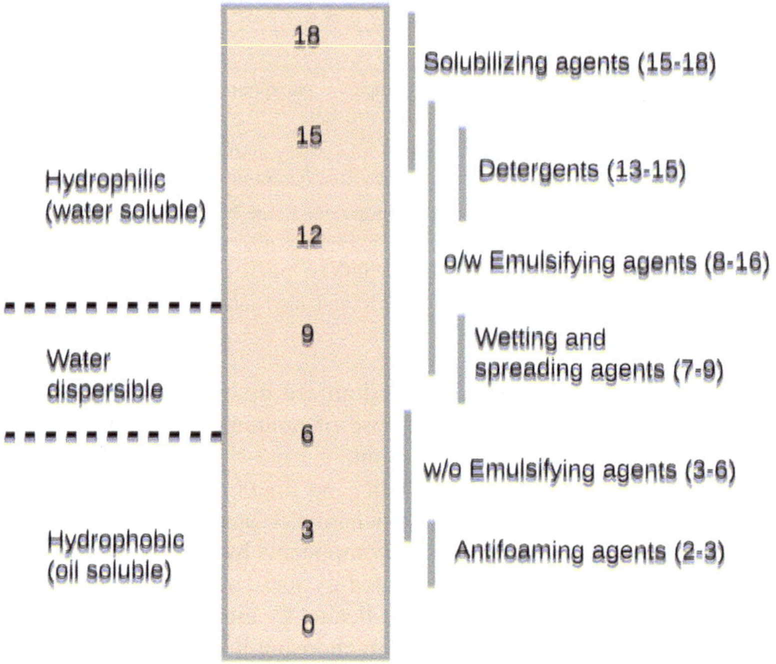

Fig. 2 HLB scale showing classification of surfactant function

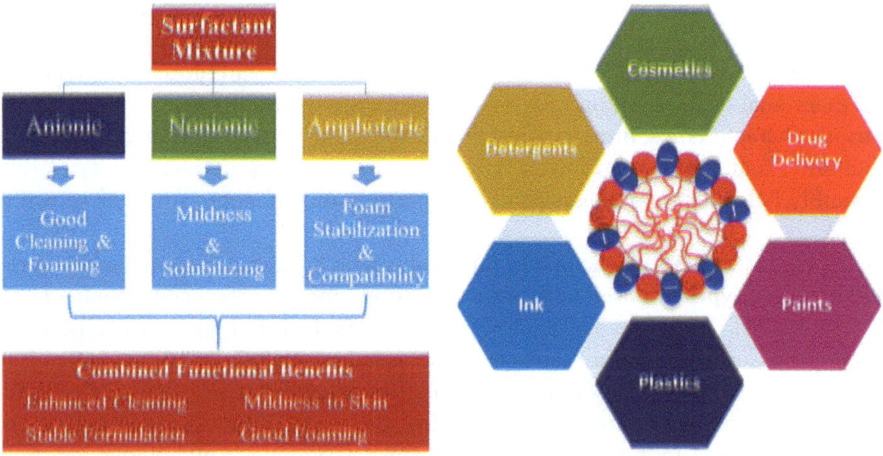

Fig. 3 Schematic representation of mixed surfactant systems in various applications

interchangeably with cleansing. Skin dryness encompasses much more than just the surface-level elimination of fluid. The biological mechanisms supporting skin health are fundamentally disrupted by surfactant interaction with lipids and proteins. Only in the past ten years have genuinely moisturizing cleansers started to appear despite the fact that mild surfactants have produced cleaners with much lower drying and harmful potential. The knowledge of mild surfactancy is expanded upon by the technology used to cleanse skin and enhance hydration, which also incorporates knowledge of skin and moisturization [21–25]. New knowledge on how humectants, emollients, and surfactants interact with skin can only result in cleaners that offer even more advantages. Therefore, moisturizing cleansers mark a major reimagining of the most fundamental cosmetic product in history.

2 Literature Review

Biosurfactants are becoming more and more popular as sustainable, safe substitutes for synthetic surfactants in cosmetics, according to recent assessments. Compared to conventional synthetic surfactants, which can irritate skin and damage the environment, biosurfactants—which are generated from microbes or plants—are prized for their mildness, biodegradability, and decreased toxicity.

Key Comparisons: Biosurfactants versus Synthetic Surfactants

Safety and Skin Compatibility: Unlike some synthetic surfactants that might irritate skin, trigger allergic reactions, or compromise skin integrity, biosurfactants are less likely to cause allergic reactions or harm the skin barrier.

Environmental Impact: Because biosurfactants are more biodegradable and less harmful to aquatic life and water systems, they function better than synthetic surfactants. Biosurfactants have a more environmentally favorable profile than synthetic surfactants, which are frequently generated from petroleum sources.

Functionality: In cosmetic formulations, biosurfactants (such glycolipids, lipopeptides, and saponins) exhibit superior emulsification, foaming, moisturizing, and antibacterial qualities that either match or surpass those of conventional synthetic surfactants.

Market Shift: Cosmetic companies are moving away from petrochemical surfactants and toward biosurfactants and other renewable substitutes due to consumer desire for natural, sustainable ingredients. These substitutes are being more widely used commercially, and continuous research is confirming their effectiveness.

A combination of surfactants is present in all commercial formulations that incorporate surfactants. An overview of surfactant mixtures in water is provided in this chapter, with particular attention to key ideas that set them apart from single-component systems. Emphasis is placed on surfactant mixtures that are crucial for applications [26–40]. In addition to cleaning and foaming, surfactants are being developed to provide other advantages including texture, emulsification, conditioning, pH regulation, etc. They can also serve many purposes (e.g.,, light cleanser and conditioner) or lessen the need for additional chemicals. A glowing skin and shiny well-groomed hair not only adds to self-confidence but also is a reflection of health and personality of that person. It is something that every person looks forward to that forms the base on which the multibillion global. Strong chemicals and surfactants cause the organization of the skin's lipid and protein layers to be disrupted, which increases transepidermal water loss (TEWL) and results in dryness, irritation, itching, and a tight feeling. Numerous studies have examined the use of surfactant mixtures in personal care products.

Research indicates that sulfate-free substitutes have been created because conventional anionic surfactants like sodium lauryl sulfate (SLS) provide better cleansing but can irritate skin. Additionally, advances in biodegradable surfactants have aided in the development of eco-friendly formulations without compromising efficacy. Natural biosurfactants are also becoming more and more popular, as evidenced by novel formulations that use enzyme-modified surfactants derived from natural sources [26, 41–44].

On the other hand, nonionic sugar-derived/ester surfactants like sucrose esters, sugar-based, alkyl polyglucosides (APGs), etc., are gaining interest due to mildness, skin friendliness, and biodegradability. Studies are looking into their phase behavior, aggregation, tailoring their hydrophile-lipophile balance (HLB). Recent research has also focused on synergistic formulations, which combine surfactant mixtures to maximize washing efficacy while minimizing irritation [45–60]. Amphoteric surfactants, for instance, have been shown to enhance mildness without sacrificing foaming ability when added to formulations containing anionic surfactants. These advancements satisfy consumer desire for environmentally friendly personal care products that balance efficacy and environmental responsibility. The literature indicates that multifunctional, mild, and sustainable surfactants that meet consumer and regulatory standards are becoming more and more popular. As research advances, new formulation methods that employ bio-based and low-irritancy surfactants are likely to influence the personal care industry's future.

According to published research, combining the various surfactant types presented in Table 2 can result in a synergistic impact, customized performance, structural polymorphism, cost and environmental advantages, the creation of sophisticated materials, and broad applicability. Variety of surfactant mixtures having different combination of surfactants are studied from among the surfactants listed below.

3 Roles of Surfactants in Personal Care

Because surfactants lower the surface tension between water and oil, they can be used as emulsifiers, detergents, foaming agents, conditioners, and solubilizers in personal care products, facilitating efficient product formulation and cleaning. They assist combine water-based and oil-based substances in lotions and creams, remove and

Table 2 Common surfactant classes and their uses in personal care products

Surfactant type	Basic characteristics	Examples	Typical personal care uses
Anionic	• Negatively charged in water • Strong cleansing and foaming	Sodium lauryl sulfate (SLS), ammonium lauryl sulfate (ALS), sodium lauryl ether sulfate (SLES), alpha olefin sulfonates (AOS), disodium laureth sulphosuccinate (DSLS), sodium cocyl glutamate (SCG), sodium cocoyl glycinate (SCGLY), sodium cocoyl methyl taurate (SCMT), sodium cocoyl isethionate (SCI), sodium lauroyl sarcocinate (SNLS)	Shampoos, body washes, cleansers, mildness
Cationic	• Positively charged • Antistatic, conditioning, antimicrobial	Quaternary ammonium compounds, Polyquaternium-10	Hair conditioners, hair dyes, rinse aids
Nonionic	• No electrical charge • Gentle, emulsifying, grease-removing, low foam	Decyl glucoside (DG), lauryl glucoside (LG), fatty alcohols, polysorbates (e.g., Tween 80) PEG-7 glyceryl cocoate, polysorbate 20, and sorbitan monostearate (SM)	Facial cleansers, shampoos, conditioners, lotions
Amphoteric	• Can be anionic or cationic depending on pH • Mild, conditioning, thickening	Cocamidopropyl betaine (CAPB), lauramidopropyl betaine (LAPB) cocamidopropyl hydroxysultaine (CAH), and sodium lauroamphoacetate (SLA)	Shampoos, conditioners, liquid soaps, skin cleansers, foams

suspend dirt, oil and filth from skin and hair, produce lather in shampoos and body washes, and aid in conditioning by coating skin and hair in protective layers.

Cleansing: Surfactants work as cleaners by reducing the surface tension of water, which enables it to mix with dirt and oils. The filth is then suspended for washing away by being encapsulated in micelles. They are vital ingredients in soaps, detergents, and shampoos because of their dual nature, which enables them to draw in both water-loving (hydrophilic) and oil-loving (hydrophobic) substances, efficiently removing stains, grease, and other pollutants from surfaces like skin and clothing (Table 3).

Table 3 Mixed surfactant systems used in cleansing

Mixed surfactant system	Applications
SCI + CAPB	Common combo in gentle facial cleansers
SLES + CAPB	Used in body washes and shampoos to balance cleansing with foam and reduced irritation
DG + LG	Used in ultra-gentle and natural formulations

Table 4 Mixed surfactant systems used in foaming

Mixed surfactant system	Applications
SNLS + CAPB	Good foaming
AES + NMEA	Foam booster

Foaming: Surfactants function as foaming agents by stabilizing the foam once it occurs and by reducing a liquid's surface tension, which facilitates the formation of bubbles. They accomplish this by producing a protective barrier at the air–liquid interface by adsorption, which stops bubbles from combining (coalescing) and slows down rapid water drainage. This procedure makes it possible to create and preserve a stable foam, which is necessary for a variety of applications, including improving oil recovery and cleaning goods (Table 4).

Emulsification: Surfactants function as emulsifying agents, stabilizing emulsions by lowering the interfacial tension between immiscible liquids like water and oil. For uses in food, cleaning goods, and industrial processes, they provide a protective barrier at the interface that keeps the dispersed phase consistently suspended inside the continuous phase and inhibits droplet agglomeration. By generating a stable emulsion, surfactants allow for the stable mixing of inherently incompatible water and oil to create stable lotions, creams, and cosmetics (Table 5).

Conditioning: The main way that surfactants function as conditioning agents is by adhering to negatively charged surfaces, such as skin and hair, where they lubricate and lower friction, creating a softer, smoother feel, and less static. Because of their strong affinity for the hair's surface and capacity to build a lubricating coating that enhances manageability and appearance, cationic surfactants are very excellent conditioning agents.

Solubilization: Above a certain concentration known as the critical micelle concentration (CMC), surfactants produce micelles and function as solubilizing agents. With their hydrophilic (loving water) and hydrophobic (hating water) components, these amphiphilic molecules organize into micellar structures. This creates a hydrophobic core that can encapsulate weakly soluble chemicals, enhancing their apparent solubility in an aqueous environment. In industries like cleaning and textiles, where surfactants are employed to dissolve and disperse hydrophobic compounds in water.

Wetting: By reducing a liquid's surface tension, surfactants function as wetting agents, facilitating more even dispersion across solid surfaces and penetration of porous materials. By altering the liquid–solid interaction, decreasing the contact

Table 5 Mixed surfactant systems used in emulsion

Mixed surfactant system	Applications
Glyceryl stearate + PEG-100 stearate	Common O/W emulsifier pair in creams
Sorbitan oleate (Span 80) + Polysorbate 80 (Tween 80)	Span-Tween pairs are classic for controlled HLB targeting

angle, and enhancing adhesion and coverage, they accomplish this. In applications where uniform spreading, penetration, and improved absorption are critical for performance, such as textile dyes, cleaning solutions, and agricultural sprays, this increased wettability is necessary (Table 6).

Thickening: By self-assembling into micelle structures that increase viscosity and improve product adhesion and performance, surfactants function as thickening agents in formulations such as shampoos and detergents. This thickening is accomplished by either interacting with other components to increase the flow resistance or by generating massive, rod-like micelles. The type and concentration of the surfactant, as well as the presence of additional substances like salt, all affect how effective this process is (Table 7).

Antimicrobial Activity: The main way that surfactants function as antimicrobial agents is by electrostatically and hydrophobically rupturing bacteria' cell membranes, which results in cell death. Cationic surfactants are especially efficient at breaking down the lipid bilayers of bacterial and fungal cells because of their charged heads and hydrophobic tails. Surfactants are employed in a variety of industries, including food and health care, to avoid contamination and infections because of their nonspecific action, which hinders the development of resistance by bacteria. By functioning as a preservative, some surfactants with bactericidal qualities can help extend the shelf life of cosmetic items (Table 8).

Table 6 Mixed surfactant systems used in wetting

Mixed surfactant system	Applications
Alkylpolyglucoside (APG) + ethoxylated fatty alcohol (AE)	Common O/W emulsifier pair in creams

Table 7 Mixed surfactant systems used in thickening

Mixed surfactant system	Applications
SLES + CAPB	Micelle growth, salt-sensitive
SLS + CAPB	More aggressive, less mild
SOS + CAPB	Synergistic thickening, works with salt

Table 8 Mixed surfactant systems used in antimicrobial activity

Surfactant	Applications
Benzalkonium chloride (BAC)	Disinfectants, hand sanitizers, eye drops
Cetyltrimethylammonium bromide (CTAB)	Antimicrobial agent, antiseptics
Dodecyltrimethylammonium chloride (DTAC)	Cosmetics, antiseptics
Chlorhexidine	Oral rinses, skin disinfectants
Sophorolipids	Mild antimicrobial used in cosmetics

4 Fundamental Mechanisms: What Drives Synergism or Antagonism

The molecular-level effects listed below dictate whether or not combined surfactants exhibit synergistic behavior, or are superior to each one alone:

Electrostatic interactions: Strong attraction may occur between two surfactants with opposite charges (cationic and anionic); however, mixed ionic–nonionic systems might occasionally benefit when nonionic species lessen repulsion between ionic heads.

Head-group compatibility: Size, charge density, hydration, etc. Mismatches can lead to less favorable interactions (steric hindrance and hydration issues).

Tail compatibility: Length, branching, saturation, rigidity. If tails are very different, packing is less good.

Counterion binding and screening: Counterions and additional salts are crucial for screening charges or adjusting micelle stability in ionic surfactants, often known as mixed ionic systems.

Hydrophobic interactions among tail parts; van der Waals forces.

5 Emerging Research Trends

Due to customer desire for softer, more environmentally friendly products, bio-based and sustainable alternatives are the focus of emerging research in personal care surfactants. Producing sulfate-free formulations to lessen skin irritation, investigating new amphoteric and nonionic surfactants for better performance and kinder applications, and producing new biosurfactants from renewable resources like oils and sugars are some of the major developments. Additionally, research uses cutting-edge methods like molecular modeling and cryo-TEM to comprehend and improve surfactant microstructure and self-assembly.

I. **Sulfated/ Sulfate-Free Mild Formulations**

Instead of harsh sulfates like SLS or SLES, milder surfactants like coco glucoside are used in sulfate-free and mild personal care formulations to efficiently cleanse without irritating the skin or removing natural oils from the hair and skin. These products are perfect for colored hair, sensitive skin, and everyday usage to keep the skin barrier and scalp healthy. For assurance in mild formulations, look for natural chemicals, mild surfactants, and certifications from reputable brands.

SLES/CMEA in Mixed Surfactant System

Anionic surfactant sodium lauryl ether sulfate (SLES) in the presence of nonionic cocamide monoethanolamine (CMEA) depicts high surface activity and foam stability [12]. In aqueous SLES-CMEA binary mixes, the introduction of NaCl can cause a sphere-to-rod micellar transition, which was attributed to screening of electrostatic repulsion inside the charged headgroups of the SLES. SLES-CMEA-NaCl entangled wormlike micelles have demonstrated extremely viscoelastic characteristics as depicted in Fig. 4. The viscosity values were in good agreement with the micellar size and shape as determined by DLS and SANS. Furthermore, it had been discovered that CMEA and NaCl can be used to modify the foamability and foam stability of SLES. The results of this study will support a viable strategy for providing pre-solubilized CMEA composition in SLES, which might be very advantageous for personal care products by allowing CMEA blending at room temperature instead of requiring high temperature mixing.

Sodium N-Lauroyl Sarcosinate in Mixed Surfactant System

Amino acid-based anionic surfactant sodium N-lauroyl sarcosinate (SNLS) finds wide application in cosmetic formulations. A mixture of SNLS, cocamide monoethanolamine (CMEA), and lauryl glucoside (LG) formed entangled wormlike micelles [13]. By adjusting the pH of the solution, a combination of SNLS, CMEA, and LG created entangled wormlike micelles with surfactant concentrations ranging from 12 to 20 weight percent. It was discovered that synergistic viscosity build in SNLS/CMEA/LG combinations was impacted by pH values of about 5.8. In some SNLS/CMEA/LG compositions, shear-thinning rheology provided support for this finding. Hydrogen bonds and electrostatic interactions between surfactant molecules decreased micellar charge density measurements when pH was lowered to the viscosity peak, increasing packing factor. Furthermore, the efficacy of mixing SNLS/CMEA/LG surfactants was confirmed by increased surface activity, improved mildness, superior foamability, and stability. It was shown that adding CMEA/LG surfactant and adjusting the pH-thickened SNLS surfactants, which were normally challenging to thicken. Formulators can choose surfactants and their ratios to maximize product performance by using this study as a guide.

Sodium N-acyl sarcosinate in Mixed Surfactant System

Three alkyl sarcosinate solutions (dodecyl, tetradecyl, and cocoyl) were subjected to systematic studies to ascertain their viscosity response at pH > 6 (ionized form),

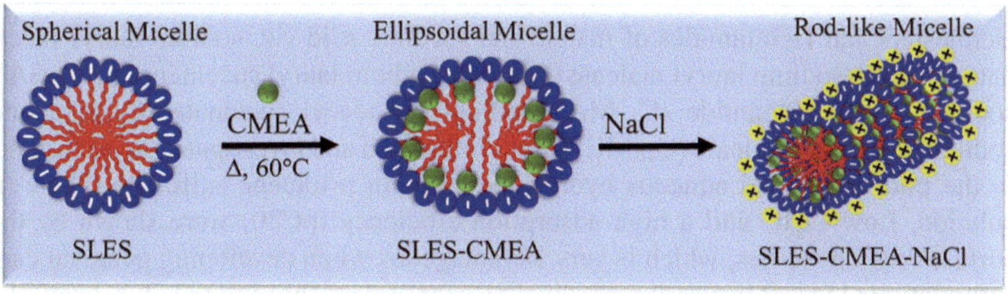

Fig. 4 Schematic representation of sphere-rod transition in SLES micelles in CMEA and NaCl

in the presence of electrolytes (NaCl and KCl). The primary goal of this research was to determine the critical elements affecting the rheological behavior of solutions containing sodium sarcosinate and their mixes with cationic, zwitterionic, and nonionic co-surfactants [52]. Although the formation of mixed micelles is restricted by steric hindrance from side methyl groups, the elongation of these micelles is responsible for the rise in viscosity when amine oxide and sulfobetaine are used. This study's practical value stems from the discovery that longer-chain alkyl sarcosinates, like tetradecyl, which was studied here, can achieve noticeably higher viscosities at lower salt concentrations.

Sodium Cocoyl Glycinate in Mixed Surfactant System

In comparison with other amino acid-based surfactants, sodium cocoyl glycinate (SCGLY), a mild anionic surfactant based on amino acids obtained from the natural fatty acid of coconut and the amino acid glycine, has a significantly lower charged head [15]. Glycinate is preferred for usage in hair shampoos, skincare products, shower gel, and shaving cream due to its remarkable surface activity and aggregation behavior, low toxicity, high biodegradability, and mildness (safe and non-irritating to skin). When nonionic surfactants such as CDEA and LG are combined with SCGLY, they co-assemble with SCGLY micellar solutions to generate mixed micelles. Surface tension experiments indicate that the ternary mixture of surfactant solution exhibits more aggregation ability and higher surface activity than SCGLY alone. This is because the presence of CDEA and LG molecules stuck among charged SCGLY head groups causes a decrease in electrostatic repulsion among SCGLY polar head groups. The small rod-like micelles grow into long rod-like micelles by adjusting solution pH. Developing mild product formulations with amino acid-based surfactants in mixed systems may greatly benefit from this knowledge. Furthermore, the combination of LG and CDEA with SCGLY results in foam that has a low liquid fraction and tiny bubble diameters, which improves foam stability and foamability. By increasing the packing at the air–liquid interface, the addition of CDEA and LG to SCGLY increases the bulk viscosity. This increases the stability of the foam by lowering the liquid drainage from the foam lamellae and avoiding coalescence and coarsening. In addition to greater cleaning, improved foamability and stability are sought in a number of applications (Fig. 5).

Surfactants Based on Maleic and Succinic acid in Mixed Surfactant Systems

Hemiesters and Hemiamides of maleic and succinic acid viz. sodium lauryl succinate ($C_{12}SE$), sodium lauryl maleate ($C_{12}ME$), sodium lauryl succinamide ($C_{12}SA$), sodium lauryl maleamide ($C_{12}MA$), sodium hexadecyl succinate ($C_{16}SE$), and sodium hexadecyl maleate ($C_{16}ME$) were synthesized and investigated as surfactants in the pure water and aqueous hydrotrope [sodium p-toluene sulfonate (NaPTs)] solution. Low CMC and a high adsorption efficiency (pC20) were shown by the surface tension studies, which is very advantageous when developing personal care formulations [11]. Because larger micelles cannot pass through the skin layer, the dynamic light scattering (DLS) approach revealed the production of larger micelles,

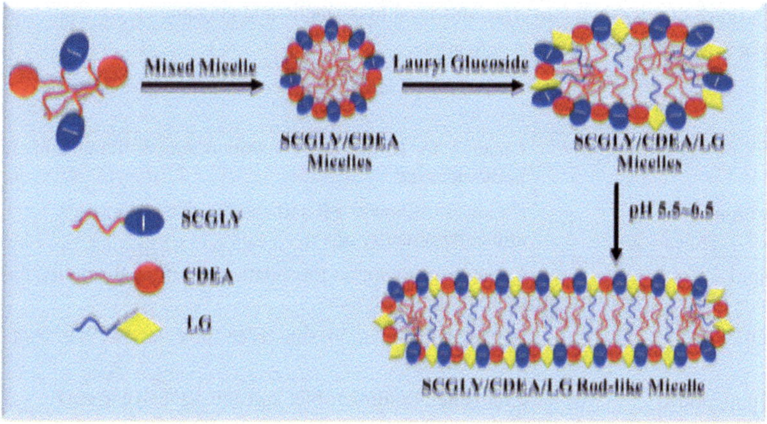

Fig. 5 Schematic representation of sphere-rod transition in a ternary mixture of SCGLY/CDEA/LG surfactants as observed by rheology measurements

which was significant for skin care. Additionally, these surfactants showed superior foamability and stability, which were attributable to their small bubble size and quicker monomer adsorption both of which were desirable for cleaning applications. Furthermore, compared to other widely used conventional surfactants, these surfactants' benign skin behavior was suggested by their minimal protein/lipid solubilization. Viscosity studies indicated that surfactants had a respectable capacity to thicken when co-surfactants such as lauramine oxide were present. The new hemiamides and hemisters appear to be attractive surfactants for enhancing a number of attributes in prospective personal care products.

II. **Bio-based Sustainable Surfactants**:

Sustainable and bio-based surfactants, which are made from microorganisms or renewable resources like sugars and fats, are environmentally beneficial substitutes for conventional petroleum-based surfactants in personal care products. They satisfy consumer demand for natural and eco-friendly beauty products by providing advantages including mildness, biodegradability, and low toxicity. In order to lessen their influence on the environment and satisfy regulatory demands for cleaner chemicals, shampoos, lotions, and other personal care products are increasingly using biosurfactants. Alkyl polyglucosides (APG) are a common bio-based surfactant found in shampoos, soaps, and lotions because of their mildness and good compatibility with other surfactants. Glycolipid biosurfactants are perfect for environmentally aware skin care because they are biodegradable, mild on the skin, and offer efficient cleansing without unpleasant side effects. Fermentation produces rhamnolipids, a type of biosurfactant that performs well in a variety of personal care applications and has a lower carbon footprint.

Bio-Based Surfactants: Derived from renewable fats, sugars, or amino acids but synthesized chemically, including methyl ester sulfonates, alkyl polyglucosides,

Table 9 Types of microbial biosurfactants used in cosmetic and personal care

Microbial biosurfactant	Applications
Rhamnolipid	Anti-aging product, cleanser in shampoos, anti-adhesive activity
Rhamnolipid/Sophorolipid	Cleanser for antidandruff shampoo, moisturizing skin cleanser, body cleanser
Sophorolipid	Cleanser in shower gel and shampoo, body washer, anti-inflammatory agent
Glycolipid	Cleanser in shampoo formulation, antifungal activity, hair care conditioning polymers
Lipopeptide	Antiacne formulation, sunscreen formulations, hair care formulation
Glycolipopeptide	Emulsifier, (antimicrobial and anti-adhesive agent)
Glycolipopeptide/Lipopeptide	Rosemary oil/water emulsions, preservative
Oligomeric biosurfactant	Conditioning agent for hair products

glycinate amino acid-based surfactants, and sugar-based surfactants. They bridge the gap between synthetic and microbial surfactants but retain some environmental and safety concerns due to chemical synthesis steps.

Microbial Biosurfactants: Produced directly by microorganisms such as bacteria and yeasts, these are secondary metabolites including glycolipids (rhamnolipids, sophorolipids), lipopeptides (surfactin), glycolipopeptides, and bioemulsifiers (phospholipids, polymeric substances). They are characterized by high biocompatibility, biodegradability, and diverse bioactivities.

Microbial biosurfactants have demonstrated abilities to reduce water surface tension significantly (down to ~ 26 mN/m), form stable emulsions, and possess antimicrobial and anti-inflammatory activities. They are promising in formulations targeted at skin and hair health, including anti-aging products, cleansers, moisturizers, sunscreens, and anti-acne treatments.

Microbial biosurfactants represent a sustainable and biocompatible alternative to synthetic and chemically derived bio-based surfactants for cosmetic and pharmaceutical applications. Their multifunctional properties, combined with lower toxicity and environmental impact, make them highly attractive for next-generation formulations. Overcoming production and regulatory challenges is key to unlocking their full potential and enabling widespread industrial use (Table 9).

6 Conclusions

Surfactants are crucial components in cosmetics and personal care products because they aid in cleaning, emulsification, and foaming. The focus of current research has been on developing mild and biodegradable surfactants to lessen skin discomfort and its effects on the environment. Mixed surfactant systems provide a compromise between safety, mildness, and performance. They are a fundamental component of contemporary personal care chemistry because of their intricate interactions, which include mixed micelle and sub-micellar production, which allow formulators to create solutions with the best cleaning, foaming, sensory attributes, and lowest risk of irritation. Compared to single surfactants, mixed surfactant systems can synergistically improve qualities including cleaning, mildness, and skin compatibility, making them essential in personal care products. By combining molecules with various architectures, mixed surfactant systems create formulations that are more effective, perform better, and frequently cause less irritation. For products like shampoos, body washes, and face cleansers, this synergy enables producers to strike a balance between mildness and cleaning effectiveness.

While mixed surfactant systems offer many advantages, there are few challenges:

- The system becomes more complex due to the addition of more parameters (such as temperature, pH, ionic strength, surfactant kinds, concentrations, etc.).
- Micelle or interfacial film composition can be difficult to predict, particularly when surfactants have widely disparate characteristics (size, charge, etc.).
- Phase behavior could be complicated, including mixed micelles, many types of micelles coexisting, mixed liquid crystalline phases, and precipitation in specific circumstances.
- Mixed surfactant systems may be less stable when exposed to changes in temperature, dilution, or salt.
- Some blended surfactants may be costlier or cause environmental issues, particularly if they contain costly or less biodegradable components.

Acknowledgements Authors thank Navrachana University for Seed Grant (NUV/Seed Grant/2023_24/02) and UGC-DAE (CRS/2023-24/03/1217) for providing the financial support.

Authors declare that they have no conflict of interest.

References

1. Kronberg B, Holmberg K, Lindman B (2014) Surface chemistry of surfactants and polymers. Wiley
2. Rosen MJ, Kunjappu JT (2012) Surfactants and interfacial phenomena, 4th edn. Wiley
3. Tadros TF (2005) Applied surfactants: principles and applications. Wiley, Weinheim
4. Banjare MK (ed) Role of surfactants in drug delivery, pesticides, antimicrobial activity and bioremediation. Deep Science Publishing
5. Dreiss CA, Feng Y (eds) (2017) Wormlike micelles: advances in systems, characterization and applications. Royal Society of Chemistry, London
6. Sakamoto K, Lochhead R, Maibach H, Yamashita Y (eds) Cosmetic science and technology: theoretical principles and applications. Elsevier
7. Ghosh S, Ray A, Pramanik N (2020) Self-assembly of surfactants: an overview on general aspects of amphiphiles. Biophys Chem 265:106429

8. Varade D, Rodríguez-Abreu C, Shrestha LK, Aramaki K (2007) Wormlike micelles in mixed surfactant systems: effect of cosolvents. J Phys Chem B 111:10438–10447
9. Acharya DP, Varade D, Aramaki K (2007) Effect of temperature on the rheology of wormlike micelles in a mixed surfactant system. J Colloid Interface Sci 315:330–336
10. Lukic M, Pantelic I, Savic S (2016) An overview of novel surfactants for formulation of cosmetics with certain emphasis on acidic active substances. Tenside Surf Det 53:7–19
11. Zhou L, Liu Z, Wang Y (2025) Molecular insights: how counterions determine surfactant aggregation. Adv Colloid Interface Sci 341:103484
12. Janni S, Rajput G, Pandya N, Subramanyam G, Varade D (2021) Interfacial properties of novel surfactants based on maleic and succinic acid for potential application in personal care. J Mol Liq 342:117484
13. Pandya N, Rajput G, Janni S, Subramanyam G, Ray D, Aswal V, Varade D (2021) SLES/CMEA mixed surfactant system: effect of electrolyte on interfacial behavior and microstructures in aqueous media. J Mol Liq 325:115096
14. Varade D, Rajput G, Janni S, Subramanyam G (2024) Rheology and surface active properties of sodium N-lauroyl sarcosinate in mixed surfactant system. Colloids Surf A Physicochem Eng Asp 697:134349
15. Zhao H, He C, Zhou Y, Yang J, Luo C, Xu B (2019) Study on foaming properties of N-acyl amino acid surfactants: sodium N-acylglycinate and sodium N-acyl phenylalaninate. Colloids Surf A Physicochem Eng Asp 567:240–248
16. Rajput G, Janni DS, Subramanyam G, Ray D, Aswal VK, Varade D (2022) Novel approach for tuning micellar characteristics and rheology of a sulfate-free anionic surfactant sodium cocoyl glycinate. Colloids Surf A: Physicochem Eng Asp 648:129426
17. Zhecheng ZW, Zhihui ZZ, Minjia MY (2025) Study on the foam properties of mixed systems based on sodium lauroyl glutamate. China Surfactant Deterg Cosmet 55:1035–1041
18. Eunseo EK, Jongsun JY, Jong-dae JJ, Heemuk HO, Hyomin HL (2025) Mixed surfactants for stabilizing and controlling water transport in double emulsions. Sci Adv 11:1–14
19. Zilong ZL, Ping PL, Yanxiao YH, Di DS, Fengzhi FG, Xue XL, Wenxiu WL, Qichao QL, Weichao WS (2024) Synergistic mechanism in anionic/zwitterionic and anionic/nonionic surfactant mixtures on improving the thermal stability of emulsions: an experimental and simulation study. Chem Eng Sci 289:119824
20. Nozomu NS (2024) Interaction parameters for the formation of mixed micelles and partitioning of solutes in them: a review. Applied Chem 4:1–14
21. Biswal NR, Paria S (2014) Interfacial and wetting behavior of natural-synthetic mixed surfactant systems. RSC Adv 4:9182–9188
22. Holmberg K, Bilén F, Bordes R (2025) Development of amino acid-based surfactants: from synthesis to applications. Curr Opin Colloid Interface Sci 75:101884
23. Veeramanoharan A, Kim S-C (2024) A comprehensive review on sustainable surfactants from CNSL: chemistry, key applications and research perspectives. RSC Adv 14:25429–25471
24. Ananthapadmanabhan KP (2019) Amino-acid surfactants in personal cleansing (review). Tenside Surf Det 56:378–386
25. Takehara M (1989) Properties and applications of amino acid based surfactants. Colloids Surf A Physicochem Eng Asp 38:149–167
26. De S, Malik S, Ghosh A, Saha R, Saha B (2015) A review on natural surfactants. RSC Adv 5:65757–65767
27. Vu T, Weaver MR, Kasting GB, Koenig P (2021) Effect of pH on the structure and dynamics of wormlike micelles in an amino acid-derived surfactant composition. Langmuir 37:4112–4120
28. Vu T, Koenig P, Cochran BM, Ananthapadmanabhan KP, Weaver M, Reeder B, Hutton HD, Kasting GB (2020) Thickening mechanisms for an amino acid-derived surfactant composition. Colloids Surf A Physicochem Eng Asp 589:124424
29. Tzocheva SS, Kralchevsky PA, Danov KD, Georgieva GS, Post AJ, Ananthapadmanabhan KP (2012) Solubility limits and phase diagrams for fatty acids in anionic (SLES) and zwitterionic (CAPB) micellar surfactant solutions. J Colloid Interface Sci 369:274–286

30. Shrestha RG, Shrestha LK, Aramaki K (2007) Formation of wormlike micelle in a mixed amino-acid based anionic surfactant and cationic surfactant systems. J Colloid Interface Sci 311:276–284
31. Nurendra NC, Suranjan SS, Moazzam MA (2025) Amino acid-based sodium N-lauroylsarcosinate as an optimised surfactant for propranolol interaction: comparative analysis of mixed micelle formation with cationic and zwitterionic surfactants. J Mol Liq 433:127899
32. Wanping WZ, Lihong LH, Kuanhong KJ, Feiyan FG, Zijia ZZ, Guangyong GZ (2025) Phase behavior and rheology of mixed systems of potassium cocoyl glycinate and fatty acid salts. J Surfactants Deterg 28:933–946
33. Wanping WZ, Jiajia JG, Xin XD, Donghu DL, Hanjun HC, Liang LC, Qianjie QZ (2025) Rheological behavior and mechanism of pH-responsive wormlike micelles based on sodium cocoyl glycinate/cocamidopropyl hydroxypropyl sulfobetaine. Colloids Surf A Physicochem Eng Asp 710:136244
34. Bin BQ, Zhao ZC, Xiaomei XP, Zhenggang ZC, Binglei BS (2025) Viscoelastic behavior of wormlike micellar solutions formed by an aspartame-based bicephalous anionic surfactant. Soft Matter 21:1583–1591
35. Longbin LL, Weixin WL, Diansong DC, Ling LM, Li LJ, Chunrui CH, Liwei LZ, Jianxin JJ (2025) Foaming and emulsification synergies in skin mild and environmentally friendly detergents based on gleditsia sinensis saponins. J Mol Liq 417:126649
36. Foluso FA, Samiul SA (2024) Stimuli-responsive bio-based surfactant-polymer gels. Colloids Surf A Physicochem Eng Asp 703:135149
37. Anirudh AS, Mukul MK, Ravi RP, Masood JM, Sandeep SK (2024) Effect of the mixed micelles of zwitterionic-anionic surfactant on efficiency of antibiotic azithromycin dehydrate. J Mol Liq 408:125317
38. Deeming L, Leaman NR, Honecker D, Chen Y, Ma K, Edler KJ (2025) Mixed micellar solutions of an oleo-furan sulfonate surfactant with anionic and non-ionic surfactants. J Colloid Interface Sci 699:138201
39. Zhou J, Huang C, Zou X, Meng T, Lai C (2025) Glycerol α-monolaurate-induced evolution of aggregate morphology and viscosity in the aqueous sodium N-lauroylsarcosinate solution. Colloids Surf A Physicochem Eng Asp 725:137671
40. Yuan YY, Meng MC, Jian JZ (2025) Synergistic effects of betaine and anionic surfactants at the oil-water interface: a molecular dynamics simulation study. J Mol Liq 437:128456
41. Feihong FW, Wenhui WF, Zhendong ZZ, Jiahao JZ, Hongyuan HW, Leping LD (2024) Coacervating behavior of amino acid anionic and amphoteric mixed micelle-polymer. Soft Matter 20:5733–5744
42. Holmberg K (2001) Natural surfactants. Curr Opin Colloid Interface Sci 6:148–159
43. Zhou C, Wang Y (2020) Structure-activity relationship of cationic surfactants as antimicrobial agents. Curr Opin Colloid Interface Sci 45:28–43
44. Cornwell PA (2018) A review of shampoo surfactant technology: consumer benefits, raw materials and recent developments. Int J Cosmetic Sci 40:16–30
45. Guo J, Sun L, Zhang F, Sun B, Xu B, Zhou Y (2022) Review: progress in synthesis, properties and application of amino acid surfactants. Chem Phys Lett 794:139499
46. Hiranuma T, Yamamoto Y, Nonomura Y (2025) Coacervate formation behavior and lubrication properties of a hair shampoo formulated with amino acid surfactants and cationized cellulose. J Surfactants Deterg 0:1–9
47. Gonçalves RA, Holmberg K, Lindman B (2023) Cationic surfactants: a review. J Mol Liq 375:121335
48. Chu Z, Dreiss CA, Feng Y (2013) Smart wormlike micelles. Chem Soc Rev 42:7174–7203
49. Li H, Lin Z, Chen Z, Cui Z, Lei L, Song B (2024) Wormlike micellar solutions formed by an anionic surfactant and a cationic surfactant with two head groups. Soft Matter 20:978–984
50. Tripathy DB, Mishra A, Clark J, Farmer TS (2018) Synthesis, chemistry, physicochemical properties and industrial applications of amino acid surfactants: a review. C R Chim 21:112–130
51. Sakai K, Nomura K, Shrestha RG, Endo T, Sakamoto K, Sakai H, Abe M (2012) Wormlike micelle formation by acylglutamic acid with alkylamines. Langmuir 28:17617–17622

52. Lorenzo LV, Giuseppe GC, Najet NM, Giovanni GF, Gerardo GP, Emiliano EF, Luigi LG (2025) Bicelle-bicelle lamellar stacking of a fatty alcohol and A C8–10 alkyl polyglycoside surfactant. J Mol Liq 435:128193
53. Rusanova DG, Stoeva M, Mitrinova Z, Pagureva N, Burdzhiev N, Tcholakova S (2025) Role of electrolytes and co-surfactants on the rheological properties of sodium N-Acyl sarcosinate solutions. J Mol Liq 434:128069
54. Zhendong ZZ, Jiahao JZ, Feihong FW, Wenhui WF, Leping LD, Hongyuan HW (2024) Synergistic effect and phase behavior of SCG-CAPB-H_2O ternary compound system. Appl Sci 14:3081
55. Jiamin JF, Jun JZ, Xiuquan XY, Liang LB, Yuan YZ, Zhiyu ZW, Ziyu ZQ (2022) Study on the properties of the sodium lauroyl glycinate and sodium lauroyl lactylate composite system. Langmuir 38:16112–16121
56. Bordes R, Holmberg K (2015) Amino acid-based surfactants—do they deserve more attention? Adv Colloid Interface Sci 222:79–91
57. Kronberg B, Mixtures S (1997) Curr Opin Colloid Interface Sci 2:456–463
58. Bajani D, Gharai D, Dey J (2018) A comparison of the self-assembly behaviour of sodium N-lauroyl sarcosinate and sodium N-lauroyl glycinate surfactants in aqueous and aqueo-organic media. J Colloid Interface Sci 529:314–324
59. Fan J, Zhang J, Yang X, Bai L, Zhou Y, Wu Z, Qin Z (2022) Study on the properties of the sodium lauroyl glycinate and sodium lauroyl lactylate composite system. Langmuir 38:16112–16121
60. Li R, Zhang G, Yu S, Deng R, Yin J, Zhang T, He Q (2025) Impact of molecular structures of lauroyl glycine derivatives on foaming properties. Front Chem 24:1563560

Self-assembly of Pluronic Surfactants and Their Applications

Nisar Ahmad Malik

Abstract This chapter examines how Pluronic surfactants self-assemble and the various ways they can be used in science and technology. Pluronics, which are triblock copolymers made of poly(ethylene oxide) and poly(propylene oxide), possess distinctive thermoresponsive and amphiphilic characteristics that promote micellization, gelation, and the creation of various nanostructures. In the chapter, the physicochemical principles that govern their self-assembly are discussed, along with the factors that influence structural transitions and the effects of concentration, temperature, and solvent conditions. Applications in drug delivery, nanomedicine, and emulsification are highlighted. Gaining an understanding of the self-assembly mechanisms of Pluronics can provide important insights for the design of sophisticated materials intended for biomedical and industrial uses.

Keywords Pluronic surfactants · Thermodynamics · Self-assembly · Types · Micellization · Drug delivery · Poloxamer · Triblock copolymers

1 Introduction

Molecular building blocks can organize themselves into ordered structures without external direction, a process known as self-assembly. This phenomenon is fundamental to the creation of micelles, vesicles, liquid crystalline phases, and numerous soft-matter structures. Pluronic polymers (also known as poloxamer or PEO–PPO–PEO triblock copolymers) are one of the most thoroughly researched types of amphiphilic polymers, due to their biocompatibility, customizable structure, and behavior that responds to stimuli [1–3].

Pluronic surfactants are utilized in various fields, including drug delivery, nanomaterials, rheological modifiers, and responsive gels (Fig. 1). Understanding how they

N. A. Malik (✉)
Department of Chemistry, Islamic University of Science and Technology, Awantipora, Pulwama, Jammu and Kashmir, India
e-mail: nisarchmjmi@gmail.com

M. Abdul Rub et al. (eds.), *Industrial Applications of Individual and Mixed Surfactant Systems—Volume 2*, Sustainable Materials and Technology,
https://doi.org/10.1007/978-981-95-9195-4_12

Fig. 1 Some of the Pluronic surfactant applications

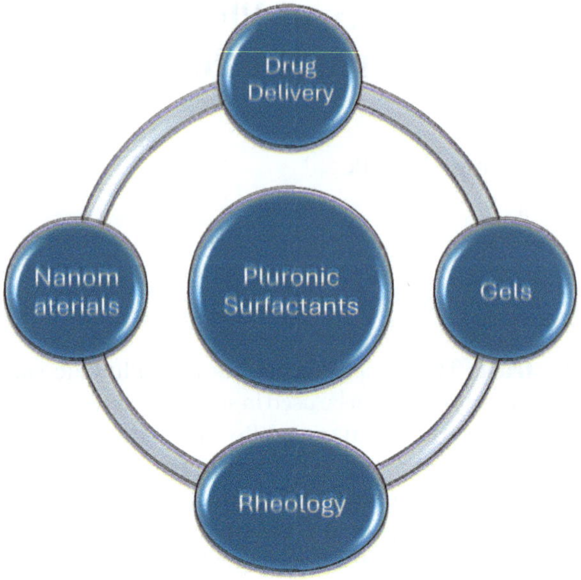

self-assemble is vital for customizing functional systems in nanomedicine, catalysis, and soft materials science. The formation, stability, and transitions of self-assembled structures are controlled by the complex interplay between hydrophilic (PEO) and hydrophobic (PPO) segments, along with factors such as temperature, concentration, additives, or salts.

2 Chemical Structure and Nomenclature

A Pluronic molecule is a type of triblock copolymer structured as PEO–PPO–PEO, with PEO representing poly(ethylene oxide) and PPO representing poly(propylene oxide). The central hydrophobic block of PPO is flanked by the hydrophilic end blocks of PEO. The hydrophilic–lipophilic balance (HLB) of the molecule is determined by the relative lengths of PEO and PPO. Pluronics intended for commercial use are usually identified by a combination of a letter and numbers (e.g., F127, P123, P85: Figs. 2 and 3.), which reflects their molecular weight and the ratio of EO to PO units [1, 4].

Fig. 2 Chemical structure of P85

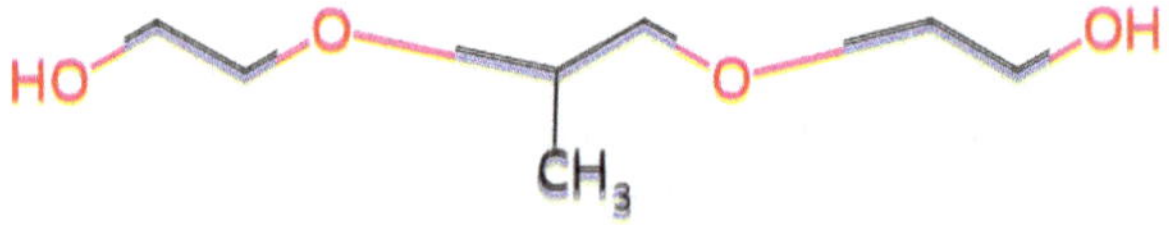

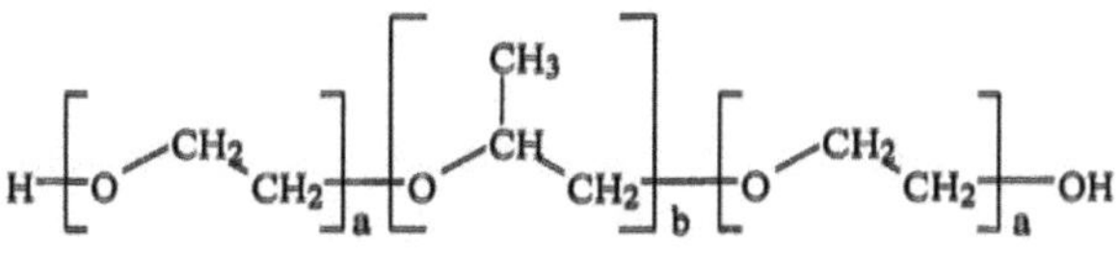

Fig. 3 Chemical structure of F127 ($a = 100$ and $b = 65$), a = number of ethylene oxide and b = propylene oxide monomers

As an illustration, Pluronic F127 has a higher PEO content, which increases its hydrophilicity, whereas Pluronic L121 contains more PPO and is thus more hydrophobic. This tunability results in the variation of critical micelle concentrations (CMCs), critical micelle temperatures (CMTs), and phase behavior among different Pluronics in aqueous media [2, 5].

Pluronic copolymers are nonionic surfactants (meaning they do not carry a formal charge), which makes their self-assembly easier to interpret compared to ionic surfactants. However, the interaction of hydrogen bonding, hydration, and entropic contributions is still complex.

3 Thermodynamics and Kinetics of Micellization

In dilute aqueous solutions, Pluronic molecules primarily exist as unimers. With rising concentration (or temperature), driving forces (the hydrophobic effect of PPO and entropy gain of water) promote aggregation into micelles [1, 3]. This process is regulated by two important thresholds: the critical micelle concentration (CMC) and critical micelle temperature (CMT) [2, 5]. As the temperature rises, the hydration level of PPO segments decreases (they become more hydrophobic), resulting in a reduction of the CMT and CMC. As a result, higher temperatures promote micellization [2–9]. When the temperature is kept constant at a value above the CMT, an increase in concentration results in an increase in the number of micelles and possibly morphological changes (spheres → rods → lamellae). Due to their amphiphilic characteristics, they can form micelles in water at concentrations and temperatures exceeding specific CMC and CMT values.

Both thermodynamic and kinetic parameters govern the process of micellization, controlling the formation, stability, and behavior of micelles. The spontaneous self-assembly process of micellization is driven by the balance of hydrophobic interactions, hydrogen bonding, and the entropic effects of water structuring. The analysis can be conducted based on changes in Gibbs free energy $\left(\Delta G_m^0\right)$, enthalpy $\left(\Delta H_m^0\right)$, and entropy $\left(\Delta S_m^0\right)$ [10–12].

Free Energy of Micellization: The standard Gibbs free energy of micellization per mole of monomer units can be expressed as:

$$\Delta G_m^0 = \mathrm{RTln(CMC)}, \tag{1}$$

where R = gas constant, T = absolute temperature, CMC = critical micelle concentration. A negative value of ΔG_m^0 Indicates that micellization is spontaneous.

Enthalpy and Entropy of Micellization: The temperature dependence of CMC provides information about enthalpy and entropy using the van't Hoff equation:

$$\frac{d\ln(\mathrm{CMC})}{d\left(\frac{1}{T}\right)} = \frac{-\Delta H_m^0}{R}, \tag{2}$$

$$\Delta S_m^0 = \frac{\left(\Delta H_m^0 - \Delta G_m^0\right)}{T}. \tag{3}$$

At temperatures below the CMT, Pluronic molecules are molecularly dispersed. Above CMT, micelles form due to dehydration of PPO segments. The relationship is:

$$\frac{\mathrm{d}\left(\Delta G_{\mathrm{m}}^0\right)}{\mathrm{d}T} = -\,\Delta S_{\mathrm{m}}^0. \tag{4}$$

Kinetics of Micellization: Micellization involves nucleation of micelles and growth to equilibrium size.

Rate of Micelle Formation: The rate can be described as:

$$\frac{\mathrm{d}C_m}{\mathrm{d}t} = k_f\left(C_1 - C_{\mathrm{eq}}\right) - k_b C_m, \tag{5}$$

$$\text{At equilibrium}: k_f\left(C_1 - C_{\mathrm{eq}}\right) = k_b C_m, \tag{6}$$

$$\text{Relaxation time}: \tau = \frac{1}{k_f + k_b}. \tag{7}$$

Temperature-Triggered Kinetics: The rate constant follows Arrhenius dependence:

$$k = Ae^{\frac{-E_a}{RT}}. \tag{8}$$

Thermodynamic and kinetic parameters help in understanding and tuning Pluronic performance. Examples include spontaneity $\left(\Delta G_m^0\right)$, heat effects $\left(\Delta H_m^0\right)$, entropy changes $\left(\Delta S_m^0\right)$, and micellization kinetics (τ, E_a) [13, 14] (Table 1).

These parameters direct the design of drug delivery systems, forecasting of solubilization capacity, refinement of stimuli-responsive systems, comprehension of nanocarrier self-assembly, and assessment of environmental stability. The process of Pluronic surfactants micellization is influenced by a combination of thermodynamic and kinetic factors, with the hydrophobic dehydration of PPO blocks and entropic effects being the main drivers. The thermodynamic parameters and kinetic

Table 1 Thermodynamic parameters and their interpretation

Parameter	Meaning	Relevance/application
(ΔG_m^0)	Spontaneity of micellization	Determines feasibility of micelle formation under given conditions
(ΔH_m^0)	Heat absorbed/released during micellization	Helps in designing temperature-sensitive drug delivery systems
(ΔS_m^0)	Entropy change (water structure)	Reflects the hydrophobic dehydration and molecular ordering
CMC, CMT	Concentration/temperature thresholds for micellization	Used to define formulation stability windows
(E_a)	Activation energy	Determines temperature sensitivity and rate of micellization
(τ)	Relaxation time	Defines response time in thermoresponsive systems

constants allow for the rational design of Pluronic-based materials used in drug delivery, nanotechnology, and biomedical applications.

4 Self-assembled Structures and Phase Behavior

Pluronic's self-assembled structures exhibit a rich variety of architectures beyond simple spherical micelles, depending on concentration and temperature, including spherical, cylindrical, and vesicular micelles [3, 6]. When concentrations are high enough, micelles can arrange themselves into cubic, hexagonal, or lamellar phases (lyotropic phases) [7] (Fig. 4). Hybrid systems utilize Pluronic micelles as templates for nanostructured materials [2, 8]. In the context of drug delivery, hydrophobic drugs are encapsulated by Pluronic micelles, whose self-assembly and release behavior is affected by interactions with drugs or biomolecules [4].

Recent studies have demonstrated that highly hydrophilic Pluronics, which typically resist aggregation, can be prompted to self-assemble in the presence of sugars acting as kosmotropes [9]. Owing to their amphiphilic nature, they can self-assemble into diverse nanostructures in water, rendering them useful for a variety of applications that include drug delivery and the synthesis of nanomaterials. Self-assembly behavior and phase transitions of Pluronics are affected by temperature, concentration, and the presence of additives. In water, Pluronic surfactants can cluster together to create micelles, with the hydrophobic PPO core protected from water by the hydrophilic PEO corona. The micelles can differ in size and shape depending on the copolymer's molecular weight and environmental conditions. The self-assembly of Pluronics is critically influenced by temperature. As an example, Pluronic F127 displays a lower critical solution temperature (LCST) characteristic, whereby a rise in temperature induces a switch from hydrophilic to hydrophobic, thereby encouraging micelle formation. The morphology is also influenced by concentration: At

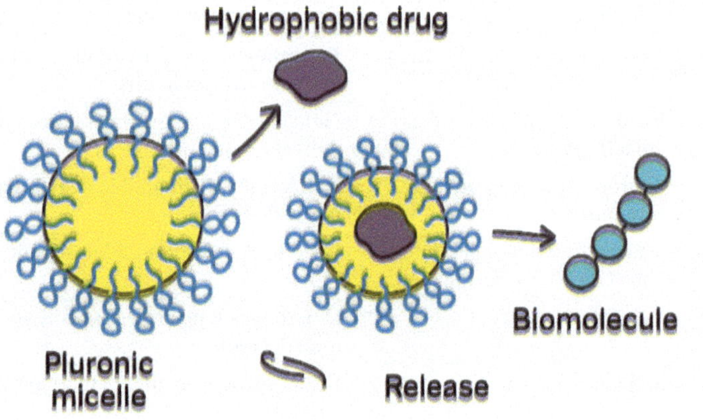

Fig. 4 Schematic representation of hydrophobic drug-pluronic assembly

elevated concentrations, Pluronics can generate more ordered structures like lamellar or hexagonal phases (Fig. 5) [15–20].

Phase diagrams can be used to map the phase behavior of Pluronic surfactants, delineating regions that correspond to various self-assembled structures. As an illustration, Pluronic P84 in water and p-xylene shows regions associated with normal micellar cubic, hexagonal, bicontinuous cubic, and lamellar phases, along with reverse phases like reverse micellar and bicontinuous cubic phases [21–23]. Due to their capability of encapsulating hydrophobic drugs, Pluronic micelles are appropriate for controlled drug delivery.

The core–shell structure enables the solubilization of hydrophobic agents within the core, while the hydrophilic shell promotes stability and biocompatibility. Pluronic surfactants act as templates for the synthesis of nanomaterials. Their self-assembled

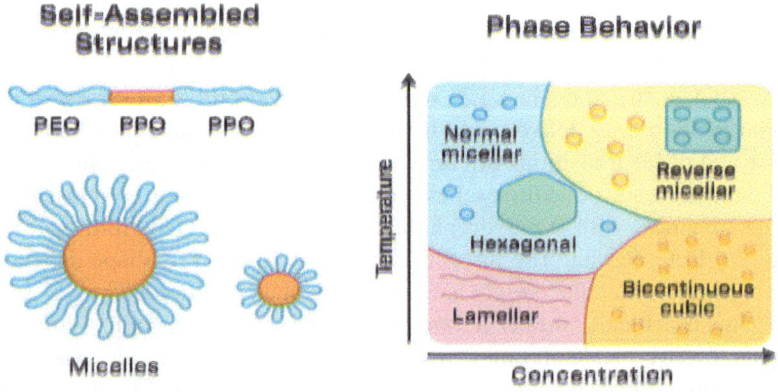

Fig. 5 Schematic representation of pluronic surfactant assembly and phase behavior

structures can direct the creation of nanoparticles, nanowires, and other nanostructures that are essential in areas such as catalysis and electronics. Due to their adjustable characteristics and compatibility with biological systems, hydrogels based on Pluronic are well suited for use as scaffolds in tissue engineering. Due to their capability of undergoing sol–gel transitions in reaction to temperature variations, they can be used in injectable formulations that solidify upon reaching body temperature. Pluronic surfactants demonstrate the capability of amphiphilic block copolymers to create self-assembled structures with a variety of morphologies.

Thus, the phase behavior understanding is essential for customizing the properties of pluronic for uses in drug delivery, nanotechnology, and biomedical engineering. Further investigation into the self-assembly mechanisms and phase transitions of Pluronics will augment the applicability of pluronics across a range of scientific and industrial fields [20–24] (Table 2).

Table 2 Commonly used pluronic surfactants and their uses

Full name (generic/trade)	Common name	Typical uses/applications	Refs.
Poly(ethylene oxide)-poly(propylene oxide)-poly(ethylene oxide) triblock copolymer	Pluronic F127/ poloxamer 407	Thermoresponsive hydrogel (in situ gelling), drug delivery/micelles, tissue engineering scaffolds, mucosal formulations, steric stabilization	[25]
Poly(ethylene oxide)-poly(propylene oxide)-poly(ethylene oxide)	Pluronic F68/ poloxamer 188	Shear-protectant excipient in cell culture/biologics, micellar drug carriers, stabilizer in formulations	[26]
Poly(ethylene oxide)-poly(propylene oxide)-poly(ethylene oxide)	Pluronic P123/ poloxamer 403	Micelle-forming surfactants for drug solubilization; soft-templating in mesoporous silica synthesis; drug delivery vehicles (micelles/vesicles)	[27]
Poly(ethylene oxide)-poly(propylene oxide)-poly(ethylene oxide)	Pluronic P85	Membrane/surface activity; used to enhance cellular uptake, chemo sensitization in MDR cancer research; micelle carriers in delivery	[28]
Poly(ethylene oxide)-poly(propylene oxide)-poly(ethylene oxide)	Pluronic L61	Hydrophobic, used as a biological sensitizer (e.g., affects drug distribution in MDR cells), component in mixed micelle formulations and formulation tuning	[29]
Poly(ethylene oxide)-poly(propylene oxide)-poly(ethylene oxide)	Pluronic P338/ poloxamer P338	Cell-culture additive/transduction enhancers and low-toxicity excipient in certain viral transduction and biologics protocols	[30]

5 Characterization Techniques

The self-assembly of pluronics requires characterization across multiple scales. Common tools encompass Small-Angle X-Ray Scattering (SAXS), Small-Angle Neutron Scattering (SANS), Dynamic Light Scattering (DLS), cryo-TEM, Rheology, and Differential Scanning Calorimetry (DSC) [3, 6, 7]. The integration of these methods allows for the construction of a detailed representation that spans from molecular packing to macroscopic rheological behavior.

6 Determination of Critical Micellization and Surface Activity

6.1 *Surface Tension/Tensiometry*

Surface tension measurements using techniques such as the Wilhelmy plate, Du Noüy ring, and pendant drop tensiometry are traditional methods for determining surface activity and estimating the critical micelle concentration (CMC) in surfactant solutions. This is accomplished by plotting surface tension against the logarithm of concentration and identifying the breakpoint. It is crucial to manage the baseline and temperature carefully for Pluronics, which typically exhibit gradual transitions due to the gradual dehydration of PPO blocks [5, 31–35].

6.2 *Fluorescence Probes (Pyrene, Nile Red)*

The fluorescence of hydrophobic probes (for instance, the pyrene I_1/I_3 ratio) is commonly employed to assess the polarity of the microenvironment and to ascertain the apparent CMC/CMT. Pyrene's partitioning into micellar cores results in spectral modifications that depend on probe concentration and solvent conditions; it is advisable to use a low probe concentration to prevent disturbance of self-assembly [36].

6.3 *Conductivity and Light Scattering Titrations*

Although it is less common to use conductivity for nonionic Pluronics, it can still indicate micellization when ionic additives are present. Static and dynamic light scattering (SLS/DLS) measurements as a function of concentration can identify the emergence of larger scattering objects at the CMC/CMT and provide estimates of hydrodynamic radius and weight-averaged molar mass [37].

6.4 Size, Polydispersity, and Morphology

6.4.1 Dynamic Light Scattering (DLS)

DLS serves as a standard technique for measuring the hydrodynamic size (R_h) and polydispersity index (PDI) of Pluronic micelles and larger aggregates. It is quick, little sample preparation is required. Limitations: high sensitivity to dust and aggregates of higher order; interpret R_h carefully for samples that are non-spherical or polydisperse. When researching temperature-dependent micellization or gelation, it is essential to control the temperature [38]. Carefully filter samples using suitable pore membranes (for instance, 0.45 μm followed by 0.22 μm if feasible), utilize multiple scattering angles if the instrument permits and verify DLS sizes with direct imaging techniques.

6.4.2 Static Light Scattering (SLS)/Multi-angle Light Scattering (MALS)

SLS (or MALS linked to SEC/AF4) offers the radius of gyration (R_g) and absolute molar mass (Mw). The ratio R_g/R_h provides information about the shape of particles (for example, it is approximately 0.775 for compact spheres). The combination of DLS and SLS forms a robust duo for estimating shape and aggregation numbers [37].

6.4.3 Transmission Electron Microscopy (TEM) and Cryo-TEM

After negative staining, conventional TEM can reveal micelle morphology; however, it carries the risk of drying artifacts and the collapse of soft micelles. The gold standard for directly visualizing micelle size, shape, and packing in gels is cryo-TEM, which preserves a near-native hydrated morphology. Although cryo-TEM entails a higher resource expenditure and demands meticulous vitrification, it yields crucial structural validation, particularly in cases where atypical morphologies are presumed [26, 38, 39].

6.4.4 Atomic Force Microscopy (AFM) and Scanning Probe Methods

AFM proves beneficial for surface-deposited Pluronic films and micellar layers. While it offers mechanical and topographical contrast, it may be affected by substrate interactions and drying, similar to TEM. Although liquid-cell AFM can minimize drying artifacts, it requires more technical expertise.

7 Structural Scattering Techniques

7.1 Small-Angle X-Ray Scattering (SAXS)

SAXS examines nanoscale structure and is highly effective for ascertaining core–shell micelle architecture, aggregation number, micelle radius, and micelle packing in concentrated systems or gels. When combined with complementary data [34, 38], model fitting (core–shell sphere, form and structure factors) produces robust geometric parameters. SAXS proves especially useful for examining Pluronic micelles that are loaded with oil or drugs [34].

7.2 Small-Angle Neutron Scattering (SANS)

SANS offers contrast variation (which is beneficial for D_2O/H_2O mixtures) and is effective for identifying PEO versus PPO domains, assessing hydration profiles, and clarifying core–shell contrast that might be unclear in SAXS. The most comprehensive depiction of internal micelle structure is offered by combined SAXS/SANS studies [34, 38].

Wide-angle X-ray Diffraction (WAXS/XRD) can be used on crystalline domains, such as ordered micelle lattices or crystallizable guest molecules. Pluronic solutions are usually amorphous, but at high concentrations, ordered phases can generate Bragg peaks that indicate packing symmetry.

8 Thermal and Rheological Characterization

8.1 Differential Scanning Calorimetry (DSC)

Thermal events related to micellization and gelation, such as endothermic peaks linked to PPO dehydration and aggregation, are detected using DSC. It produces transition temperatures (Tonset, Tpeak, Tendset) and enthalpies that assist in quantifying the energetics of micellization. For dilute solutions, Micro-DSC provides greater sensitivity [35].

8.2 *Thermogravimetric Analysis (TGA)*

TGA serves as a valuable tool for analyzing the stability and composition of dried samples or formulations (for instance, to measure water content or thermal degradation of additives).

8.3 *Rheology*

Gelation (storage modulus G', loss modulus G"), yield stress, thixotropy, and viscoelastic properties of Pluronic gels are characterized by oscillatory and steady shear rheology. Through temperature sweeps, one can identify the sol–gel transition temperatures and the microstructure that dictates macroscopic mechanics. The combination of rheology with SAXS/SANS during temperature ramps provides valuable insights into the structural basis of mechanical transitions [26, 35–39].

9 Spectroscopic and Chemical Characterization

9.1 *Fourier Transform Infrared Spectroscopy (FTIR)*

FTIR reports on the vibrational modes of PEO/PPO and its hydrogen-bonding interactions with solvents or additives. Changes in characteristic bands (e.g., C–O–C stretches) may suggest dehydration of PPO and alterations in chain conformation during micellization.

9.2 *Nuclear Magnetic Resonance (NMR)*

Proton NMR can investigate chain mobility and micelle core–shell exchange: signals from PPO protons in the core broaden or weaken during micellization, while PEO signals maintain relative mobility. DOSY NMR provides diffusion coefficients for various species and can distinguish free polymer from micelle-associated chains in favorable cases [40].

9.3 *UV–Vis and Fluorescence Spectroscopies*

When chromophoric drugs or probes are present, UV–Vis proves beneficial for quantified loading/release studies. Fluorescence techniques (both intrinsic and extrinsic probes) uncover details about microenvironment polarity, guest partitioning, and dynamic exchange kinetics (time-resolved fluorescence) [36].

10 Interfacial, Electrokinetic, and Binding Measurements

10.1 Zeta Potential and Electrophoretic Mobility

Pluronics are neutral nonionic polymers, but when they are combined with charged drugs, surfactants, or polyelectrolytes, zeta potential measurements provide insights into the complexation and stability of the resulting mixed systems.

10.2 Isothermal Titration Calorimetry (ITC)

ITC assesses thermal variations during the formation of micelles, the binding of polymer to drug, or the combination with additives. It offers binding stoichiometry, association constants, and thermodynamic parameters (ΔH, ΔS, ΔG). ITC holds special importance for assessing the thermodynamics of drug loading in poloxamer micelles [41].

10.3 Surface Plasmon Resonance (SPR) and Quartz Crystal Microbalance (QCM)

Adsorption of Pluronic molecules on surfaces (such as biomaterials) can be probed using SPR and QCM, allowing for mass uptake quantification and investigation of antifouling properties [42].

11 Challenges and Future Directions

Despite decades of study, many challenges remain in the self-assembly of Pluronic systems [1, 3, 7]. These include precise control over morphology transitions, influence of additives, kinetic trapping, and the need for responsive or "smart" assemblies. Recent critical reviews highlight the importance of correlating rheological behavior with microstructure and the effect of salt and blending strategies [1]. Pluronic surfactants are extensively utilized in pharmaceuticals, nanotechnology, cosmetics, and biotechnology because of their amphiphilic characteristics, compatibility with biological systems, and remarkable self-assembly properties. Even though they are versatile, a number of scientific, technological, and regulatory issues still restrict their wider potential and require concentrated focus in upcoming research. The intricate and extremely sensitive self-assembly behavior of Pluronics in water is one of the main difficulties. Micellization, cloud point, and gel formation can be significantly affected by factors like temperature, concentration, molecular weight, block length ratio, and the presence of electrolytes or co-solutes. Due to this sensitivity, it is challenging to attain consistent and reproducible performance in areas like drug delivery, where exact control of the critical micelle concentration (CMC) and critical micelle temperature (CMT) is crucial. Even slight changes to the formulation conditions can result in instability, early drug release, or diminished therapeutic efficacy, which represent a significant obstacle to clinical translation.

Another significant challenge is the restricted capacity for drug loading for certain hydrophobic and bulky molecules. Despite the ability of Pluronic micelles to encapsulate hydrophobic drugs within their PPO core, the low encapsulation efficiency and rapid dissociation that often occur when diluted in biological fluids can be attributed to the small volume of the core and the dynamic nature of the micelles. This may lead to burst release and diminished bioavailability of the therapeutic agent. Furthermore, contact with plasma proteins like albumin may result in the breakdown or modification of micelle structure, making their in vivo performance even more complex.

Although toxicity and biocompatibility are generally favorable, they still raise concerns, particularly with long-term or high-dose exposure. There have been reports indicating that certain grades of Pluronic can lead to oxidative stress, alterations in cellular signaling pathways, and membrane perturbation. Research is also being conducted into how they affect mitochondrial activity and efflux transporters like P-glycoprotein, as these effects can be helpful or harmful depending on the circumstances. Furthermore, Pluronics have a comparatively slow rate of biodegradation in physiological conditions. This may result in accumulation and unexpected biological effects, particularly in chronic applications.

The synthesis and purification of Pluronic surfactants are complex and energy-intensive from an industrial standpoint, leading to high production costs. Variability in molecular weight distribution and block composition from one batch to another can result in performance inconsistencies, posing challenges for large-scale production, particularly in regulated industries such as pharmaceuticals. Moreover, the ecological ramifications of these synthetic polymers—especially their durability in soil and water—are posing a new challenge with sustainability taking center stage in the field of materials science.

There are considerable prospects for addressing these challenges via molecular engineering and advanced formulation strategies in the future. Stability, drug-loading capacity, and environmental degradability could be improved by designing Pluronic derivatives for the next generation that have modified block lengths, incorporate biodegradable linkages, or feature functional end groups. By integrating Pluronics with other polymers, lipids, peptides, or inorganic nanoparticles, hybrid systems may enhance structural integrity and multifunctionality. This advancement could lead to more intelligent drug delivery platforms and responsive materials capable of adapting to stimuli like pH, temperature, light, or magnetic fields.

Another promising avenue is the incorporation of Pluronic surfactants into personalized medicine, allowing for the creation of customized formulations tailored to individual patient needs and specific disease profiles. It is anticipated that developments in computational modeling, molecular simulation, and machine learning will be instrumental in forecasting self-assembly behavior and refining formulations with enhanced accuracy, thereby diminishing the dependence on trial-and-error methods. Simultaneously, it will be essential to conduct thorough research into the molecular and cellular biological interactions of Pluronics in order to determine their long-term safety and gain regulatory approval.

References

1. Di Spirito NA, Grizzuti N, Pasquino R (2024) Self-assembly of Pluronics: a critical review and relevant applications. Phys. Fluids 36(11)
2. Kerkhofs S, Willhammar T, van den Noortgate H, Kirschhock CE, Breynaert E, van Tendeloo G, Bals S, Martens JA (2015) Self-assembly of Pluronic F127, silica spherical core-shell nanoparticles in cubic close-packed structures. Chem Mater 27(15):5161–5169
3. Song HM, Zink JI (2019) Ag (i)-mediated self-assembly of anisotropic rods and plates in the surfactant mixture of CTAB and Pluronics. RSC Adv 9(8):4380–4389
4. Prasanthan P, Kishore N (2021) Self-assemblies of Pluronic micelles in partitioning of anticancer drugs and effectiveness of this system towards target protein. RSC Adv 11(36):22057–22069
5. Naharros-Molinero A, Caballo-González MÁ, de la Mata FJ, García-Gallego S (2022) Direct and reverse Pluronic micelles: design and characterization of promising drug delivery nanosystems. Pharmaceutics 14(12):2628
6. Kumar SS, Harikrishnan KK, Urmila SP, Gauri V, Saritha A, Gangopadhyay M (2023) Comprehensive review of Pluronic® polymers of different shapes with prominent applications in photodynamic therapy. Eur Polym J 22(200):112534
7. Qutub M, Tatode A, Premchandani T, Taksande J, Mane D, Umekar M (2024) Blending induced variations in Poloxamer's/Pluronic's® gelation: thermodynamic and rheological perspectives. JCIS Open 1(16):100126
8. Patel D, Tripathi N, Ray D, Aswal VK, Kuperkar K, Bahadur P (2023) Self-assembly generation triggered in highly hydrophilic Pluronics® by sugars/polyols. J Mol Liq 15(378):121614
9. Park SI, Song HM (2021) Bottom-up self-assembly of nanofibers in the surfactant mixture of CTAB and Pluronics. AIP Adv 11(12)
10. Alexandridis P, Hatton TA (1995) Poly (ethylene oxide) poly (propylene oxide) poly (ethylene oxide) block copolymer surfactants in aqueous solutions and at interfaces: thermodynamics, structure, dynamics, and modeling. Colloids Surf A 96(1–2):1–46
11. Fripiat JJ, Gatineau L, Van Damme H (1986) Multilayer physical adsorption on fractal surfaces. Langmuir 2(5):562–567
12. Kobayashi M, Yoshioka T, Kozasa T, Tashiro K, Suzuki J, Funahashi S, Izumi Y (1994) Structure of physical gels formed in syndiotactic polystyrene/solvent systems studied by small-angle neutron scattering. Macromolecules 27(6):1349–1354
13. Kabanov AV, Batrakova EV, Alakhov VY (2002) Pluronic® block copolymers for overcoming drug resistance in cancer. Adv Drug Deliv Rev 54(5):759–779
14. Katz LE, Hayes KF (1995) Surface complexation modeling: I. strategy for modeling monomer complex formation at moderate surface coverage. J Colloid Interface Sci 170(2):477–490

15. Ghosh S, Ray A, Pramanik N (2020) Self-assembly of surfactants: an overview on general aspects of amphiphiles. Biophys Chem 1(265):106429
16. Munafò I, Costa D, Milano G, Munaò G (2025) Self-assembly of Pluronics L121 in water and added drugs with different hydrophobicity. Polymer 6:128871
17. Rosenholm JB (2014) Phase equilibriums, self-assembly and interactions in two-, three-and four medium-chain length component systems. Adv Colloid Interface Sci 1(205):9–47
18. Piculell L (2013) Understanding and exploiting the phase behavior of mixtures of oppositely charged polymers and surfactants in water. Langmuir 29(33):10313–10329
19. Karlstroem G, Carlsson A, Lindman B (1990) Phase diagrams of nonionic polymer-water systems: experimental and theoretical studies of the effects of surfactants and other cosolutes. J Phys Chem 94(12):5005–5015
20. Baccile N, Seyrig C, Poirier A, Alonso-de Castro S, Roelants SL, Abel S (2021) Self-assembly, interfacial properties, interactions with macromolecules and molecular modelling and simulation of microbial bio-based amphiphiles (biosurfactants). A tutorial review. Green Chem 23(11):3842–3944
21. Hammouda B (2010) SANS from Pluronic P85 in d-water. Eur Polym J 46(12):2275–2281
22. Jang JD, Yoon YJ, Jeon SW, Han YS, Kim TH (2021) Molecular weight-dependent, flexible phase behaviors of amphiphilic block copolymer/additive complexes in aqueous solution. Polymers 13(2):178
23. Atkin R, De Fina LM, Kiederling U, Warr GG (2009) Structure and self-assembly of Pluronic amphiphiles in ethylammonium nitrate and at the silica surface. J Phys Chem B 113(36):12201–12213
24. Yeredla N, Kojima T, Yang Y, Takayama S, Kanapathipillai M (2016) Aqueous two-phase system assisted self-assembled PLGA microparticles. Sci Rep 6(1):27736
25. Pitto-Barry A, Barry NP (2014) Pluronic® block-copolymers in medicine: from chemical and biological versatility to rationalisation and clinical advances. Polym Chem 5(10):3291–3297
26. Di Spirito NA, Grizzuti N, Lutz-Bueno V, Urciuoli G, Auriemma F, Pasquino R (2024) Pluronic F68 micelles as carriers for an anti-inflammatory drug: a rheological and scattering investigation. Langmuir 40(2):1544–1554
27. Tan G, Wang Y, He Y, Miao G, Li Y, Wang X (2022) Bioinspired poly (cation-π) micelles drug delivery platform for improving chemotherapy efficacy. J Control Release 1(349):486–501
28. Goliaei A, Lau EY, Adhikari U, Schwegler E, Berkowitz ML (2016) Behavior of P85 and P188 poloxamer molecules: computer simulations using united-atom force-field. J Phys Chem B 120(33):8631–8641
29. Venne A, Li S, Mandeville R, Kabanov A, Alakhov V (1996) Hypersensitizing effect of pluronic L61 on cytotoxic activity, transport, and subcellular distribution of doxorubicin in multiple drug-resistant cells. Can Res 56(16):3626–3629
30. Peña Corona SI, Borbolla-Jiménez FV, Duarte-Peña L, Moreno A, Pérez-Caltzontzin LE, Del Prado-Audelo ML, Romero-Montero A, González-Torres M, Cortés H, Hernández-Parra H, Sharifi-Rad J (2025) Repurposing poloxamers as antimicrobial agents: a comprehensive review of mechanisms and applications. Drug Dev Res 86(5):e70130
31. Kabanov AV, Batrakova EV, Alakhov VY (2002) Pluronic® block copolymers as novel polymer therapeutics for drug and gene delivery. J Control Release 82(2–3):189–212
32. Constantinou AP, Nele V, Doutch JJ, S. Correia J, Moiseev RV, Cihova M, Gaboriau DC, Krell J, Khutoryanskiy VV, Stevens MM, Georgiou TK (2022) Investigation of the thermogelation of a promising biocompatible ABC triblock terpolymer and its comparison with pluronic F127. Macromolecules 55(5):1783–1799
33. Bodratti AM, Alexandridis P (2018) Formulation of poloxamers for drug delivery. J Funct Biomater 9(1):11
34. Grillo I, Morfin I, Prévost S (2018) Structural characterization of Pluronic micelles swollen with perfume molecules. Langmuir 34(44):13395–13408
35. Shriky B, Kelly A, Isreb M, Babenko M, Mahmoudi N, Rogers S, Shebanova O, Snow T, Gough T (2020) Pluronic F127 thermosensitive injectable smart hydrogels for controlled drug delivery system development. J Colloid Interface Sci 1(565):119–130

36. Perumal S, Atchudan R, Lee W (2022) A review of polymeric micelles and their applications. Polymers 14(12):2510
37. Ghezzi M, Pescina S, Padula C, Santi P, del Favero E, Cantù L, Nicoli S (2021) Polymeric micelles in drug delivery: an insight of the techniques for their characterization and assessment in biorelevant conditions. J Control Release 10(332):312–336
38. da Silva LC, Borges AC, de Oliveira MG, de Farias MA (2020) Visualization of supramolecular structure of Pluronic F127 micellar hydrogels using cryo-TEM. MethodsX 1(7):101084
39. Mortensen K, Talmon Y (1995) Cryo-TEM and SANS microstructural study of Pluronic polymer solutions. Macromolecules 28(26):8829–8834
40. Vinarov Z, Katev V, Burdzhiev N, Tcholakova S, Denkov N (2018) Effect of surfactant–bile interactions on the solubility of hydrophobic drugs in biorelevant dissolution media. Mol Pharm 15(12):5741–5753
41. Wu G, Lee KY (2009) Interaction of poloxamers with liposomes: an isothermal titration calorimetry study. J Phys Chem B 113(47):15522–15531
42. Santander-Ortega MJ, Jódar-Reyes AB, Csaba N, Bastos-González D, Ortega-Vinuesa JL (2006) Colloidal stability of pluronic F68-coated PLGA nanoparticles: a variety of stabilisation mechanisms. J Colloid Interface Sci 302(2):522–529

An Overview of the Current State of Ionic Liquid-Based Surfactants for Biological Applications

Devanshu Pathak and Sumit S. Bhawal

Abstract Ionic liquid-based surfactants (ILBSs) represent a new class of hybrid amphiphilic systems that integrate the physicochemical properties of ionic liquids (ILs)—such as high ionic conductivity, negligible vapor pressure, thermal stability, and structural tunability—with the self-assembly behavior of classical surfactants. By rational molecular design of cationic (imidazolium, pyridinium, ammonium, phosphonium), anionic, or zwitterionic headgroups coupled with hydrophobic chains or functional tails, ILBS can be engineered to display tailored interfacial activity, aggregation behavior, and biological interactions. Their ability to form micelles, vesicles, liquid-crystalline phases, and nanoarchitectures enables efficient encapsulation, solubilization, and controlled release of hydrophobic and hydrophilic biomolecules. Recent advances demonstrate that ILBS can offer enhanced biocompatibility, intrinsic antimicrobial and membrane-active properties, improved DNA/RNA complexation efficiency, and enhanced enzyme stabilization under stress environments. These multifunctional molecular assemblies are increasingly explored for applications in drug and gene delivery, gene transfection, protein refolding, tissue engineering, biosensing, biocatalysis, and bio-separation systems. Moreover, tunable physicochemical parameters such as alkyl chain length, headgroup charge density, hydrogen-bonding capability, and counter-ion identity play critical roles in governing cytotoxicity, bio-membrane interactions, biodegradability, and functional performance. Despite promising biological activity and self-assembly versatility, key challenges remain in controlling long-term toxicity, ensuring biodegradability, improving colloidal stability in physiological media, and achieving scalable, cost-efficient synthesis. Structure–property–bioactivity relationships, guided by computational modeling, molecular dynamics simulations, and high-throughput design strategies, are expected to accelerate the development of next-generation ILBS. Future research efforts will focus on designing stimulus-responsive, task-specific ILBS with precision biocompatibility, minimal environmental impact, and tailored functionalities to advance their translation into biomedicine and biotechnology.

D. Pathak · S. S. Bhawal (✉)
School of Science, Navrachana University, Vasna-Bhayli Road, Vadodara, Gujarat 391410, India
e-mail: sumitb@nuv.ac.in

M. Abdul Rub et al. (eds.), *Industrial Applications of Individual and Mixed Surfactant Systems—Volume 2*, Sustainable Materials and Technology,
https://doi.org/10.1007/978-981-95-9195-4_13

Graphical Abstract

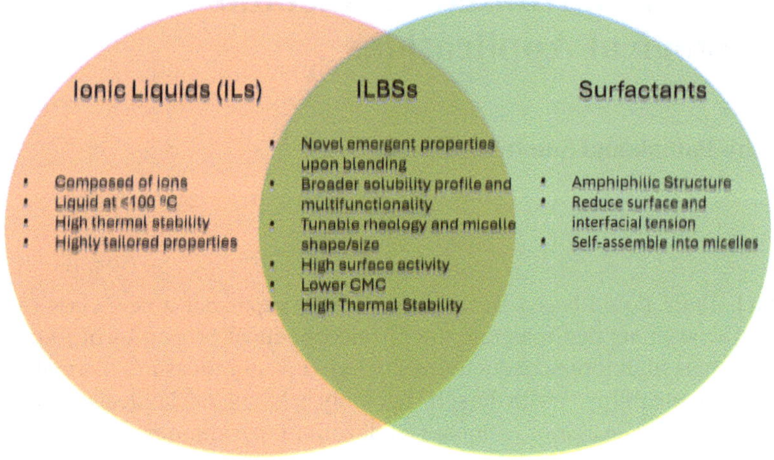

Keywords Ionic liquids · Surface-active ionic liquids · Biocompatibility · Structure–function correlation · Tuneability

1 Introduction

Ionic liquid-based surfactants (ILBSs), also known as surface-active ionic liquids (SAILs), are an emergent new family of functional materials that bridge/combine the distinctive properties of ionic liquids (ILs) with the well-established functionality of conventional surfactants [1, 2]. Briefly, the ILBS structure is an amphiphilic molecule with a hydrophilic ionic headgroup and hydrophobic alkyl tail [3]. The reason the field of ILBS differs from that of classic surfactants is the vast chemical composition tunability, enabling systematic molecular structure design with precisely engineered properties [4, 5]. The modular structure of the ILBS, whereby the cation and the anion can be independently tuned, allows unprecedented freedom in the design of desired material characteristics like critical micelle concentration (CMC), surface activity, solubility, thermal stability, and crucially, biocompatibility [6, 7].

An important design element in the development of ILBSs for biological applications is the choice of cation–anion pairs to balance surface activity with low toxicity [8, 9]. Commonly employed cationic moieties are the imidazolium, pyridinium, ammonium, and phosphonium units, while the counter-anions range from the simplest halide and tetrafluoroborate to the bio-based fatty acid derivatives [10–13]. The hydrophobic tail is typically of the linear or branched alkyl chains (C8–C18), but including renewable or natural building blocks will enhance the sustainability profile of the final surfactant [14]. The choice of the ion pair is crucial: While the hydrophobicity dictates the micellization and surface activity, the headgroup and counter-anion

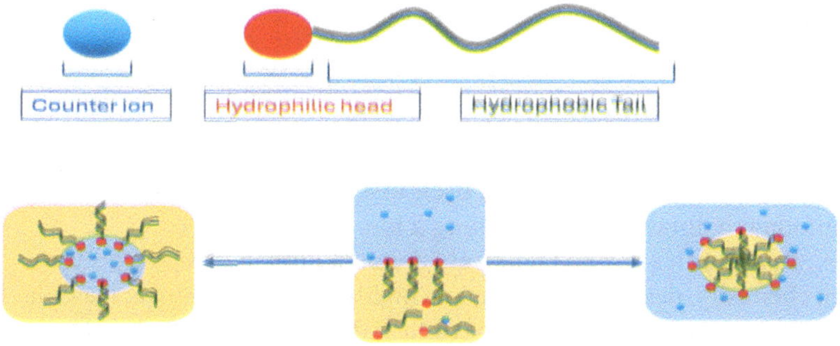

Fig. 1 Schematic of micellization and aggregation pathways for ionic liquid surfactants

strongly influence interactions with biological membranes, cell uptake, and overall biocompatibility [15].

The ILBS inherently displays low CMC values and broader interfacial activity windows compared with conventional surfactants because of stronger ionic interaction and unique packing and aggregation behavior, allowing effective surface modification and self-assembly at low concentrations under physiologically relevant conditions [16] (Fig. 1). This is particularly advantageous in biomedical applications, in which decreased dosage reduces cytotoxicity [17]. The ability of the ILBS to self-organize into nanoscale assemblies such as micelles, vesicles, and lamellar phases enables access to unique nanostructures like micro-segregated domains, bicontinuous, and super-aggregate structures. Thus, their modular ionic architecture and highly tunable aggregation behavior enable diverse applications in drug delivery, antimicrobial drugs, protein stabilization, and the creation of new biomaterials [18, 19].

However, despite their growing interest, the large-scale use of ILBS in biological applications is impeded by concerns related to cost and potential toxicity, the latter being stimulated largely because of the chemical nature of the select pair [20, 21]. Rational selection and design of low-toxicity, biocompatible ions based on renewable or natural origins, therefore, represent the primary challenge and opportunity in this field [22]. Through a focus on the systematic development of ion pairs that match surface activity with favorable biological profiles, this chapter aims to provide an in-depth overview of the synthesis of ILBS, their physicochemical characterization, as well as the latest advances in their applications in biology and biotechnology. By leveraging molecular modeling (which includes molecular dynamics (MD) simulation, quantum chemical calculations, and docking studies) as a powerful tool for designing and optimizing ILBSs, the development process becomes more efficient, predictive, and tailored to specific applications, thereby accelerating innovation in this rapidly evolving field.

2 Ionic Liquid-Based Surfactants

2.1 Historical Context

The idea of ionic liquids (ILs) dates back to 1914, when Paul Walden synthesized ethylammonium nitrate, a salt exhibiting a low melting point of 12 °C [23, 24]. For decades, IL research remained relatively dormant until there was a growing interest in green chemistry, designer solvents, and highly tunable functional materials fueled a surge in the late-twentieth and early-twenty-first centuries [25–27]. The discovery that long-chain ILs (Fig. 2), particularly imidazolium-based salts, exhibit surface-active properties in aqueous solutions led to the emergence of ionic liquid-based surfactants (ILBSs) around the early 2000s [28, 29].

ILBS combines the modular structure of ILs (i.e., independently tunable cation and anion) with amphiphilic characteristics, providing an exceptional opportunity to systematically modulate their physicochemical and biological properties [30–34]. From the 2020s onward, increased environmental awareness and regulatory pressure have driven efforts toward the development of bio-derived and biodegradable ILBS, with key attention directed toward tailoring the chemical structure of ion pairs to optimize surface activity, biocompatibility, and multifunctionality [35, 36].

The modular characteristics of the ILs (i.e., independently adjustable cation and anion) are combined with amphiphilic behavior in the ILBS, which provides an exceptional opportunity of systematic variation in the physicochemical and biological properties [37, 38]. Since the 2020s, environmental issues and regulating pressure

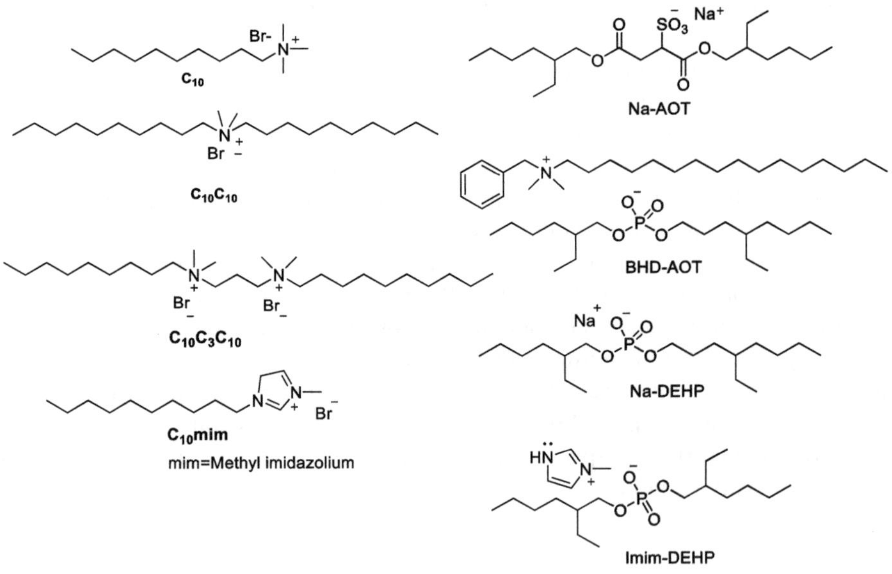

Fig. 2 Some common ILBSs [40–45]

have motivated the development in the direction of bio-derived and biodegradable ILBS, with core attention being given to tailoring the chemical structure of ion pairs to optimize surface activity, biocompatibility, and multifunctionality [39].

2.2 Classification of ILBS

ILBS is classified based on their molecular structure and functionality.

- **Cationic ILBS**: Cationic ILBS has so far been the most researched type, typically comprising imidazolium, pyridinium, ammonium, or phosphonium cations paired with halide, tetrafluoroborate, or organic counter-ions. The amphiphilic design includes a hydrophilic ionic headgroup and a hydrophobic alkyl chain (C8–C18) [46].
  - **Structure—Property Relationship**: Increasing the alkyl chain length systematically lowers the critical micelle concentration (CMC) and enhances interfacial activity due to increased hydrophobic interactions driving micelle formation [30].
  - A significant development includes amino acid-based cationic headgroups (e.g., histidine-derived), enhancing biodegradability and cytocompatibility without compromising on self-assembly behavior or antimicrobial function [47]. These modifications also enable efficient endosomal escape and cargo release, demonstrating how subtle chemical modifications at the headgroup level have a direct impact on biological function.
- **Anionic ILBS**: Anionic ILBS is constructed by pairing cations like imidazolium with anionic headgroups like dodecyl sulfate or carboxylates [48].
  - **Structure–Property Relationship**: The balance of hydrophilic headgroup size versus alkyl tail length dictates aqueous solubility and aggregation behavior. For example, $[C_4MIm][C_{12}SO_4]$ showed remarkable performance as a pseudo-stationary phase in micellar electrokinetic chromatography (PubMed, 2019), owing to favorable electrostatic interactions and micelle stability [28, 49].
  - The limited exploration of anionic ILBS creates a crucial gap in the current knowledge of correlating ion pair selection with biocompatibility and specific surface activity profiles relevant to drug delivery or protein stabilization.
- **Gemini ILBS**: Gemini ILBS consists of two amphiphilic units joined with a spacer, giving rise to enhanced interfacial properties [50].
  - **Structure–Property Relationship**: The presence of two head–tail groups decreases the CMC considerably and accelerates surface adsorption kinetics. Spacer length and flexibility significantly influence micellar geometry and interfacial packing density [51].
  - Applications such as Cu_2O nanoparticle morphology and enhanced oil recovery reflect how tailored ion pair combinations affect particle templating

and interfacial tension reduction. Rational pair design is possible with Gemini so that they can offer high-end functionalities under biological conditions, while adjustable cytotoxicity is maintained [52, 53].
  - These examples highlight how Gemini ILBS acts not only as surfactants but as functional modulators of interfacial phenomena, a property that traditional surfactants rarely achieve.

- **Task-Specific ILBS**: Task-specific ILBS comprises stimuli-responsive ion pairs that react to pH, redox conditions, light, or temperature [54, 55].
  - **Structure–Property Relationship**: Functional groups added to the cation or anion offer switchable hydrophilicity or micellization behavior, enabling on-demand assembly and disassembly [56, 57].
  - For example, the use of ILs as orientational solvents for low-temperature desalination (Durling et. al.) inspires the use of ILBS in precision medicine, whereby the correct choice of the ion pair would modulate micellar stability and cargo release profiles in response to specific biological stimuli[58, 59].
- **Bio-Based ILBS**: Bio-based ILBS contains naturally derived cations or anions such as amino acids, sugars, or fatty acids [60].
  - **Structure–Property Relationship**: The choice of bio-derived ions will result in lower toxicity and improved biodegradability while maintaining amphiphilic self-assembly [61, 62].
  - Despite the previous evidence of high antimicrobial activity and biodegradability, much work remains to correlate some molecular features (e.g., hydrophilic headgroup size, hydrophobic tail length, functional groups) to both biological activity and surface performance.
- **Magnetic ILBS**: Magnetic ILBS features magnetic counter-ions (e.g., Fe^{3+}-based anions) that offer magneto-responsive behavior [63].
  - **Structure–Property Relationship**: The magnetic functionality depends on the incorporation of the paramagnetic cations and the structural rigidity of the ionic liquid backbone, which together influence micellar stability and responsiveness, whose combination dictates the micellar stability and response in the presence of an external magnetic field [64].
  - The challenge remains to engineer ion pairs that provide both strong magnetic properties and high biocompatibility, enabling magnetically targeted drug delivery or separations.
- **Polymeric ILBS**: Polymeric ILBS grafts IL moieties onto polymer backbones or side chains [65].
  - **Structure–Property Relationship**: Polymer chain length and density of ionic groups affect micelle size, mechanical stability, and release kinetics [66, 67].
  - These systems offer long-term stability, but ion pair selection is critical so as to prevent cytotoxicity while preserving surface activity and stimuli-responsiveness.

- **Zwitterionic ILBS**: Zwitterionic ILBS contains both positive and negative charges within the same molecular structure [68].
  - **Structure–Property Relationship**: The internal charge balance causes exceptional salt tolerance, enzyme stabilization, and biocompatibility [69].
  - A 2024 study demonstrated a zwitterionic Gemini ILBS forming solid-like monolayers at the oil–water interface, enabling ultra-stable emulsions and 3D liquid patterning [70].
  - Further development entails thorough research on zwitterionic ion pairs that synergistically promote stability, low toxicity, and structural functionality in biological applications (Table 1).

3 Synthesis of ILBS

The synthesis of ILBS is typically a two-step process: (i) quaternization of amines or phosphines by alkyl halides or alkyl sulfates (commonly through SN_2 mechanism, e.g., Menshutkin reaction), followed by (ii) anion exchange, if necessary, to yield the desired ILBS salt. The order of reactivity of alkyl halides generally follows: RI > RBr > RCl, while the reaction rate decreases with increasing alkyl chain length [75] (Scheme 1).

3.1 General Synthetic Approaches

- **Quaternization (Cation Formation)**

The cationic headgroup is attached via N-alkylation of nitrogen or phosphorous-containing precursors such as imidazoles, pyridines, trialkyls or aryl phosphines, or tertiary amines using long-chain alkyl halides.

Example (Scheme 2):

Typically, the yields are high (85–90%), and the reaction is carried out in polar aprotic solvents (e.g., acetonitrile, THF, DMF, etc.) under reflux.

- **Anion Exchange (Metathesis)**

Metathesis is applied by reacting the halide ILBS intermediate with a salt of the target anion (e.g., $NaBF_4$, KPF_6, sodium carboxylate) when halide counter-ions need to be different. The precipitation of the insoluble and stable inorganic salt (NaBr, KBr) is the thermodynamic driving force for the process, which enables efficient exchange.

Example (Schemes 3 and 4):

- **Green Synthetic Strategies**

Emerging approaches prioritize atom economy, sustainability, and reduced toxicity:

Table 1 Structural classes of ILBS and their physicochemical parameters [71–76]

Class	Type	Surfactant name/abbreviation	Cationic structure	Counterion	CMC (mM)	γCMC (mN/m)
Imidazolium	Monocatenary	1-Decyl-3-methylimidazolium bromide	$[C10mim]^+$	Br^-	66.8	38.8
	Monocatenary	1-Dodecyl-3-methylimidazolium bromide	$[C12mim]^+$	Br^-	15.6	36.3
	Monocatenary	1-Tetradecyl-3-methylimidazolium bromide	$[C14mim]^+$	Br^-	3.8	34.1
	Monocatenary	1-Hexadecyl-3-methylimidazolium bromide	$[C16mim]^+$	Br^-	0.9	32.5
	Monocatenary	1-Dodecyl-3-methylimidazolium chloride	$[C12mim]^+$	Cl^-	17.5	39
	Monocatenary	1-Tetradecyl-3-methylimidazolium chloride	$[C14mim]^+$	Cl^-	4.6	36.5
	Monocatenary	1-Hexadecyl-3-methylimidazolium chloride	$[C16mim]^+$	Cl^-	1.2	34
Pyridinium	Monocatenary	N-Dodecylpyridinium chloride	$[C12py]^+$	Cl^-	16	38.5
	Monocatenary	N-Tetradecylpyridinium chloride	$[C14py]^+$	Cl^-	4	36
	Monocatenary	N-Hexadecylpyridinium chloride	$[C16py]^+$	Cl^-	0.9	33.5
	Monocatenary	N-Dodecylpyridinium bromide	$[C12py]^+$	Br^-	14.8	37.9
Cholinium	Monocatenary	Choline laurate	$[(CH_3)_3N^+CH_2CH_2OH]$	Laurate $[C_{11}H_{23}COO]^-$	8.5	29.1
	Monocatenary	Choline oleate	$[(CH_3)_3N^+CH_2CH_2OH]$	Oleate $[C_{17}H_{33}COO]^-$	1.2	31.5
Amino acid	Monocatenary (Anionic)	Sodium N-lauroylsarcosinate	Na^+	$[C_{11}H_{23}CON(CH_3)CH_2COO]^-$	14.6	28
	Monocatenary (Anionic)	Sodium cocoyl glycinate	Na^+	$[Cocoyl\text{-}NHCH_2COO]^-$	4.3	29.5
Gemini (quaternary ammonium)	Dicatenary	1,2-Bis(dodecyl-dimethyl-ammonium)ethane dibromide (12-2-12)	$[(C_{12}H_{25}(CH_3)_2N^+)^-(CH_2)_2{}^-(N^+(CH_3)_2C_{12}H_{25})]$	$2Br^-$	0.98	37
	Dicatenary	1,3-Bis(dodecyl-dimethyl-ammonium)propane dibromide (12-3-12)	$[(C_{12}H_{25}(CH_3)_2N^+)^-(CH_2)_3{}^-(N^+(CH_3)_2C_{12}H_{25})]$	$2Br^-$	0.48	35.5
	Dicatenary	1,4-Bis(dodecyl-dimethyl-ammonium)butane dibromide (12-4-12)	$[(C_{12}H_{25}(CH_3)_2N^+)^-(CH_2)_4{}^-(N^+(CH_3)_2C_{12}H_{25})]$	$2Br^-$	0.23	34

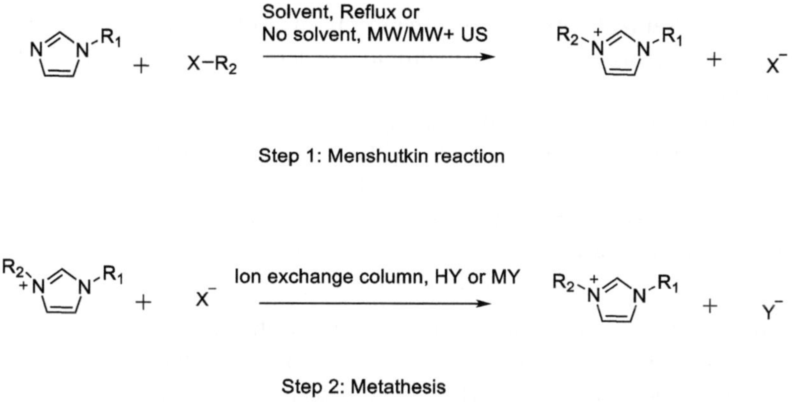

Scheme 1 General synthetic route of imidazolium ionic liquids via Menshutkin reaction and metathesis

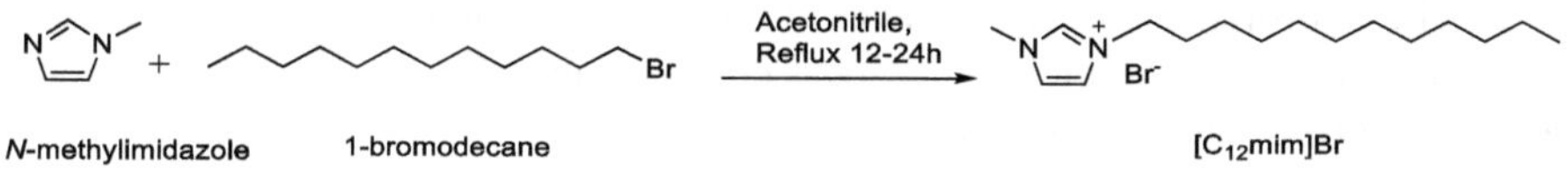

Scheme 2 Synthesis of 1-dodecyl-3-methylimidazolium bromide ([C_{12}mim][Br]) via quaternization reaction [76]

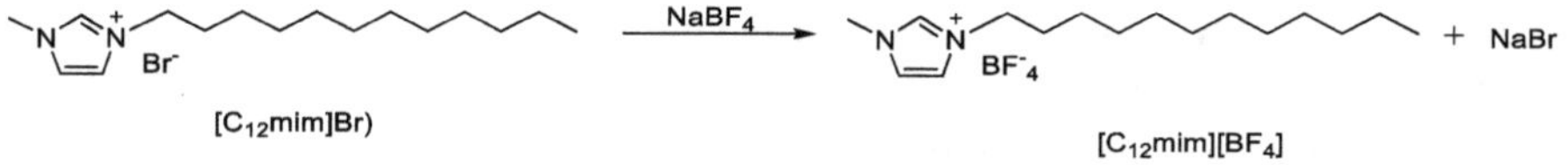

Scheme 3 Preparation of [C_{12}mim][BF_4] via anion metathesis with $NaBF_4$

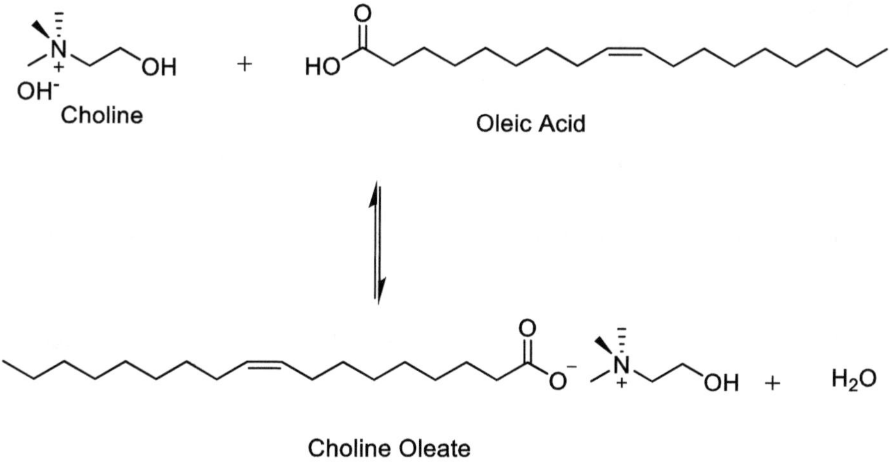

Scheme 4 Choline oleate synthesis: Acid–base reaction between choline hydroxide and oleic acid. Redrawn from Campos [77]

- **Acid–Base Neutralization**: Proton transfer from bio-based acids (e.g., fatty acids) to basic precursors (e.g., choline hydroxide) generates ILBS and water as the only byproduct:
- **Mechanochemistry (Solvent-Free Synthesis)**: Ball-milling process enables solid-state quaternization reactions without solvents, allowing for minimized waste and energy use with similar yields [78, 79].
- **Microwave-Assisted Synthesis**: The kinetics of reactions reduces the time of reactions and increases yields without relying on bulk solvents [80–82].
- Bio-Derived and Functional ILBS Design

Innovative synthesis increasingly integrates renewable feedstocks (e.g., choline, carbohydrates, amino acids, fatty acids) to yield **biodegradable, non-toxic ILBS** with inherent functionality [83, 84]. Structural design incorporates:

- **Circular Economy Concept**: ILBS engineered to degrade progressively under physiological or environmental conditions (through systematic introduction of inbuilt hydrolyzable ester linkages or pH-labile moieties), allowing for total molecular disintegration into harmless components [85–87].
- Ionic Gelation and Polymer-Based Approaches

Alternative methods include the synthesis of ILBS through **ionic gelation** or polymer-supported ionic liquids, where IL moieties functionalized polymer backbones are employed, such that hierarchically formed nanostructure and extended stability are realized. Such a system yields material having controllable viscosity, controlled release kinetics, and superior biocompatibility [88–90].

Synthetic routes for some more ILBSs are given in Schemes 5 and 6.

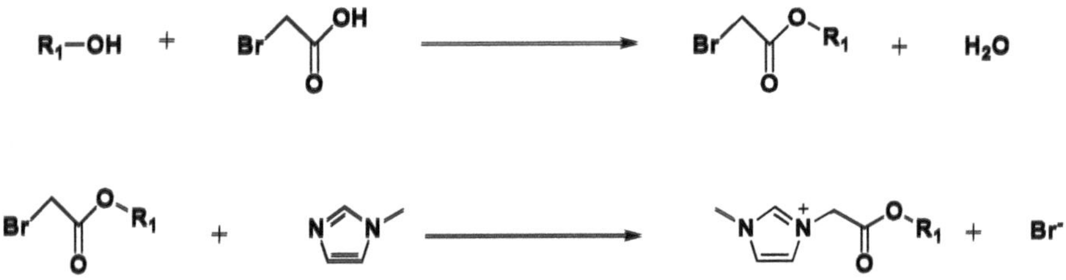

Scheme 5 Synthetic route for ILBSs via ester/amide side chain. Redrawn from Kanjilal [91]

Scheme 6 Synthetic route of Gemini ILBSs via dihaloalkanes. Redrawn from Baltazar [92]

3.2 Synthesis Challenges

ILBS preparation involves inherent challenges from molecular tunability balance to practical necessities for scalable, green, and reproducible synthesis [93]. While classical synthetic routes involve a two-step quaternization followed by anion exchange (metathesis), offering structural flexibility to a certain extent, they are prone to persistent issues that hinder practical implementation [94].

One of the primary challenges is impurity contamination present in the quaternization and metathesis steps. Such resultant halides are likely to significantly impact the surface activity, interfacial properties, and physicochemical stability of ILBS, even in trace amounts, due to their use in pharmacy and biology, where purity is a requirement. Removal of such impurities typically requires laborious purification steps such as extensive washing, ion-exchange resins, or chromatography, which increase cost, energy consumption, and environmental footprint [95–97].

Additionally, the drive for **bio-based feedstocks** to achieve biodegradability, low toxicity, and sustainable sourcing adds complexity in synthesis. The natural starting materials, like fatty acids, amino acids, and derivatives of choline, inherently exist as mixtures with different chain lengths and functional groups. Such a feature makes reproducibility and molecular designs difficult, hence presenting a significant bottleneck for scale-up on an industrial scale [98].

To address these challenges, recent literature highlights several promising solutions:

- **Process Intensification**: Continuous-flow reactors are utilized for enhancing reaction kinetics, heat, and mass transfer and for process control in the steps of quaternization because their small channels and high-surface area-to-volume ratio enable faster mixing and more efficient temperature control. Such processes reduce byproduct formation, lower the duration of reactions, and enhance yield consistency [99].
- **Advanced Purification Technologies**: Scalable liquid–liquid wash technologies and robotic adsorption columns have been introduced for high-performance removal of halide impurities and other unwanted byproducts. These are applied for the high-purity production of ILBS for use in biology [100].
- **Standardization of Bio-feedstocks**: Agro-industrial supplier strategic collaboration facilitates acquisition of higher-quality, standardized renewable precursors. This mitigates feedstock variability, ensuring reproducible performance of the resulting ILBS across batches [100, 101].
- **Mechanochemical and Solvent-Free Approaches**: To enhance sustainability further and reduce processing complexity, mechanochemistry through ball milling eliminates the use of solvents completely and minimizes energy input, while delivering similar yields to classic methods.

4 Physicochemical Properties of ILBS

4.1 Critical Micelle Concentration (CMC)

Ionic liquid-based surfactants (ILBSs) possess very low critical micelle concentrations (CMCs), typically 10–100 times lower than conventional surfactants, often in the range 0.01–1 mM. This enhanced effectiveness is due to the cumulative effect of the strong electrostatic interactions among the ionic headgroups and the large hydrophobic contribution by long alkyl chains that cause self-assembly at very small concentration. For instance, the CMC for 1-hexadecyl-3-methylimidazolium bromide ($[C_{16}mim]Br$) is approximately 0.1 mM, far lower than the 8.2 mM achieved for sodium dodecyl sulfate (SDS) at comparable conditions. This property makes ILBS very effective in systems that seek low loadings of surfactants [102, 103].

4.2 Thermal and Chemical Stability

ILBS demonstrates remarkable thermal stability (> 300 °C) and is active in high-salt systems (> 5 M NaCl), where standard surfactants such as SDS precipitate or degrade. This behavior results from the inherent stability of the ionic framework and the strong solvation of charges in polar media. Such stability is critical for applications in extreme conditions, such as the enhanced recovery of oil and high-temperature catalysis, and in biological systems where stability at a range of pH and ionic strength is critical [104]. The ability of ILBS to retain interfacial activity under extreme conditions distinguishes them from conventional surfactants.

4.3 Biocompatibility and Degradability

Bio-derived ILBS and particularly those synthesized through natural precursors such as amino acids, organic acids, and choline possess noticeably lower cytotoxicities than their synthetic counterparts [87]. For example, cholinium-amino acid ILBS possesses low-toxicity profiles (LC_{50} > 100 μM in mammalian cell lines) and is promising candidates for drugs, delivery agents, and other biomedical applications. [105, 106]. These bio-based ILBS has features like hydrolyzable ester bonds or pH-labile functionalities, enabling controlled degradation into non-toxic byproducts. This molecular design promotes sustainable applications in drug delivery systems, antimicrobial formulations, and protein stabilization platforms, while addressing regulatory and environmental imperatives. Such biocompatibility and the potential to be degradable make the bio-based ILBS sustainable counterparts in delicate applications.

4.4 Emerging Insights

Recent studies highlight the possibility of further ILBS property adjustment via controlled cation–anion pairing to enable systematic adjustment of interfacial tension, micellar shape, and degradation profiles [107, 108].

Moreover, high-throughput screening and sophisticated computational modeling approaches today assist in predicting structure-related physicochemical behavior, accelerating the determination of the functionally optimal ILBS lead compounds with specifically engineered biocompatibility and functionality (Table 2).

Table 2 Comparison of physicochemical properties of conventional surfactants and SAIL [76, 105, 112]

Property	Conventional surfactants	Surface-active ionic liquids (SAILs)
Thermal stability	Generally low; prone to decomposition at elevated temperatures	High; often stable up to 300–400 °C, inheriting stability from the IL core
Chemical stability	Variable; can be sensitive to extreme pH or high salinity, leading to hydrolysis or precipitation	High; robust across a wide range of pH and salinity conditions
Volatility	Generally non-volatile, but some smaller components can be semi-volatile	Negligible vapor pressure; non-volatile by design
Tunability	Limited; primarily through modification of the hydrophobic tail and headgroup type	Extremely high; properties can be fine-tuned by combinatorial variation of cation, anion, and substituents
Critical micelle concentration (CMC)	Varies widely, but often higher for a given alkyl chain length	Characteristically lower than conventional counterparts, indicating higher efficiency
Performance in harsh conditions	Often poor; efficacy is reduced by high temperature, pressure, and salinity	Excellent; maintain performance under harsh conditions where conventional surfactants fail
Potential for biodegradability/ toxicity	Varies; many are derived from petrochemicals and can be toxic and persistent (e.g., cationic biocides)	A key area of design; bio-based SAILs and those with cleavable linkages are being developed for improved biodegradability and lower toxicity

5 Biological Applications of ILBS

5.1 *Drug Delivery*

ILBS represents a significant change in drug delivery by combining solubilization, permeation improvement, stability, and response to stimuli into one platform. Their very-low CMC (e.g., ~ 0.1 mM compared to ~ 8 mM for SDS) allows for efficient nanoscale self-assembly into micelles, vesicles, or microemulsions that can dissolve poorly water-soluble APIs (e.g., paclitaxel) through hydrophobic partitioning, hydrogen bonding, and ion–dipole interactions [110]. Importantly, ILBS (e.g., imidazolium or pyridinium derivatives) improves drug permeability by fluidizing epithelial membranes and adjusting tight junction proteins. Structural tunability enables tumor-specific, pH- or enzyme-triggered breakdown, which reduces off-target toxicity [111, 112]. The shift toward bio-derived ILBS (e.g., choline-fatty acid systems) overcomes legacy toxicity challenges, advancing them from passive excipients to active, smart delivery agents for biologics and chemotherapeutics.

5.2 *Ionic Liquid-Based Drugs (ILBDs) with Surface Activity: Multifunctional Therapeutics*

Ionic liquid-based drugs (ILBDs) with inherent surface activity represent a cutting-edge class of therapeutics showing pharmacological activity along with aquatic amphiphilicity, allowing a better solubilization, facile membrane interaction, a controlled delivery [113].

5.2.1 Dual Role: API and Surfactant

In contrast to single-compound drugs that rely on separately added surfactants or excipients, ILBDs uniquely integrate both the therapeutic activity of the API and its surfactant properties directly into the ionic liquid framework of the drug itself. This dual functionality arises from their amphiphilic architecture, in which a hydrophilic ionic head group, such as imidazolium, pyridinium, or choline, is covalently linked to a hydrophobic alkyl tail. This structural arrangement facilitates spontaneous self-assembly into organized nanostructures, including micelles, vesicles, or ionogels, even at very low concentrations, owing to their remarkably low critical micelle concentration (CMC) in the micromolar range [113–115].

5.2.2 Advantages of Surface Activity of ILBDs

- **Enhanced Solubilization of Poorly Soluble APIs:** The hydrophobic environment of ILBDs provides a nano-environment that enhances the solubility of lipophilic drugs (e.g., paclitaxel, acyclovir) and overcomes the solubility-permeability barrier of BCS class II/IV compounds [116].
- **Membrane Fluidization and Permeability Enhancement:** Surface-active ILBDs promote drug permeation by fluidizing the lipid bilayers and interacting with tight junction proteins (e.g., occludin, ZO-1), which leads to an increase in transcellular and paracellular transport [117].
- **Intrinsic Antimicrobial Activity:** The cationic headgroups facilitate direct interaction with microbial membranes and cause depolarization of the inner membrane potential, disrupting phospholipid bilayers and inhibiting biofilm formation, providing an adjuvant therapeutic action that is especially valuable in wound treatment or device coatings [118, 119].
- **Stimuli-Responsive Drug Release**: ILBDs can be engineered with sensitive ionic bonds that can hydrolyze in the tumor microenvironment (TME) conditions (pH ~6.5, overexpression of enzymes), offering site-specific drug release without exogenous inductions [120, 121].

5.2.3 Polymorphism Elimination and Formulation Stability

ILBDs exist as liquids or amorphous phases, eliminating the problem of polymorphism observed for crystalline drugs [122]. The surface activity of the polymers ensures that they self-assemble into predetermined nanostructures, providing homogeneous formulations with no separation of phase or erratic crystal growth during storage or use.

An ILBD assembled with an antitumor anion (e.g., ibuprofenate or paclitaxelate) and an imidazolium-based cation establishes stable micelles to improve transdermal and oral bioavailability. At small concentrations (<1 mM), these ILBD micelles exhibit higher drug loading and better permeation than classical formulations and are also able to protect wound sites against microorganisms [123]. ILBDs with surface activity blur the line between drug and excipient, offering a new class of multifunctional therapeutics:

- Solubilizer → Permeation enhancer → Antimicrobial agent → Controlled release platform.

Such rational design has been extended beyond classic formulations and led to the development of programmable, bio-responsive, and sustainable drug carriers, which may revolutionize the next-generation drugs in terms of better efficacy, fewer adverse effects, and multifunctional properties [124].

5.3 Antimicrobial Agents

ILBS serves as next-generation antimicrobials that disrupt microbial membranes via cationic amphiphilicity, offering potent, resistance-averse action against planktonic cells and biofilms. The antimicrobial strength is very sensitively adjustable by alkyl chain length, cation, and anion, and programmable prokaryote/eukaryote selective toxicity is obtained. Importantly, bio-derivative ILBS, such as those with amino acid or choline moieties, exhibits low hemolytic activity alongside high bactericidal effectiveness, allowing coating for topical antiseptics, wound dressing, and antimicrobial coating. ILBS also improves synergy with antibiotics through penetration of biofilm EPS barriers, hence expanding their use against MDR pathogens [125, 126].

5.4 Protein Stabilization

ILBS stabilizes proteins by combining thermodynamic exclusion effects and steric protection within micellar nano-environments, preventing denaturation, aggregation, and surface adsorption. ILBS, by strategically selecting hydrogen-bond-basicity anions (e.g., acetate) and biocompatible cations (e.g., choline), strengthens proteins without perturbing the hydrophobic cores. This is a key requirement to enable long-term stability of biologics like monoclonal antibodies and vaccines subjected to thermal and mechanical stress, placing ILBS as programmable, rather than passive, stabilizers [127].

5.5 Biomaterials and Ionogels

ILBS enables the design of advanced biomaterials and ionogels that integrate antimicrobial action, mechanical robustness, and electrochemical functionality. As active components, the ILBS can prevent phase separation in ionogels and bestow them high with high ionic conductivity (>10 mS/cm), stretchability (>500% strain), and self-healing properties, meeting the requirements for wearable biosensors or implantable devices [128]. The choline and amino acid ILBS can improve the biocompatibility, biodegradability, enabling green ionogels with programmable degradation profiles and efficient drug release. They also facilitate high-quality bio interfaces for ECG/EMG recording, marking a significant step toward intelligent and environmentally cooperative bioelectronics [129].

5.6 *Emerging Applications*

- **Vaccine Adjuvants**: ILBS improves antigen stability, cellular uptake, and depot formation at the injection sites, which enhances both the humoral as well as cellular immune responses [130, 131].
- **Bioimaging**: ILBS is used as support capping for QDs and GNPs, having better photostability, low cytotoxicity, and preferred targeting specificity [132–134].
- **Enzyme Catalysis**: ILBS-assisted micellar systems constitute protective nano-environments that improve the activity of the enzyme, thermal stability, and recyclability for p-nitroacetophenone in non-aqueous media [135].

In summary, ILBS is the multifunctional active agents toward bridging material science and biomedicine, offering the possibility for rational design of smart delivery systems, antimicrobial surfaces, biosafe ionogel, and enzyme-friendly reaction environments. The revolutionary concept is to elevate ILBS from a passive drug delivery matrix toward programmable, bio-active therapeutic principles with intrinsic safety and ecological benefits.

ILBS offers several advantages over conventional surfactants:

- **Recyclability**: Low volatility enables ILBS recovery and reuse in formulations and decreases the generation of wastewater and process costs.
- **Eco-friendliness**: Renewable-resource-derived bio-based ILBS reduces environmental exposure and is compatible with the tenets of green chemistry.
- **High-Salinity Tolerance**: ILBS maintains colloidal stability and functionality in high-ionic-strength media (e.g., physiological fluids, brine) and preserve effectiveness when conventional surfactants fail to function adequately.
- **Enhanced Bioavailability**: ILBS increases the solubilization of drugs, penetrability at the membrane level, and delivery efficiency to obtain enhanced therapeutic performance.
- **Versatility**: The tunable physicochemical properties permit ILBS to be engineered for different applications from drug delivery to biomass conversion.

These advantages position ILBS as a sustainable and efficient alternative to traditional surfactants, particularly in biomedical contexts.

6 Challenges in ILBS Development

6.1 *Toxicity Concerns*

The toxicity of ILBSs is highly structure-sensitive, and the length of the alkyl chain and the nature of the ionic headgroup both influence biocompatibility. Larger chain lengths, e.g., C_{16}-imidazolium-containing ILBSs, not only exhibit excellent antimicrobial activity but also increase the level of cytotoxicity and acute fish toxicity by several orders of magnitude, with LC_{50} values below 10 μM for zebrafish embryos reported [136]. Even when the amino acid-, choline-, or sugar-headgroup-containing bio-derived ILBSs are used and exhibit much lower toxicity, the studies indicate that the degradation processes inevitably produce intermediate products whose long-term impacts are not known [137]. Moreover, the failure to conduct systematic in vivo and chronic exposure studies inhibits translation to biomedical and pharmaceutically applicable areas. Accordingly, comprehensive toxicological profiling well beyond the single-cell level remains a necessity in order to permit safe incorporation of ILBSs in biological systems.

6.2 Scalability and Cost

Mass production of ILBSs is prevented by inhibitory costs deriving mainly from multi-step syntheses, expensive ionic liquid precursors, and purification requirements. As a typical example, the production cost of $[C_{12}mim][BF_4]$ had been evaluated at \$50–100 g^{-1}, while the conventional surfactant SDS costs only \$1–2 g^{-1} [138]. Such prohibitions in cost prevent industry adoption despite the functional merits of ILBSs. Recent studies present the development of scalable "one-pot" syntheses (e.g., by using renewable feedstocks such as glycerol and amino acids) and solvent-free processes that potentially offer lower energy inputs and lower wastes. However, areas for innovation relate to achieving high purity at low cost, while retaining the structural tunability that warrants ILBSs' attractiveness.

6.3 Environmental Impact

Although marketed as greener alternatives, many synthetic ILBSs contain low rates of biodegradability and may ultimately end up remaining in water ecosystems, potentially contributing to bioaccumulation and ecotoxicity. Research shows that imidazolium- and pyridinium-type ILBSs may disrupt populations of microorganisms and stop enzymatic activity in wastewater treatment processes [139]. As a countermeasure for the resulting effects, ILBS development with biodegradable anions (e.g., lactate or acetate) and natural cations (e.g., cholinium or amino acid-derived cations) shows promise and exhibits > 60% rates of biodegradation through OECD testing protocols. However, life-cycle analysis shows that even bio-based ILBSs can carry significant environmental burdens if energy-intensive production takes place [140]. Thus, the combination of molecular design and green chemistry techniques such as enzymatic production, renewable feedstocks, and closed-loop recycling must take place for long-term ecological risk mitigation.

7 Recent Advancements

7.1 Bio-based ILBS

Bio-derived ILBS (choline/amino-acid/sugar/fatty-acid counter-ions) reached maturity from proof-of-concept to reproducible formulations: Choline-carboxylate SAILs (e.g., choline laurate) reliably show low micromolar → millimoral cytotoxicity in cell assays and pass OECD-style 28-day ready-biodegradability screens (most choline-C_{12} analogues approach or exceed the 60% 28-day biodegradation pass threshold), and as such the choline- and amino acid-based ILBS is gaining enthusiasm as the long-awaited "third-generation" biocompatible surfactants [141, 142]. The literature, however, holds important qualifications: (i) the biodegradation profiles are heavily chain-length and ester/amide linkage dependent (esterified tails increase mineralization rates, long tails (C_{16}–C_{18}) increase aquatic toxicity), (ii) most toxicity data are available in vitro (e.g., HeLa/SK-MEL cells, fibroblasts) and the necessary in vivo chronic/toxicokinetic data are few and far between, and (iii) the formulated ILBS performance (CMC, micellar stability, protein interactions) is highly sensitive to counter-ion choice, so translation requires standardized in vivo toxicology, environment fate studies and regulatory harmonization [143].

7.2 Stimuli-Responsive ILBS

Overt stimuli-responsive motifs (pH-sensitive headgroups, thermoresponsive linkers or amide/ester spacers) on SAILs now reproducibly self-assemble to form catanionic vesicles, reversible micelle ↔ vesicle transitions, and organogels whose aggregation state and cargo release can be tuned by pH, temperature, or light; recent advances have provided pH-responsive catanionic vesicles and amide-functionalized SAILs displaying reversible temperature-induced micelle ↔ vesicle ↔ gel transitions. Parallel but independent research uses non-surfactant ILs (e.g., CAGE and other cholinium-derivative ILs) as protein stabilizers or pore-forming agents (technically not ILBS), and the IL-based systems displayed improved insulin stabilization and increased transdermal/oral absorption in preclinical systems. The practical takeaway: SAIL chemistries already offer precise, polymer-free control over aggregate morphology for on-demand release; however, clinical evidence for the release of protein-scale therapeutics, such as insulin, at physiological temperature (37 °C) remains sparse and warrants rigorous, systematic validation. Furthermore, many of the most promising insulin delivery outcomes reported to date originate from broader IL-mediated systems rather than SAILs specifically, and direct head-to-head in vivo comparisons with dedicated SAIL formulations are still lacking.

7.3 Magnetic ILBS

Magnetic ionic liquids (MILs) and magnetic SAIL variants (e.g., Fe-containing anions such as $[FeCl_4]^-$ and long-chain imidazolium cations) permit the stoichiometric inclusion of magnetic responsiveness at the subunit level in the surfactants; MIL-based surfactants yield magnetically responsive microemulsions or micelles separable by exogenous fields. Reviews and recent experiments describe viable magnetic microemulsion assembly using such entities as $[C_{12}mim][FeCl_4]$ and demonstrate clearly improved in vitro targeting under applied fields; animal-level evidence using SAIL-based carriers exists nonetheless only marginally, and translational barriers are not small (iron release, oxidative Fenton chemistry, colloidal stability at physiological ionic strength, and the regulatory safety of the Fe-containing anions) [144, 145]. Briefly, magnetic ILBS constitutes a promising route to carriers directed by remote control, but quantitative in vivo targeting and safety information remain to be determined.

7.4 Polymeric and Zwitterionic ILBS

Polymerized ionic-liquid surfactants (PILS/polymeric ILBS) and zwitterionic ILBS address two typical formulation needs: long-term colloidal stability and enzyme/biomolecule compatibility. ILBS-containing polymer matrices offer mechanically stable nanoparticles or ion-dense polymeric micelles with high-capacity and slow-release rates; various recent articles (including polymeric IL matrices and preliminary articles on PIL-stabilized nimodipine nanoparticles) show multifold solubility enhancement and significantly greater physical stability than small-molecule surfactants [146]. Zwitterionic ILBS (sulfobetaine-type surfactants) is particularly attractive for biocatalysis since they retain enzyme structure and interfacial functionality: SAIL-stabilized hydrophobic-IL bicontinuous microemulsions and zwitterionic SAIL systems improved enzymatic conversion and operating stability in model systems (effects depend on enzyme and substrate but are reproducible across lipase and ester hydrolysis systems) [147]. Significant Limitations: Polymeric ILBS carries greater molar mass and potential clearance/toxicity risk and must be evaluated for biodegradability and immunogenicity; zwitterionic formats avoid protein adsorption, but scale and cost optimization remain.

7.5 Ionogels and API-ILs

Hybrid ionogels (SAIL/ILBS entrapped in biopolymers such as gelatin, silk, or gelatin-like matrices) combine the mechanical control of a gel with the solubilization and interfacial properties of ILBS; recent reviews and experimental ionogel formulations show clear advantages for sustained, stimuli-tunable dermal/transdermal release and for stabilizing labile APIs. Separately, API-IL strategies (converting the API into an ionic-liquid form) remain one of the most effective routes to massive solubility gains [148]. Methotrexate-based API-ILs and other API-ILs have reported orders-of-magnitude solubility increases in water/biorelevant media, and ionogel-API-IL composites have already shown improved permeation and controlled delivery in vitro [149]. Cautionary Notes: API-ILs change drug solid-state properties (dose form, stability, and PK) and so must be validated for activity, toxicity, and regulatory compatibility; ionogel mechanical/transport properties depend critically on ILBS diffusion and polymer crosslinking, so formulation-level optimization is essential.

8 Future Prospects

The future of SAILs in biology will be directed by continued innovation in molecular design and integration of other advanced technologies. Several key trends are shaping the future generation of materials.

8.1 Multifunctional and "Smart" SAILs

The research trajectory extends beyond single-function molecules to the creation of highly integrated, multifunctional SAILs. These are single molecules engineered to perform various functions concurrently, e.g., both antimicrobial activity and delivery and transport of a synergistic antibiotic. The integration of stimuli-response functionalities will yield "smart" SAILs delivered by order on demand through the use of exterior agents such as light or local pH shift and provide highly selective and controlled therapeutic function.

8.2 Incorporation with Nanotechnology and Nanomedicine

The integration of SAILs with nanotechnology is a very rich aspect of research. The integration of SAILs with various nanomaterials, including metallic nanoparticles, quantum dots, or carbon nanotubes, forms new hybrid systems through the exploitation of the new and emergent properties arising through the integrative combination [150]. In the field of nanomedicine, SAILs are seen as central elements in the construction of advanced delivery platforms such as liposomes, solid lipid nanoparticles, and polymeric micelles [151].

8.3 Sophisticated Diagnostics and Personalized Medicine

The ideal tunability of SAILs makes them the ideal instrument for the era of personalized medicine. It may be used for the creation of patient-specific preparations of drugs whose solubility and bioavailability must be optimized to complement the individual body in consideration. Apart from that, unique properties are also being exploited for the development of advanced platforms for diagnostics, e.g., organ-on-a-chip systems, in which they can be used to create stable and biologically relevant microenvironments amenable to cell culture and high-throughput screening [152].

8.4 Artificial Intelligence (AI) and Machine Learning in SAIL Design

The combinatorial chemical space of viable SAIL architecture far exceeds by many orders of magnitude the number accessible by experimental syntheses and testing one at a time by hand. AI and machine learning are becoming indispensable new tools to help expedite the discovery process [153]. By training machine-learning models on available sets of experimental data, predictive models can be built that can precisely predict the structure–property-activity relationships of new, computer-designed SAILs. This in-silico screening procedure will permit the ready identification of promising candidates for a specified biological challenge and will reduce development time and cost by many orders of magnitude [154].

9 Conclusion

Surface-active ionic liquids are at the intersection of life sciences and materials chemistry. They are something greater than incremental improvements on the traditional surfactants, but a new molecular family whose transformational potential derives directly from the unprecedented molecular level designability. By virtue of delivering a platform to tightly control and integrate such properties as thermal stability, interfacial activity, biocompatibility, and selective biological interactions, SAILs offer solutions to long-standing and most intransigent questions for the fields of medicine and biotechnology. The prospects are not free from difficulty. The central difficulties are the production cost becoming excessive, the scalability, and the specifications for approval by the regulatory body must be addressed systematically. The region must also continue to adhere to the tenets of green chemistry and proceed strongly on the path to the development of SAILs, not only high-performance and safe for human handling, but also environmentally friendly. The "sustainability trilemma" functionality vs. safety vs. biodegradability will be the overarching challenge and the guiding ethos for future research. Despite such challenges, the future direction for the area is one of unprecedented promise. Continued development by rational molecular design, increasingly guided by computation and machine-learning methods, will yield new SAIL generations with progressively more advanced and selective properties. By the development of the sort of interdisciplinary research that traverses chemistry and biology through engineering and medicine, surface-active ionic liquids will be converted from the object of lab curiosity to the indispensable agents that will remake domains ranging from advanced delivery and biocatalysis through intelligent biomaterials and personalized medicine.

References

1. Pandya I, Mishra S, Patel T, Keppeler N, Kumar S, Aswal VK, Kumar Kailasa S, El Seoud O, Malek NI (2024) 5-Fluoro uracil mediated transformation of surface-active ionic liquid based micellar nanoaggregates into pH-responsive vesicular nanoaggregates as promising drug delivery vehicle. J Mol Liq 409:125443. https://doi.org/10.1016/J.MOLLIQ.2024.125443
2. Asadabadi S, Saien J, Kharazi M (2024) Enhanced interfacial activity by maximizing synergy between long-chain ionic liquid and conventional surfactant for enhanced oil recovery. RSC Adv 14:18942–18949. https://doi.org/10.1039/D4RA02092H
3. Phan CM (2024) Relationship between molecular structure and surface activity of ionic surfactants. J Phys Chem B 128:11839–11845. https://doi.org/10.1021/ACS.JPCB.4C05527/SUPPL_FILE/JP4C05527_SI_001.PDF
4. Khachatrian AA, Mukhametzyanov TA, Salikhov RZ, Klimova AE, Gafurov ZN, Kantyukov AO, Yakhvarov DG, Garifullin BF, Larionov RA, Voloshina AD, Solomonov BN (2025) Interaction between newly synthesized surface-active ionic liquids with pharmaceutically active anion and bovine serum albumin. Int J Biol Macromol 286:138431. https://doi.org/10.1016/J.IJBIOMAC.2024.138431
5. Bagheri M, Ghaffari F, Shekaari H, Mokhtarpour M, Golmohammadi B (2025) Molecular interactions between surface-active ionic liquids based on 2-hydroxyethylammonium laurate with gabapentin: electrical conductivity and surface tension studies. RSC Adv 15:26–37. https://doi.org/10.1039/D4RA06670G
6. Xing Y, Hu Y, Zhang X, Zheng D, Ma G, Diao Y, Yue H, Wei W, Zhang S (2025) Cationic alkyl chain length and nanoaggregate form of ionic liquids dominate biocompatibility and toxicity. Nat Commun 16:1–15. https://doi.org/10.1038/s41467-025-62206-x
7. Zhen D, Zhang M, Liu F, Peng Q (2024) Research progress of surface-active ionic liquids for drug delivery. J Mol Liq 400:124546. https://doi.org/10.1016/J.MOLLIQ.2024.124546
8. Wojcieszak M, Kaczmarek DK, Karolak M, Pałkowski Ł, Lewandowska A, Marcinkowska A, Dopierała K, Materna K (2024) Surface-active ionic liquids and surface-active quaternary ammonium salts from synthesis, characterization to antimicrobial properties. Molecules 29:443. https://doi.org/10.3390/MOLECULES29020443/S1
9. Wojcieszak M, Syguda A, Materna K (2024) The influence of herbicidal anions on chemical shifts in NMR, phytotoxicity and surface properties of pyrrolidinium surface-active ionic liquids. Tenside, Surfactants, Deterg 61:146–157. https://doi.org/10.1515/TSD-2023-2571/ASSET/GRAPHIC/J_TSD-2023-2571_FIG_006.JPG
10. Gryniukova A, Borysko P, Myziuk I, Alieksieieva D, Hodyna D, Semenyuta I, Kovalishyn V, Metelytsia L, Rogalsky S, Tcherniuk S (2024) Anticancer activity features of imidazole-based ionic liquids and lysosomotropic detergents: in silico and in vitro studies. Mol Divers 28:3817–3833. https://doi.org/10.1007/S11030-023-10779-4/METRICS
11. Jahirul MI, Nath A, Khadiza N, Kumer A, Barua P, Sarkar A, Israt N, Abedin M, Islam MM, Khan MW (2025) Synthesis, characterization, antimicrobial activity and in silico study of

pyridinium based ionic liquids. RSC Adv 15:25009–25018. https://doi.org/10.1039/D5RA03648H
12. Marzullo P, Gruttadauria M, D'Anna F (2024) Quaternary ammonium salts-based materials: a review on environmental toxicity, anti-fouling mechanisms and applications in marine and water treatment industries. Biomolecules 14:957. https://doi.org/10.3390/BIOM14080957
13. Madej M, Kowalczyk J, Kowalski M, Grabowski P, Wernik J (2024) Tribological properties of selected ionic liquids in lubricated friction nodes. Materials 2025 18:18. https://doi.org/10.3390/MA18010018
14. Wojcieszak M, Zięba S, Dubis AT, Karolak M, Pałkowski Ł, Marcinkowska A, Skrzypczak A, Putowska A, Materna K (2024) Impact of alkyl spacer and side chain on antimicrobial activity of monocationic and dicationic imidazolium surface-active ionic liquids: experimental and theoretical insights. Molecules 29:5743. https://doi.org/10.3390/MOLECULES29235743/S1
15. Fernandes MM, Carvalho EO, Correia DM, Esperança JMSS, Padrão J, Ivanova K, Hoyo J, Tzanov T, Lanceros-Mendez S (2022) Ionic liquids as biocompatible antibacterial agents: a case study on structure-related bioactivity on *Escherichia coli*. ACS Appl Bio Mater 5:5181–5189. https://doi.org/10.1021/ACSABM.2C00615/ASSET/IMAGES/LARGE/MT2C00615_0007.JPEG
16. Pathan T, Girase M, Patel K, Pillai SA, Pathan S, Parekh P, Kuperkar K, Patel VI (2024) Synergistic self-assembly in a surface-active ionic liquid and an anionic surfactant mixed system: a comprehensive physicochemical analysis. Ind Eng Chem Res 63:9688–9700. https://doi.org/10.1021/ACS.IECR.4C00668/ASSET/IMAGES/MEDIUM/IE4C00668_0008.GIF
17. Priyanka VP, Harikrishna AS, Kesavan V, Gardas RL (2024) Synergistic interaction and antibacterial properties of surface-active mono- and di-cationic ionic liquids with ciprofloxacin. J Mol Liq 399:124359. https://doi.org/10.1016/J.MOLLIQ.2024.124359
18. Deng Z, Tang S, Ou M, Wang Z (2025) Ionic liquids self-assembled micelles for transdermal delivery: characterization, functions and applications. Int J Pharm X 10. https://doi.org/10.1016/J.IJPX.2025.100350
19. Jain M, Ukani H, Kumar S, Aswal VK, Kailasa SK, El Seoud OA, Malek NI (2024) Unveiling the potential of surface active ionic liquids based catanionic vesicles for tailored delivery of curcumin and its enhanced therapeutic efficacy. ChemistrySelect 9:e202401144. https://doi.org/10.1002/SLCT.202401144
20. Sadeghi A, Mohammad Jasour A, Kowsari E, Gheibi M, Hadi Ghasemi M, Ramakrishna S (2024) Comprehensive viewpoint on ionic liquids applications in sustainable pharmaceutical technology (experiments, simulations, and managerial insights). J Mol Liq 404:124991. https://doi.org/10.1016/J.MOLLIQ.2024.124991
21. Arakelyan LA, Arkhipova DM, Seitkalieva MM, Vavina AV, Sahharova LT, Kurbanalieva SK, Posvyatenko AV, Egorova KS, Ananikov VP (2024) A comprehensive dataset on cytotoxicity of ionic liquids. Sci Data 2024 11:1–11. https://doi.org/10.1038/s41597-024-04190-3
22. Zheng D, Jiang S, Qin S, An B, Gao L (2025) Sustainable ionic liquid design for enhanced solubility and safety of Tamoxifen: a computational exploration. Sustain Chem Pharm 45:102046. https://doi.org/10.1016/J.SCP.2025.102046
23. Benedetto A, Galla HJ (2018) Editorial of the "ionic liquids and biomolecules" special issue. Biophys Rev 10:687–690. https://doi.org/10.1007/S12551-018-0426-3/FIGURES/1
24. Emel'Yanenko VN, Boeck G, Verevkin SP, Ludwig R (2014) Volatile times for the very first ionic liquid: understanding the vapor pressures and enthalpies of vaporization of ethylammonium nitrate. Chem–Euro J 20:11640–11645. https://doi.org/10.1002/CHEM.201403508
25. Plechkova NV, Seddon KR (2007) Applications of ionic liquids in the chemical industry. Chem Soc Rev 37:123–150. https://doi.org/10.1039/B006677J
26. Shiflett MB, Magee JW, Tuma D (2020) Important developments in the history of ionic liquids from academic curiosity to commercial processes and products, pp 3–29. https://doi.org/10.1007/978-3-030-35245-5_1
27. Matuszek K, Piper SL, Brzęczek-Szafran A, Roy B, Saher S, Pringle JM, Macfarlane DR, Matuszek K, Roy B, Saher S, Macfarlane DR, Piper SL, Pringle JM, Brz A (2024) Unexpected

energy applications of ionic liquids. Adv Mater 36:2313023. https://doi.org/10.1002/ADMA.202313023

28. Dong B, Li N, Zheng L, Yu L, Inoue T (2007) Surface adsorption and micelle formation of surface active ionic liquids in aqueous solution. Langmuir 23:4178–4182. https://doi.org/10.1021/LA0633029/ASSET/IMAGES/MEDIUM/LA0633029N00001.GIF
29. Chen Y, Zhao Y, Chen J, Zhuo K, Wang J (2011) Aggregation behavior of a model ionic liquid surfactant in monosaccharide + water solutions. J Colloid Interface Sci 364:388–394. https://doi.org/10.1016/J.JCIS.2011.08.052
30. Huang G, Sun L, Li L, Pei L, Xue W, Wang R, Wang Y (2023) Exploring the effect mechanism of alkyl chain lengths on the tribological performance of ionic liquids. ACS Omega 9:3184–3192. https://doi.org/10.1021/ACSOMEGA.3C01885/ASSET/IMAGES/LARGE/AO3C01885_0012.JPEG
31. Zakharova LY, Pashirova TN, Doktorovova S, Fernandes AR, Sanchez-Lopez E, Silva AM, Souto SB, Souto EB (2019) Cationic surfactants: self-assembly, structure-activity correlation and their biological applications. Int J Mol Sci 20:5534. https://doi.org/10.3390/IJMS20225534
32. Brown P, Butts CP, Eastoe J, Grillo I, James C, Khan A (2013) New catanionic surfactants with ionic liquid properties. J Colloid Interface Sci 395:185–189. https://doi.org/10.1016/J.JCIS.2012.12.046
33. Zuo Y, Lv J, Wei N, Chen X, Tong J (2023) Effect of anions and cations on the self-assembly of ionic liquid surfactants in aqueous solution. J Mol Liq 375:121342. https://doi.org/10.1016/J.MOLLIQ.2023.121342
34. Brown P, Butts CP, Eastoe J, Fermin D, Grillo I, Lee HC, Parker D, Plana D, Richardson RM (2012) Anionic surfactant ionic liquids with 1-Butyl-3-methyl-imidazolium cations: characterization and application. Langmuir 28:2502–2509. https://doi.org/10.1021/LA204557T/SUPPL_FILE/LA204557T_SI_001.PDF
35. Nazar M, Ul Hassan Shah M, Ahmad A, Goto M, Zaireen Nisa Yahya W, Moniruzzaman M (2023) Aggregation, toxicity, and biodegradability study of an ionic liquid-based formulation for effective oil spill remediation. Chemosphere 344:140412. https://doi.org/10.1016/J.CHEMOSPHERE.2023.140412
36. Pérez L, Pinazo A, Morán MC, Pons R (2020) Aggregation behavior, antibacterial activity and biocompatibility of catanionic assemblies based on amino acid-derived surfactants. Int J Mol Sci 21:8912. https://doi.org/10.3390/IJMS21238912
37. Wei Y, He X, Zhang X, Xie S, Dong J, Wei D (2025) Amino acid anions and antimicrobial performance correlation of amino acid-based ionic liquids. J Environ Chem Eng 13:119203. https://doi.org/10.1016/J.JECE.2025.119203
38. Moshikur RM, Ali MK, Moniruzzaman M, Goto M (2021) Recent advances in surface-active ionic liquid-assisted self-assembly systems for drug delivery. Curr Opin Colloid Interface Sci 56:101515. https://doi.org/10.1016/J.COCIS.2021.101515
39. Xu L, Huang Z, Deng Z, Du Z, Sun TL, Guo ZH, Yue K, Transparent A (2021) Highly stretchable, solvent-resistant, recyclable multifunctional ionogel with underwater self-healing and adhesion for reliable strain sensors. Adv Mater 33:2105306. https://doi.org/10.1002/ADMA.202105306
40. Figueira-González M, Francisco V, García-Río L, Marques EF, Parajó M, Rodríguez-Dafonte P (2013) Self-aggregation properties of ionic liquid 1,3-didecyl-2-methylimidazolium chloride in aqueous solution: from spheres to cylinders to bilayers. J Phys Chem B 117:2926–2937. https://doi.org/10.1021/JP3117962
41. Brown P, Butts CP, Eastoe J, Fermin D, Grillo I, Lee HC, Parker D, Plana D, Richardson RM (2012) Anionic surfactant ionic liquids with 1-butyl-3-methyl-imidazolium cations: characterization and application. Langmuir 28:2502–2509. https://doi.org/10.1021/LA204557T
42. Villa CC, Correa NM, Silber JJ, Falcone RD (2019) Catanionic reverse micelles as an optimal microenvironment to alter the water electron donor capacity in a SN2 reaction. J Org Chem 84:1185–1191. https://doi.org/10.1021/ACS.JOC.8B02492

43. Villa CC, Silber JJ, Correa NM, Falcone RD (2014) Effect of the cationic surfactant moiety on the structure of water entrapped in two catanionic reverse micelles created from ionic liquid-like surfactants. ChemPhysChem 15:3097–3109. https://doi.org/10.1002/CPHC.201402307
44. Cheng N, Ma X, Sheng X, Wang T, Wang R, Jiao J, Yu L (2014) Aggregation behavior of anionic surface active ionic liquids with double hydrocarbon chains in aqueous solution: experimental and theoretical investigations. Colloids Surf A Physicochem Eng Asp 453:53–61. https://doi.org/10.1016/J.COLSURFA.2014.03.083
45. Andrada HE, Silva OF, Morales GM, Correa NM, Falcone RD (2020) Spontaneous formation of unilamellar vesicles based on the surfactant 1-methylimidazolium bis-(2-ethylhexyl) phosphate, evaluated as a function of pH and in saline solution. Colloids Surf A Physicochem Eng Asp 606:125435. https://doi.org/10.1016/J.COLSURFA.2020.125435
46. El Seoud OA, Keppeler N, Malek NI, Galgano PD (2021) Ionic liquid-based surfactants: recent advances in their syntheses, solution properties, and applications. Polymers (Basel) 13:1100. https://doi.org/10.3390/POLYM13071100/S1
47. Polidori I, To D, Kali G, Bernkop-Schnürch A (2024) Histidine-based ionizable cationic surfactants: novel biodegradable agents for hydrophilic macromolecular drug delivery. Drug Deliv Transl Res 14:2370–2385. https://doi.org/10.1007/S13346-023-01511-8
48. Javadian S, Nasiri F, Heydari A, Yousefi A, Shahir AA (2014) Modifying effect of imidazolium-based ionic liquids on surface activity and self-assembled nanostructures of sodium dodecyl sulfate. J Phys Chem B 118:4140–4150. https://doi.org/10.1021/JP5010049
49. Chen B, Du Y (2010) Evaluation of the enantioseparation capability of the novel chiral selector clindamycin phosphate towards basic drugs by micellar electrokinetic chromatography. J Chromatogr A 1217:1806–1812. https://doi.org/10.1016/J.CHROMA.2010.01.043
50. Bhadani A, Singh S (2009) Novel Gemini pyridinium surfactants: synthesis and study of their surface activity, DNA binding, and cytotoxicity. Langmuir 25:11703–11712. https://doi.org/10.1021/LA901641F/SUPPL_FILE/LA901641F_SI_001.PDF
51. Cao Y, Yang W, Jiang Y, Wang Y, Ju H, Geng T (2021) Studies on physicochemical properties of three Gemini surfactants with different spacer groups. J Mol Liq 325:115039. https://doi.org/10.1016/J.MOLLIQ.2020.115039
52. Roshdy K, Mohamed HI, Ahmed MH, El-Dougdoug WI, Abo-Riya MA (2023) Gemini ionic liquid-based surfactants: efficient synthesis, surface activity, and use as inducers for the fabrication of Cu_2O nanoparticles. RSC Adv 13:31128–31140. https://doi.org/10.1039/D3RA04646J
53. Saien J, Shokri B, Kharazi M (2024) Effective synergistic action of benzimidazolium nano Gemini ionic liquid and conventional surfactant for chemical enhanced oil recovery. ACS Omega 9:22336–22344. https://doi.org/10.1021/ACSOMEGA.4C01768/ASSET/IMAGES/MEDIUM/AO4C01768_0010.GIF
54. Lu B, Zhou G, Xiao F, He Q, Zhang J (2020) Stimuli-responsive poly(ionic liquid) nanoparticles for controlled drug delivery. J Mater Chem B 8:7994–8001. https://doi.org/10.1039/D0TB01352H
55. Subraveti SN, Nader MG, AziziHariri P, John VT, Lamichhane N, Raghavan SR (2024) Vesicle–micelle transitions driven by ROS, light and heat. Nanoscale 16:16942–16951. https://doi.org/10.1039/D4NR01543F
56. Kumari S, Aggrawal R, Sonu, Halder S, Sundar G, Saha SK (2017) Effect of hydrophobicity of tails and hydrophilicity of spacer group of cationic Gemini surfactants on solvation dynamics and rotational relaxation of coumarin 480 in aqueous micelles. ACS Omega 2:5898–5910. https://doi.org/10.1021/ACSOMEGA.7B00818/SUPPL_FILE/AO7B00818_SI_001.PDF
57. Tian M, Chen X, Zou X, Qian Y, Liu Z, Fan Y (2021) pH-responsive rheological properties and microstructure transition in mixture of anionic Gemini/Cationic monomeric surfactants. Molecules 26:5013. https://doi.org/10.3390/MOLECULES26165013
58. Durling GM, Milazzo T, Marin-Rimoldi E, Ouimet JA, Phillip WA, Maginn EJ, Luo T, Ashfeld BL (2025) Tuning the directional solubility of ionic liquids through multicomponent ions for low-temperature desalination. J Am Chem Soc 147:20294–20301. https://doi.org/10.1021/JACS.4C16294

59. Ristroph K, Salim M, Wilson BK, Clulow AJ, Boyd BJ, Prud'homme RK (2021) Internal liquid crystal structures in nanocarriers containing drug hydrophobic ion pairs dictate drug release. J Colloid Interface Sci 582:815–824. https://doi.org/10.1016/J.JCIS.2020.08.045
60. Toledo Hijo AAC, Silva EK, Meirelles AAD, Cunha RL, Meirelles AJA (2023) Novel naturally derived encapsulation agents in the ionic liquid form for sustainable emulsion-based products. Sustain Food Technol 1:275–279. https://doi.org/10.1039/D2FB00054G
61. Wu S, Li F, Zeng L, Wang C, Yang Y, Tan Z (2019) Assessment of the toxicity and biodegradation of amino acid-based ionic liquids. RSC Adv 9:10100–10108. https://doi.org/10.1039/C8RA06929H
62. García MT, Bautista E, Pérez L, Vázquez S (2025) Self-assembly, antimicrobial properties and biodegradability of ester-functionalized choline-based surface-active ionic liquids. Molecules 30:1280. https://doi.org/10.3390/MOLECULES30061280/S1
63. Huang Z, Yi M, Liu Y, Qi P, Song A, Hao J (2020) Magnetic polymerizable surfactants: thermotropic liquid crystal behaviors and construction of nanostructured films. New J Chem 44:16537–16545. https://doi.org/10.1039/D0NJ03029E
64. Kulshrestha A, Gehlot PS, Kumar A (2020) Magnetic proline-based ionic liquid surfactant as a nano-carrier for hydrophobic drug delivery. J Mater Chem B 8:3050–3057. https://doi.org/10.1039/D0TB00176G
65. Niesyto K, Keihankhadiv S, Mazur A, Mielańczyk A, Neugebauer D (2024) Ionic liquid-based polymer matrices for single and dual drug delivery: impact of structural topology on characteristics and in vitro delivery efficiency. Int J Mol Sci 25:1292. https://doi.org/10.3390/IJMS25021292/S1
66. Torres-Luna C, Hu N, Domszy R, Fan X, Yang J, Briber RM, Wang NS, Yang A (2021) Effect of carbon chain length ionic strength, and pH on the in vitro release kinetics of cationic drugs from fatty-acid-loaded contact lenses. Pharmaceutics 13:1060. https://doi.org/10.3390/PHARMACEUTICS13071060
67. Lunagariya J, Kumar NS, Asif M, Dhar A, Vekariya RL (2017) Dependency of anion and chain length of imidazolium based ionic liquid on micellization of the block copolymer F127 in aqueous solution: an experimental deep insight. Polymers 9:285. https://doi.org/10.3390/POLYM9070285
68. Chen X, Xin Y, Yu F, Lan Z (2017) Synthesis and surface properties of novel carboxylic ester-containing imidazolium-based zwitterionic surfactants: monoalkyl 2-(3-methylimidazolium-1-yl) succinate inner salts. J Surfactants Deterg 20:1121–1127. https://doi.org/10.1007/S11743-017-1987-Z/METRICS
69. Huet G, Araya-Farias M, Alayoubi R, Laclef S, Bouvier B, Gosselin I, Cézard C, Roulard R, Courty M, Hadad C, Husson E, Sarazin C, Nguyen Van Nhien A (2020) New biobased-zwitterionic ionic liquids: efficiency and biocompatibility for the development of sustainable biorefinery processes. Green Chem 22:2935–2946. https://doi.org/10.1039/D0GC01002B
70. Yan F, Hu L, Ji Z, Lyu Y, Chen S, Xu L, Hao J (2024) Highly interfacial active Gemini surfactants as simple and versatile emulsifiers for stabilizing lubricating and structuring liquids. Angew Chem Int Ed 63:e202318926. https://doi.org/10.1002/ANIE.202318926
71. Inoue T, Ebina H, Dong B, Zheng L (2007) Electrical conductivity study on micelle formation of long-chain imidazolium ionic liquids in aqueous solution. J Colloid Interface Sci 314:236–241
72. Dong B, Li N, Zheng L, Yu L, Inoue T (2007) Surface adsorption and micelle formation of surface active ionic liquids in aqueous solution. Langmuir 23:4178–4182
73. Łuczak J, Hupka J, Thöming J, Jungnickel C (2008) Self-organization of imidazolium ionic liquids in aqueous solution. Colloids Surf A 329:125–133
74. Goodchild I, Collier L, Millar SL et al (2007) Structural studies of the phase, aggregation and surface behaviour of 1-alkyl-3-methylimidazolium halide + water mixtures. J Colloid Interface Sci 307:455–468
75. Kaur R, Kumar H, Singla M (2022) Micellization studies of selected imidazolium based ionic liquid in aqueous solution and in presence of different additives: a review. J Ionic Liquids 2(2):100036

76. Wahyuningrum D, Nur Pasya N, Eduardus E, Handayani N, Damayanti S (2024) An experimental and computational approach to the microwave-assisted synthesis of 2'-hydoxychalcone derivatives in 1-dodecyl-3-methylimidazolium bromide ([DDMIm]Br) ionic liquids. ChemistrySelect 9:e202303211. https://doi.org/10.1002/SLCT.202303211
77. Campos WF, Silva EC, Oliveira TJ, Oliveira JM, Tubino M, Pereira C, Vila MMDC, Balcão VM (2020) Transdermal permeation of bacteriophage particles by choline oleate: potential for treatment of soft-tissue infections. Future Microbiol 15:881–896. https://doi.org/10.2217/FMB-2019-0290
78. Pattanayak P, Saha S, Chatterjee T, Ranu BC (2024) Sustainable and solvent-free synthesis of molecules of pharmaceutical importance by ball milling. Chem Commun 61:247–265. https://doi.org/10.1039/D4CC05127K
79. Go ES, Hong EJ, Lee JY, Stolar T, Peterson GI, Emmerling FL, Kim K, Kim JG (2025) Insights into mechanochemical solid-state ball-milling reaction: monitoring transition from heterogeneous to homogeneous conditions. JACS Au 5:2720–2727. https://doi.org/10.1021/JACSAU.5C00322/SUPPL_FILE/AU5C00322_SI_001.PDF
80. Tiwari G, Khanna A, Mishra VK, Sagar R (2023) Recent developments on microwave-assisted organic synthesis of nitrogen- and oxygen-containing preferred heterocyclic scaffolds. RSC Adv 13:32858. https://doi.org/10.1039/D3RA05986C
81. Wang Y, Hou Q, Ju M, Li W (2019) New developments in material preparation using a combination of ionic liquids and microwave irradiation. Nanomaterials 9:647. https://doi.org/10.3390/NANO9040647
82. Zhang Y, Liu Z, Li Y, Chi R (2014) Optimization of ionic liquid-based microwave-assisted extraction of isoflavones from Radix *Puerariae* by response surface methodology. Sep Purif Technol 129:71–79. https://doi.org/10.1016/J.SEPPUR.2014.03.022
83. Ribeiro SF, Braga HS, Santiago-Aguiar RS (2025) Protic ionic liquids as eco-friendly surfactants for oil spill remediation: synthesis, characterization, and performance evaluation. ACS Omega 10:27069–27081. https://doi.org/10.1021/ACSOMEGA.5C02170
84. Yazdani A, Sivapragasam M, Levêque J-M, Moniruzzaman M (2016) Microbial biocompatibility and biodegradability of choline-amino acid based ionic liquids. J Microb Biochem Technol Open Access J 8:415–421. https://doi.org/10.4172/1948-5948.1000318
85. Stachowiak W, Wysocki M, Guazzelli L, Niemczak M (2025) Glycine betaine-functionalized ionic liquids: design of versatile bioplatform for incorporating biologically active moieties into cations. ACS Med Chem Lett 16:1756–1761. https://doi.org/10.1021/ACSMEDCHEMLETT.5C00280
86. Patil HV, Badgujar NP, Suresh, Kulkarni RD, Nagaraj K (2025) Sustainable bio-based surfactants: advances in green chemistry and environmental applications. Mater Today Commun 48:113583. https://doi.org/10.1016/J.MTCOMM.2025.113583
87. Alreshidi MA, Yadav KK, Gunasekaran S, Gacem A, Sambandam P, Subbiah G, Bhutto JK, Palanivel S, Fallatah AM, Abo El-Khair MA, Almalawi JF, Alam MW, Perumal T, Annadurai S (2025) A review on the evolution of ionic liquids: sustainable synthesis, applications, and future prospects. Mater Today Sustain 31:101160. https://doi.org/10.1016/J.MTSUST.2025.101160
88. Kuddushi M, Bin Xu B, Malek N, Zhang X (2024) Review of ionic liquid and ionogel-based biomaterials for advanced drug delivery. Adv Colloid Interface Sci 331:103244. https://doi.org/10.1016/J.CIS.2024.103244
89. Abdel-Hameed R, Ashmawy A, Abourashed NM, Hegazy A, Huwaimel B, Fargally O, Ali AA, Tantawy AH, Mohamed HI (2025) Ionic liquid-based cationic Gemini surfactants with polymeric spacer as inducers for Cu_2O nanoparticles production: green synthesis, characterization, and surface activity. J Mol Liq 426:127306. https://doi.org/10.1016/J.MOLLIQ.2025.127306
90. Fernandes LC, Correia DM, Costa CM, Lanceros-Mendez S (2025) Recent advances in ionic liquid-based hybrid materials for electroactive soft actuator applications. Macromol Mater Eng 310:2400279. https://doi.org/10.1002/MAME.202400279

91. Kanjilal S, Sunitha S, Reddy PS, Kumar KP, Murty KPK, Prasad RBN (2009) Synthesis and evaluation of micellar properties and antimicrobial activities of imidazole-based surfactants. Eur J Lipid Sci Technol 111:941–948. https://doi.org/10.1002/EJLT.200800292
92. Baltazar QQ, Chandawalla J, Sawyer K, Anderson JL (2007) Interfacial and micellar properties of imidazolium-based monocationic and dicationic ionic liquids. Colloids Surf A Physicochem Eng Asp 302:150–156. https://doi.org/10.1016/J.COLSURFA.2007.02.012
93. Ubale S, Patil H, Waghmare J (2025) Ionic liquid surfactants: a comprehensive review of their synthesis, properties, and emerging application. Int J Innov Sci Res Technol 10:185–192. https://doi.org/10.38124/IJISRT/25MAR262
94. Mansour EM, Hosny R, Mohamed AS, Abdelhafiz FM (2025) Synthesis and evaluation of amino acid ionic liquid for enhanced oil recovery: experimental and modeling simulation studies. Sci Rep 15:1–19. https://doi.org/10.1038/S41598-025-85560-8;SUBJMETA
95. Sharma P, Sharma S, Kumar H (2024) Introduction to ionic liquids, applications and micellization behaviour in presence of different additives. J Mol Liq 393:123447. https://doi.org/10.1016/J.MOLLIQ.2023.123447
96. Almeida C, Pedro AQ, Tavares APM, Neves MC, Freire MG (2023) Ionic-liquid-based approaches to improve biopharmaceuticals downstream processing and formulation. Front Bioeng Biotechnol 11:1037436. https://doi.org/10.3389/FBIOE.2023.1037436/FULL
97. Tang F, Wu K, Ding L, Yuan J, Liu Q, Nie L, Yao S (2008) Purification of undiluted ionic liquids from trace-colored impurities for spectroscopy by octadecylsilyl solid-phase extraction. Sep Purif Technol 60:245–250. https://doi.org/10.1016/J.SEPPUR.2007.08.018
98. Pham TD, Pham TT, Phan MN, Ngo TMV, Dang VD, Vu CM (2020) Adsorption characteristics of anionic surfactant onto laterite soil with differently charged surfaces and application for cationic dye removal. J Mol Liq 301:112456. https://doi.org/10.1016/J.MOLLIQ.2020.112456
99. García-Verdugo E, Altava B, Burguete MI, Lozano P, Luis SV (2015) Ionic liquids and continuous flow processes: a good marriage to design sustainable processes. Green Chem 17:2693–2713. https://doi.org/10.1039/C4GC02388A
100. Verdía Barbará P, Choudhary H, Nakasu PS, Al-Ghatta A, Han Y, Hopson C, Aravena RI, Mishra DK, Ovejero-Pérez A, Simmons BA, Hallett JP (2025) Recent advances in the use of ionic liquids and deep eutectic solvents for lignocellulosic biorefineries and biobased chemical and material production. Chem Rev 125:5461–5583. https://doi.org/10.1021/ACS.CHEMREV.4C00754
101. Farias CBB, Almeida FCG, Silva IA, Souza TC, Meira HM, de Soares da Silva RCF, Luna JM, Santos VA, Converti A, Banat IM, Sarubbo LA (2021) Production of green surfactants: market prospects. Elect J Biotechnol 51:28–39. https://doi.org/10.1016/J.EJBT.2021.02.002
102. Buettner CS, Cognigni A, Schröder C, Bica-Schröder K (2022) Surface-active ionic liquids: a review. J Mol Liq 347:118160. https://doi.org/10.1016/J.MOLLIQ.2021.118160
103. Ni C, Hassan PA, Kaler EW (2005) Structural characteristics and growth of pentagonal silver nanorods prepared by a surfactant method. Langmuir 21:3334–3337. https://doi.org/10.1021/LA046807C
104. Li B, Elsaid M, Ge H (2022) Transition-metal-catalyzed site-selective γ- and δ-C(sp3)–H functionalization reactions. Chem 8:1254–1360. https://doi.org/10.1016/J.CHEMPR.2022.04.018
105. Bae E, Beil S, König M, Stolte S, Escher BI, Markiewicz M (2024) The mode of toxic action of ionic liquids: narrowing down possibilities using high-throughput, in vitro cell-based bioassays. Environ Int 193:109089. https://doi.org/10.1016/J.ENVINT.2024.109089
106. Egorova KS, Gordeev EG, Ananikov VP (2017) Biological activity of ionic liquids and their application in pharmaceutics and medicine. Chem Rev 117:7132–7189. https://doi.org/10.1021/ACS.CHEMREV.6B00562
107. Sharma B, Pérez-García L, Chaudhary GR, Kaur G (2025) Innovative approaches to cationic and anionic (catanionic) amphiphiles self-assemblies: synthesis, properties, and industrial applications. Adv Colloid Interface Sci 337:103380. https://doi.org/10.1016/J.CIS.2024.103380

108. Khan IA, Gnezdilov OI, Wang YL, Filippov A, Shah FU (2020) Effect of aromaticity in anion on the cation-anion interactions and ionic mobility in fluorine-free ionic liquids. J Phys Chem B 124:11962–11973. https://doi.org/10.1021/ACS.JPCB.0C08421
109. Nandwani SK, Malek NI, Chakraborty M, Gupta S (2020) Insight into the application of surface-active ionic liquids in surfactant based enhanced oil recovery processes–a guide leading to research advances. Energy Fuels 34:6544–6557. https://doi.org/10.1021/ACS.ENERGYFUELS.0C00343
110. Xie B, Liu Y, Li X, Yang P, He W (2024) Solubilization techniques used for poorly water-soluble drugs. Acta Pharm Sin B 14:4683. https://doi.org/10.1016/J.APSB.2024.08.027
111. Sidat Z, Marimuthu T, Kumar P, du Toit LC, Kondiah PPD, Choonara YE, Pillay V (2019) Ionic liquids as potential and synergistic permeation enhancers for transdermal drug delivery. Pharmaceutics 11:96. https://doi.org/10.3390/PHARMACEUTICS11020096
112. Khan O, Bhawale R, Vasave R, Mehra NK (2024) Ionic liquid-based formulation approaches for enhanced transmucosal drug delivery. Drug Discov Today 29:104109. https://doi.org/10.1016/J.DRUDIS.2024.104109
113. Zhuo Y, Cheng HL, Zhao YG, Cui HR (2024) Ionic liquids in pharmaceutical and biomedical applications: a review. Pharmaceutics 16:151. https://doi.org/10.3390/PHARMACEUTICS16010151
114. Sangiorgi S, Albertini B, Bertoni S, Passerini N (2025) An overview on the role of ionic liquids and deep eutectic solvents in oral pharmaceuticals. Pharmaceutics 17:300. https://doi.org/10.3390/PHARMACEUTICS17030300
115. Tourné-Péteilh C, Coasne B, In M, Brevet D, Devoisselle JM, Vioux A, Viau L (2014) Surfactant behavior of ionic liquids involving a drug: from molecular interactions to self-assembly. Langmuir 30:1229–1238. https://doi.org/10.1021/LA404166Y
116. Shamshina JL, Cojocaru OA, Kelley SP, Bica K, Wallace SP, Gurau G, Rogers RD (2017) Acyclovir as an ionic liquid cation or anion can improve aqueous solubility. ACS Omega 2:3483. https://doi.org/10.1021/ACSOMEGA.7B00554
117. Beaven E, Kumar R, An JM, Mendoza H, Sutradhar SC, Choi W, Narayan M, kyu Lee Y, Nurunnabi M (2023) Potentials of ionic liquids to overcome physical and biological barriers. Adv Drug Deliv Rev 204:115157. https://doi.org/10.1016/J.ADDR.2023.115157
118. Nikfarjam N, Ghomi M, Agarwal T, Hassanpour M, Sharifi E, Khorsandi D, Ali Khan M, Rossi F, Rossetti A, Nazarzadeh Zare E, Rabiee N, Afshar D, Vosough M, Kumar Maiti T, Mattoli V, Lichtfouse E, Tay FR, Makvandi P (2021) Antimicrobial ionic liquid-based materials for biomedical applications. Adv Funct Mater 31:2104148. https://doi.org/10.1002/ADFM.202104148
119. Venkata Nancharaiah Y, Reddy GKK, Lalithamanasa P, Venugopalan VP (2012) The ionic liquid 1-alkyl-3-methylimidazolium demonstrates comparable antimicrobial and antibiofilm behavior to a cationic surfactant. Biofouling 28:1141–1149. https://doi.org/10.1080/08927014.2012.736966
120. Datta D, Colaco V, Bandi SP, Dhas N, Janardhanam LSL, Singh S, Vora LK (2025) Stimuli-responsive self-healing ionic gels: a promising approach for dermal and tissue engineering applications. ACS Biomater Sci Eng 11:1338–1372. https://doi.org/10.1021/ACSBIOMATERIALS.4C02264
121. Zhang W, Guo Y, Yang J, Tang G, Zhang J, Cao Y (2023) Prodrug based on ionic liquids for dual-triggered release of thiabendazole. ACS Omega 8:3484. https://doi.org/10.1021/ACSOMEGA.2C07511
122. Wu X, Zhu Q, Chen Z, Wu W, Lu Y, Qi J (2021) Ionic liquids as a useful tool for tailoring active pharmaceutical ingredients. J Control Release 338:268–283. https://doi.org/10.1016/J.JCONREL.2021.08.032
123. Lin R, Qian J, Zhang J, Li H, Wu Y (2025) Ionic liquids as transdermal drug carriers: a review of recent advancements. J Mol Liq 429:127595. https://doi.org/10.1016/J.MOLLIQ.2025.127595
124. Huang W, Wu X, Qi J, Zhu Q, Wu W, Lu Y, Chen Z (2020) Ionic liquids: green and tailor-made solvents in drug delivery. Drug Discov Today 25:901–908. https://doi.org/10.1016/J.DRUDIS.2019.09.018

125. Zunita M, Batara B (2023) Antibacterial-based ionic liquids for environmental wastewater treatment. Case Stud Chem Environ Eng 7:100341. https://doi.org/10.1016/J.CSCEE.2023.100341
126. Siddiquee MA, Patel R, Saraswat J, Khatoon BS, ud din Parray M, Wani FA, Khan MR, Busquets R (2021) Interfacial and antibacterial properties of imidazolium based ionic liquids having different counterions with ciprofloxacin. Colloids Surf A Physicochem Eng Asp 629:127474. https://doi.org/10.1016/J.COLSURFA.2021.127474
127. Veríssimo NV, Vicente FA, de Oliveira RC, Likozar B, de Oliveira RPS, Pereira JFB (2022) Ionic liquids as protein stabilizers for biological and biomedical applications: a review. Biotechnol Adv 61:108055. https://doi.org/10.1016/J.BIOTECHADV.2022.108055
128. Gao Y, Zhang W, Li L, Wang Z, Shu Y, Wang J (2023) Ionic liquid-based gels for biomedical applications. Chem Eng J 452:139248. https://doi.org/10.1016/J.CEJ.2022.139248
129. Pedro AQ, Castro LS, Coutinho JAP, Freire MG (2024) Ionogels as advanced materials for overcoming challenges in wound healing and drug delivery. Nano Mater Sci. https://doi.org/10.1016/J.NANOMS.2024.06.010
130. Fang T, Cao X, Wang L, Chen M, Deng Y, Chen G (2024) Bioresponsive and immunotherapeutic nanomaterials to remodel tumor microenvironment for enhanced immune checkpoint blockade. Bioact Mater 32:530–542. https://doi.org/10.1016/J.BIOACTMAT.2023.10.023
131. Mohammadi K, Sani MA, Azizi-Lalabadi M, McClements DJ (2022) Recent progress in the application of plant-based colloidal drug delivery systems in the pharmaceutical sciences. Adv Colloid Interface Sci 307:102734. https://doi.org/10.1016/J.CIS.2022.102734
132. Sun H, Wu L, Gao N, Ren J, Qu X (2013) Improvement of photoluminescence of graphene quantum dots with a biocompatible photochemical reduction pathway and its bioimaging application. ACS Appl Mater Interfaces 5:1174–1179. https://doi.org/10.1021/AM3030849
133. Pereira J, Souza R, Moita A (2024) A review of ionic liquids and their composites with nanoparticles for electrochemical applications. Inorganics 12:186. https://doi.org/10.3390/INORGANICS12070186
134. Yang F, Yang H, Liu S, He R, Song W, Zhao X, Lu X, Wang Y (2025) The synthesis and application of functional materials strengthened by ionic liquids. J Mol Liq 429:127493. https://doi.org/10.1016/J.MOLLIQ.2025.127493
135. Potdar MK, Kelso GF, Schwarz L, Zhang C, Hearn MTW (2015) Recent developments in chemical synthesis with biocatalysts in ionic liquids. Molecules 20:16788–16816. https://doi.org/10.3390/MOLECULES200916788
136. Roy R, Chick P, York E, Rawling T (2024) Cytotoxicity of acridinium-based ionic liquids: structure-activity relationship and mechanistic studies. Chem Biol Interact 396:111042. https://doi.org/10.1016/J.CBI.2024.111042
137. Wang Y, Zhang Y, Li X, Sun M, Wei Z, Wang Y, Gao A, Chen D, Zhao X, Feng X (2015) Exploring the effects of different types of surfactants on zebrafish embryos and larvae. Sci Rep 5:1–10. https://doi.org/10.1038/SREP10107;TECHMETA
138. Tan VWG, Abuhasheesh Y, Tang DYY, Zhang Y, Ma Z, Chen L, Krishnamoorthy R, Show PL (2025) A critical assessment of ionic liquid-based aqueous biphasic systems for biomolecules extraction and CO2 absorption. Chem Eng J Adv 21:100670. https://doi.org/10.1016/J.CEJA.2024.100670
139. Docherty KM, Kulpa CF (2005) Toxicity and antimicrobial activity of imidazolium and pyridinium ionic liquids. Green Chem 7:185–189. https://doi.org/10.1039/B419172B
140. Flieger J, Flieger M (2020) Ionic liquids toxicity—benefits and threats. Int J Mol Sci 21:6267. https://doi.org/10.3390/IJMS21176267
141. El Mohamad M, Han Q, Dyett B, Yu H, Edgecomb S, Pride MC, Chism CM, Roberts A, Jones D, Tanner EEL, Drummond CJ, Greaves TL, Zhai J (2024) Cytotoxicity and cell membrane interactions of choline-based ionic liquids: comparing amino acids, acetate, and geranate anions. Chemosphere 364:143252. https://doi.org/10.1016/J.CHEMOSPHERE.2024.143252
142. Singh M, Singh G, Kaur H, Muskan N, Kumar S, Aswal VK, Kang TS (2024) Self-assembly of choline-based surface-active ionic liquids and concentration-dependent enhancement in the enzymatic activity of cellulase in aqueous medium. Phys Chem Chem Phys 26:16218–16233. https://doi.org/10.1039/D4CP01236D

143. García MT, Bautista E, de la Fuente A, Pérez L (2023) Cholinium-based ionic liquids as promising antimicrobial agents in pharmaceutical applications: surface activity. Antibact Act Ecotoxicological Profile Pharm 15:1806. https://doi.org/10.3390/PHARMACEUTICS15071806/S1
144. Li L, Wang J, Hu G, Yang S, Wang F, You X, Zhang H, He M (2025) Preparation of magnetic ionic liquid microemulsion collector by combining simulation and experiment and study of its flotation performance. Sep Purif Technol 354:128792. https://doi.org/10.1016/J.SEPPUR.2024.128792
145. Chen Y, Li J (n.d.) Development of a novel ionic liquid based microemulsion system for gemcitabine hydrochloride and in vitro evaluation in human cervical cancer HeLa cells. Indian J Pharm Educ Res 55. https://doi.org/10.5530/ijper.55.3.140
146. Ji L, Chen M, Zhang W, Nian B, Hu Y (2024) Comprehensive applications of ionic liquids in enzyme immobilization: current status and prospects. Mol Catal 552:113675. https://doi.org/10.1016/J.MCAT.2023.113675
147. Mansour AS, Abo-Aly MM, Rizk SA, Abd El Rahman AMM, Ghaly NF, Ragab AA, Alsabagh AM (2025) Synthesis of new zwitterionic surfactants and investigation of their surface active and thermodynamic properties. Sci Rep 15:1–19. https://doi.org/10.1038/S41598-025-97814-6;SUBJMETA
148. Shmool TA, Martin LK, Jirkas A, Matthews RP, Constantinou AP, Vadukul DM, Georgiou TK, Aprile FA, Hallett JP (2023) Unveiling the rational development of stimuli-responsive silk fibroin-based ionogel formulations. Chem Mater 35:5798. https://doi.org/10.1021/ACS.CHEMMATER.3C00303
149. Moshikur RM, Ali MK, Wakabayashi R, Moniruzzaman M, Goto M (2021) Methotrexate-based ionic liquid as a potent anticancer drug for oral delivery: *In vivo* pharmacokinetics, biodistribution, and antitumor efficacy. Int J Pharm 608:121129. https://doi.org/10.1016/J.IJPHARM.2021.121129
150. Hagarová I, Nemček L (2021) Application of metallic nanoparticles and their hybrids as innovative sorbents for separation and pre-concentration of trace elements by dispersive micro-solid phase extraction: a minireview. Front Chem 9:672755. https://doi.org/10.3389/FCHEM.2021.672755/BIBTEX
151. Wang Y, Zhang M, Yan Z, Ji S, Xiao S, Gao J (2024) Metal nanoparticle hybrid hydrogels: the state-of-the-art of combining hard and soft materials to promote wound healing. Theranostics 14:1534. https://doi.org/10.7150/THNO.91829
152. Rani S, Bagchi D, Pal U, Kumari M, Sharma M, Bera A, Shabir J, Pal SK, Saha-Dasgupta T, Mozumdar S (2020) The role of imidazolium-based surface-active ionic liquid to restrain the excited-state intramolecular h-atom transfer dynamics of medicinal pigment curcumin: a theoretical and experimental approach. ACS Omega 5:25582–25592. https://doi.org/10.1021/ACSOMEGA.0C02438/SUPPL_FILE/AO0C02438_SI_002.AVI
153. Koutsoukos S, Philippi F, Malaret F, Welton T (2021) A review on machine learning algorithms for the ionic liquid chemical space. Chem Sci 12:6820. https://doi.org/10.1039/D1SC01000J
154. Racki A, Paduszyński K (2025) Recent advances in the modeling of ionic liquids using artificial neural networks. J Chem Inf Model 65:3161–3175. https://doi.org/10.1021/ACS.JCIM.4C02364

Biosurfactants in Novel Drug Delivery: A Natural Approach to Enhanced Therapeutics

Mehak Awasthi, Pankaj Popli, Keshav Dhiman, and Rajan Swami

Abstract Biosurfactants are natural amphiphilic molecules obtained from natural sources that have gained significant attention as sustainable and biocompatible compared to synthetic surfactants in novel drug delivery systems. Their ability to reduce surface and interfacial tension, self-assemble into micelles or vesicles, and stabilize emulsions makes them valuable tools for enhancing the solubility, stability, and bioavailability of therapeutic agents. Structurally diverse classes such as glycolipids, lipopeptides, phospholipids, and polymeric biosurfactants exhibit various physicochemical and biological properties. When incorporated into novel systems, biosurfactants improve encapsulation efficiency and controlled release. Their antimicrobial, anticancer, and wound-healing properties further extend their biomedical relevance. As natural, eco-friendly excipients, biosurfactants represent a bridge between biotechnology and pharmaceutical innovation, providing a pathway toward safer and sustainable drug delivery platforms.

Keywords Biosurfactants · Microorganisms · Glycolipids · Liposomes · Antimicrobial

1 Introduction

In the new era, sustainability has become an important part of innovation. As researchers and industry personnel, they are moving toward eco-friendly alternatives to synthetic molecules. This event is clearly visible in the worldwide surfactant area, which continues to grow at a consistent rate of 4.5–5% annually. The market, valued at around USD 58.3 billion in 2024, is projected to rise to approximately USD 81.7 billion by 2030 [1]. An upward trend showed enhanced demand for safer and economical formulations worldwide [2].

M. Awasthi · P. Popli · K. Dhiman · R. Swami (✉)
Chitkara College of Pharmacy, Chitkara University, Rajpura, Punjab, India
e-mail: drrjnswami@gmail.com; rajan.2127@chitkara.edu.in

M. Abdul Rub et al. (eds.), *Industrial Applications of Individual and Mixed Surfactant Systems—Volume 2*, Sustainable Materials and Technology,
https://doi.org/10.1007/978-981-95-9195-4_14

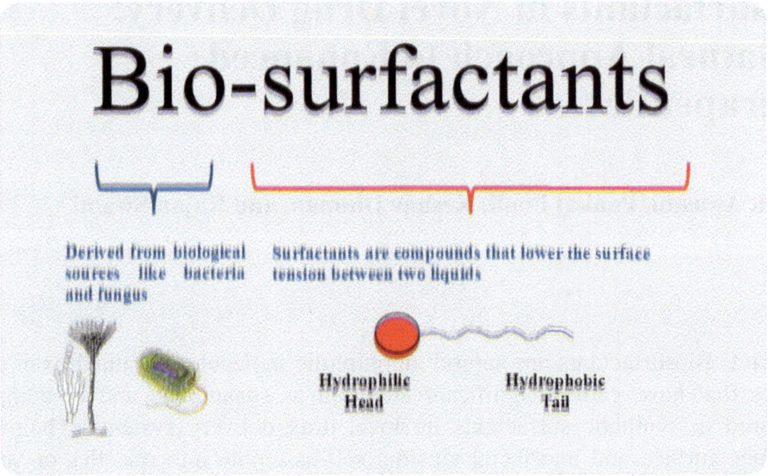

Fig. 1 Diagrammatic representation of biosurfactants illustrating their amphiphilic structure and biological origin

Biosurfactants are obtained from natural biological sources and have been demonstrated as a potential alternative to synthetic surfactants. These molecules are produced by numerous organisms such as bacteria, fungi, plants, and animals [3]. Moreover, their environmental compatibility, biodegradability, and low toxicity make them a suitable alternative to other synthetic surfactants. Common naturally occurring biosurfactants are dipalmitoyl phosphatidylcholine, obtained from mammalian lungs, and plant-derived lecithin [4, 5]. Based on structure, biosurfactants demonstrate the amphiphilic nature, containing both hydrophilic and hydrophobic domains. This dual nature enables them to self-assemble into micelles and make themselves at interfaces, making them effective in stabilizing emulsions, dispersing particles, and lowering interfacial tension, as illustrated in Fig. 1 [6, 7].

Biosurfactants are categorized into two groups, on the basis of their origin and production mode. Origin-based biosurfactants, often known as first-generation biosurfactants, such as, saponins, alkanolamines, esters of sugars, and other surface-active molecules, are derived from plants or animals [8]. On the other hand, the second-generation biosurfactants, called as microbial or green biosurfactants, are generated by microbial fermentation [9]. They include glycolipids such as rhamnolipids and sophorolipids. Moreover, microbial-derived biosurfactants have certain properties, including better stability and ease in production [2, 10]. Their biocompatibility property with biological systems often makes them appropriate choice to utilize [11]. Moreover, due to their structural property, biosurfactants can be utilized in numerous sectors such as pharmaceuticals, food technology, cosmetics, textiles, detergents, and even petrochemical processes [12, 13]. Biosurfactants often demonstrated significant pharmacological potential. Numerous lipopeptide and glycolipid biosurfactants have shown antimicrobial, antiviral, and anticancer activities. Moreover, their amphiphilic nature and ability to form micelles enhance solubility and ameliorate bioavailability of the drug, ensuring controlled drug release [14]. With the properties of biological safety, biosurfactants not only show efficient performance but also represent a meaningful move toward environmentally focused solutions. This chapter explores their emerging role in novel drug delivery systems and highlights their growing importance in pharmaceutical innovation.

2 Classification of Biosurfactants

Biosurfactants are majorly amphiphilic molecules that are synthesized by numerous microorganisms. They have the property of lower surface and interfacial tension. On the basis of molecular weight, these biomolecules are divided into low molecular weight (LMW) and high molecular weight (HMW). LMW biosurfactants lie in the 500–1500 Da range and are effective in lowering the surface tension; they are used for properties requiring strong interfacial activity [15]. On the other hand, HMW, including polymeric and particulate, exhibits higher emulsifying and stabilizing properties due to their large size and complexity in structure. On the basis of structural and functional characteristics, biosurfactants are divided into numerous subclasses, each recognized by its own chemical features and origin. Their structural heterogeneity is demonstrated in Fig. 2, which provides an overview of the major groups [16].

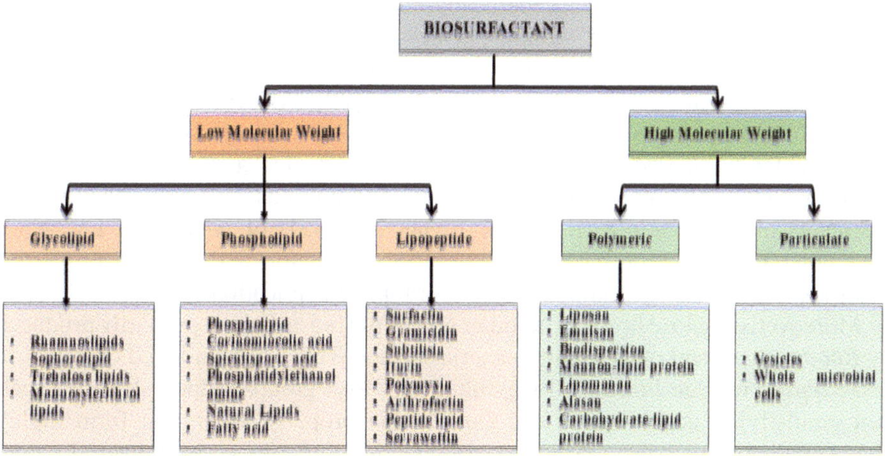

Fig. 2 Classification of biosurfactants

2.1 Low-Molecular-Weight Biosurfactants

LMW biosurfactants are less than 1 kDa biomolecules, recognized for their exceptional ability to lower surface and interfacial tension. Owing to their small size, they can quickly migrate and align at interfaces, which significantly enhances the wetting, spreading, and emulsifying behavior of liquid systems. Among the most prominent members of this category are glycolipids and lipopeptides, which play a crucial role in pharmaceutical applications [16, 17].

2.1.1 Glycolipids

Glycolipids are comprised of sugar units, which are covalently linked to long-chain fatty acids [17]. Numerous yeast and bacterial species produce the amphiphilic molecules and are considered biodegradable and non-toxic [18]. Among the glycolipids, rhamnolipids, which are extensively synthesized by *Pseudomonas* species, have received the utmost attention due to their structure; they consist of one or two rhamnose sugar residues glycosidically linked to β-hydroxy fatty acids. Their surface properties enable them to lower the surface tension of water, while their high biodegradability and minimal toxicity are prominent properties for bioremediation, detergency, and environmental cleanup applications [19].

Moreover, sophorolipids are another category of biosurfactants synthesized by non-petrogenic yeasts such as *Starmerella bombicola.* According to their structure, they have a sophorose disaccharide headgroup linked to a hydroxylated fatty acid tail and have two structural variants, which are lactonic and acidic. These variants have shown surface-active properties and also demonstrated biological activities, including antiviral and anticancer effects [20]. As reported by Adu et al., acidic sophorolipids and mono-rhamnolipids showed no cytotoxicity in human keratinocytes compared to conventional surfactants such as sodium lauryl ether sulfate and other glycolipids. Furthermore, di-rhamnolipids have been seen to lower the levels of interleukin-8 (IL-8) and higher the levels of interleukin-1 receptor antagonist (IL-1RA) expression, ensuring anti-inflammatory activity. Such properties make sophorolipids a potential approach for topical applications [21].

Moreover, trehalose lipids are another glycolipid biosurfactant primarily produced by *Rhodococcus* and other Actinobacteria species. They are comprised of trehalose disaccharide residue linked to long-chain fatty acyl groups, which showed excellent emulsifying capabilities [22]. Due to this property, these can form highly stable emulsions and also make them valuable in oil recovery and industrial procedures. Trehalose lipids also exhibit notable antimicrobial and antiviral properties [23]. For instance, Songdech et al. successfully created *Meyerozyma guilliermondii* MX to biosynthesize potent antimicrobial glycolipids using xylose and palm oil as substrates. Through strategic antibiotic regulation and fermentation optimization, the researchers achieved an impressive yield of 159.42 g/L, accompanied by substantial biofilm inhibition against *Candida albicans.* These findings emphasize the antifungal efficacy [24].

2.1.2 Phospholipid

Phospholipid-based biosurfactants have a major group of amphiphilic molecules that include microbial membrane lipids and numerous fatty acid derivatives, demonstrating surfactant properties [25]. Corinomocolic acids, also known as corynomycolic acids, are structurally intricate 3-hydroxy-α-branched fatty acids (R-CH(OH)-CH(R')-COOH) obtained from *Actinobacteria,* for instance, *Nocardia erythropolis* and *Corynebacterium lequs* [26]. These molecules have the properties of hydroxyl and alkyl groups, along with long hydrocarbon chains, which offer surface activity and the ability to lower water surface tension. Notably, their emulsifying efficiency remains consistent across a wide pH range (2–10), underscoring their chemical stability [27]. Due to their biodegradability property and higher emulsifying efficiency, corinomocolic acids are increasingly recognized as valuable candidates for bioremediation, detoxification of the environment, and formulation of green cosmetics and pharmaceuticals [28]. Their properties offer an appealing choice for biocompatible surfactant systems. Moreover, spiculisporic acid is another type of phospholipid which has γ-butenolide lipid (4,5-dicarboxy-4-pentadecanolide) characterized by a decyl (C_{10}) hydrocarbon chain, a lactone ring, and two terminal carboxyl groups [27, 29]. Obtained from *Penicillium spiculisporum,* it has also been procured from relatable fungi such as *Talaromyces trachyspermus* and *Aspergillus* species [16, 27]. Owing to its strong surface property, biocompatibility, and low skin irritation potential, spiculisporic acid has gained attention for its antibacterial activity and stabilized the oil-in-water emulsions and metal ion complexes [29]. These properties make it a potential biosurfactant for heavy metal remediation and as a mild emulsifying agent in cosmetic formulations.

Furthermore, another most commonly utilized component in nanocarrier formulation is phospholipid-based biosurfactant, specifically phosphatidylethanolamine (PE), a zwitterionic glycerophospholipid featuring a glycerol backbone, two fatty acid chains, and a phosphate-ethanolamine headgroup [30]. In microbial cell membranes, PE are naturally produced. PE is released in extracellular vesicles by bacteria such as *Acinetobacter* and *Rhodococcus pp.* Moreover, utilizing these surfactants can offer a stable microemulsion, reducing the hexadecane–water interfacial tension to below 1mN/m, and offer stability across wide pH and salinity ranges. Consequently, PE-based biosurfactant demonstrated potential in oil spill remediation and drug delivery formulations [16, 31]. For instance, Nwaguma et al. isolated *Klebsiella pneumoniae* IVN51 from hydrocarbon-contaminated Nigerian soil and found that it produced a PE biosurfactant capable of emulsifying petrol, diesel, kerosene, and other hydrocarbons. Maximum activity was achieved at 30–35 °C and pH 7–9 [32]. On the other hand, Rojweska et al. found the therapeutic property of PE demonstrated that purified liquorice saponin interacted effectively with DPPE membranes than crude extracts, ameliorating both monolayer stability and surfactant property.

These saponins targeted the phospholipid headgroups, modulating zeta potential without penetrating deeper bilayer, providing their utility in biomembrane-specific drug delivery [33].

Some microorganisms also originate with no-charge lipid biosurfactants, composed mainly of triacylglycerols, wax esters, or fatty acyl esters. For instance, *Clostridium pasteurianum* secretes a non-charge lipid having a property of reducing water surface tension from approximately 72 to 55 mN/m. These no-charge amphiphilic mixtures make stable hydrocarbon emulsions and are co-produced with glycolipids under anaerobic or saline stress conditions. Their stability and emulsification efficiency make them an attractive option for oil recovery and bioremediation processes [34]. Further understanding this, Bellebcir et al. reported the secretion of *Streptomyces thinghirensis* 7SDS from oil-contaminated soil, capable of releasing the phospholipid biosurfactant termed "Streptingfactin," which is generated from water substrates. This molecule reduced surface tension to 27.96 mN/m, demonstrated activity under extreme pH or temperature, and exhibited both antimicrobial and hydrocarbon-emulsifying activity, confirming its promising approach for industrial feasibility [35]. Moreover, long-chain fatty acids (C_{12}–C_{20}), such as palmitic and oleic acids, have the property of natural biosurfactants. Generated by microbial strains such as *Bacillus* and *Pseudomonas*, these saturated and unsaturated fatty acids demonstrated a mild reduction in surface tension and ability of emulsifying. However, alone its activity is limited, they also act synergistically with other biosurfactants, demonstrating the activity of oil displacement, surface wetting, and biofilm dispersion. Due to their biocompatibility property, these fatty acids hold significant potential environmentally friendly surfactants in food, cosmetic, and nanoformulations [27, 34].

2.1.3 Lipopeptides

Lipopeptides are biosurfactants comprised of a short chain of peptide functionalized with fatty acid tails. Many molecules are cyclic, having the lactone or lactum ring structure that is responsible for strong amphiphilic properties [36]. Moreover, surfactin is one of the most studied cyclic lipo-heptapeptides, produced by Bacillus subtilis. According to their structure, it consists of a seven-residue peptide which linked with a lactone ring to a β-hydroxy fatty acid [37]. These biosurfactants demonstrate surface activity properties, with a significant reduction in hexadecane–water interfacial tension to approximately 1 mN/m. Its amphipathic nature properties demonstrated efficient in forming micelles and particle aggregation at low concentration, and these properties have significant potential in bioremediation and enhanced oil recovery. Additionally, surfactin has a remarkable property of antimicrobial and anti-inflammatory [38].

Furthermore, iturins, another class of cyclic lipopeptides, are comprised of α-amino acid heptapeptide linked with a β-amino acid and are generated by Bacillus species [39]. Iturin A, one of the types of iturins, exhibits antifungal property by demolishing the fungal membrane integrity, with minimum toxicity. Iturin molecules have a potent inhibitory effect against plant-pathogenic fungi and bacteria, making them a remarkable biopesticide. Furthermore, they also reduce surface tension and help in emulsion formation, which is utilized in agricultural, food preservation, and drug delivery formulation [40]. For instance, Phulpoto et al. demonstrated the use of lipopeptide biosurfactants derived from Bacillus subtilis CW2 on biohydrogen production from lignocellulosic biomass under anaerobic dark fermentation. Reactors supplemented with 25–50 mg/L biosurfactant showed 30–36% enhancement in hydrogen yield compared to controls. SEM data demonstrated substrate porosity and microbial accessibility. Illumina sequencing revealed microbial shifts favoring Firmicutes, notably Ruminiclostridium and Bacillus, associated with cellulose degradation and hydrogen evolution. The study supports biosurfactant-mediated modulation of microbial consortia to improve lignocellulose valorization in biological applications [41]. Mnif et al. characterized lipopeptide biosurfactants produced by *Bacillus subtilis* ZNI5, isolated from hydrocarbon-contaminated soil. HPLC-MS analysis identified surfactin, iturin, licheniformin, and novel isoforms. The biosurfactant reduced surface tension to 32 mN/m (CMC: 350 mg/L) and dispersed motor oil up to 80 mm. It demonstrated high thermal, pH, and salt stability, with performance comparable to synthetic surfactants. The study supports its application in oil spill remediation and industrial processes requiring eco-friendly surface-active agents [42].

2.2 *High-Molecular-Weight Biosurfactants*

HMW biosurfactants consist of large polymeric or particulate biomolecules that primarily stabilize emulsions rather than markedly reducing surface or interfacial tension [16]. They are broadly classified into two categories: polymeric and particulate biosurfactants, each contributing to emulsification, dispersion, or enhanced bioavailability through various mechanisms.

2.2.1 Polymeric Biosurfactants

Polymeric biosurfactants are macromolecular complexes, often exceeding 100 kDa, composed of polysaccharides, proteins, or lipoprotein conjugates. These compounds exhibit remarkable emulsifying, stabilizing, and film-forming capabilities, making them highly valuable in environmental and biomedical applications (31). For instance, emulsan is generated from *Acinetobacter calcoaceticus* strain RAG-1. Emulsan is a heterolipopolysaccharide comprised of an anionic polysaccharide backbone covalently coupled with fatty acids with esters and amide bonds. With molecular weights approaching 1000 kD, emulsan is highly efficient at low concentrations and is extensively studied for its ability to stabilize oil-in-water emulsions, and its exceptional performance has also inspired applications in drug delivery systems and nanoparticle formulations [43]. Another prominent is alasan, secreted by

Acinetobacter radioresistens; it is an anionic heteropolysaccharide protein complex composed of an acidic polysaccharide coupled with an alanine-rich 45 kDa proteins. It effectively emulsifies aromatics and crude oil, highlighting its potential in bioremediation and enhanced oil recovery (31). Furthermore, Liposan is derived from *Candida lipolytica* and various mannoprotein complexes from yeast species. These macromolecules are characterized for biocompatibility, biodegradability, and having the property of natural emulsifiers in pharmaceutical and cosmeceutical formulations [3].

2.2.2 Particulate Biosurfactant

These differ from polymeric types in that they are composed of whole cellular structures or membrane-derived vesicles rather than isolated biomolecules. Their amphiphilic properties are produced by the heterogeneous distribution of hydrophilic and hydrophobic molecules on microbial surfaces, which effectively stabilize emulsions and interact with hydrophobic substrates. Membrane vesicles, with a size of 20 to 200 nm in diameter, have the outer membrane lipids, lipopolysaccharides, and proteins. These vesicles can be loaded with hydrophobic molecules, which improves their solubilization [16]. For instance, *Acinetobacter* species grown on hydrocarbon compounds can produce phospholipid—or protein-rich vesicles with a size range of 20–50 nm that help in forming the stable microemulsions and enhancing substrate uptake. In addition to vesicles, whole microbial cells can function as particulate biosurfactants [27]. Certain bacteria, such as *Acinetobacter calcoaceticus* 2CA2, possess highly hydrophobic cell surfaces capable of emulsifying hydrocarbons without additional biosurfactant secretion. Thus, both extracellular vesicles and intact microbial cells play integral roles in surface activity, contributing significantly to bioprocessing, hydrocarbon degradation, and environmental remediation efforts [16, 27]. They are composed of cellular components rather than isolated biomolecules. These include membrane vesicles and intact microbial cells, which display amphiphilic surface features. Membrane vesicles are typically 20–200 nm in diameter and consist of outer membrane lipids, lipopolysaccharide, and protein. They can encapsulate hydrophobic compound and aid in their solubilization. For example, Acinetobacter species grown on hydrocarbons secrete vesicles (20–50 nm) enriched in phospholipids, protein, and LPS, which help form stable microemulsion and enhance substrate update. In addition, whole microbial cells may act as particulate biosurfactant. Their surfaces contain both hydrophobic and hydrophilic domains, enabling them to stabilize emulsions and reduce surface tension. Acinetobacter calcoaceticus 2CA2 is noted for its highly hydrophobic cell surface, capable of emulsifying hydrocarbons independently of additional biosurfactant production. As a result, both extracellular vesicles and microbial cells contribute meaningfully to surfactant activity in bioprocessing and environmental cleanup applications.

3 Physicochemical Properties of Biosurfactants Governing Nanocarriers

3.1 Reduction in Surface and Interfacial Tension

Reduction in surface tension is a fundamental physicochemical application that enables biosurfactants to facilitate the formation of nanoscale drug delivery systems. For instance, surfactin, a cyclic lipopeptide derived from *Bacillus* species, reduces the surface tension of water from 72 mN/m to approximately 27 mN/m at low concentrations. This property enhanced the production of droplets and micelles, thereby ameliorating the solubility of poorly water-soluble drugs. Lowering interfacial tension further enhances encapsulation efficiency, making biosurfactants essential for stabilizing micellar, liposomal, and nanoemulsion formulations [44].

3.2 Critical Micelle Concentration

The ability of biosurfactants to form micelles at low critical micelle concentrations enhances their efficiency as natural emulsifiers. Below the CMC, biosurfactant molecules remain as single monomers; beyond the threshold, they spontaneously self-assemble into micelles capable of encapsulating lipophilic drugs. For example, biosurfactants derived from *Lactobacillus* species exhibit CMC values around 2–3 mg/ml, above which micelles readily solubilize hydrophobic drug molecules. A low CMC is particularly advantageous in drug delivery, as it allows stable carrier formation with minimal surfactant use, reducing toxicity and cost of formulation, as represented in Fig. 3 [45].

3.3 Hydrophilic–Lipophilic Balance (HLB)

Biosurfactant HLB essentially determines whether it leans toward water or oil, which dictates the type of emulsion that will form. By nudging the HLB calculation of the carriers, high-HLB biosurfactants drift toward oil-in-water systems which solubilize hydrophobic drugs while low-HLB tend to form water-in-oil systems or micelles. Selecting a biosurfactant that lands on the HLB spot paves the way for stable nanoemulsions [3].

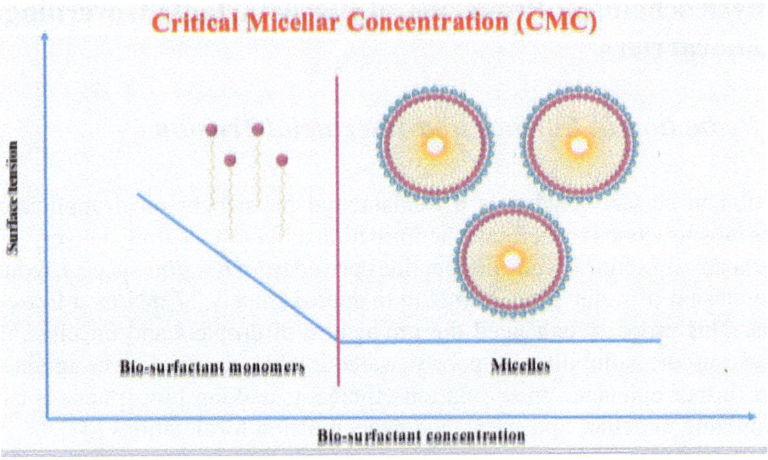

Fig. 3 Demonstrating CMC showing biosurfactant monomer to micelle transition

3.4 *Interfacial Tension and Emulsification Efficiency*

The property of biosurfactants to minimize the interfacial tension is utilized in nanoformulations. For instance, surfactin can minimize interfacial tension to 3–4 mN/m, even in extreme conditions of pH or temperature. This enables the production of a nanoemulsion, thereby ameliorating the encapsulation of hydrophobic drugs in lipid-based formulations. This approach enhanced the solubility of the drug and also minimized the particle aggregation and long-term stability [46].

3.5 *Zeta Potential and Colloidal Stability*

Zeta potential is one of the major parameters that ensure the stability of the formulation, which is influenced by the physicochemical properties of the biosurfactant. Many biosurfactants have anionic functional groups, demonstrating the negative charge on the surface, which is around − 20 mV that producing the electrostatic repulsion and preventing particle aggregation. In nanoformulation, zeta potential is responsible for suspension stability and interactions with biological membranes. On the other hand, highly negative charges ameliorate dispersion and shelf life, which can be helpful in cellular uptake [47].

3.6 *Influence of Molecular Weight and Structural Architecture*

Molecular weight and structural complexity of biosurfactants showed self-made behavior. LMW type of biosurfactants, such as glycolipids and lipopeptides, produces micelles and demonstrate lower interfacial tension, enhancing the solubilization of hydrophobic drugs. On the other hand, HMW type of biosurfactants offers stabilization in emulsions mainly by the steric hindrance, producing durable interfacial films. Difference in length of chain, polarity of head group, and saturation level impact the hydrophobicity, aggregation, and release of drug through nanoformulations [16].

3.7 *Self-Assembly and Colloidal Organization*

Above their CMC, biosurfactants self-assemble into a uniform structure such as micelles, vesicles, or liquid-crystalline phases; these structures are utilized in drug delivery. Moreover, the morphology of particles depends upon the structure of molecules and environmental conditions such as pH, ionic strength, and temperature [48]. Furthermore, amphiphilic nature of biosurfactants showed electrostatic and steric stability, demonstrating colloidal stability. While the thermodynamic stability of microemulsions requires external energy input, biosurfactants can form kinetically stable nanoemulsions that demonstrate long shelf life [49].

3.8 *Rheological and Environmental Stability*

Rheological properties of biosurfactants-based formulations such as shear-thinning behavior and viscoelasticity, offer additional advantages in drug delivery. These systems flow easily under applied stress; it could be injection or mixing, and yet reform their structure, ensuring controlled release and physical stability [50]. Furthermore, biosurfactants show environmental stability across a wide range of parameter such as pH, temperature, and salinity. Glycolipids and lipopeptides demonstrate the surface activity under acidic or alkaline conditions and remain effective in saline environments [51].

4 Biosurfactant-Enabled Nanocarrier Systems

Nanotechnology is an advanced drug delivery system that lies between the size ranges of 1–100 nm. In drug delivery systems, it ameliorates the solubility of poorly water-soluble drugs, protects from drug degradation, and offers targeted or controlled release, thereby enhancing bioavailability and reducing systemic toxicity [52]. Biosurfactants are capable of self-assembling into micelles or interfacial films that encapsulate and solubilize hydrophobic drugs [46]. Utilizing biosurfactants in nanocarriers makes them biocompatible and stable drug delivery options with higher drug loading. Numerous platforms have been produced using biosurfactants.

4.1 Lipid-Based Nanocarriers

4.1.1 Liposomes

Liposomes are bilayer vesicles composed of phospholipids, capable of encapsulating both hydrophilic and hydrophobic drugs. Designing liposomes with biosurfactants can replace synthetic surfactants and ameliorate the properties of liposomes [53]. For instance, glycolipid biosurfactants like rhamnolipids and sophorolipids have been used to produce liposomes. Rhamnolipid-modified liposomes (loaded with curcumin) demonstrated ameliorated vesicle formation and loading efficiency, while sophorolipid-coated liposomes maintained dispersion and offered ameliorated drug content [4].

Biosurfactants mixed with lipids form bilayers, which enhance the membrane rigidity and reduce aggregation. For instance, Dias-Souza et al. engineered liposome using rhamnolipid, meropenem, and chloramphenicol to evaluate their antibiofilm efficacy. The findings demonstrated particle size of liposome is 100 nm, and the rhamnolipid-loaded liposomes achieved up to 92% biofilm reduction against *Stenotrophomonas maltophilia*, while meropenem liposomes reduced *Pseudomonas aeruginosa* biofilms by 85%, and chloramphenicol liposomes demonstrated inhibition by 60%. After 90 days of storage, all formulations confirmed good stability [54]. In another study, Sanches et al. developed rhamnolipid-based liposomes to investigate the antibacterial activity of the ParELC3 peptide, which normally lacks bacterial permeability. Rhamnolipids extracted from *Pseudomonas aeruginosa* were combined with POPC and cholesterol to form stable, nano-sized vesicles with a size range of 93 to 212 nm and encapsulation efficiency of 63%, demonstrating sustained release up to 15 h. Confocal imaging demonstrated that encapsulation enabled the peptide to penetrate *E. coli* and *S. aureus* cells, overcoming membrane barriers. The liposomal ParELC3 showed markedly reduced minimum inhibitory concentration values of 11 μmol L^{-1} compared to the free peptide > 100 μmol L^{-1}. Hemolysis and cytotoxicity assays demonstrated no toxicity in HepG2 cells, indicating excellent biocompatibility [55]. Moreover, Zhou et al. engineered chitosan-coated rhamnolipid liposomes to enhance the stability and bioavailability of curcumin. The positively charged chitosan layer formed an electrostatic interaction with anionic rhamnolipid, forming a uniform size range of 413 nm, with a zeta potential of + 19.7 mV. Utilizing chitosan coating ameliorated acid and heat stability, effectively reducing curcumin degradation at 80 °C with a 94.4% retention compared to 78.8% in uncoated liposomes. Under simulated gastrointestinal conditions, chitosan-coated liposomes showed slower curcumin release, enhanced transformation, and greater bioavailability due to improved protection and stability, whereas chitosan reduced bioaccessibility [56].

Furthermore, Chen and Zhang developed a sophorolipid-functionalized piperine-loaded liposome. Incorporation of the biosurfactants demonstrated the reduction in particle size and PDI, increased entrapment efficiency, and ameliorated stability at 4 °C. The modified liposomes also showed stability under varying NaCl and pH conditions and exhibited sustained drug release [57]. This underscores the significant potential of biosurfactant-engineered liposome formulation in drug delivery [4, 54].

4.1.2 Solid Lipid Nanoparticles (SLNs)

Solid lipid nanoparticles are the lipid particulate system, consisting of a solid lipid core (e.g., triglycerides or waxes) stabilized by surfactants [58]. They provide sustained drug release and protect labile actives. In biosurfactant-enabled SLNs, a microbial surfactant often serves as the emulsifier [59]. For instance, Asadpour Panbechouleh et al. developed rhamnolipid-based SLNs encapsulating the antioxidant menadione sodium bisulfite (MSB) for topical anti-pigmentation therapy. The optimized formulation exhibited a mean particle size of 117.26 nm, zeta potential − 6.28 mV, and entrapment efficiency of 83.34%, indicating good uniformity and stability. The rhamnolipid-stabilized SLNs provided a controlled release profile and significantly enhanced skin penetration compared to free MSB. Biological evaluation showed strong melanin inhibition through 95.14% L-DOPA oxidation suppression and high cell viability (85.58%), confirming biocompatibility. Skin irritation studies further verified their safety for topical application, suggesting rhamnolipid-based SLNs as an effective and eco-friendly platform for antioxidant and skin-lightening formulations [60]. This indicates that the biosurfactant both stabilizes the lipid matrix and enhances skin permeation. In general, SLNs combine the advantages of lipids (biocompatible GRAS components) with nanoparticle functionality [61].

4.1.3 Nanostructured Lipid Carriers (NLCs)

NLCs are another generation of lipid particulate system comprising mixed solid/liquid lipids; the use of biosurfactants makes them the potential to use in drug delivery [62]. In one study, Azevedo et al. formulated rhamnolipid-stabilized nanostructured lipid carriers (NLCs) for vitamin D_3 delivery using mixed solid (glycerol monostearate) and liquid (medium-chain triglycerides) lipids. The NLCs showed particle sizes between 92.1 and 108.1 nm, low PDI, and remained stable for 60 days at 4 °C. Structural analyses indicated that increasing liquid lipid content reduced crystallinity and improved matrix flexibility. Cytotoxicity tests confirmed that rhamnolipid concentrations ≤ 0.25 mg/mL were biocompatible with Caco-2 cells. In vitro digestion studies demonstrated that formulations with higher solid lipid content and smaller particle size enhanced lipolysis and bioaccessibility of vitamin D_3 [63]. In a separate investigation, Azevedo et al. developed nanostructured lipid carriers (NLCs) using rhamnolipids as natural biosurfactants and evaluated the effect of lipid composition on their properties. The NLCs showed spherical morphology, particle sizes of 97–120 nm, and good uniformity (PDI 0.20–0.26). Increasing liquid lipid content reduced crystallinity and altered melting behavior, enhancing flexibility. The optimized formulation (0.05% rhamnolipids; 10:90 solid-to-liquid lipid ratio) remained stable at 4 °C for 30 days and retained its structure after freeze-drying, confirming rhamnolipids as effective stabilizers for biocompatible, long-term lipid nanocarriers [64].

4.1.4 Core–Shell Nanocapsules

Core–shell nanocapsules using biosurfactants are an emerging class of nanocarriers where a biosurfactant acts as a key structural or stabilizing component, enabling the encapsulation and delivery of therapeutic agents with improved efficiency and biocompatibility. Katiyar et al. developed lipid—and biosurfactant-based core–shell nanocapsules for efficient paclitaxel (PTX) delivery in breast cancer therapy. The nanocapsules exhibited a uniform size (250 nm), high drug loading (20%), and biphasic release—an initial burst followed by sustained Fickian-controlled release. Biosurfactant incorporation enhanced stability, cellular uptake, and cytotoxicity in MCF-7 cells compared to free PTX. In vivo studies demonstrated improved pharmacokinetics, higher therapeutic efficacy, and reduced systemic and organ toxicity versus commercial formulations (Intaxel® and Nanoxel®) [65].

4.1.5 Niosomes

Niosomes are surfactant-based vesicular systems. By utilizing biosurfactants, it enhanced the biocompatibility and stability [66]. These nanocarriers ameliorated the encapsulation of the drug, sustained release, and reduction in toxicity. For instance, Mazzotta et al. identified *ammonium glycyrrhizate* (AG), a triterpenoid saponin derived from liquorice root, as a natural biosurfactant. AG's amphiphilic structure enabled the self-assembled structure into uniform vesicles. These vesicles demonstrated nanosize, negative zeta potential, and high colloidal stability. Importantly, AG niosomes effectively encapsulated both fluorescein sodium and curcumin drug and demonstrated enhanced skin penetration. The formulation showed no cytotoxic or hemolytic effects and demonstrated anti-inflammatory activity confirmed by albumin denaturation assays [67]. Moreover, Shiri et al. formulated berberine-loaded niosomes using saponin as a natural biosurfactant to ameliorate antibacterial performance. Formulation demonstrated high encapsulation efficiency of 93.7% with a nano-sized particle. In vitro studies demonstrated sustained release pattern and exhibited antibacterial activity against *Staphylococcus aureus* and *Escherichia coli* [68].

4.1.6 Emulsion-Based System

Nanoemulsions are submicron oil-in-water emulsions (droplets < 100 nm) stabilized by surfactants; microemulsions are similar but form spontaneously due to surfactant/co-surfactant mixtures. Both serve to solubilize hydrophobic drugs and enhance absorption. Biosurfactants can act as the emulsifier (or co-emulsifier) in these systems [69].

Microemulsions

Microemulsions are thermodynamically stable systems that enable nanoscale solubilization of actives, offering enhanced antimicrobial and physicochemical performance. For example, Luo et al. developed a sophorolipid (SL)-based microemulsion (ME) system to enhance the delivery of xanthohumol (XN), a poorly soluble flavonoid from hops. By partially replacing the synthetic surfactant Tween 80 with the natural biosurfactant SL, the researchers achieved a dramatic 4000-fold increase in XN solubility and extended its storage half-life to over 150 days. The SL incorporation preserved the microemulsion's structural integrity while improving solubilization efficiency, storage stability, and antioxidant activity. In vitro release studies indicated non-Fickian diffusion behavior and enhanced digestive stability and bioaccessibility of XN. This biosurfactant-based formulation provided an eco-friendly and efficient strategy for stabilizing and delivering hydrophobic bioactives in food and pharmaceutical applications [70]. Shen et al. formulated rhamnolipid-stabilized essential oil microemulsions to enhance the stability and antimicrobial performance of volatile oils for fruit preservation. The optimized microemulsion demonstrated a particle size of 16.8 nm, with high transmittance and long-term stability, with an extremely slow release of 24.4% over 35 days. It showed significant antimicrobial activity against *E. coli*, *S. aureus*, *C. albicans*, *S. cerevisiae*, and *A. niger*, with a markedly lower MIC for *S. aureus* of 39 mg/L compared to free essential oils of 1250 mg/L. When applied to blueberries, the rhamnolipid-based microemulsion significantly prevented microbial decay and preserved fruit quality, notably reducing anthocyanin loss [71].

Nanoemulsions

Nanoemulsions are another type of lipid-based system that require surfactant to get stabilized; however, the use of biosurfactants demonstrates ameliorated droplet uniformity, kinetic stability, and surface properties, offering promising delivery of BCS class II drug. For instance, Al-Sakkaf and Onaizi formulated crude O/W nanoemulsions stabilized by rhamnolipid biosurfactants and evaluated their stability under varying pH (3–12) and salinity conditions. The optimized nanoemulsions demonstrated a particle size of 38 nm with a negative zeta potential and kinetic stability. An increase in pH and salinity raised the viscosity due to the formation of smaller and closely packed droplets [72]. Moreover, Zanotto et al. utilized the mannosylerythritol lipid as a biosurfactant for stabilizing O/W emulsions containing essential oils *Thymus vulgaris*, *Lippia sidoides*, and *Cymbopogon citratus*. Utilizing the biosurfactant enhanced the emulsion's physicochemical stability. Formulation exhibited ameliorated antimicrobial potency against a range of bacterial and fungal strains. Furthermore, biosurfactants significantly demonstrated the antioxidant activity and the enhancement in radical-scavenging activity by 13.7% [73]. Similarly, Yin et al. formulated a rhamnolipid-coated cinnamaldehyde nanoemulsion utilizing spontaneous emulsification. The optimized formulation demonstrated a particle size of 20.5 nm, spherical morphology, and high encapsulation efficiency. Formulation caused membrane disruption and degradation of biofilm, reducing extracellular polysaccharides and protein levels by 70% and 50%, respectively. In vivo studies demonstrated a significant reduction in bacterial load and inflammatory biomarkers (TNF-α and IL-6) and showed nearly doubled the wound-closure rate [74]. Furthermore, Haddaji et al. formulated a nanoemulsion utilizing a lipopeptide biosurfactant from Bacillus species to stabilize the formulation at a concentration of 1–3.3%. The formulation demonstrated the uniform droplet size and high physical stability. It exhibited antibacterial and antibiofilm properties against multidrug-resistant *E. coli* and *S. aureus*, significantly disrupting preformed biofilms and inhibiting bacterial adhesion. When it is combined with standard antibiotics, the system re-sensitizes resistant strains, enhancing antibiotic susceptibility [69]. Moreover, Mottola et al. optimized the use of these biosurfactants in agriculture by formulating nanoemulsions by incorporating synthetic rhamnolipids to increase the plant immunity. The formulation exhibited small particle size, high colloidal stability, and strong surface activity, ensuring uniform leaf coverage. When applied to *Arabidopsis thaliana*, the formulation demonstrated defense response and efficiently resisted against pathogenic infection without causing phytotoxicity or disrupting soil microbial communities. Compared to natural rhamnolipids, formulation offered ameliorated protection [75].

Multiple Emulsions (E.G., W/O/W)

Biosurfactants stabilize complex multiemulsions (W/O/W) and co-encapsulate the hydrophilic and hydrophobic drug, enabling sustained release and improved drug delivery efficiency. For example, Lee et al. designed rhamnolipid-coated water-in-oil-in-water (W/O/W) double emulsion nanoparticles for the co-delivery of doxorubicin (DOX) and erlotinib (ERL) to achieve synergistic cancer therapy. The rhamnolipid shell enabled the encapsulation of both hydrophilic (DOX) and hydrophobic (ERL) agents with high loading efficiency and improved colloidal stability. The dual-drug nanoparticles exhibited enhanced cellular uptake and sustained drug release, leading to greater cytotoxicity against cancer cells than single-drug or free-drug treatments. In vivo studies confirmed superior tumor accumulation, potent tumor suppression, and minimal systemic toxicity, validating the biosurfactant-coated double emulsion system as a safe and effective nanoplatform for combination cancer therapy [76].

4.2 Hybrid Nanocarrier

Nanoscale complexes co-formulated from surfactants and dendrimers for co-delivery. For example, Malavia et al. engineered green surfactant–dendrimer aggreplexes by integrating a histidine–lauric acid-based biosurfactant with poly(amidoamine) (PAMAM) dendrimers for the co-delivery of docetaxel (DTX) and SIRT1 shRNA. The self-assembled aggreplexes formed via electrostatic interactions exhibited a particle size of 262 nm, PDI of 0.25, and encapsulation efficiency of 70.56%, with spherical morphology confirmed by TEM. The system enabled biphasic, sustained DTX release and was hemocompatible, making it suitable for intravenous administration. Cellular studies in MDA-MB-231 breast cancer cells revealed 1.87-fold higher uptake, a 4.16-fold reduction in IC_{50}, and a 3.57-fold increase in apoptotic index compared to free DTX. Additionally, the aggreplexes showed enhanced gene transfection efficiency relative to naked shRNA and commercial reagents, demonstrating synergistic chemo-gene therapeutic potential. These results underscore the value of amino acid-based "green" surfactants as biocompatible, dual-function materials for safe and effective combination cancer therapy [77]. Moreover, nanosystems combine lipid vesicles with metal nanoparticles and biosurfactant. For instance, Kannan et al. developed multilamellar liposomes co-encapsulating a lipopeptide biosurfactant from *Paenibacillus thiaminolyticus* and copper nanoparticles (CuNPs) to combat biofilm-forming pathogens. The formulation (153 nm, PDI 0.20) showed synergistic antibacterial effects, significantly lowering MIC values against *Pseudomonas aeruginosa* (89 μg/mL) and MRSA (105 μg/mL). It effectively inhibited virulence factors such as pyocyanin and staphyloxanthin, reduced EPS and biofilm formation (> 80%), and induced strong ROS-mediated bacterial death. In vivo studies using *C. elegans* confirmed enhanced survival and reduced pathogen colonization, demonstrating the formulation's strong therapeutic and anti-biofilm potential [78].

4.3 Polymer-Based Nanocarriers

4.3.1 Natural Polymer Nanoparticles

Polymer nanoparticles stabilized through biosurfactant offer enhanced stability, modulated surface charge, and ameliorated antimicrobial activity, offering therapeutic delivery. For instance, Karamchandani et al. formulated the fungal chitosan nanoparticles derived from *Cunninghamella echinulata* for sustainable agricultural use with rhamnolipid biosurfactant. The formulation demonstrated a mean particle size of 144 nm and a zeta potential of + 45.6 mV. It exhibited antimicrobial and antifungal activities, significantly reducing MIC and inhibiting fungal spore germination by up to 90% [79]. Arif et al. developed chitosan and PMLA nanoparticles loaded with rhamnolipids. The formulation demonstrated 73.1% encapsulation efficiency, with an average size of 217 nm. It exhibited 99% inhibition of *H.*

pylori biofilm while remaining biocompatible with fibroblasts. TEM confirmed the spherical formation of the formulation. The formulation demonstrated enhanced antimicrobial efficacy [80]. Moreover, Bettencourt et al. engineered chitosan-based nanoparticles utilizing the sophorolipids and rhamnolipids. Prepared formulations were stable and demonstrated the mean sizes of 210 ± 15.7 nm of sophorolipids formulation and 329.6 ± 8.0 nm for rhamnolipids formulation with an encapsulation efficiency of 41.1 ± 8.8% and 74.2 ± 1.3%, respectively. TEM confirmed the uniform morphology of the formulation. Rhamnolipid formulation demonstrated enhanced antimicrobial property against *Staphylococcus aureus* and also achieved a minimal inhibitory concentration of 130 μg/mL and 99% biofilm inhibition compared to the sophorolipids formulation. The formulation demonstrated the controlled release, while the cytocompatibility test confirmed no adverse effects on human dermal fibroblasts [81].

4.3.2 Synthetic Polymers

Synthetic polymeric nanoparticles have demonstrated the benefit of biosurfactant inclusion by ameliorating emulsification, lowering interfacial tension, and enabling dual-drug loading. Rana et al. found the in vivo toxicity obtained from *Candida parapsilosis* loaded in the biodegradable poly(lactide-co-glycolide) (PLGA)-PEG nanoparticles. The prepared formulation was administered intravenously to mice at a concentration of 20 μg/mL, whereas the control groups received no treatment. Hematological, biochemical, and histopathological evaluations were conducted on days 7, 14, and 21. The results demonstrated no significant changes in blood parameters and in histological reports compared to controls. Mild variation in biochemical parameters was observed. Hence, the research group demonstrated that the formulation was well tolerated and showed no toxicity [82]. Moreover, Soares et al. extracted a lipopeptide biosurfactant from *Aspergillus welwitschiae* and used it in formulating the polycaprolactone nanoparticle. The formulation demonstrated the uniform, spherical nanoparticles with mean diameters of 179 to 181 nm, with a low polydispersity index of 0.14 ± 0.04, and a negative zeta potential of − 29.10 ± 8.70 mV. Structural and morphological analyses confirmed uniform particle shape, while the biosurfactant property of amphiphilic nature helped in ameliorating dispersion and stability [83]. Furthermore, Wadhawan et al. engineered PLA-PEG nanoparticles loaded with a lipopeptide biosurfactant from Candida parapsilosis. Formulation demonstrated 84.9% entrapment efficiency, 28.3% drug loading, and sustained release over 120 h. With the help of folic acid conjugation, the formulation was efficiently taken up and induced apoptosis in MDA-MB-231 breast cancer cells [84] Haggag et al. engineered PEGylated PLGA nanocapsules loaded with sophorolipids for targeting colon cancer. Utilizing 10% PEG ameliorated systemic circulation and enhanced cellular uptake and accumulation in CT26 colon carcinoma cells. In in vivo studies, the formulation demonstrated an inhibition of 57% in tumor growth [85].

4.4 Nanofibers Platforms

Electrospun nanofiber systems also utilize the surfactants to reduce fiber diameter, while biosurfactants came out as another approach to ameliorate the reduction in fiber diameter and also demonstrated the significant therapeutic property, such as antimicrobial activity. For instance, Baccile et al. found the self-assembly property of palmitic acid sophorolipids under acidic conditions (pH < 6), demonstrating that their transformation into significant ordered fibrillar nanostructures forms robust thixotropic hydrogels. These hydrogels demonstrated a quick structural recovery and notable elasticity, with storage capability surpassing 10 kPa. SAXS and cryo-TEM revealed a reversible micelle-to-fiber transition with a 8.6 nm diameter, denoting a dynamic yet stable supramolecular organization [86]. Moreover, Afsharipour et al. engineered electrospun nanofibers loaded with lipopeptide biosurfactant from *Acinetobacter junii* B6, formulated with CMC, PVA, and poloxamer for wound healing. Prepared formulation with the concentration of 2% CMC, 9% PVA, 10% poloxamer, and 5% LPB demonstrated uniform, thermally stable nanofibers with a size of 590 nm with ameliorated spreadability. SEM images revealed smooth, uniform fibers. The composite scaffold is stable under temperature variations and promotes cell interaction, moisture retention, and healing efficiency [87].

Furthermore, De Rosette et al. developed PVDF electrospun nanofiber membranes decorated with rhamnolipid biosurfactant and metal oxides (TiO_2, CuO) for advanced wastewater treatment. Rhamnolipid decoration ameliorated hydrophilicity and antifouling, while TiO_2 improved oil removal up to 95%. CuO addition enhanced flux recovery while slightly lowering oil rejection. The optimized membranes achieved > 99.9% microplastic removal [88].

4.5 Metallic Nanoparticles

In metallic nanoparticles synthesis, surfactants were also utilized to formulate, but rather than using biosurfactants, they promote antimicrobial properties and ameliorate colloidal stability. Deshmukh et al. engineered the silver nanoparticles using rhamnolipid biosurfactant obtained from *Pseudomonas aeruginosa* strain LTR1, which was extracted from oil-contaminated soil. The formulation demonstrated spherical, crystalline shape with a particle size of 34 nm. The rhamnolipids prevented the nanoparticle aggregation and demonstrated excellent stability for over 45 days. Antimicrobial testing demonstrated potent inhibition against E. coli, S. aureus, B. subtilis, P. vulgaris, S. typhi, and C. albicans, with improved efficacy [89].

4.6 *Carbon-Based Nanostructures*

Carbon nanoparticles incorporated with biosurfactants demonstrated enhanced dispersion, reduced toxicity, and ameliorated bioactivity. For example, Bansal et al. formulated a theranostic nanoconjugate by incorporating the biosurfactant, which is obtained from *Candida parapsilosis* with the functionalization of graphene quantum dots and folic acid for breast cancer treatment. The formulation exhibited excellent aqueous stability and photoluminescence. Fluorescence microscopy and in vitro assays confirmed enhanced cellular uptake and accumulation in MCF-7 breast cancer cells. With the help of functionalization, formulation demonstrated higher cytotoxicity against tumor cells [90].

5 Biosurfactants in Biomedical and Therapeutic Applications

Biosurfactants are biocompatible and biodegradable, which makes them an attractive choice for pharmaceutical applications. Biosurfactants offer numerous advantages; due to their biodegradability, they have lower toxicity and often demonstrate therapeutic benefits. For example, many microbial glycolipids and lipopeptides are effective emulsifiers and foaming agents. On the other hand, synthetic surfactants can cause irritation or allergic reactions in cosmetic and drug formulations. Biosurfactants can replace harmful synthetic surfactants in creams, lotions, and drug coating, ameliorating product safety [91]. Moreover, numerous applications with key mechanisms are represented in Table 1.

5.1 *Antimicrobial and Antibiofilm Activity*

Biosurfactants produced from *Lactiplantibacillus plantarum* exhibit higher antimicrobial properties by inhibiting $\geq 90\%$ of *S.aureus* and *E. coli* at low concentrations. They also efficiently harm and prevent biofilm formation, enhancing the efficacy of antibiotics. Obtained from probiotic microbes, these biosurfactants are non-toxic and biocompatible, making them appropriate for use in food, cosmetics, and medical applications. Biosurfactants possess antimicrobial and antibiofilm properties, making them safe and effective alternatives [92, 93].

Table 1 Several biosurfactants with their applications and mechanisms

Biosurfactant	Application	Mechanism
Lactiplantibacillus-derived biosurfactants	Antimicrobial and antibiofilm	Disrupts biofilms, kills pathogens, enhances antibiotic action, non-toxic
Sophorolipids	Anticancer properties	Induces apoptosis via oxidative stress Enhances nanoparticle drug delivery
Rhamnolipids, surfactin, sophorolipids	Wound healing and tissue repair	Exhibits antioxidant/anti-inflammatory activity, clears biofilms, and promotes cellular regeneration
MEL-A, sophorolipids, surfactin, rhamnolipids,	Immunomodulation and anti-inflammation	Modulates cytokine production, polarizes T-helper cells, and inhibits phospholipase activity
Sophorolipids, micellar formulations	Antiviral activity	Destabilizes viral envelope Blocks entry, simulates immune response
MEL-A, rhamnolipids, surfactin	Gene and drug delivery systems	Assembles into nanostructures for targeted delivery, facilitates membrane fusion, and improves cellular uptake
Surfactin, bacillus-derived lipopeptides	Vaccine adjuvant	Activates TLR2 pathways, enhances antigen presentation, and promotes adaptive immunity

5.2 *Anticancer Properties*

Biosurfactants such as sophorolipids can induce apoptosis in cancer cells by increasing oxidative stress and triggering caspase activation. Additionally, when used in nanoparticle formulation, they enhance the delivery and effectiveness of chemotherapeutic agents like doxorubicin and resveratrol, especially in drug-resistant cancer types. This dual function improves drug stability, promotes tumor cell uptake, and helps suppress resistance mechanism [94].

5.3 *Wound Healing and Tissue Repair*

Biosurfactants aid wound healing by controlling infection, reducing inflammation, and supporting tissue regeneration. Their antimicrobial, anti-inflammatory, and antioxidant properties help clear biofilms, lower bacterial load, and neutralize toxins, promoting faster repair. Especially useful in chronic wounds, biosurfactants also suppress harmful cytokines and scavenge free radicals. Being non-toxic and biodegradable, they are ideal for use in safe, bioactive dressing like hydrogel, foams, or sprays [93].

5.4 Immunomodulatory and Anti-Inflammatory Effects

Sophorolipids and rhamnolipids suppress pro-inflammatory mediators such as IL-1β, TNF-α, and nitric oxide while enhancing regulatory cytokines like IL-10 and TGF-β1 in urine models of dextran-induced colitis. Dietary sophorolipids reduced inflammation and regulated mucosal protectant including MUC2, IL-10, and TGF-β. Microbial surfactants, particularly glycolipids and lipopeptides, exhibit potent immunomodulatory and anti-inflammatory properties by interacting with immune receptors and altering membrane dynamics. Mannosylerythritol lipid-A (MEL-A) binds immunoglobulins and lectins, triggering cytokine release and immune cell activation [4, 95]. Sophorolipids and rhamnolipids suppress pro-inflammatory mediators such as IL-1β, TNF-α, and nitric oxide while enhancing regulatory cytokines like IL-10 and TGF-β1 [96]. In murine models of dextran-induced colitis, dietary sophorolipids reduced inflammation and upregulated mucosal protectants including MUC2, IL-10, and TGF-β [4]. Surfactin, a cyclic lipopeptide, inhibits phospholipase A_2 and dampens excessive immune activation [96]. Additionally, biosurfactants modulate T-helper cell polarization (Th2 to Th1) and promote wound-relevant cell differentiation, positioning them as promising candidate for treating inflammatory and auto-immune conditions.

5.5 Antiviral Activity

Biosurfactants show broad antiviral potential, particularly against enveloped viruses, by disrupting their lipid membrane. Sophorolipids can neutralize HIV-1 and HSV by modifying viral glycosides and stimulating immune responses. Against SARS-CoV-2, micellar formulations destabilize the viral envelope, blocking cell entry and promoting disassembly. Structural trait like acetylated sugar moieties further improves their efficacy. These features position biosurfactants as viable candidate for antiviral therapy, supporting both direct viral and inactivation and modulation of hyperinflammatory response in severe infections [96].

5.6 Gene and Drug Delivery Systems

Biosurfactants are a promising approach in drug and gene delivery by forming stable nanostructures. For instance, biosurfactants are utilized in micelles, liposomes, and nanoemulsions to enhance solubility, cellular interaction, and therapeutic properties. In antifungal applications, surfactin micelles loaded with itraconazole reduce hemolysis while improving efficacy. Rhamnolipid-based liposomes, engineered for pH sensitivity using DOPE lipids, enable targeted drug release in acidic endosomal environments. For gene delivery, mannosylerythritol lipid demonstrated efficient DNA transfection by binding lectins and promoting membrane fusion, with engineered fatty acid profiles ameliorating gene transfer. Thus, biosurfactant incorporation ameliorates drug loading, cellular uptake, and site-specific delivery, which may help address multidrug resistance—making them promising agents for advanced therapeutic platforms [95].

5.7 *Vaccine Adjuvant*

Certain microbial biosurfactants particularly lipopeptides from non-pathogenic Bacillus strains demonstrate potential as vaccine adjuvants by activating Toll-like receptor 2 and stimulating both innate and adaptive immunity without triggering pyrogenic effects. These molecules improve antigen presentation and support cytotoxic T-cell activation via MHC-I pathways. Surfactin's emulsifying properties further enhance antigen delivery and immune priming. Compared to conventional adjuvants, biosurfactant-based systems can elicit robust antibody and T-cell responses with reduced toxicity. Their natural origin and biodegradability make them well suited for next-generation vaccine formulations [4].

6 Regulatory Framework of Biosurfactants

Regulatory agencies such as FDA, EMA, and ICH treat microbial biosurfactants as novel pharmaceutical excipients (i.e., substance used in drug for the first time) and accordingly require full characterization of their manufacture, composition, and analytical controls, along with comprehensive safety data [97]. The EMA's "Excipients in the Dossier" guideline explicitly states that novel excipients must have full manufacturing details, analytical specifications, and cross-referenced toxicological data. Consistent with this, new biosurfactant excipients must undergo comprehensive GLP-compliant nonclinical testing [98]. Recommended studies include acute and repeat-dose toxicity in two species (rodent and nonrodent via the intended route), ICH-S7A safety-pharmacology evaluations, pharmacokinetic/ADME assessments per ICH-S3A/S3B, and a standard genotoxicity battery per ICH S2B. FDA guidance further emphasizes that any proposed new suspect requires appropriate justification of its concentration and a complete toxicology program, reinforcing the need for a clear rationale beyond safety data alone. Similarly, EMA mandates that the drug application includes an explanation of each surfactant's function, concentration, and impact on product performance, adding a performance-based dimension to the regulatory evaluation [99]. For chronic or reproductive use, ICH S5/S1A studies are also advised. In addition, impurity limits must follow ICH guidance: ICH Q3D(R1) establishes risk-based limits on elemental (metal) impurities in excipients, and ICH Q6B (along with USP monographs) governs residual bioprocess impurities (e.g., host cell proteins or DNA) in biologically derived excipients [100]. Because biosurfactants originate from microbial fermentation, agencies emphasize controlling bioprocess impurities, underscoring the importance of robust purification and analytical validation. Regulatory acceptance may be streamlined if the biosurfactant is produced by a GRAS-status organism; many biosurfactant producers (e.g., Bacillus or GRAS yeast species) are already GRAS-listed strains, such as Saccharomyces or Candida, and FDA guidance notes that demonstrated prior human use on GRAS status may qualify an excipient without full new testing [99]. Thus, strict adherence to FDA/EMA/ICH requirements for novel excipients, including rigorous GLP toxicology (with ADME and genotoxicity studies) and impurity control, is essential to achieving regulatory acceptance of biosurfactant in drug delivery [97, 98].

7 Future Outlook of Biosurfactants in Novel Drug Delivery System

Biosurfactants are amphiphilic compounds produced by microorganisms that can self-assemble into carriers such as micelles, liposomes, and nanoemulsions, which encapsulate hydrophobic drugs to enhance their solubility and bioavailability. Their natural origin provides advantages, including excellent biodegradability, low toxicity, and the capacity to remain effective under harsh conditions, making them highly biocompatible excipients in pharmaceutical applications [4, 46]. Lipid—and peptide-based biosurfactants such as lipopeptides, glycolipids, and mannosylerythritol lipids are increasingly incorporated into nanocarrier systems to improve drug loading, formulation stability, and cellular uptake [101]. For example, biosurfactant-coated liposomes have been developed that remain intact at neutral pH but release their contents selectively in acidic endosomal environments [95]. Looking ahead, research is increasingly focused on developing tailor-made biosurfactants and implementing sustainable manufacturing processes. Advances in metabolic and genetic engineering, including the application of CRISPR tools and synthetic promoters, are enabling precise structural modifications and improving production yields. Additionally, machine learning and systems biology approaches are being used to explore microbial diversity and optimize biosurfactant biosynthesis pathways to create novel variants with specialized function [102, 103] These innovations are expected to expand the repertoire of natural surfactants available for therapeutic applications. Overall, biosurfactant-based drug delivery systems represent a promising strategy to improve dissolution, targeting, and therapeutic efficacy while minimizing adverse effects.

8 Conclusion

Biosurfactants are a class of natural amphiphiles that can benefit advanced drug delivery systems. Their ability to self-assemble into stable nanostructures, enhancing solubility and enabling targeted release, offers a promising approach compared to synthetic surfactants. When utilized with lipidic or polymeric carriers, they ameliorate therapeutic efficacy, stability, and safety. Moreover, when used as excipients, many biosurfactants demonstrate therapeutic properties such as antimicrobial, anti-inflammatory, and anticancer properties. Continued advances in metabolic engineering and green production processes will further optimize their functionality, establishing biosurfactants as key components in upcoming eco-friendly pharmaceutical and biomedical formulations.

References

1. Miao Y, To MH, Siddiqui MA, Wang H, Lodens S, Chopra SS et al (2024) Sustainable biosurfactant production from secondary feedstock—recent advances, process optimization and perspectives. Front Chem 12:1327113
2. Kadam V, Dhanorkar M, Patil S, Singh P (2024) Advances in the co-production of biosurfactant and other biomolecules: statistical approaches for process optimization. J Appl Microbiol 135(2):lxae025
3. Romero Vega G, Gallo SP (2025) Bio-based surfactants and biosurfactants: an overview and main characteristics. Molecules 30(4):863
4. Ceresa C, Fracchia L, Sansotera AC, De Rienzo MAD, Banat IM (2023) Harnessing the potential of biosurfactants for biomedical and pharmaceutical applications. Pharmaceutics 15(8):2156
5. Akiyode OA (2017) The physico-chemical and biological analysis of microbial biosurfactants (BSs) for applications towards cancer therapies, wound healing dressings and drug delivery. University of Greenwich
6. Bhattacharya B, Ghosh TK, Das N (2017) Application of bio-surfactants in cosmetics and pharmaceutical industry. Sch Acad J Pharm 6(7):320–329
7. Raji MI, Bello S (2024) Biosurfactant therapeutic potentials: a review. Fudma J Sci 8(3):43–56
8. Lourenço M, Duarte N, Ribeiro IA (2024) Exploring biosurfactants as antimicrobial approaches. Pharmaceuticals 17(9):1239
9. Mallik T, Banerjee D (2022) Biosurfactants: the potential green surfactants in the 21st century. J Adv Sci Res 13(01):97–106
10. Shakeri F, Babavalian H, Amoozegar MA, Ahmadzadeh Z, Zuhuriyanizadi S, Afsharian MP (2021) Production and application of biosurfactants in biotechnology. Biointerface Res Appl Chem 11(3):10446–10460
11. Purohit GK, Dubey D, Rath S, Subhadarsini SS, Tripathi D, Behera S et al (2024) Biosurfactants: moving towards healthcare applications. Int J Adv Life Sci Res 7(1):24–37
12. Gürkök S, Özdal M (2021) Microbial biosurfactants: properties, types, and production. Anatol J Biol 2(2):7–12
13. Sari C, Hertadi R, Gozan M, Roslan A (2019) Factors affecting the production of biosurfactants and their applications in enhanced oil recovery (EOR). a review. IOP Conf Ser: Earth Environ Sci
14. Nayak NS, Joshi DS, Rathi YS Biosurfactants-A boon for therapeutics
15. Schultz J, Rosado AS (2020) Extreme environments: a source of biosurfactants for biotechnological applications. Extremophiles 24:189–206

16. Sharma J, Sundar D, Srivastava P (2021) Biosurfactants: potential agents for controlling cellular communication, motility, and antagonism. Front Mol Biosci 8:727070
17. Adu SA, Twigg MS, Naughton PJ, Marchant R, Banat IM (2023) Glycolipid biosurfactants in skincare applications: challenges and recommendations for future exploitation. Molecules 28(11):4463
18. Bellanthudawa B, Nawalage N, Handapangoda H, Suvendran S, Wijayasenarathne K, Rathnasuriya M et al (2023) A perspective on biodegradable and non-biodegradable nanoparticles in industrial sectors: applications, challenges, and future prospects. Nanotechnol Environ Eng 8(4):975–1013
19. Kabeil SS, Darwish AM, Abdelgalil SA, Shamseldin A, Salah A, Taha HA et al (2025) Rhamnolipids bio-production and miscellaneous applications towards green technologies: a literature review. PeerJ 13:e18981
20. Daverey A, Dutta K, Joshi S, Daverey A (2021) Sophorolipid: a glycolipid biosurfactant as a potential therapeutic agent against COVID-19. Bioengineered 12(2):9550–9560
21. Adu SA, Twigg MS, Naughton PJ, Marchant R, Banat IM (2023) Characterisation of cytotoxicity and immunomodulatory effects of glycolipid biosurfactants on human keratinocytes. Appl Microbiol Biotechnol 107(1):137–152
22. Kuyukina MS, Ivshina IB (2019) Production of trehalolipid biosurfactants by Rhodococcus. In: Biology of rhodococcus. Springer, pp 271–298
23. Andreolli M, Villanova V, Zanzoni S, D'Onofrio M, Vallini G, Secchi N et al (2023) Characterization of trehalolipid biosurfactant produced by the novel marine strain Rhodococcus SP. SP1d and its potential for environmental applications. Microb Cell Factories 22(1):126
24. Songdech P, Jayasekara L, Watchaputi K, Butkinaree C, Yingchutrakul Y, Soontorngun N (2025) Elucidating a novel metabolic pathway for enhanced antimicrobial glycolipid biosurfactant production in the yeast Meyerozyma guilliermondii. Sci Rep 15(1):1–17
25. Master NG, Markande AR (2023) Importance of microbial amphiphiles: interaction potential of biosurfactants, amyloids, and other exo-polymeric-substances. World J Microbiol Biotechnol 39(11):320
26. Burkovski A (2018) The role of corynomycolic acids in Corynebacterium-host interaction. Antonie Leeuwenhoek 111(5):717–725
27. Ahmadi-Ashtiani H-R, Baldisserotto A, Cesa E, Manfredini S, Sedghi Zadeh H, Ghafori Gorab M et al (2020) Microbial biosurfactants as key multifunctional ingredients for sustainable cosmetics. Cosmetics 7(2):46
28. Kumari V (2022) Biosurfactant as an antimicrobial and biodegradable agent a review. In: Microbial surfactants. CRC Press, pp 139–157
29. Moriwaki-Takano M, Asada C, Nakamura Y (2021) Production of spiculisporic acid by Talaromyces trachyspermus in fed-batch bioreactor culture. Bioresour Bioprocess 8:1–11
30. Liu Y, Zhao J, Qiao W, Yang B, Liu Q, Chen L (2024) Effect of infant formula production processes on phospholipid composition and structure of milk fat globules. LWT 195:115808
31. Gayathiri E, Prakash P, Karmegam N, Varjani S, Awasthi MK, Ravindran B (2022) Biosurfactants: potential and eco-friendly material for sustainable agriculture and environmental safety—a review. Agronomy 12(3):662
32. Nwaguma IV, Chikere CB, Okpokwasili GC (2016) Isolation, characterization, and application of biosurfactant by Klebsiella pneumoniae strain IVN51 isolated from hydrocarbon-polluted soil in Ogoniland, Nigeria. Bioresour Bioprocess 3:1–13
33. Rojewska M, Smułek W, Grzywaczyk A, Kaczorek E, Prochaska K (2023) Study of interactions between saponin biosurfactant and model biological membranes: phospholipid monolayers and liposomes. Molecules 28(4):1965
34. Wenjie X, Li Y, Ping W, Jianlong X, Hanping D (2012) Characterization of a thermophilic and halotolerant Geobacillus pallidus H9 and its application in microbial enhanced oil recovery (MEOR). Ann Microbiol 62:1779–1789
35. Bellebcir A, Merouane F, Chekroud K, Bounabi H, Vasseghian Y, Kamyab H et al (2023) Bioprospecting of biosurfactant-producing bacteria for hydrocarbon bioremediation: optimization and characterization. Korean J Chem Eng 40(10):2497–2512

36. Mnif I, Ghribi D (2015) Review lipopeptides biosurfactants: mean classes and new insights for industrial, biomedical, and environmental applications. Pept Sci 104(3):129–147
37. Tseng CC, Bruner SD, Kohli RM, Marahiel MA, Walsh CT, Sieber SA (2002) Characterization of the surfactin synthetase C-terminal thioesterase domain as a cyclic depsipeptide synthase. Biochemistry 41(45):13350–13359
38. Théatre A, Cano-Prieto C, Bartolini M, Laurin Y, Deleu M, Niehren J et al (2021) The surfactin-like lipopeptides from bacillus spp.: natural biodiversity and synthetic biology for a broader application range. Front Bioeng Biotechnol 9:623701
39. Dini S, Oz F, Bekhit AEDA, Carne A, Agyei D (2024) Production, characterization, and potential applications of lipopeptides in food systems: a comprehensive review. Compr Rev Food Sci Food Saf 23(4):e13394
40. Yaraguppi DA, Bagewadi ZK, Patil NR, Mantri N (2023) Iturin: a promising cyclic lipopeptide with diverse applications. Biomolecules 13(10):1515
41. Phulpoto IA, Bobo W, Qazi MA, Yu Z (2024) Lipopeptide biosurfactants enhanced biohydrogen evolution from lignocellulose biomass and shaped the microbial community and diversity. Int J Energy Res 2024(1):3084702
42. Mnif I, Rajhi H, Bouallegue A, Trabelsi N, Ghribi D (2022) Characterization of lipopeptides biosurfactants produced by a newly isolated strain Bacillus subtilis ZNI5: potential environmental application. J Polym Environ 30(6):2378–2391
43. Simões CR, da Silva MWP, de Souza RFM, Hacha RR, Merma AG, Torem ML et al (2024) Biosurfactants: An overview of their properties, production, and application in mineral flotation. Resources 13(6):81
44. Dini S, Bekhit AE-DA, Roohinejad S, Vale JM, Agyei D (2024) The physicochemical and functional properties of biosurfactants: a review. Molecules 29(11):2544
45. Abruzzo A, Parolin C, Corazza E, Giordani B, di Cagno MP, Cerchiara T et al (2021) Influence of lactobacillus biosurfactants on skin permeation of hydrocortisone. Pharmaceutics. 13(6):820
46. Bjerk TR, Severino P, Jain S, Marques C, Silva AM, Pashirova T et al (2021) Biosurfactants: properties and applications in drug delivery, biotechnology and ecotoxicology. Bioengineering 8(8):115
47. Sharma V, Singh D, Manzoor M, Banpurkar AG, Satpute SK, Sharma D (2022) Characterization and cytotoxicity assessment of biosurfactant derived from lactobacillus pentosus NCIM 2912. Braz J Microbiol 53(1):327–340
48. Singh N, Hu X-H, Kumar V, Solanki MK, Kaushik A, Singh VK et al (2024) Microbially derived surfactants: an ecofriendly, innovative, and effective approach for managing environmental contaminants. Front Bioeng Biotechnol 12:1398210
49. Hellweg T, Oberdisse J, Sottmann T (2023) Mini review: recent advances in biosurfactant-based association colloids–formation of microemulsions. Front Soft Matter 3:1260211
50. Fragal EH, Poirier A, Bleses D, Silva YFG, Baccile N, Rharbi Y (2025) Microbial biosurfactant hydrogels with tunable rheology for precision 3D printing of soft scaffolds. Soft Matter 21(22):4476–4487
51. Phulpoto IA, Yu Z, Hu B, Wang Y, Ndayisenga F, Li J et al (2020) Production and characterization of surfactin-like biosurfactant produced by novel strain bacillus nealsonii S2MT and it's potential for oil contaminated soil remediation. Microb Cell Factories 19:1–12
52. Silva DF, Melo AL, Uchôa AF, Pereira GM, Alves AE, Vasconcellos MC et al (2023) Biomedical approach of nanotechnology and biological risks: a mini-review. Int J Mol Sci 24(23):16719
53. Yadav D, Sandeep K, Pandey D, Dutta RK (2017) Liposomes for drug delivery. J Biotechnol Biomater 7(04):276
54. Dias-Souza MV, Haq IU, Pagnin S, Veiga AA, Dos Santos VL (2025) Liposome-encapsulated antibiotics and biosurfactants: an effective strategy to boost biofilm eradication in cooling towers. Microb Cell Factories 24(1):1–12
55. BCP S, Rocha CA, Martin Bedoya JG, VLd S, PBd S, Fusco-Almeida AM et al (2021) Rhamnolipid-based liposomes as promising nano-carriers for enhancing the antibacterial

activity of peptides derived from bacterial toxin-antitoxin systems. Int J Nanomedicine:925–939
56. Zhou W, Cheng C, Ma L, Zou L, Liu W, Li R et al (2021) The formation of chitosan-coated rhamnolipid liposomes containing curcumin: stability and in vitro digestion. Molecules 26(3):560
57. Chen H, Zhang Q (2023) Surface functionalization of piperine-loaded liposomes with sophorolipids improves drug loading and stability. J Pharm Innov 18(2):747–755
58. Dobreva M, Stefanov S, Andonova V (2020) Natural lipids as structural components of solid lipid nanoparticles and nanostructured lipid carriers for topical delivery. Curr Pharm Des 26(36):4524–4535
59. Rana S, Handa S, Aggarwal Y, Puri S, Chatterjee M (2023) Role of Candida in the bioremediation of pollutants: a review. Lett Appl Microbiol 76(9):ovad103
60. Panbehchouleh FA, Amani H, Saeedi M (2024) Menadione sodium bisulfite loaded Rhamnolipid based solid lipid nanoparticle as skin lightener formulation: a green production beside in vitro/in vivo safety index evaluation. Adv Pharm Bulletin 14(3):623
61. Pandey S, Shaikh F, Gupta A, Tripathi P, Yadav JS (2021) A recent update: solid lipid nanoparticles for effective drug delivery. Adv Pharm Bulletin 12(1):17
62. Rehman M, Tahir N, Sohail MF, Qadri MU, Duarte SO, Brandão P et al (2024) Lipid-based nanoformulations for drug delivery: an ongoing perspective. Pharmaceutics 16(11):1376
63. Azevedo MA, Cerqueira MA, Gonçalves C, Amado IR, Teixeira JA, Pastrana L (2023) Encapsulation of vitamin D3 using rhamnolipids-based nanostructured lipid carriers. Food Chem 427:136654
64. Azevedo MA, Cerqueira MA, Fucinos P, Silva BF, Teixeira JA, Pastrana L (2021) Rhamnolipids-based nanostructured lipid carriers: effect of lipid phase on physicochemical properties and stability. Food Chem 344:128670
65. Katiyar SS, Ghadi R, Kushwah V, Dora CP, Jain S (2020) Lipid and biosurfactant based core–shell-type nanocapsules having high drug loading of paclitaxel for improved breast cancer therapy. ACS Biomater Sci Eng 6(12):6760–6769
66. Jadhav A, Nagaraj K (2025) Surfactant-enabled Nanocarriers in breast cancer therapy: targeted delivery and multidrug resistance reversal. Pharmaceutics 17(6):779
67. Mazzotta E, Romeo M, Hafidi Z, Perez L, Perrotta ID, Muzzalupo R (2024) Ammonium Glycyrrhizate: a natural biosurfactant for the design of innovative nanocarriers. J Mol Liq 398:124291
68. Shiri S, Gharanjig K, Tahghighi A, Hosseinnezhad M, Etezad M (2025) Formulation and characterization of BBR loaded niosomes using saponin as a nonionic biosurfactant investigating synergistic effects to enhance antibacterial activity. Sci Rep 15(1):5231
69. Haddaji N, Bahloul B, Bahia W, Bechambi O, Mahdhi A (2023) Development of nanotechnology-based drug delivery systems for controlling clinical multidrug-resistant Staphylococcus aureus and Escherichia coli associated with aerobic vaginitis. Pharmaceutics 15(8):2133
70. Luo J, Yang B, Yang X, Ji S, Guo Z, Liu Y et al (2023) Sophorolipid-based microemulsion delivery system: multifaceted enhancement of physicochemical properties of xanthohumol. Food Chem 413:135631
71. Shen C, Li J, Meng Q, Xu L, Zhang G (2024) Rhamnolipids stabilized essential oils microemulsion for antimicrobial and fruit preservation. Food Chem 457:140167
72. Al-Sakkaf MK, Onaizi SA (2023) Crude oil/water nanoemulsions stabilized by rhamnolipid biosurfactant: effects of acidity/basicity and salinity on emulsion characteristics, stability, and demulsification. Fuel 344:128052
73. Zanotto AW, Kanemaru MYS, de Souza FG, Duarte MCT, de Andrade CJ, Pastore GM (2023) Enhanced antimicrobial and antioxidant capacity of Thymus vulgaris, Lippia sidoides, and Cymbopogon citratus emulsions when combined with mannosylerythritol a lipid biosurfactant. Food Res Int 163:112213
74. Yin L, Guo Y, Xv X, Dai Y, Li L, Sun F et al (2024) Cinnamaldehyde nanoemulsion decorated with rhamnolipid for inhibition of methicillin-resistant Staphylococcus aureus biofilm formation: in vitro and in vivo assessment. Front Microbiol 15:1514659

75. Mottola M, Bertolino MC, Kourdova LT, Valdivia Pérez JA, Bogino MF, Nocelli NE et al (2023) Nanoemulsions of synthetic rhamnolipids act as plant resistance inducers without damaging plant tissues or affecting soil microbiota. Front Plant Sci 14:1195718
76. Lee Y, Lee D, Park E, Jang S-y, Cheon SY, Han S et al (2021) Rhamnolipid-coated W/O/W double emulsion nanoparticles for efficient delivery of doxorubicin/erlotinib and combination chemotherapy. J Nanobiotechnol 19(1):411
77. Malavia N, Ghadi R, Kuche K, Date T, Bhargavi N, Chaudhari D et al (2021) Green surfactant-dendrimer aggreplexes: an ingenious way to launch dual attack on arch-enemy cancer. Colloids Surf B: Biointerfaces 204:111821
78. Kannan S, Solomon A, Krishnamoorthy G, Marudhamuthu M (2021) Liposome encapsulated surfactant abetted copper nanoparticles alleviates biofilm mediated virulence in pathogenic Pseudomonas aeruginosa and MRSA. Sci Rep 11(1):1102
79. Karamchandani BM, Maurya PA, Dalvi SG, Waghmode S, Sharma D, Rahman PK et al (2022) Synergistic activity of rhamnolipid biosurfactant and nanoparticles synthesized using fungal origin chitosan against phytopathogens. Front Bioeng Biotechnol 10:917105
80. Arif M, Sharaf M, Samreen KS, Chi Z, Liu C-G (2021) Chitosan-based nanoparticles as delivery-carrier for promising antimicrobial glycolipid biosurfactant to improve the eradication rate of helicobacter pylori biofilm. J Biomater Sci Polym Ed 32(6):813–832
81. Bettencourt AF, Tome C, Oliveira T, Martin V, Santos C, Gonçalves L et al (2021) Exploring the potential of chitosan-based particles as delivery-carriers for promising antimicrobial glycolipid biosurfactants. Carbohydr Polym 254:117433
82. Rana S, Singh J, Wadhawan A, Khanna A, Singh G, Chatterjee M (2021) Evaluation of in vivo toxicity of novel biosurfactant from Candida parapsilosis loaded in PLA-PEG polymeric nanoparticles. J Pharm Sci 110(4):1727–1738
83. Soares AR, JCd A, Lacerda CD, Azevedo SG, MTM P, Manzato L et al (2024) Synthesis of polymeric nanoparticles using fungal biosurfactant as stabilizer. PRO 12(12):2739
84. Wadhawan A, Singh J, Sharma H, Handa S, Singh G, Kumar R et al (2022) Anticancer biosurfactant-loaded PLA–PEG nanoparticles induce apoptosis in human MDA-MB-231 breast cancer cells. ACS Omega 7(6):5231–5241
85. Haggag Y, Elshikh M, El-Tanani M, Bannat IM, McCarron P, Tambuwala MM (2020) Nanoencapsulation of sophorolipids in PEGylated poly (lactide-co-glycolide) as a novel approach to target colon carcinoma in the murine model. Drug Deliv Transl Res 10(5):1353–1366
86. Baccile N, Ben Messaoud G, Le Griel P, Cowieson N, Perez J, Geys R et al (2021) Palmitic acid sophorolipid biosurfactant: from self-assembled fibrillar network (SAFiN) to hydrogels with fast recovery. Phil Trans R Soc A 379(2206):20200343
87. Afsharipour S, Kavianipoor S, Ranjbar M, Bagheri AM, Najafi ML, Banat IM, et al., (eds) (2023) Fabrication and characterization of lipopeptide biosurfactant-based electrospun nanofibers for use in tissue engineering. In: Annales pharmaceutiques françaises, Elsevier
88. de Rosset A, Torres-Mendieta R, Pasternak G, Yalcinkaya F (2025) Synergistic effects of natural biosurfactant and metal oxides modification on PVDF nanofiber filters for efficient microplastic and oil removal. Process Saf Environ Prot 194:997–1009
89. Deshmukh N, Kadam D, Deshmukh P, Kathwate G (2022) Utilization of biosurfactant produced by Pseudomonas aeruginosa strain LTR1 for the synthesis of the silver nanoparticles and their application as antimicrobial agents. J Appl Biol Biotechnol 10:35–41
90. Bansal S, Singh J, Kumari U, Kaur IP, Barnwal RP, Kumar R et al (2019) Development of biosurfactant-based graphene quantum dot conjugate as a novel and fluorescent theranostic tool for cancer. Int J Nanomedicine 14:809–818
91. Karnwal A, Shrivastava S, Al-Tawaha ARMS, Kumar G, Singh R, Kumar A et al (2023) Microbial biosurfactant as an alternate to chemical surfactants for application in cosmetics industries in personal and skin care products: a critical review. Biomed Res Int 2023(1):2375223
92. Thakur B, Kaur S, Dwibedi V, Albadrani GM, Al-Ghadi MQ, Abdel-Daim MM (2024) Unveiling the antimicrobial and antibiofilm potential of biosurfactant produced by newly isolated Lactiplantibacillus plantarum strain 1625. Front Microbiol 15:1459388

93. Ohadi M, Forootanfar H, Dehghannoudeh N, Banat IM, Dehghannoudeh G (2023) The role of surfactants and biosurfactants in the wound healing process: a review. J Wound Care 32(Sup4a):xxxix–xlvi
94. Singh SB, Kuniyal K, Rawat A, Bisht A, Shah V, Daverey A (2025) Sophorolipids as anticancer agents: progress and challenges. Discov Oncol 16(1):507
95. Wang X, An J, Cao T, Guo M, Han F (2024) Application of biosurfactants in medical sciences. Molecules 29(11):2606
96. Sarangi MK, Padhi S, Patel L, Rath G, Nanda SS, Yi DK (2022) Theranostic efficiency of biosurfactants against COVID-19 and similar viruses-a review. J Drug Delivery Sci Technol 76:103764
97. Ismail R, Baaity Z, Csóka I (2021) Regulatory status quo and prospects for biosurfactants in pharmaceutical applications. Drug Discov Today 26(8):1929–1935
98. Bami MS, Khazaeli P, Lahiji SF, Dehghannoudeh G, Banat IM, Ohadi M (2024) Potential of biosurfactant as green pharmaceutical excipients for coating of microneedles: a mini review. AIMS Microbiol 10(3):596
99. FDA U. Guidance for Industry Nonclinical Studies for the Safety Evaluation of Pharmaceutical Excipients 2005. Available from: https://www.fda.gov/media/72260/download#:~:text=consider%20factors%20such%20as%20use,with%20the%20prior%20use%20may
100. Liu F (2024) Safety assessment of drug impurities for patient safety: a comprehensive review. Regul Toxicol Pharmacol 153:105715
101. Aguirre-Ramírez M, Silva-Jiménez H, Banat IM, Díaz De Rienzo M (2021) Surfactants: physicochemical interactions with biological macromolecules. Biotechnol Lett 43(3):523–535
102. Chabhadiya S, Acharya D, Mangrola A, Shah R, Pithawala EA (2024) Unlocking the potential of biosurfactants: innovations in metabolic and genetic engineering for sustainable industrial and environmental solutions. Biotechnol Notes 5:111–119
103. Vigil TN, Felton SM, Fahy WE, Kinkeade MA, Visek AM, Janiga AR et al (2024) Biosurfactants as templates to inspire new environmental and health applications. Frontiers in Synthetic Biology 2:1303423

Preparation, Characteristics, and Applications of Zwitterionic Surfactants

Chirayu Lunkad, Gauri Pawaskar, Susmita Paranjpe, Prasad Sanap, and Amit Pratap

Abstract Zwitterionic surfactants occupy a unique position in colloid and interface science as they unite both cationic and anionic functionalities within a single molecular framework. This makes them extremely adaptable in terms of surface activity, solubility shift with pH and ionic strength, enabling performance in environments where conventional surfactants fail. This chapter presents a comprehensive yet accessible overview of zwitterionic surfactants, covering their molecular framework, synthesis methods, categorization, interfacial properties, functional performance across diverse applications along with the safety and toxicity profile. From an industrial standpoint, their utility extends across multiple sectors, spans, including household and personal care formulations, pharmaceuticals, biomedicine, enhanced oil recovery, and environmental remediation.

Keywords Zwitterionic surfactants · Amphoteric surfactants · Betaines · Sulfobetaines · Imidazoline derivatives · Surface activity · Biocompatibility · Biodegradability · Enhanced oil recovery · Environmental Remediation

---

C. Lunkad · S. Paranjpe · P. Sanap · A. Pratap (✉)
Department of Oils, Oleochemicals and Surfactants Technology, Institute of Chemical Technology (University Under Section 3 of UGC Act 1956, Formerly UDCT/ UICT), Matunga (East), Mumbai, India
e-mail: amitpratap2001@gmail.com

C. Lunkad
e-mail: 22oilcs.lunkad@ug.ictmumbai.edu.in

S. Paranjpe
e-mail: oil24ss.paranjpe@phd.ictmumbai.edu.in

P. Sanap
e-mail: oil21ps.sanap@pg.ictmumbai.edu.in

G. Pawaskar · A. Pratap
Department of Specialty Chemicals Technology, Institute of Chemical Technology (University Under Section 3 of UGC Act 1956, Formerly UDCT/ UICT), Matunga (East), Mumbai, India
e-mail: 23dyegn.pawaskar@ug.ictmumbai.edu.in

M. Abdul Rub et al. (eds.), *Industrial Applications of Individual and Mixed Surfactant Systems - Volume 2*, Sustainable Materials and Technology,
https://doi.org/10.1007/978-981-95-9195-4_15

1 Introduction

Surfactants (surface-active agents) are molecules having a dual nature, consisting of a hydrophobic tail and hydrophilic head. Surfactants exhibit surface activity, meaning they can reduce the free energy associated with surfaces and interfacial regions. The hydrophilic head might be polar or ionic, whereas the hydrophobic tail is usually a hydrocarbon or fluorocarbon chain. The amphiphilic nature of surfactants allows them to lower surface tension and interact with different phases of a solution. The term originates from Greek, where amphi means "both" and phil means "like," signifying molecules that have affinity for both polar and nonpolar environments. The hydrophilic head can be either polar or ionic, while the hydrophobic tail is typically a hydrocarbon or fluorocarbon chain. This amphiphilic structure allows surfactants to reduce surface tension and interact with different phases in a solution [1].

Surfactants in aqueous solutions self-assemble form micelles, which adsorb at the interfaces between phases. Their chemical structure generally includes an alkyl chain containing 8–22 carbon atoms, which forms the hydrophobic segment, and a functional group that serves as the hydrophilic head. This distinct composition accounts for their vast range of applications, including emulsification, detergency, and wetting, making them indispensable in industries such as pharmaceuticals, cosmetics, and food processing [2].

In water-based systems, surfactants spontaneously organize into micelles and accumulate at the boundaries separating different phases. This unique composition is responsible for their broad spectrum of uses such as emulsification, detergency, and wetting, making them essential in industries such as pharmaceuticals, cosmetics, and food processing [1].

1.1 General Classification of Surfactants

Surfactants are conventionally grouped based on the nature of their hydrophilic head, with four principal categories: anionic, cationic, nonionic, and zwitterionic (amphoteric) (Fig. 1) [1].

Most surfactants have a molecular weight below 500 Da; however, nonionic surfactants with lengthy polyoxyethylene tails may have a higher molecular

Fig. 1 Types of surfactants

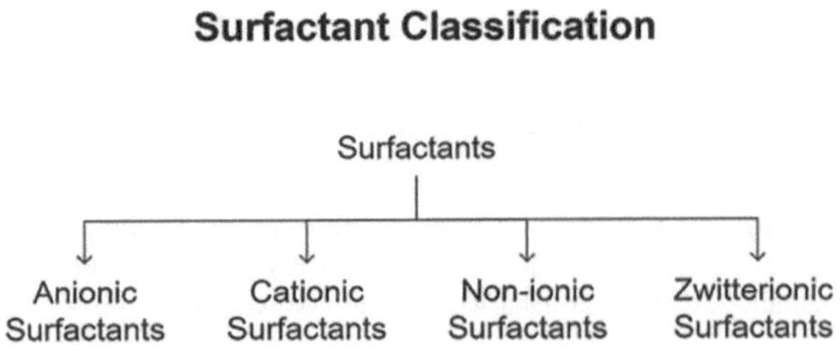

weight. Structurally with the low-molecular weight macromolecules containing both hydrophilic and hydrophobic parts can be referred to as polymeric surfactants. Several natural polymers are known to be polymeric surfactants, e.g., proteins are not considered from this point of view, but in multiple cases, proteins are found in natural systems as emulsion stabilizers [3]. Surfactants present in nature are referred to as polar lipids.

Surfactants are broadly sourced either from oleochemical routes, primarily using vegetable oils, or from petrochemical routes, which rely on precursors like ethylene obtained through naphtha cracking. In certain cases, both approaches yield products with comparable properties [4]. Since the homolog distribution is mostly controlled by the distillation process, both approaches produce straight-chain alcohols with little variation, and each pathway is employed on a large industrial scale.

In addition to the multiple variations in molecular structure and origin, surfactants can also be classified based on the nature of their ionic character in aqueous solutions. Zwitterionic surfactants are the ones which contain both acidic and basic functional groups and can exhibit cationic or anionic behavior depending on the pH. Due to their unique characteristics and their distinct physicochemical behavior and broad applicability, they merit an extensive evaluation.

Amphoteric surfactants represent a distinct class of surface-active agents that contain both acidic and basic functional groups within the same molecule. Their ionic character is strongly influenced by the pH of the surrounding medium: they can behave as cationic surfactants in acidic conditions, anionic in alkaline environments, or remain electrically neutral near their isoelectric point. Within this broader group, zwitterionic surfactants occupy a special position. Unlike general amphoterics that may switch charge states with pH, zwitterionic molecules bear both positive and negative charges simultaneously in their hydrophilic head groups across a wide pH range, resulting in a net neutral structure. This inherent dual charge endows them with superior interfacial behavior, biocompatibility, and stability compared to conventional amphoterics, making them especially valuable in advanced applications.

1.2 Zwitterion

A zwitterion is a molecule that contains both a positive and a negative charge within the same structure yet remains electrically neutral overall. This unique property arises from the presence of functional groups with differing acid–base characteristics, leading to internal charge separation [4].

Unlike simple ions that are either positively or negatively charged, zwitterions possess an intramolecular balance of charges, making them highly polar and water-soluble while exhibiting characteristics of both cationic and anionic species.

A typical zwitterion consists of:

- A cationic moiety such as ammonium, imidazolium, or pyridinium.
- An anionic moiety such as carboxylate, sulfonate, or phosphate.

The distance and flexibility between these charge centers influence the overall solubility, reactivity, and physicochemical properties of the zwitterion.

Common zwitterionic groups include:

- Amino acids (e.g., glycine, lysine)–containing amine ($–NH_3^+$) and carboxylate ($–COO^-$) groups.
- Betaine surfactants (e.g., coco-betaine)–containing quaternary ammonium ($–N^+R_3$) and carboxylate ($–COO^-$) groups.
- Sulfobetaine surfactants–containing sulfonate ($–SO_3^-$) instead of carboxylate groups for enhanced water solubility.

Unlike simple salts, where charges exist on separate ions, zwitterions maintain intramolecular charge balance, leading to unique solution behavior and interactions with polar and nonpolar substances.

2 Zwitterionic Surfactant

Zwitterionic surfactants possess both cationic and anionic moieties within the same headgroup, separated by a short spacer segment, commonly a methylene chain ($–CH_2–)_n$. This arrangement generates a strong dipolar field, the magnitude of which does not increase linearly with spacer length (n), as the mobility of the intercharge spacer allows for conformational adjustments [5, 6].

Zwitterionic surfactants represent the smallest of the four main classes of surfactants and are characterized by molecules consisting of both charges an anionic and cationic; they are also known as amphoteric surfactants. In most cases, the cationic moiety of zwitterionic surfactants consists of quaternary ammonium, imidazolium, pyridinium, or phosphonium groups, while the anionic counterpart is typically a carboxylate, sulfonate, or sulfate group [7]. Zwitterionic (amphoteric) surfactants are used in significantly lower volumes than conventional anionic, nonionic, and simple cationic surfactants, primarily because their commercial use has been concentrated in mild personal care and specialty formulations rather than bulk detergent applications [8]. Their more elaborate synthesis routes and higher production costs, relative to commodity anionic and nonionic types, further restrict their adoption in cost-sensitive high-tonnage markets [8]. Despite the large variety of zwitterionic surfactants, in practice only a small group of products are used because of factors such as cost, performance, availability of raw materials, and the complexity of the reactions involved; moreover these products have high procurement and production costs and are ususally produced as dilute, low active formulations that sell at a premium and command higher margins. These factors collectively contribute to the relatively limited industrial uptake of zwitterionic surfactants.

One of their most notable properties is pH-dependent ionization, wherein they change their state of ionization according to their pH. At low pH values, zwitterions gain protons and exist predominantly as positively charged cations, whereas at high pH, they lose protons and convert into negatively charged anionic forms [1].

Compared to other surfactants, zwitterionic surfactants exhibit only minor influences from pH, temperature, and added electrolytes [9].

Surfactant science experienced a significant transformation beginning in 1930, with the mid-twentieth century marking the advent of synthetic surfactants and the modern era of surfactant development. Notably, the 1950s saw A.E. Long and E.B. Pullman of DuPont synthesize lauryl dimethylamine oxide, the first zwitterionic surfactant, initially applied in industrial processes for paper, textiles, and detergents.

The unique properties of zwitterionic surfactants, characterized by their amphoteric nature and the equilibrium between hydrophilicity and hydrophobicity defined by the Hydrophilic-Lipophilic Balance (HLB), led to their categorization as emulsions, detergents, wetting, and foaming agents [10, 11]. By the 1970s, their gentle nature and low irritation potential made them suitable for personal care formulations such as shampoos and body washes, leading to wider use beyond industrial applications. Today, they serve a broad range of roles in personal care, household detergents, cleaning agents, industrial operations, pharmaceuticals, biotechnology, the food sector, and agriculture. Because they can interact with many different substances, these surfactants have become a significant class in modern chemistry and industry, demonstrating versatile properties and diverse applications.

Zwitterionic surfactants, the smallest among the four main classes, are characterized by having both positive and negative ionic groups within their head group, with their overall charge depending on the pH of the system. Often classified as specialty, secondary or co-surfactants which are used to complement and create a synergistic effect with primary surfactants, particularly in formulations requiring foam stabilization and salt-induced thickening. This is especially necessary due to the increased use of surfactants in personal and home care products; skin contact direct or indirect skin is a common factor and food preparation surfaces are involved. Consequently, consumer safety is the priority and the most important concern in the development and application of these versatile compounds [12]. In 2024, the global market for zwitterionic surfactants was estimated at around USD 0.9–1.2 billion, with projections indicating a compound annual growth rate (CAGR) of 6–8.5% between 2026 and 2033. The market is growing due to increase in demand of bio-based materials, this shift toward renewable sources has led to the development of bio-based zwitterionic surfactant which are sourced from castor oil and methyl oleate [13, 14]. Their versatility also extends to life sciences, chemical sciences, and even vaccine formulations where they are being used as non-cytotoxic carriers for nano emulsions which increase the drug delivery efficiency (Fig. 2) [15, 16].

Figure 1 illustrates that with a hydrophilic head that has both positive (+) and negative (−) charges and a hydrophobic chain, the dual nature of a zwitterionic surfactant molecule. These areas are connected by a methylene tether (acts as a spacer), which makes it easier for the molecule to cross water and oil interfaces.

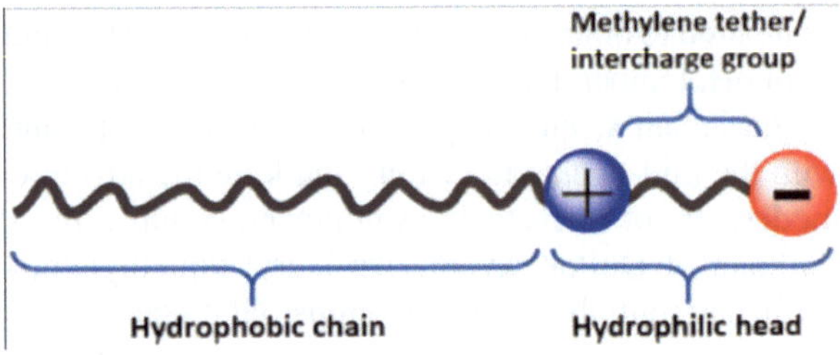

Fig. 2 Presence of positive and negative charge in a zwitterionic surfactant molecule

3 Classification of Zwitterionic Surfactants

Zwitterionic surfactants can be categorized based on their chemical structure and their reaction to pH variations. Their classification provides a systematic approach to understanding their behavior, functionality, and suitability for specific applications. The following sections detail the two primary classification methods: one based on chemical structure and the other on pH sensitivity (Table 1).

3.1 Chemical Structures

The structural differentiation of zwitterionic surfactant occurs from the presence of different functional groups present. They are classified into betaines, amino acid-based type surfactants and imidazoline derivatives, depending on the core molecular framework of the molecule. Each of this class exhibits distinct sets of characteristic and properties which influence the surfactants characteristics such as: wetting, foaming/defoaming, emulsification/demulsification (both macro- and micro-emulsions), dispersion and aggregation of solids, solubility and solubilization (hydrotropic properties), adsorption, micellization and detergency (which is a complex combination of several of these properties).

3.1.1 Betaines

According to Kruger, betaine or (trimethylammonium) acetate (I) is a natural product which was first isolated from beets (*Beta vulgaris*) by Scheibler [17]. The name

Table 1 Classification of surfactants based on its properties

1. Based on chemical structure (a) Betaines (b) Amino acid (c) Imidazoline derivatives	2. Based on the pH (potential of hydrogen) (a) pH sensitive (b) pH insensitive

betaine was proposed by Bruhl to cover all the compounds which are similar in structure to the natural product.

The distinction between betaine-type amphoteric surfactants and other amphoteric surfactants lies in the presence of quaternary ammonium nitrogen, which prevents them from existing as anionic surfactants in alkaline environments. Depending on the pH range, betaine-type amphoteric surfactants can exist either as zwitterionic or cationic species. In the isoelectric zone, amphoteric surfactants are not as prone to a sharp decrease in solubility like other amphoteric surfactants with weakly basic nitrogen.

Betaine-type amphoteric surfactants are also different from cationic surfactants. Some researchers believe that it should be classified as a "quaternary ammonium salt amphoteric surfactant" or "quaternary ammonium salt surfactant". Unlike cationic surfactants such as "external quaternary ammonium salt surfactants", betaine-type amphoteric surfactants can be used in combination with anionic surfactants and will not form "electrically neutral" compounds [18, 19]. Betaines are the zwitterionic surfactants who are gaining attention due to their toxicological profile in the personal and household care industries. The presence of a functional group of a quaternary ammonium or in some cases a sulfonium (II) or phosphonium (III), ion and the carboxylic group provide betaine with the unique set of surface-active properties [17]. The zwitterionic characteristic of the betaine headgroup causes self-screening, enabling the headgroups to pack tightly together [20].

Characterization of betaines is done by a fully quaternized nitrogen atom and their ability to not exhibit anionic properties in alkaline solutions or also demonstrate reduced water solubility in their isoelectric range, thus meaning they exist only as zwitterions [12]. At neutral pH, they remain uncharged however they get protonated and form a cation due to their amphoteric nature. This characteristic causes betaines to behave more like nonionic surfactants rather than ionic ones, making them compatible with anionic surfactants within the pH range commonly found in personal and household care products (Table 2) [12].

Classification of Betaines

(I) They are classified into alkyl betaines (carboxybetaines), alkyl amido betaines, sulfobetaines, phosphoniobetaines, and other phosphorous containing betaines [20].

Table 2 Difference between betaine type of zwitterionics

Carboxylate ion surfactants	Sulfonate ion surfactants
Form external salts with strong acids	External salts are not obtained
Have achieved commercial importance	Have not penetrated the marketplace
Similarity: Both show outstanding solubility in strong electrolyte solutions and are insensitive to hard water	

(1) Alkyl Betaines

They are a class of zwitterionic surfactants with an amphoteric structure, combining a cationic quaternary ammonium group and an anionic carboxylate group within a single molecule. Alkyl betaines are sometimes referred to as carboxybetaines, to indicated unambiguously that they contain a carboxylate group [20].

Their core structure is 2-(alkyldimethylammonio) acetate, where the alkyl chain length varies from C8 to C22 [21], commonly derived from coconut oil (C12–C14) or tallow (C16–C18). They primarily function as general-purpose amphoteric surfactants, primarily used as secondary surfactants to enhance foam stability and mildness in formulations. Their zwitterionic nature allows them to remain stable across a wide pH range, with the hydrophobic alkyl tail and hydrophilic dual-charge head enabling effective surfactant properties. Due to their compatibility with anionic surfactants and low irritation potential, alkyl betaines are widely used in personal care products (shampoos, body washes), household cleaners, and industrial applications.

The straight-chain alkyl groups are typically derived from natural sources such as coconut or palm kernel oils. The most common chain lengths are C12–C14, though mixtures ranging from C8 to C18 may be used. These surfactants demonstrate excellent biodegradability due to their linear structure [20].

There are various types of Alkyl Betaines classified on their alkyl chain length and origin

1. Lauryl Betaine (C12): It is predominantly used in shampoos and cleansers for its foaming and mildness. It features a fully saturated C12 alkyl chain and is occasionally called lauryl dimethylglycine. This surfactant is regarded as mild because it does not induce skin irritation, such as dermatitis or inflammation [22].
2. Coco-Betaine: A mixture of C8–C18 chains derived from coconut oil, often confused with cocamidopropyl betaine (CAPB) but different structurally.

   The surfactant is a blend of molecules with varying alkyl chain lengths and is synthesized from a mixture of fatty alcohols originally obtained from coconut oil [12].

   The CMC without salt was reported as 1.9×10^{-3} mol/kg at 25 °C and 1.8×10^{-3} mol/kg at 35 °C [23].
3. Myristic Betaine (C14): Enhances viscosity and foam stability in rinse-off products. The synergistic effect between alkyl betaines and anionic surfactants gives rise to increased packing efficiency at the interface which increases surface shear viscosity which results into better foam stability [24].
4. Behenyl Betaine (C22): Used in conditioning products for its thickener, emulsifier and stabilizing properties. It is high molecular weight; saturated fatty alcohol derivative gives it the above-mentioned properties.

$$\text{R-OH} + (\text{CH}_3)_2\text{NH} \xrightarrow[\text{Reduction}]{\text{Catalytic}} \text{R-N(CH}_3)_2 + \text{H}_2\text{O}$$

Fig. 3 The reaction above represents reductive amination of a fatty alcohol with dimethylamine to yield a tertiary amine which serves as a precursor for quaternary ammonium surfactants and betaines

$$\text{R-N(CH}_3)_2 + \text{ClCH}_2\text{COO}^-\text{Na}^+ \xrightarrow[\text{pH 8-10}]{\text{80-90°C}} \text{R-N}^+(\text{CH}_3)_2\text{-CH}_2\text{COO}^- + \text{NaCl}$$

Fig. 4 Here, the tertiary dimethylalkylamine undergoes SN2 attach on the primary carbon of sodium chloroacetate which displaces chloride ion and produces N-carboxymethyl quaternary ammonium salt with NaCl as a by-product. Reaction conditions are moderate to keep the carboxylate group deprotonated. This is an efficient route to produce alkyl betaines however requires downstream removal of NaCl and unreacted substances to achieve high purity

5. Oleyl Betaine (C18:1): Features an unsaturated chain for improved solubility in formulations. Due to this, they are used in liquid and semi-solid formulations enhancing formulation flexibility and performance [25].

Preparation of Alkyl Betaines

The synthesis of alkyl betaines involves a multistep process that begins with the derivation of bio-based fatty alcohols from natural sources such as coconut oil or palm kernel oil. The overall process consists of the following key stages:

Alkyl betaines, zwitterionic surfactants derived from natural fatty alcohols, are synthesized via a two-step quaternization process:

(1) Tertiary Amine Formation

In the first step, fatty alcohols (C_8–C_{18}) react with dimethylamine under reductive conditions to produce dimethyl alkylamines. The tertiary amine's lone pair attacks the electrophilic carbon in chloroacetate replacing the chloride ion (Cl^-) in a SN2 nucleophilic substitution and zwitterion of the quaternary ammonium group (N^+) and carboxylate group (–COO) is formed. The resulting dimethyl alkylamines are purified by extraction and distillation, and their structures are confirmed by NMR spectroscopy (Fig. 3) [26].

Step 2: Quaternization with Sodium Chloroacetate (SCA)

The tertiary amine undergoes nucleophilic substitution with SCA to form alkyl betaines. The product is obtained in admixture with sodium chloride. Takaku and Imai found that the reaction yield can be increased by applying the Le-Chatliers principle, adding a small amount of betaine is added prior to adding the bulk of reactants (Fig. 4) [27].

Optimal conditions for the reaction:

$$C_{12}H_{25}\text{-}N(CH_3)_2 + ClCH_2COO^{-}Na^{+} \longrightarrow C_{12}H_{25}\text{-}N^{+}(CH_3)_2\text{-}CH_2COO^{-} + NaCl$$

Fig. 5 The reaction above is SN2 quaternization of lauryl dimethylamine with sodium chloroacetate to produce lauryl betaine, a zwitterionic surfactant

- Molar ratio should be between 1:1.03 and 1:1.3 mol (Amine: SCA) to ensure complete conversion and minimize residual amine [28]
- Temperature is kept to 80–90 °C during the reaction, as heat accelerates the reaction
- pH is maintained at 8–10 using NaOH to prevent protonation of the amine. Alkaline conditions ensure that there are no side reactions ensuring zwitterion formation.

Purification Step

This involves a targeted thermal and chemical treatment to eliminate residual halogenated compounds ensuring high purity in the final product. Thermal degradation of residual halogenated compounds: The reaction mixture is heated at 100–150 °C for 1–4 h. This step is designed to degrade compounds as mono halo acetic acid, thereby reducing their presence in the final product [12, 27].

In post-reaction treatment, the betaine solution is adjusted to a pH range of 5–12 and the solution is heated again. This stage specifically targets the reduction of organically bonded halogen impurities, such as chloroacetate derivatives, lowering their concentration to below 10 ppm [28].

The final step is high-temperature hydrolysis at 100–150 °C. This treatment further removes any residual halogenated impurities, including unreacted chloroacetate, ensuring that the final alkyl betaine meets stringent purity standards. For scaling up the process, continuous production is required where tubular reactors with suitable modifications are to be made to factor in chloroacetate's solid state [29].

Example: Synthesis of Lauryl Betaine (C_{12}) (Fig. 5).

CMC Affecting Alkyl Betaines

Despite their extensive use in formulations and commercial products, the physical and chemical characteristics of alkyl betaines have not been thoroughly documented in scientific literature.

Research indicates that while alkyl betaines are industrially significant, comprehensive scientific studies are limited, with lauryl betaine receiving the most attention [2, 30]. Chorro et al. determined the CMC of LB to be 1.9×10^{-3} mol/kg at 25 °C and 1.8×10^{-3} mol/kg at 35 °C, demonstrating a minimal temperature dependence. Notably, the presence of salts significantly reduces the CMC, highlighting the role of charge interactions in the aggregation of these zwitterionic surfactants [23]. Divalent salts, such as $CaCl_2$, exhibit a more pronounced effect compared to monovalent salts like NaCl, emphasizing the influence of ion valency on CMC (Table 1) [23]. Furthermore, salt presence alters LB adsorption onto silica, increasing adsorption

Table 3 CMC susceptibility of alkyl betaine with respect to chain length

CMCs with respect to the chain length of alky betaine [22, 31]	
Alkyl chain length of the betaine	CMC (uM)
8	1.86×10^5
10	1.80×10^4
12	1.84×10^3
14	186
16	20.1
18	4.78

at lower surfactant concentrations. The observed stronger binding of LB to anions (Cl^-) relative to cations (Na^+, Ca^{2+}) further underscores the importance of charge interactions in its physicochemical behavior [23].

Extending the discussion to other alkyl betaines, Birnie et al. [31] CMC values that align with Chorro et al.'s findings, showing a consistent trend of CMC decreasing approximately tenfold for every two methylene units added to the alkyl chain. This trend adheres to Stauff-Kleven's rule, with a notable exception for octadecyl betaine, which displays a higher-than-expected CMC [31]. Beyond CMC, the foaming properties of LB are also significant. Syed et al. explored the synergism of LB with nanoparticles and other surfactants, revealing increased viscosity when combined with alpha-olefin sulfonate (AOS) or sodium dodecyl sulfate (SDS), but not with silicon dioxide (SiO_2). The most stable foam was observed with a 9:1 AOS to LB ratio [22]. These findings collectively highlight the complex interplay of factors, including salt presence, alkyl chain length, and synergistic effects, in determining the physicochemical properties of alkyl betaines (Table 3).

(2) Alkyl Amido Betaines

Alkyl amidopropyl betaines are zwitterionic surfactants that combine a hydrophobic alkyl chain, an amide bond, and a quaternary ammonium-carboxylate headgroup. They utilize fatty acids or fatty acid esters as their hydrophobic precursor instead of fatty alcohols. These amphoteric molecules adapt to pH changes, functioning as cationic surfactants in acidic conditions, and zwitterionic forms at neutral pH. Their unique structure gives them the following properties: mildness, foaming synergy, and compatibility with other surfactants has made them indispensable in personal care, detergents, and industrial formulations.

They feature an amide linkage between a fatty alkyl chain and the betaine head group, yielding enhanced mildness and foam stability relative to non-amidated analogous [32, 33]. Their synthesis typically involves fatty acids or esters undergoing amidation with an appropriate amine linker, followed by quaternization, resulting in a zwitterionic structure [16, 34]. Studies have shown that chain length, degree of unsaturation, and presence of branching can influence their interfacial properties, foam generation, and overall effectiveness [16, 22, 35]. The literature also highlights reduced irritation, making them attractive for personal care applications where mildness is paramount [9, 16] (heads or tails).

The alkyl amido betaines are commercially the most important class of zwitterionic surfactants in use today, specifically cocamidopropyl betaine (CAPB). Cocamidopropyl betaine (CAPB) is manufactured on a large industrial scale, and the resulting economies of scale have significantly reduced its production cost, enabling it to supersede other classes of zwitterionic surfactants in commercial formulations over time.

In the 1950s, CAPB was first synthesized by Henkel Corporation as a gentle alternative to harsh anionic surfactants like sodium lauryl sulfate (SLS), initially finding application in textiles and household cleaners for its softening properties (https://yeserchem.com/history-and-evolution-of-cocamidopropyl-betaine). As consumer awareness grew and the need for low irritation products became evident, the 1970s and 1980s witnessed a surge in the adoption of CAPB in personal care products, particularly shampoos, where its ability to synergize with anionic surfactants resulted in improved foam stability and enhanced viscosity. Entering the 2000s and continuing to the present day, CAPB has become a dominant ingredient in formulations such as baby shampoos and sulfate-free products.

Types of Alkylamidopropyl Betaines

They are classified based on alkyl chain length and fatty acid source:

(A) By alkyl chain length

(1) Short-Chain Variants (C8–C10)

E.g., Caprylamidopropyl Betaine (C8):

- Derived from caprylic acid (octanoic acid).
- Properties: High water solubility, low viscosity.
- Applications: Mild cleansers (e.g., micellar waters, baby products).

(2) Mid-Chain Variants (C12–C14)

E.g., Lauramidopropyl Betaine (C12):

- Primary component in coconut-derived mixtures.
- Properties: Superior foaming, viscosity modulation.
- Applications: Shampoos, body washes

E.g., Myristamidopropyl Betaine (C14):

- Derived from myristic acid.
- Properties: Enhanced foam stability.

(3) Long-Chain Variants (C16–C18)

E.g., Palmitamidopropyl Betaine (C16):

- Sourced from palm oil.

- Properties: Thickening agent, conditioning.

E.g., Stearamidopropyl Betaine (C18):

- Derived from stearic acid.
- Properties: High emulsification, used in heavy-duty cleansers.

(B) By Fatty Acid Source

(a) Coconut/Palm Derived (C8–C18 Mixture)

E.g., Cocamidopropyl Betaine (CAPB):

- CAS: 61789-40-0 (inner salt), 70851-07-9 (sodium salt).
- Composition: Mixture of C8–C18 chains, predominantly C12.
- Applications: Gold standard in personal care (shampoos, facial cleansers)13.

(b) Tallow-Derived (C16–C18)

E.g., Tallow amidopropyl Betaine:

- Sourced from hydrogenated animal fats.
- Properties: High stability in hard water, used in industrial detergents.

Synthesis of Alkyl Amido Betaines

They are synthesized by multistep processes based on coconut or palm kernel oil; the process begins with the derivation of hydrophobes from the feedstocks. The process is a meticulously controlled one, involving amidation and quaternization with attention to impurity management and reaction conditions.

1. Amidation: It is the reaction between coconut fatty acids and DMAPA from an amide bond. This reaction creates the amidoamine intermediate, a key precursor for synthesizing alkyl amido propyl betaines.

   Reactants:

   - Fatty acids: Derived from coconut oil (C8–C18 chains, predominantly C12 lauric acid) via saponification or methyl esterification.
   - DMAPA (3-dimethylaminopropylamine): Synthesized by reacting dimethylamine with acrylonitrile to form dimethyl aminopropionitrile (DMAPN), followed by hydrogenation.

     The reaction temperature is 150–175 °C to drive amide bond formation between DMAPA's primary amine and fatty acids/methyl esters. Due to this high temperature, certain by-products like color bodies (e.g., oxidized amines) are generated, therefore distillation is required for purification (Fig. 6).

2. Quaternization: This is a terminating step where the amidoamine intermediate reacts with sodium chloroacetate (SCA), generated in situ from monochloroacetic

$$R\text{–}COOH + H_2NCH_2CH_2CH_2N(CH_3)_2 \xrightarrow{\Delta} R\text{–}CONHCH_2CH_2CH_2N(CH_3)_2 + H_2O$$

Fig. 6 Direct amidation of a fatty acid derived from coconut oil with DMAPA. An amide bond is formed after a nucleophile attack from primary amine of DMAPA which also releases water as a condensation by-product

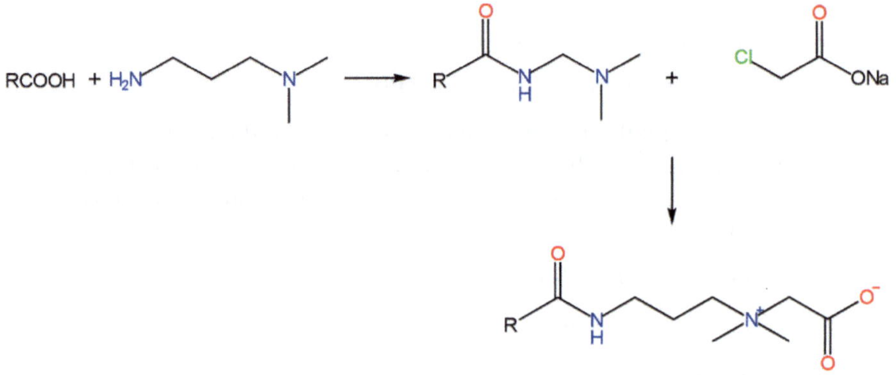

Fig. 7 The reaction above represents quaternization of amidoamine intermediate with sodium chloroacetate to yield alkylamidopropyl betaine. The reaction is driven by SN2 quaternization which forms the zwitterionic structure

acid (MCA) and sodium hydroxide to form a zwitterionic betaine structure. The molecule is converted into the final alkyl amido propyl betaine surfactant.

Reactants:

- Sodium chloroacetate (SCA): Generated in situ by neutralizing monochloroacetic acid (MCA) with NaOH

The pH of the reaction is maintained at 9–10 pH using NaOH to suppress SCA hydrolysis (which produces HCl and sodium glycolate, reducing yield). The temperature is 80–100 °C for 2–8 h (Fig. 7).

Control of Impurities and Product Integrity

The synthesis of alkyl amido propyl betaines like cocamidopropyl betaine (CAPB) demands rigorous impurity control to meet safety and formulation benchmarks across sensitive applications. The main points of control are.

(1) The use of ultra-high purity monochloroacetic acid (MCA, $\geq$ 99%) to reduce dichloroacetic acid (DCA) a suspected carcinogen to negligible levels.
(2) A residual 3-dimethylaminopropylamine (DMAPA), a skin irritant, is restricted to $<$ 5 ppm through post-synthesis treatments like sodium metabisulfite neutralization or ethanol washing.[36]

Purification and Final Composition

After post synthesis, the product undergoes purification to meet the industry's specifications. The salt content is ~ 6% sodium chloride (NaCl) which is reduced to < 1% via ethanol/methanol washing or vacuum evaporation enhancing compatibility in delicate formulations. The impurity targets include < 30 ppm aminopropyl dimethyl amino betaine (a contributor to off taste in oral care) and < 0.5% free fatty acids. The final product is supplied as a 30–40% active aqueous solution with trace amounts of glycolic acid, NaCl ($\leq$5%), and glycerol esters depending on the preliminary feedstock.

(3) Sulfobetaines

Sultaines or sulfobetaines are electroneutral salts which contain two ionic centers with different charges. They are chemically analogous with betaines, with a sulfonate group present in place of betaine carboxylate within their headgroup and an alkyl tail. They are also structurally like alkyl betaines, where the carboxylic acid groups are replaced by a sulfonate group. It appears that "sultaine" is a contraction of "sulfobetaine," and the terms "sultaine" and "sulfobetaine" are used interchangeably.

The presence of a spacer and the functional group that connects the headgroup and tail group allow molecules to be further subdivided into different groups based on their structure. For example, amidosultaines and hydroxysultaines [20].

Key Characteristics of Sulfobetaine

They are a group of electroneutral salts with two ionic centers of different charges in the same molecule. They have no net charge [37] and are pH insensitive which make them stable across neutral, acidic, and alkaline pH ranges [38]. They are also known for their hardwater tolerance and resistance to degradation by oxidizing and reducing agents [38]. Due to their non-toxic, non-irritating skin, and eyes, they do not cross the skin's natural barrier which provides the formulation with required dermal compatibility. They also show minimum dependence on temperature changes and electrolyte addition.

The specific structure of sulfobetaines also allows them to exhibit excellent functional properties like strong detergency, emulsification, foaming, biodegradability, and excellent solubility in water. Due to this, they have found use not only in cosmetic products but also in domestic detergent formulations [1, 12–14]. Additionally, they do not transport substances across the skin and are combined with cationic or anionic surfactants offering high surface activity.

Given their ability to solubilize proteins and membrane receptors, sulfobetaines may be employed in structural biology research [15]. There is also a possibility that they are also used for preparation of ultrafiltration membranes via a phase inversion method. Longer operating lifetimes and an efficient decrease in membrane fouling across a broad pH range are the benefits of employing these materials. Increasing the quantity of sulfobetaine groups in the membrane could enhance its hydrophilicity and antifouling characteristics.

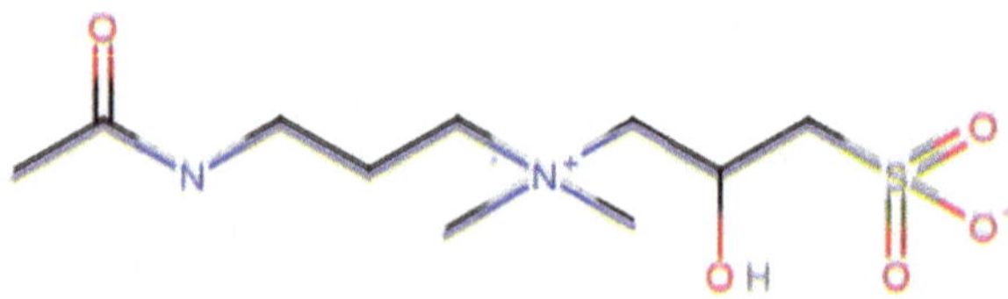

Fig. 8 Structure of amido propyl hydroxysultaine

Types of Sulfobetaines

(I) Hydroxysultaines

They are a type of sultaine that has a hydroxyl group (-OH) present in the part of a molecule. They are the most commonly known and widely used sultaines. They also possess a headgroup which is more hydrophilic than amidopropyl betaines due to the synergistic presence of sulfate anionic group and hydroxyl group (Fig. 8).

Properties

- High water solubility (e.g., lauryl hydroxysultaine: 680 g/L at 20 °C) which facilitates high-active formulations.
- Broad pH and electrolyte tolerance thus it is ideal for formulations exposed to multiple water quality and chemical conditions.
- It acts as an effective foam booster and rheology modifier, particularly when combined with primary surfactants in personal care.
- It is dermatologically mild with a low irritation index thus suitable for leave on and rinse-off formulations even at 1–5% active use levels.
- It also creates electrostatic discharge (ESD) reduction due to zwitterionic nature, offering dual functionality in industrial settings.

Applications

- Personal Care: Used in low irritation cleansing systems, enhancing foam quality, density, and stability in shampoos, body washes, and facial cleansers.
- Industrial and Institutional Cleaning (I&I): It is employed in alkaline degreasers, metal cleaners and hard surface cleaners for its high wetting, detergency, and antistatic effects.
- Petroleum and Mining: It is applied in drilling fluids and enhanced oil recovery for its stability in brine and multivalent ion systems.

Safety

- Low acute and chronic toxicity (e.g., lauryl hydroxysultaine demonstrates a high margin of safety in dermal and ocular tests).
- Readily biodegradable and compliant with global eco-labels (e.g., EU Ecolabel, Safer Choice), aligning with green and sustainable formulation goals.
- It is found to be non-sensitizing and non-irritant in human repeat insult patch testing (HRIPT), making it suitable for sensitive skin applications.

(II) Amidosultaines

They are zwitterionic surfactants which are characterized by a sulfonate group (negative charge) along with a quaternary ammonium group (positive charge) within the same molecule. An amide linkage (–CONH–) is found between their hydrophobic tail and the hydrophilic head. This structural characteristic provides them with superior thermal, chemical, and oxidative stability, along with better biodegradability.

Structural overview
A hydrophobic alkyl chain, typically C12–C18
A tertiary amine connected via a propyl spacer to a sulfonate group
A central amide linkage formed by reacting fatty acids or esters with 3-(dimethylamine) propylamine (DMAPA), followed by sulfoalkylation with 1,3-propane sultone or sodium 3-chloropropanesulfonate

Key examples include:

- Cocoamidopropyl hydroxysultaine
- Lauryl amidopropyl hydroxysultaine
- Stearylamidodimethylpropylsultaine

(III) Phosphobetaines

Phosphobetaines are particularly noteworthy because of their unique structural motif, characterized by a covalent attachment of a quaternary phosphonium cation to an anionic functional group, typically a carboxylate, sulfonate, or phosphate moiety. They are structurally and functionally analogous to biological zwitterions like phosphatidylcholine (a major phospholipid component of cellular membranes). This resemblance has imparted some biomimetic properties to phosphobetaines, making them highly relevant in biomedical and environmental applications.

Structural Characteristics

The general structure of phosphobetaines is represented as follows:

$$R_3P^+-(CH_2)_n-Z^-$$

R_3P^+: Quaternary phosphonium group
$(CH_2)_n$: Hydrocarbon Spacer (These methylene units provide flexibility and spatial separation

Properties

The properties were studied by scientists Michaelis and Kohler who showed that when phosphonium salts with the structure $Ph_3P^+CH_2COR\ X^-$ (where R is methyl or phenyl) were treated with aqueous alkali, stable halogen-free compounds were obtained. They termed these compounds phosphobetaines and initially assigned them the structure [39].

Criteria	Theory
Stability	• Depends on the substitution, fully phenylated phosphoranes (e.g., Ph_3P:CHCOR) are water-stable at room temp • Replacing phenyl with methyl ($MenPh_{3-n}P$: CHCOR) reduces water stability, especially at higher methyl content • Early members ($n = 0, 1$) hydrolyze rapidly at room temp
Wittig reaction reactivity	General behavior • Most phosphobetaines react with aldehydes to give α, β-unsaturated products • No reaction with ketones in most cases, due to failure in the initial nucleophilic attack
	Specific examples • Ph_3P: CHCOR (R = Me, Ph): unreactive with ketones • Ph_3P: CHR (R = CN, CHO, $CONH_2$): reactive with aldehydes only • Formylmethylenetriphenylphosphorane (VI): gives cinnamaldehyde with benzaldehyde; does not react with ketones • α-Formylethylidenetriphenylphosphorane: aldehyde-selective, forms α-methylcinnamaldehyde • Methyl substitution (Me_nPh_{3-n}) on phosphorus: slightly improved ketone reactivity (e.g., fluorenone, cyclohexanone), but not with COR-type phosphoranes
	In situ generated and forcing reactions • *(Ethoxycarbonylmethyl)trimethyl phosphonium bromide* reacted with cyclohexanone (3% yield) • *Methylenetrimethylphosphorane* reacted with benzophenone (40% yield) and gave poor yields (15%) with cyclohexanone under forcing conditions
Reduction	• When treated with Zn + AcOH: phosphobetaines convert to triphenylphosphine and its corresponding ketone (~95% yield)

Synthesis Methods

(I) **Synthesis of Phosphobetaines via a Mannich-Like Reaction Involving amid amines and Phosphite Esters** (U.S. Patent No. 7371497)

Mannich-like condensation reaction involves amid amines, phosphite esters, and formaldehyde. It is one of the most efficient and synthetically versatile methods for the preparation of phosphobetaines. This method is used widely due to its operational simplicity, compatibility with a range of functional groups, and high atom economy. The reaction typically involves three key components [40].

(1) Amid amines: $R^1R^2N{-}CH_2CH_2{-}NR^3R^4$
(2) Phosphite esters: $(RO)_2P(O)H$
(3) Formaldehyde: It is added as an aqueous solution or pre-dissolved in a protic solvent

The reaction involves a one-step reaction between amid amines, phosphite esters, and formaldehyde.

Reaction Mechanism: This reaction is like a Mannich reaction wherein spontaneous quaternization of the tertiary nitrogen atom which forms the betaine-type surfactant. For by-product removal, reagents are distilled under reduced pressure.

Example: A reaction of *N, N-dimethylaminopropylamine* along with *diethyl phosphite* and formaldehyde yields a phosphobetaine.

Steps

(1) The reactants of amidaamine and phosphite esters are dissolved in a polar type of solvent.
(2) Formaldehyde is added to the mixture when the temperature is 60–140 °C under reflux conditions. The reaction takes about 2 to 5 h.

- Molar ratios are as follows: 1.1–1.2 mol of phosphite ester per tertiary amine group.
- 1.5–2 mol of formaldehyde per tertiary amine group.

(II) **Phosphation of Alkoxylated Esters Followed by Quaternization:**

This method is an industrial method where the surfactant manufactured is of the tuneable properties and environmental compatibility. This approach is widely used to produce zwitterionic compounds from personal care to metalworking fluids [41, 42]. The reaction is carried in two steps.

(1) Phosphating of Alkoxylate Esters: This is a chemical process in which phosphate groups are added to esters which have been modified with spacer units like EO or PO. This step is done so that the polar and hydrophilic phosphate groups are introduced improving the surfactants' water solubility and cleaning properties. It is widely used in making eco-friendly detergents, personal care products, and emulsifiers.

Reactants: Alkoxylated Esters and a Phosphating reagent (P_2O_5 or Polyphosphoric acid)

Procedure:

(I) Alkoxylate alcohol is heated to 60 °C, and a phosphating reagent is added is added slowly with continuous mixing.
(II) The temperature increases to 90–100 °C for 3–5 h to form a product consisting of mono and di-esters. This product is an intermediate of phosphate ester.

The molar ratio of P_2O_5 to alkoxylate: 1:1.2–1.5 to minimize diester formation.

(2) Quaternization with Epoxy derivatives: In this process, the amines are reacted with phosphate esters to form quaternary ammonium compounds. Due to this, a positive charge is added in the molecule; this adds to the antibacterial activity and compatibility with both water and oil phases.

Quarternizing agents used are DMS (as it forms stable quaternary ammonium groups), Methyl Carbonate, and Epoxy derivatives. The temperature for the reaction is 60–70 °C to prevent ester hydrolysis and the pH is 8.5–9.5.

The reaction is monitored continuously: (a) Free amine content less than 2% indicates completion and (b) 30–40% solid content for handling viscosity.

During the reaction, Diester formation is managed carefully by controlling the stoichiometry and the rate of P2O5 addition and the quaternization of phosphate esters is generally carried mildly alkaline conditions (pH 8.5–9.5) so that the integrity of the ester bonds.

3.1.2 Imidazoline

Imidazoline surfactants are a prominent class of nitrogen-containing heterocyclic compounds characterized by a five-membered ring structure with two nitrogen atoms. These compounds are formed primarily through the reaction of long-chain fatty acids with polyamines, such as diethylenetriamine (DETA), leading to the creation of amphiphilic molecules with a hydrophobic alkyl tail and a cationic head group [43]. Their structure resembles imidazole but is partially saturated, conferring distinct physicochemical properties such as increased reactivity and thermal stability.

Substituted imidazolines typically have three key positions on the ring for substitution: a fatty acid aminoethyl chain at the 1-position, a long hydrophobic alkyl chain at the 2-position, and a methyl or hydrogen substituent at the 3-position [44]. This substitution pattern significantly influences their solubility, surface activity, and compatibility with different formulation environments.

These surfactants are classified mainly into cationic and amphoteric types, depending on their ionization behavior. Cationic imidazolines possess a permanent positive charge on the nitrogen atom, making them highly effective for adsorption onto negatively charged surfaces like cellulose, glass, or metals [45]. Amphoteric derivatives, by contrast, exhibit both positive and negative charges depending on the pH, rendering them suitable for applications in personal care and cleansing formulations [43].

Owing to their amphiphilic structure, imidazoline surfactants exhibit excellent surface and interfacial properties, including emulsification, corrosion inhibition, and antistatic behavior [44]. Their cationic nature makes them particularly useful in textile softeners, where they reduce fiber friction and impart a smooth, fluffy texture even after multiple laundering cycles [45]. Furthermore, imidazoline derivatives exhibit superior thermal stability, viscosity control, and storage resilience, making them ideal for industrial formulations such as lubricants, agrochemical emulsifiers, and dewatering agents [46].

The recent molecular modeling has shown that imidazoline molecules orient themselves on surfaces in a way that their polar head groups interact with the substrate while the hydrocarbon tails extend outward, forming ordered monolayers. This arrangement enhances surface protection and contributes to the durability of treated materials (Fig. 9) [43].

The imidazoline ring typically undergoes substitution at the 1-, 2-, and 3-positions:

Position 1: Often substituted with a fatty acid aminoethyl chain.

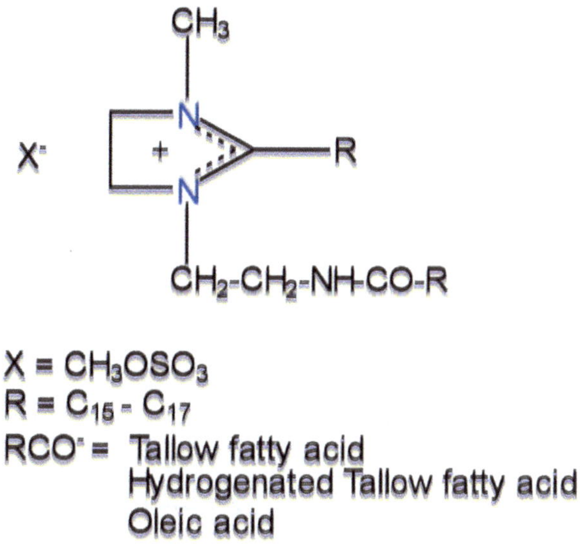

Fig. 9 Structure of imidazoline

Position 2: Commonly bears a long alkyl substituent.
Position 3: May have a methyl group.

These substitutions influence the physical properties of the molecule, such as melting point and fluidity. For instance, unsaturated C18 chains in the cis configuration lower melting points and enhance fluidity, whereas saturated C18 chains increase rigidity and raise melting temperatures.

Types of Imidazoline

(1) Cationic Imidazolines: Cationic imidazoline surfactants contain a positively charged nitrogen atom within a heterocyclic five-membered ring structure and remain cationic over a wide pH range, making them particularly suitable for applications in textile processing, corrosion inhibition, and surface conditioning [44]. Their permanent charge ensures high affinity for negatively charged substrates such as cotton, synthetic fibers, metals, and minerals [45].

Subtypes:

Hydroxyethyl Imidazolines: Hydroxyethyl imidazolines are synthesized via a 1:1 condensation reaction between fatty acids and aminoethyl ethanolamine. These derivatives are employed as rheology modifiers and adhesion promoters in lubricants and also find use in construction-related materials like road-making emulsions and industrial coatings (Fig. 10) [43].

Aminoethyl Imidazolines: Aminoethyl imidazolines, on the other hand, are formed from the reaction of fatty acids with diethylenetriamine in equimolar

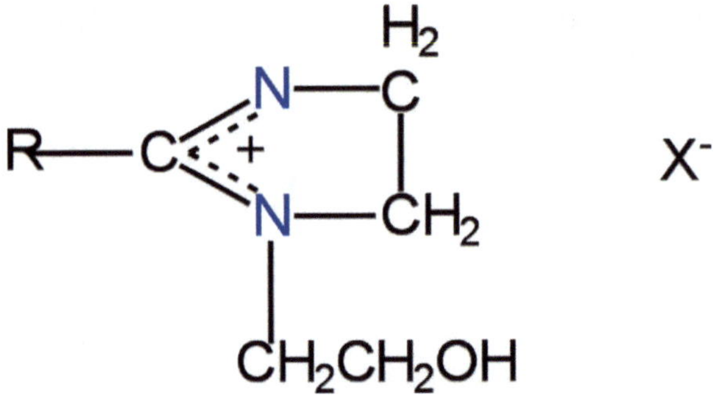

Fig. 10 Structure of hydroxyl ethyl imidazoline

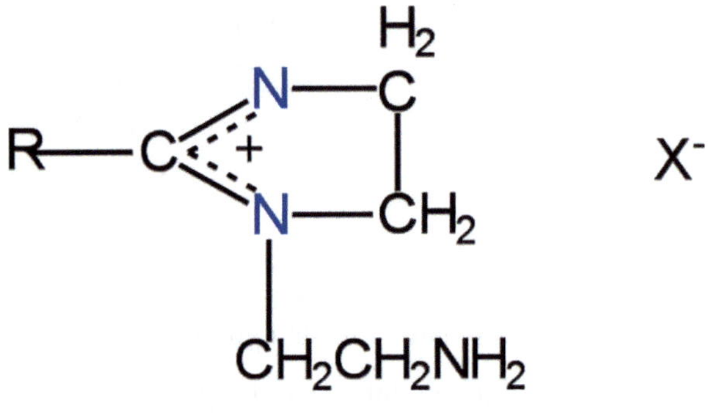

Fig. 11 Structure of aminoethyl imidazoline

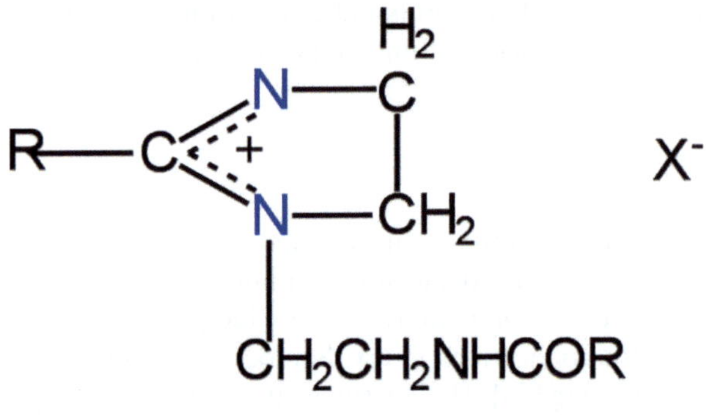

Fig. 12 Structure of amidoethyl imidazoline

ratios. These compounds are extensively used as oilfield corrosion inhibitors due to their strong adsorption onto metal surfaces and film-forming capability, as well as in dispersants and emulsifiers for petroleum-based systems (Fig. 11) [46].

Amidoethyl Imidazolines: Amidoethyl imidazolines, which are also derived from fatty acid and diethylenetriamine reactions, are typically used in fabric softeners, acid-stable detergents, and flocculating agents for water treatment and agricultural chemicals (Fig. 12) [44].

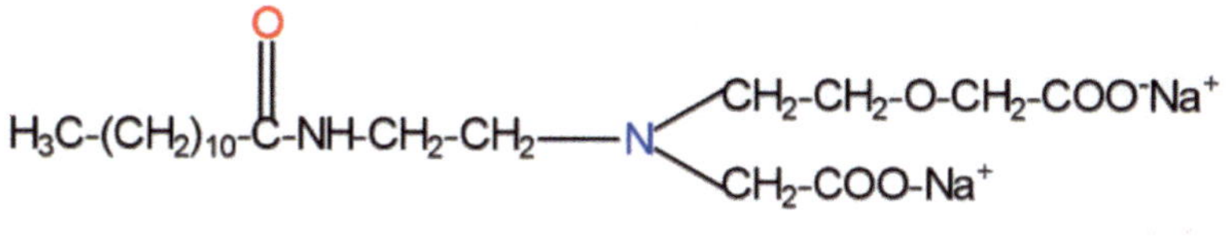

Fig. 13 Structure of disodium lauroaphodiacetate

Fig. 14 Structure of disodium lauroamphodipropionate

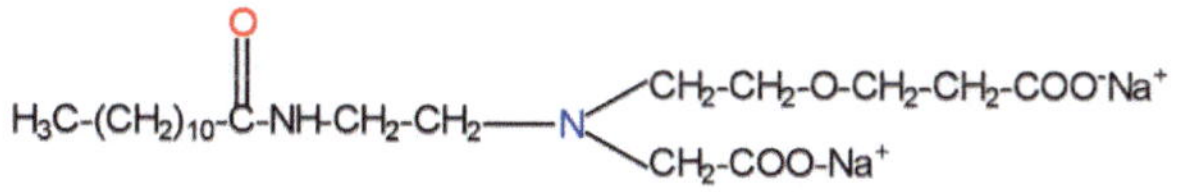

Fig. 15 Structure of sodium lauriminodipropionate

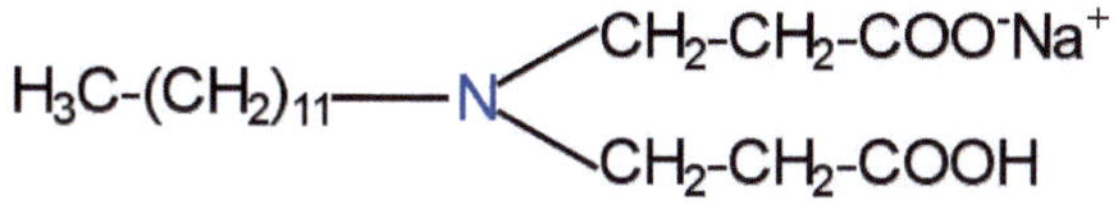

(2) Amphoteric Imidazolines: Amphoteric imidazoline surfactants differ from their cationic counterparts in that the imidazoline ring forms only transiently during synthesis and does not persist in the final molecule [43]. The final compounds feature both anionic and cationic groups, which ionize differently depending on the pH, thereby exhibiting zwitterionic behavior in aqueous environments. This imparts exceptional dermatological mildness, high compatibility with other surfactants, and reduced skin and eye irritation, making these surfactants suitable for personal care formulations such as shampoos, baby washes, and facial cleansers [44].

Subtypes:

Alkylamphodiacetates: Alkylamphodiacetates are synthesized by reacting methyl acrylate with fatty acid/amine intermediates. A commonly used example is disodium lauroamphodiacetate, which provides gentle cleansing and excellent foaming characteristics in mild shampoos and body washes (Fig. 13) [44].

Alkyliminopropionates: Alkylamphodiacetates are synthesized by reacting methyl acrylate with fatty acid/amine intermediates. A commonly used example is disodium lauroamphodiacetate, which provides gentle cleansing and excellent foaming characteristics in mild shampoos and body washes (Fig. 14) [44].

Alkylamphopropionates: Alkylamphopropionates, synthesized by combining acrylic acid with fatty acid-derived amines, such as sodium lauriminodipropionate, deliver desirable foaming, mildness, and biodegradability—especially in sulfate-free or eco-labeled formulations (Fig. 15) [44].

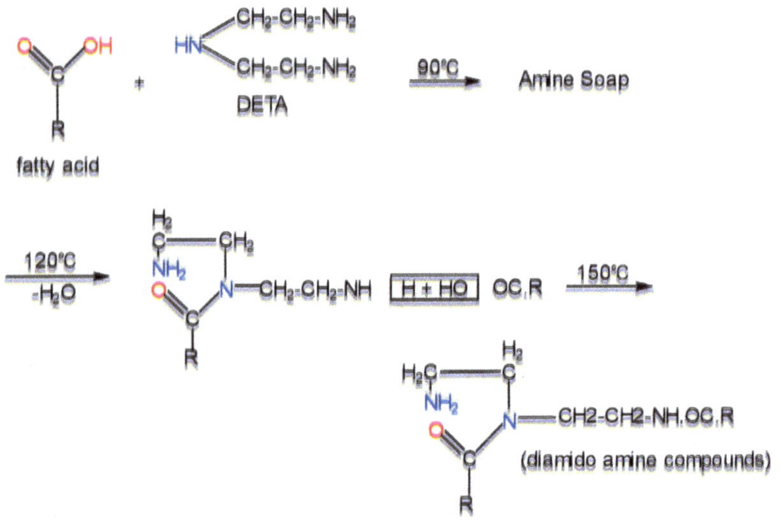

Fig. 16 Synthesis of diamidoamine compounds

Synthesis (Imidazoline and Its Derivatives: An Overview)

Imidazoline surfactants are commonly synthesized via a two-step process starting with the condensation of long-chain fatty acids and polyamines such as diethylenetriamine (DETA). The first step forms a diamidoamine intermediate, followed by intramolecular cyclization to yield the imidazoline ring structure [43]. This synthetic route is widely employed due to its efficiency and compatibility with various fatty acid feedstocks, including oleic and stearic acids.

In the first step, the fatty acid reacts with both the primary and secondary amine groups of DETA under thermal conditions, typically between 150 and 180 °C, forming a diamidoamine through two consecutive condensation reactions with the release of water [44]. This intermediate adopts a waxy, semi-solid consistency, and the yield is generally high under atmospheric conditions. The resulting product has the general structure R–CONH–CH_2CH_2–NH–CH_2CH_2–NH–COR, where R represents the hydrophobic fatty chain (Fig. 16).

Scheme 1

The second step involves ring closure via intramolecular cyclization of the diamidoamine. This reaction is conducted under reduced pressure (vacuum of 0.2–0.5 mmHg) and elevated temperatures (up to 260 °C), which drives off the water formed during the reaction and pushes the equilibrium toward cyclization [43]. The cyclization step typically takes 2–6 h and results in yields of imidazoline surfactants as high as 90–92%[46]. The efficiency of this process depends heavily on the water removal rate and reaction temperature (Fig. 17).

Fig. 17 Synthesis of imidazoline

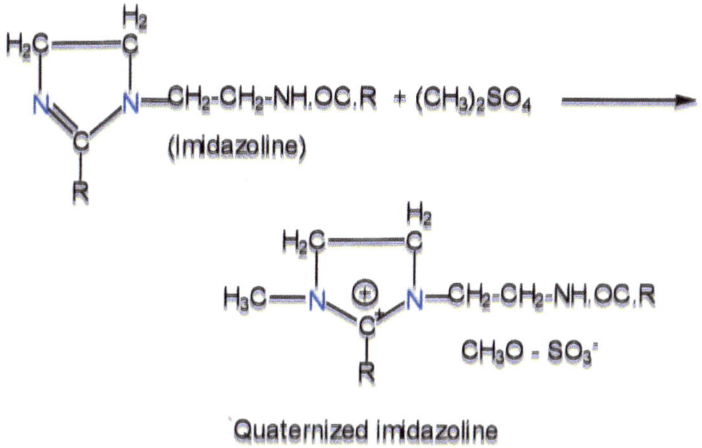

Fig. 18 Synthesis of quaternary imidazoline

An optional third step is the quaternization of the imidazoline product to improve water solubility and cationic strength. This step typically involves the alkylation of the imidazoline nitrogen using reagents such as methyl chloride, dimethyl sulfate, or benzyl chloride in alcoholic solvents like ethanol or isopropanol at 80–90 °C for several hours [44]. The resulting quaternary ammonium salts, also called imidazolinium surfactants, are more hydrophilic and perform better in aqueous formulations, especially in applications such as textile softeners and antistatic agents (Fig. 18) [43].

Alternative synthetic routes have been explored to utilize renewable feedstocks such as triglycerides. In this triglyceride-based method, animal fats like tallow are directly reacted with DETA at 180 °C to generate the diamidoamine, followed by a high-temperature vacuum step for cyclization [44]. However, this route generates glycerol as a by-product, which is hydrophilic and can compromise the final surfactant's water repellency unless removed by additional processing [46].

Another variation involves the use of methyl esters such as methyl tallowate instead of free fatty acids. This methyl ester route provides a cleaner starting material, avoiding glycerol formation. However, it suffers from slower reaction kinetics, requiring up to 13 h for completion, and the presence of water generated during ring closure can inhibit imidazoline formation [46]. Infrared spectroscopy studies show delayed ester disappearance and incomplete amine conversion in such systems, requiring optimized reaction control.

Overall, the synthesis of imidazoline surfactants is flexible, scalable, and adaptable to both petrochemical and bio-based feedstocks, making them attractive for diverse industrial applications, particularly where softening, corrosion inhibition, or emulsification are required (Fig. 19) [43].

Properties

Physical Appearance

Long-chain imidazolinium salts (those with alkyl chains C8 and above) typically appear as white, crystalline solids, or colorless compounds, depending on their purity and saturation level. In contrast, unsaturated variants—especially those with cis-configured double bonds—or benzyl-containing imidazolines are obtained as viscous liquids with hues ranging from pale yellow to brown [43].

Melting Behavior

Imidazoline derivatives show broad and relatively lower melting point ranges between 45 and 90 °C, which is significantly less than that of structurally similar amides that typically melt at 98–112 °C. For example, the oleic acid derivative of imidazoline remains liquid at ambient temperature, while its saturated isomer (elaidic acid-based) solidifies around 46–47 °C [43].

Solubility

The solubility of imidazoline surfactants is largely dictated by the chain length and the polarity of substituents. Short-chain imidazolines (C8–C14) tend to be water-soluble, whereas longer chains typically exhibit poor aqueous solubility but dissolve well in nonpolar solvents such as hydrocarbons or mineral oils. Furthermore, the presence of hydrophilic functional groups like hydroxyl or ether linkages significantly enhances solubility in polar organic solvents [43].

Thermal and Chemical Stability

These compounds exhibit excellent thermal stability due to the robustness of the imidazoline ring system, allowing them to be used in high-temperature industrial processes such as metalworking and textile curing [46]. However, their chemical stability in the free base form is more sensitive to moisture. For instance, a formulation with 38% imidazoline content degraded to 6% in 72 h in the presence of only 2% water, highlighting the need for immediate application post-synthesis or stabilization via salt formation [43].

Surface Activity

Due to their strong cationic character, imidazoline surfactants exhibit high affinity for negatively charged surfaces such as cellulose fibers, glass, and metal oxides.

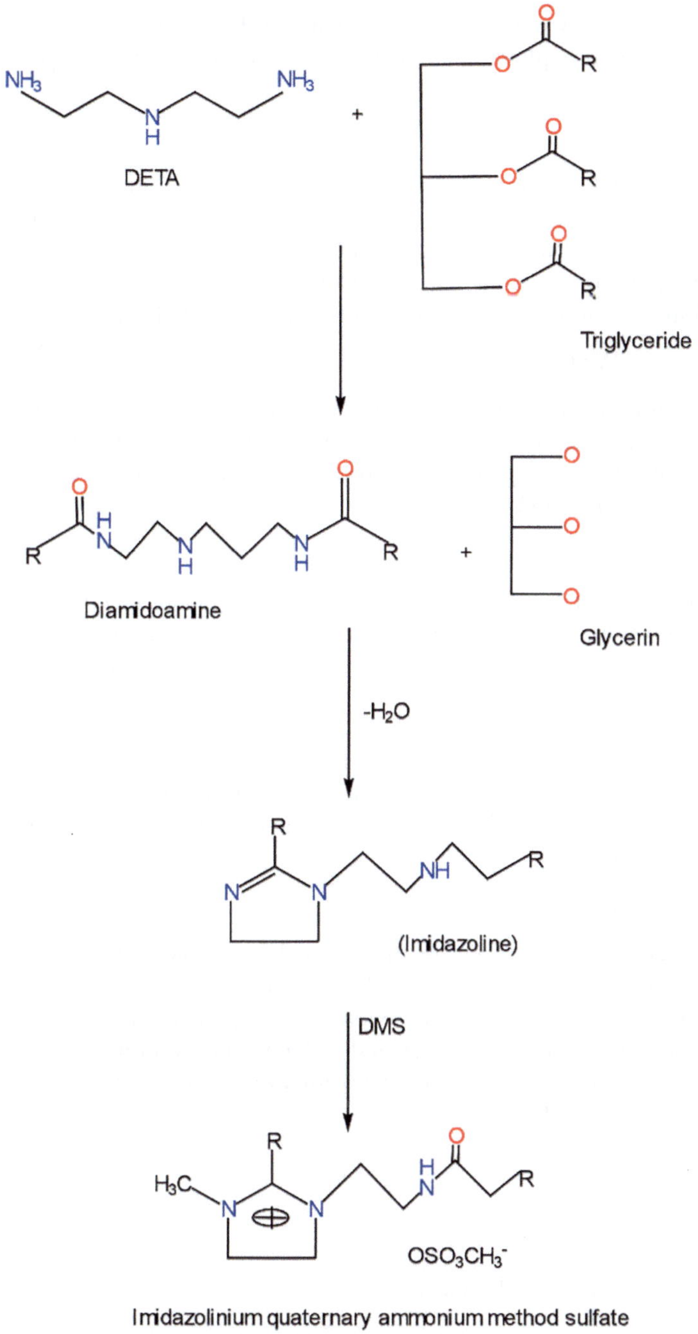

Fig. 19 Imidazolinium quaternary ammonium metho sulfate

Upon adsorption, they render hydrophilic surfaces hydrophobic by forming aligned monolayers, which explains their widespread use in textile softeners and water-repellent coatings [45].

Hydrophobicity

The hydrophobicity of imidazoline derivatives correlates directly with molecular weight, particularly the length of the alkyl chain. Compounds with longer chains exhibit higher contact angles and better water repellency. However, the presence of cis-unsaturation (as in oleyl derivatives) results in decreased hydrophobicity, due to increased molecular disorder and reduced surface packing [43].

Detergency and Wetting

Imidazolinium salts are more hydrophilic than their non-ionized free base counterparts and demonstrate excellent detergency and wetting behavior in acidic and neutral pH conditions. These properties are especially valuable in fabric conditioning and cleaning formulations [45].

Water Repellency

Both the free base and salt forms of imidazoline compounds can impart water resistance to diverse substrates, including wood, plastics, concrete, and textile fibers. This function is critical in coatings and surface treatment industries [43].

Lubricity

Imidazoline-based surfactants are noted for their superior lubrication, particularly when complexed with bentonite clays. Oleic acid salts of imidazolines are employed in metalworking fluids, wire drawing compounds, and tube lubricants to reduce friction and wear.

Spectral Properties

The characteristic infrared (IR) absorption bands of imidazoline surfactants include strong N–H stretching near 3.03 μm, a shoulder around 3.15 μm, and C=N stretching at approximately 6.03 μm. In the ultraviolet (UV) region, imidazolines exhibit broad C=N $\pi \rightarrow \pi^*$ transitions between 220 and 250 nm, with a prominent absorption peak at 230 nm. These features allow their detection even at trace concentrations of 1 ppm in complex mixtures [47].

Applications

Textile Softeners and Conditioners

Cationic imidazolines are widely used in fabric softeners for cotton and synthetic fibers due to their strong electrostatic interaction with negatively charged fabric surfaces. This interaction forms a monolayer that reduces inter-fiber friction, improves hand feel, and imparts bulkiness and antistatic properties [45]. Tallow-

and oleyl-derived imidazolines are commonly employed in rinse-cycle softeners and dryer sheets because of their high substantivity and thermal stability [43].

Personal Care and Mild Cleansing Agents

Amphoteric imidazoline derivatives such as disodium lauroamphodiacetate and disodium lauroamphodipropionate are frequently included in shampoos, baby washes, and facial cleansers. Their zwitterionic nature allows excellent skin compatibility, low irritation potential, and effective foam stability over a broad pH range and in hard water conditions. These compounds are also used in sulfate-free formulations to improve mildness without compromising performance [43].

Corrosion Inhibitors

Imidazoline surfactants, particularly aminoethyl and hydroxyethyl types, serve as effective corrosion inhibitors in oilfields and industrial piping systems. They absorb on metallic surfaces and form hydrophobic films that inhibit ionic transport and electrochemical corrosion processes, especially under acidic and high-salinity conditions [46]. Their performance is enhanced when used in blends with solvents or oils, which facilitate co-adsorption and longer-lasting protection.

Emulsifiers and Dispersing Agents

Imidazoline-based surfactants act as co-surfactants and stabilizers in emulsions and suspensions used in paints, agrochemicals, and polymer latexes. Their amphiphilic character promotes interfacial stabilization and improves the dispersion of pigments or solid particles, particularly in oil-in-water and water-in-oil systems. The structural diversity of the hydrophobic chain allows for tailored HLB values to meet specific formulation needs.

Water Treatment and Flocculants

Cationic imidazoline salts are employed in industrial wastewater treatment to flocculate suspended solids by neutralizing negatively charged colloids. They are particularly effective for removing anionic surfactants, dyes, and inorganic particles from textile and paper mill effluents [43]. Their effectiveness is comparable to or greater than conventional polyquaternary flocculants under acidic and neutral conditions.

Metalworking and Dewatering Agents

In the metal fabrication industry, imidazoline compounds—especially their oleic acid salts—are used as lubricity enhancers and dewatering agents. When applied to steel or brass surfaces, they displace trace moisture and form a hydrophobic film that reduces corrosion and facilitates processing operations such as drawing, rolling, or cutting [44].

Coatings and Water-Repellent Applications

Due to their ability to absorb porous substrates and reduce surface energy, imidazoline surfactants are used in coatings for concrete, wood, and textiles to impart water repellency. Both the free base and salt forms contribute to long-lasting hydrophobicity without significantly affecting substrate breathability [43].

Environmental Compatibility

Modern imidazoline surfactants are designed to be readily biodegradable and exhibit low aquatic toxicity, making them suitable for use in eco-labeled consumer and industrial products. Biodegradation studies indicate that these compounds break down efficiently in both aerobic and anaerobic conditions, especially those derived from naturally occurring fatty acids [46].

4 Properties of Amphoteric Surfactants

Zwitterionic surfactants are amphiphilic molecules that contain both positive and negative charges within the same hydrophilic headgroup, rendering a net neutral character. This dual-charge configuration imparts unique interfacial, physicochemical, biological, and environmental properties that are distinct from conventional surfactant types.

Structural and Molecular Features

These surfactants typically incorporate a quaternary ammonium or imidazolium-based cationic group and an anionic moiety such as carboxylate, sulfonate, or phosphate, bridged by an alkylene spacer (–$(CH_2)_n$–) [8]. The internal charge balance leads to high water solubility and effective charge shielding, while the spacer length plays a critical role in determining micellar behavior, with longer spacers often increasing the critical micelle concentration [48].

Physicochemical Properties

The internal ion pairing reduces repulsion between headgroups, allowing micellization to occur at low concentrations. CMC values typically range between 0.05 and 1 mM, depending on the surfactant's chain length and headgroup nature [49]. Surface tension values below 30 mN/m are common, contributing to excellent detergency and emulsification [50]. In oil–water systems, certain structures yield ultralow interfacial tension down to 0.037 mN/m [51, 52].

These surfactants remain stable across broad pH ranges, with some (like sulfobetaines) retaining zwitterionic character between pH 2 and 10. Thermal stability extends to temperatures up to 120 °C, making them suitable for high-stress applications [8].

Interfacial and Rheological Behavior

At the air–water interface, zwitterionic surfactants form compact adsorption layers due to minimal electrostatic repulsion. Their micelles exhibit low aggregation numbers and high hydration, supporting the formation of wormlike micelles and thermoresponsive gels when mixed with co-surfactants [53]. These systems demonstrate high surface elasticity and dilational modulus, particularly when the surfactants include branched chains or polyoxyethylene segments [54].

Surface tension is influenced by hydrophobic chain length and spacer size. Increasing chain length reduces surface tension at CMC (γ_{cmc}), increases surface excess (Γ_{max}), and decreases the molecular area (A_{min}) [55]. Spacer groups of intermediate length ($n \approx 3$–4) tend to yield optimal interfacial packing and reduced γ_{cmc} [56].

Salt addition enhances micellization and lowers CMC by screening electrostatic repulsion, with multivalent salts (e.g., $AlCl_3$) being more effective than monovalent types [57, 58].

Ion Binding and Micellar Charge Regulation

The presence of salts influences ion-binding behavior and zeta potential. Weakly hydrated anions such as PF_6^- and ClO_4^- preferentially associate with the micellar surface, a phenomenon known as the chameleon effect, contributing to more negative surface charges [59]. Cation binding follows the order: trivalent > divalent > monovalent. The Hofmeister and Pearson series governs anion interaction trends, with strongly hydrophilic anions binding less effectively than weakly solvated counterparts [60].

Foaming and Synergistic Performance

Foaming behavior improves with concentration and salt addition, as zeta potential becomes more negative and interfacial film rigidity increases. Foam stability benefits from divalent and trivalent cations ($Al^{3+} > Ca^{2+} > Na^+$), which compress the electrical double layer and promote adsorption [61]. Oils destabilize foams by penetrating and thinning the lamellae, often through Marangoni-driven flow [62].

Mixtures of amphoteric and ionic or nonionic surfactants show strong synergism, characterized by reduced CMC, higher surface activity, and improved adsorption at the interface. This behavior is quantified using the interaction parameter βm, where values < 0 and exceeding $|\ln(CMC_1/CMC_2)|$ confirm synergy [63]. The strength of interaction follows the trend: nonionic < cationic ≈ anionic, with sulfobetaine–SDS systems showing exceptional performance [64].

Biological and Environmental Compatibility

Biocompatibility is a key advantage. The zwitterionic headgroup reduces interaction with biological membranes, lowering risks of irritation and hemolysis. These surfactants are routinely incorporated in personal care and biomedical products for their mildness [8].

In drug delivery, structures like phosphorylcholine-based surfactants support enzyme stabilization, nanoparticle dispersion, and low-immunogenicity micelle formation [65, 66].

From an environmental perspective, zwitterionic surfactants are often biodegradable and meet global ecotoxicity standards. They degrade readily in microbial systems and do not bioaccumulate, aligning with REACH and EPA Safer Choice criteria [67].

5 Analysis Techniques for Zwitterionic Surfactant

To understand the unique behavior and properties of zwitterionic surfactants, an analytical approach consisting of both molecular-level details and bulk property evaluation is necessary. These surfactants are characterized by their dual-charged head groups and thus exhibit complex interactions in solution and at interfaces making their analysis both challenging and essential.

By integrating various analytical techniques, a proper understanding of zwitterionic surfactants can be done on how they behave in various environments. Note that each analytical technique offers a different view into the surfactant's performance. The following techniques and methods are used for the gaining a deeper understanding.

5.1 Molecular and Structural Characterization

These are the foundational processes in chemistry, materials sciences and biochemistry which allows researchers to understand and decode the composition, spatial organization of substances. With the help of these methods, we can get information on the atomic level and macroscopic level, thus giving us a better understanding of substances.

Molecular characterization focuses on identifying the chemical composition, bonding patterns and the identification of the functional groups within the molecule (Table 4).

5.2 Surface and Interfacial Behavior

This refers to the way molecules and their functional groups orient and interact at the boundary between the two phases such as oil and water or air and liquid. These methods evaluate the surfactants and their surface properties and are key to practical use. They focus on measuring properties like surface tension, interfacial tension, contact angle, and adsorption behavior which helps us understand how these compounds reduce energy at interfaces. These studies are essential for understanding and optimizing their function as surfactants, emulsifiers, or wetting agents (Table 5).

Table 4 Analytical techniques used to characterize zwitterionic surfactants

Technique	Objective	Application	References
NMR	To determine detailed molecular structure and purity and characterization of molecular conformations and conformational exchanges	Determines purity, investigation of the self-assembly mechanisms and phase behavior	[68]
FTIR	To identify functional groups and molecular interactions in zwitterionic surfactants	Used to confirm the presence of characteristic functional groups like quaternary ammonium and sulfonate in surfactants	[69]
MS	To determine the molecular weight and structure of surfactants	Essential for characterizing new surfactant molecules and confirming their purity	[70]
SAXS (Small Angle X-Ray scattering)	To analyze the nanoscale structure and aggregation behavior of surfactants	Provides insights into the packing and arrangement of surfactant molecules at interfaces	Tran et al. (2022)
HPLC	To separate and quantify surfactant components in mixtures	Useful for analyzing complex formulations and ensuring consistency in product quality	[69]
GC	To separate volatile components of surfactants	Applied in the analysis of surfactant degradation products and volatile impurities	[69]

Table 5 Analytical methods for identifying surface characteristics

Method	Objective	Application	Citation
Surface tension	To measure the interfacial properties of surfactants	Critical for evaluating the efficiency of surfactants in reducing surface tension and enhancing wetting properties	[70]
Contact angle			
Interfacial rheology	To study the viscoelastic properties of surfactant films	Important for understanding the stability of emulsions and foams	[71]

5.3 *Biological and Environmental Compatibility*

See Table 6.

Table 6 Biological assays for compatibility

Method	Objective	Application	References
Ecotoxicity	To assess the environmental impact and safety of surfactants	Ensures that surfactants are safe for use in sensitive environments	Tran et al. (2022)
Endocrine disruption screening			
Biodegradability	To evaluate the environmental persistence of surfactants	Determines the suitability of surfactants for eco-friendly applications	[70]

6 Application Areas and Derivatization

Zwitterionic surfactants exhibit a unique balance of hydrophilic and hydrophobic characteristics due to the coexistence of both positive and negative charges within a single molecule. This structural feature imparts high surface activity, excellent water solubility, low toxicity, biodegradability, and thermal as well as pH stability, making them versatile agents across numerous fields including petroleum recovery, personal care, analytical chemistry, nanotechnology, and environmental remediation [8].

6.1 Enhanced Oil Recovery

Amphoteric surfactants are powerful agents in chemical enhanced oil recovery (CEOR), especially under extreme salinity and temperature conditions. Their zwitterionic headgroups confer salinity tolerance (up to 20%) and thermal stability (up to 120 °C). They reduce interfacial tension (IFT) to ultralow levels and convert rock wettability from oil-wet to water-wet, thereby increasing the capillary number and improving displacement efficiency [51]. Modified sulfobetaines containing ethylene oxide (EO) groups improve water solubility and eliminate the need for co-solvents, while amidoamine-functionalized surfactants further increase thermal resistance and adsorption onto mineral surfaces like Bakken shale [8].

They form stable micro-emulsions at low CMC and display strong synergistic behavior with anionic or nonionic surfactants, lowering surfactant dosage requirements. When combined with silica nanoparticles, they also form supercritical CO_2-stable foams suitable for high-pressure reservoir flooding [8]. Bio-based amphoteric surfactants synthesized from lignin, vegetable oils, or waste oils provide low toxicity and biodegradability while maintaining performance in alkali-free flooding systems [8].

6.2 *Cosmetic and Personal Care Applications*

Zwitterionic surfactants such as sulfobetaines are widely used in cosmetic formulations due to their mildness, skin compatibility, and excellent foaming behavior. Incorporation of these compounds into shampoo formulations has been shown to improve dynamic viscosity (6000–21,000 mPa s), yield stress (~10 Pa), and foaming properties, while effectively lowering surface tension, enhancing cleansing action without forming hydrophobic films [53]. They are stable across broad pH ranges and exhibit low irritant potential, making them suitable for sensitive skin applications.

6.3 *Analytical Chemistry, Catalysis, and Separation Techniques*

Amphoteric surfactants hold a unique position in modern analytical and separation sciences due to their zwitterionic nature, tunable charge density, and excellent interfacial behavior. In chromatographic techniques, they are extensively employed in dynamic and permanent coatings for capillaries and stationary phases, improving resolution and selectivity. In ion chromatography and zwitterionic ion chromatography (ZIC), surfactants like sulfobetaines and carboxybetaines provide charge-balanced interfaces that enable simultaneous separation of anionic and cationic species. Their ability to form Donnan potential-regulated layers enhances peak shape and stability, even in high-salt or variable pH conditions [60].

In micellar electrokinetic chromatography (MEKC), amphoteric surfactants act as pseudo-stationary phases, forming micelles that differentiate analytes based on their charge and hydrophobicity. Their zwitterionic nature reduces interactions with capillary walls, minimizes band broadening, and improves electroosmotic flow control, making them especially useful for separating peptides, proteins, and polar small molecules [60].

Amphoteric surfactants demonstrate considerable efficacy in cloud point extraction (CPE) systems due to favorable thermoresponsive behavior, biocompatibility, and low environmental impact. Surfactants like SB3-9 and SB3-10 selectively preconcentrate thermolabile analytes through temperature-induced phase separation while maintaining enzymatic activity under denaturing conditions such as the presence of urea [8].

They also play catalytic roles in nanoscience and biphasic reactions, supporting enzyme activity in microreactors, and enhancing reproducibility in biosensor platforms. CHAPS and SB3-12 have been reported to co-stabilize enzymes in alcohol dehydrogenase systems, serving both structural and catalytic roles [8].

6.4 Enzyme Stabilization and Biocatalysis

In non-aqueous or micellar systems, zwitterionic surfactants stabilize enzymes by modulating interfacial environments. They reduce conformational strain and enhance water retention in reverse micelles, improving both catalytic activity and thermal stability of enzymes such as α-chymotrypsin. Their dual-charge nature aids in maintaining enzymatic integrity while enabling efficient substrate interaction [60].

6.5 Nanotechnology and Catalysis

Zwitterionic surfactants serve as templating and stabilizing agents in nanoparticle synthesis. Controlled formation of branched and hollow gold nanostructures, as well as stabilized palladium nanoparticles, has been achieved through sulfobetaine-mediated pathways. These surfactant-nanoparticle systems demonstrate enhanced catalytic efficiency, particularly in the reduction of nitroarenes and hydrogenation reactions, due to their ability to prevent aggregation and modulate surface activity [60].

They stabilize Pd and Au nanoparticles for catalytic reactions and assist in the preparation of hydroxyapatite for biomedical use. These systems exhibit pH-sensitive stabilization and high activity in hydrogenation and electrocatalytic processes [8].

6.6 Electrochemical and Battery Materials

Amphoteric surfactants contribute significantly to uniform metal deposition (Ni, Cu, Pt) during electroless plating. Their self-assembled monolayers enable electrostatic regulation of metal ion migration, improving adhesion and corrosion resistance. These systems are vital in forming Ni–P and Ni–B coatings [8]. They are also used in electrochemical sensors by enhancing adsorption on electrodes and reducing surface tension, which is particularly useful for detecting analytes like triclosan [8].

6.7 Environmental Applications

Amphoteric surfactants are used to functionalize montmorillonite clays, enhancing their capacity to adsorb phenols, heavy metals, and radioactive ions (Cs^{+}, Sr^{2+}, Co^{2+}). These surfactant-modified clays display increased thermal stability and interlayer spacing [8]. They are also integrated into nanofiltration membranes to reduce fouling while preserving high organic rejection rates, making them efficient for sustainable wastewater treatment [8].

6.8 Biomedicine, Gene Therapy, and Drug Delivery

Amphoteric surfactants enhance bioavailability and biocompatibility in nano emulsion systems (~100 nm), aiding the delivery of poorly soluble drugs and antigens. Their mucoadhesive properties make them ideal for intranasal and oral vaccine systems.

CnDMAOs demonstrate reversible pH-sensitive charge behavior, facilitating smart DNA capture and release for gene therapy vectors [9]. They also support protein solubilization in biosensors and antimicrobial surface coatings.

Surfactants like cocamidopropyl betaine and SB3-12 prevent protein aggregation and enhance enzymatic stability under stress. They support α-chymotrypsin and alcohol dehydrogenase in micellar and reverse micellar systems, contributing to robust catalytic performance in industrial and environmental applications [9, 72].

6.9 Textile and Personal Care Industries

Amphoteric surfactants are widely used in cosmetics and personal care due to their mildness, foaming, and skin compatibility. Novel sulfobetaines improve shampoo viscosity, cleansing, and foam without forming hydrophobic residues [9].

Mixtures like CAPB and LS generate viscoelastic wormlike micelles without requiring polymers, offering a gentle and stable formulation [9] exhibit reduced irritation compared to SLES and SDS and are ideal for frequent-use products.

6.10 Synergistic Formulations

Amphoteric surfactants exhibit synergism when combined with anionic, cationic, or nonionic surfactants. Mixtures like SDS/SB3-12 reduce CMC and surface tension, enhancing micelle formation and interfacial performance. These systems are commonly used in detergents, emulsifiers, and industrial cleaners [9].

7 Safety Profile and Toxicity Analysis (Environmental Impact)

Zwitterionic surfactants (ZIs), from a human and occupational safety perspective, experimental evidence suggests that the physicochemical structure of ZIs—particularly alkyl chain length—plays a crucial role in dermal absorption and systemic exposure. Short-chain compounds such as C12BET demonstrate significant skin penetration, with approximately 50% absorbed within 24 h, leading to higher systemic availability. In contrast, long-chain variants like C16BET show much lower dermal absorption (around 1.3%) but greater retention in skin tissue (approximately 15%). ZIs can disrupt the stratum corneum lipids via micelle-mediated solubilization, thereby enhancing transdermal permeability. This property, while potentially useful in targeted drug delivery systems, raises concern in industrial and personal care applications where co-contaminant penetration may occur. The extent of this effect is influenced by the surfactant's concentration and its critical micelle concentration (CMC) [72].

Toxicological data further support the generally safe profile of ZIs when used appropriately. The oral LD_{50} for C12BET in rats is approximately 71 mg/kg, while C16BET has a significantly higher threshold at 1620 mg/kg. The dermal LD_{50} values are 1300 mg/kg for C12BET and greater than 16,000 mg/kg for C16BET. For intraperitoneal exposure, LD_{50} values are 53 mg/kg and 150 mg/kg, respectively. While ZIs present low acute dermal toxicity overall, high concentrations may cause irritation, and some cases of eczema have been reported, suggesting possible dermal sensitization in susceptible individuals [24].

ZIs also interact significantly with soil and sludge due to their dual-charge characteristics and strong surface activity. Their mobility within environmental matrices is governed by factors such as pH, ionic strength, hydrophobicity, and the organic matter content of soil. Because of their strong adhesion to soil particles, especially in cases where treated sludge is applied to land, there is an increased potential for residual accumulation over time [73].

Despite the favorable biodegradability of many ZIs, some classes—particularly perfluorinated variants like PFOS and PFOA—are known for their environmental persistence. These substances resist microbial breakdown and are capable of bioaccumulating in living organisms. Such compounds have been detected in environmental matrices across the globe, including soil and surface water. Moreover, due to their typically low environmental concentrations and the complexity of surfactant mixtures, accurate monitoring of ZIs remains challenging. This necessitates the use of advanced analytical techniques, such as solid-phase extraction (SPE) combined with liquid chromatography–mass spectrometry (LC–MS), for reliable detection [73].

In terms of aquatic toxicity, ZIs have been evaluated using test organisms such as Photobacterium phosphoreum and Daphnia magna. The reported EC_{50} values range from 5 to over 200 mg/L. Toxicity tends to increase with the length of the alkyl chain. Among the compounds tested, imidazoline derivatives were found to be the least toxic, followed by alkylamido betaines, and then alkyl betaines. According to the OECD classification, amido-betaines and alkyl betaines fall under Class II acute toxicity (moderately hazardous), while imidazoline derivatives fall under Class III (slightly hazardous). Compared to cationic and nonionic surfactants such as alcohol ethoxylates and alkyl trimethyl ammonium compounds, ZIs are generally less toxic and are not typically associated with disruptions to aquatic metabolic or reproductive systems at environmentally relevant concentrations [72].

On a microbial level, zwitterionic and amphoteric surfactants have been shown to disrupt microbial activity through mechanisms such as membrane depolarization, interference with nutrient uptake and respiration, inhibition of nitrification, and disruption of photosynthetic processes in aquatic species. Prolonged exposure in wastewater and hospital effluents may also contribute to antimicrobial resistance (AMR) by promoting the selection of resistance genes and altering microbial membranes [73].

From a human and occupational safety standpoint, experimental evidence indicates that the alkyl chain length of ZIs significantly affects dermal absorption. Short-chain compounds such as C12BET show higher systemic absorption, with approximately 50% penetrating the skin within 24 h. In contrast, long-chain compounds like C16BET exhibit lower dermal absorption (approximately 1.3%) but greater retention in skin tissue (around 15%). ZIs can enhance transdermal penetration by disrupting stratum corneum lipids through micelle-mediated solubilization, which increases the potential for co-contaminant penetration. This is particularly relevant in industrial and personal care contexts, where such compounds are widely used. The penetration effect is closely related to the surfactant's concentration and critical micelle concentration (CMC) [34].

Industrially, ZIs are prevalent in a wide range of applications, including shampoos, body washes, conditioners, and deodorants in the personal care sector, as well as detergents and foaming agents in household cleaners. They also serve as emulsifiers and antistatic agents in various industrial formulations. Their widespread use results in significant presence in wastewater and biosolids. However, existing wastewater treatment plants are not specifically designed to remove ZIs, and current treatment processes may result in only partial removal. The absence of comprehensive ecotoxicological and environmental fate data for many ZIs further complicates environmental risk assessments [74].

Given their broad application, modest toxicity, and potential for environmental persistence, enhanced monitoring of ZIs in environmental systems is essential. There is a growing need for greener alternatives with improved biodegradability and lower potential for bioaccumulation. Regulatory frameworks should incorporate ZIs into safety evaluations for dermal, aquatic, and soil exposures [75]. Additionally, further research is needed to investigate the long-term ecotoxicity, bioaccumulation, and chronic exposure effects associated with these compounds.

8 Future Scope and Conclusion

Zwitterionic surfactants (ZIs), especially those of the betaine-type amphoteric class, have emerged as a focus of modern surfactant research due to their simultaneous possession of anionic and cationic groups, which renders them highly versatile in both formulation and application. These surfactants are characterized by excellent environmental compatibility, high aqueous solubility, surface activity, and reduced skin and eye irritation, making them suitable for use in personal care, pharmaceuticals, and enhanced oil recovery (EOR) sectors [35].

In the field of EOR, cocamidopropyl betaine (CAPB), a commonly studied ZI, has demonstrated superior CO_2 foam stability under high temperature and salinity conditions, outperforming conventional anionic surfactant systems and showing promise for improving oil sweep efficiency in heterogeneous reservoirs. Furthermore, ZIs exhibit reservoir-specific adsorption behavior with high adsorption on sandstone and low on carbonate formations highlighting the need for tailored application strategies based on rock type [76].

In biomedical and pharmaceutical fields, biodegradable ZIs such as CAPB and coco-betaine are recognized for forming stable oil-in-water nano emulsions useful in drug delivery of hydrophobic compounds. These systems showed high kinetic stability, maintained for over 30 days, and compatibility with therapeutic omega fatty acids such as oleic and linoleic acids, which have additional biological benefits. Their mildness and biocompatibility, coupled with scalability via co-surfactant-free emulsification methods, further reinforce their relevance in pharmaceutical and cosmeceutical applications.

From an environmental standpoint, ZIs show favorable biodegradability and low aquatic toxicity. For example, Wu et al. emphasized the role of microbial strains like Eubacterium limo sum and Sporomusa ovata in the biodegradation of betaine-type ZIs, and called for development of cost-effective, bio-based synthetic routes to improve their industrial acceptance. In a separate ecotoxicological study, it was found that ZIs had no harmful effects on freshwater microalgae (*Raphidocelis subcapitata*) and marine bacteria (*Allivibrio fischeri*), with EC_{50} values exceeding 100 mg/L, classifying them as non-hazardous under GHS standards [67]. These compounds were significantly less toxic than ionic liquids with analogous functional groups, underscoring their suitability for green formulations in detergent, environmental, and personal care industries.

Collectively, these findings position ZIs as next-generation surfactants that are not only functionally diverse and high-performing, but also environmentally benign, biodegradable, and suitable for multidomain industrial applications.

References

1. Benvenuto MA (2014) Industrial chemistry. Walter de Gruyter GmbH & Co. KG
2. Handbook of detergents
3. Raffa P, Wever DAZ, Picchioni F, Broekhuis AA (2015) Polymeric surfactants: synthesis, properties, and links to applications. Chem Rev 115:8504–8563
4. Shah J, Arslan E, Cirucci J et al (2016) Comparison of Oleo- vs Petro-sourcing of fatty alcohols via cradle-to-gate life cycle assessment. J Surfactants Detergent 19:1333–1351. https://doi.org/10.1007/s11743-016-1867-y
5. Laughlin RG, Laughlin RGJ, Zhao F, et al (1991) Tensioactive properties of zwitterionic carboxybetaine amphiphiles
6. Chevalier Y, Germanaud L, Le Perchec P (1988) Micellar properties of zwitterionic phosphobetaine amphiphiles in aqueous solution: influence of the intercharge distance. Colloid Polym Sci

7. Ward RS, Davies J, Hodges G, Roberts DW (2002) Synthesis of quaternary alkylammonium sulfobetaines. Synthesis 2002(16):2431–2439
8. Sarkar B et al (2021) Properties and applications of amphoteric surfactant: a concise review. J Surfactants Deterg 24(6):1–23. https://doi.org/10.1002/jsde.12542
9. Zieba M, Wieczorek D, Klimaszewska E et al (2019) Application of new synthesized zwitterionic surfactants as hair shampoo components. J Dispers Sci Technol 40:1189–1196. https://doi.org/10.1080/01932691.2018.1503545
10. Myers D (1988) Surfactant science and technology. John Wiley & Sons
11. Tadros TF (2006) Applied surfactants: principles and applications. John Wiley & Sons
12. Clendennen SK, Boaz NW (2019) Betaine amphoteric surfactants-synthesis, properties, and applications. In: Biobased surfactants: synthesis, properties, and applications. Elsevier, pp 447–469
13. Toan M, Choi J, Ngo HT et al (2024) Synthesis of novel zwitterionic surfactants: achieving enhanced water resistance and adhesion in emulsion polymer adhesives. Polymers (Basel) 16:3504–3504. https://doi.org/10.3390/POLYM16243504
14. Wang W, Liang MY, Lang JQ et al (2023) A new bio-based zwitterionic surfactant with strong interfacial activity at high temperature for enhanced oil recovery. Green Mater 12:1–10. https://doi.org/10.1680/JGRMA.22.00070
15. Cheng Y, Yang Y, Niu C et al (2019) Progress in synthesis and application of zwitterionic Gemini surfactants. Front Mater Sci 13:242–257. https://doi.org/10.1007/S11706-019-0473-0
16. Gerola AP, Costa PFA, Quina FH et al (2017) Zwitterionic surfactants in ion binding and catalysis. Curr Opin Colloid Interface Sci 32:39–47
17. Amphoteric surfactants—Google Books. https://books.google.co.in/books/about/Amphoteric_surfactants.html?id=x9SJzgEACAAJ&redir_esc=y. Accessed 27 May 2025
18. Ernst R, Gonzales CJ, Arditti J (1983) Biological effects of surfactants: Part 6—Effects of anionic, non-ionic and amphoteric surfactants on a green alga (Chlamydomonas). Environ Pollut Ser A Ecol Biol 31:159–175. https://doi.org/10.1016/0143-1471(83)90074-0
19. Beckett AH, Woodward RJ (1963) Surface-active betaines: *N*-Alkyl-*Nn*-dimethylglycines and their critical micelle concentrations. J Pharm Pharmacol 15:422–431. https://doi.org/10.1111/j.2042-7158.1963.tb12809.x
20. Kelleppan VT, King JP, Butler CSG et al (2021) Heads or tails? The synthesis, self-assembly, properties and uses of betaine and betaine-like surfactants. Adv Colloid Interf Sci 297
21. Bergfeld WF, Donald FACP, Belsito V et al (2014) Safety assessment of alkyl betaines as used in cosmetics. In: The 2014 Cosmetic Ingredient Review Expert Panel members are: Chairman
22. Syed AH, Yekeen N, Padmanabhan E et al (2020) Characterization of lauryl betaine foam in the Hele-Shaw cell at high foam qualities (80%–98%). Pet Sci 17:1634–1654. https://doi.org/10.1007/S12182-020-00470-W/FIGURES/5
23. Chorro M, Kamenka N, Faucompre B et al (1996) Micellization and adsorption of a zwitterionic surfactant: N-dodecyl betaine—effect of salt. Colloids Surf A Physicochem Eng Asp 110:249–261. https://doi.org/10.1016/0927-7757(95)03444-7
24. Burnett CL, Bergfeld WF, Belsito DV et al (2018) Safety assessment of alkyl betaines as used in cosmetics. Int J Toxicol 37:28S-46S. https://doi.org/10.1177/1091581818773354/ASSET/AA84EA4F-4DCA-4554-ACDA-039DC15A4C86/ASSETS/IMAGES/LARGE/10.1177_1091581818773354-FIG1.JPG
25. Pazdur L (2018) Functionalization of long chain olefins and fatty acid derivatives via boron intermediates
26. Geppert-Rybczyńska M, Mrozek-Wilczkiewicz A, Rawicka P, Bartczak P (2024) A study of the micellar formation of N-alkyl betaine ethyl ester chlorides based on the physicochemical properties of their aqueous solutions. Molecules 29:1844. https://doi.org/10.3390/MOLECULES29081844/S1
27. (1982) Process for the preparation of betaines
28. CA2141259A1—Process for the preparation of betaines—Google Patents. https://patents.google.com/patent/CA2141259A1/en. Accessed 10 Apr 2025
29. Baan (2002) Process for the continuous quaternization of tertiary amines with an alkyl halide

30. Lunstroot K, Baeten L, Nockemann P et al (2009) Luminescence of $LaF_3:Ln^{(3+)}$ nanocrystal dispersions in ionic liquids. J Phys Chem C 113:13532–13538. https://doi.org/10.1021/JP9015118
31. Birnie CR, Malamud D, Schnaare RL (2000) Antimicrobial evaluation of N-alkyl betaines and N-Alkyl-N,N-dimethylamine oxides with variations in chain length. Antimicrob Agents Chemother 44:2514–2517. https://doi.org/10.1128/AAC.44.9.2514-2517.2000
32. Kelleppan VT, Moore JE, McCoy TM et al (2018) Self-assembly of long-chain betaine surfactants: effect of tailgroup structure on wormlike micelle formation. Langmuir 34:970–977. https://doi.org/10.1021/ACS.LANGMUIR.7B02830
33. McCoy TM, Valiakhmetova A, Pottage MJ et al (2016) Structural evolution of wormlike micellar fluids formed by erucyl amidopropyl betaine with oil, salts, and surfactants. Langmuir 32:12423–12433. https://doi.org/10.1021/ACS.LANGMUIR.6B01735
34. Ridout G, Hinz RS, Hostynek JJ et al (1991) The effects of zwitterionic surfactants on skin barrier function. Fundam Appl Toxicol 16:41–50. https://doi.org/10.1016/0272-0590(91)90133-O
35. Waglewska E, Bazylińska U (2021) Biodegradable amphoteric surfactants in titration-ultrasound formulation of oil-in-water nanoemulsions: rational design, development, and kinetic stability. Int J Mol Sci 22:11776. https://doi.org/10.3390/IJMS222111776
36. CN109134287B
37. Staszak K, Wieczorek D, Zielinsk R (2013) Synthesis and interfacial activity of novel sulfobetaines in aqueous solutions. Tenside Surfactants Deterg 50:45–51. https://doi.org/10.3139/113.110233/MACHINEREADABLECITATION/RIS
38. Antibacterial activity of selected surfactants. https://www.researchgate.net/publication/329644185_Antibacterial_activity_of_selected_surfactants. Accessed 7 Sept 2025
39. Trippett BS, Walker DM (1961) The phosphobetaines: preparation and properties
40. Giersberg et al (1991) Phosphobetaines, methods for their synthesis and their use especially for industrial purposes. United States Patent (19)
41. US6117915
42. O'lenick AJ (2004) Ester based phosphobetaine 6
43. Tyagi R, Tyagi VK, Pandey SK (2007) Imidazoline and its derivatives: an overview. J Oleo Sci 56(5):211–222. https://doi.org/10.5650/jos.56.211
44. Bajpai D, Tyagi VK (2006) Fatty imidazolines: chemistry, synthesis, properties and their industrial applications. J Oleo Sci 55(7):319–329. https://doi.org/10.5650/jos.55.319
45. Puchta R (1984) Cationic surfactants in laundry detergents and laundry aftertreatment aids. J Am Oil Chem Soc 61(2):367–376. https://doi.org/10.1007/BF02678796
46. Linfield WM, Bistline RG, Noble WR, Smith FD (1977) Synthesis and properties of fatty imidazolines and their N-(2-aminoethyl) derivatives. J Am Oil Chem Soc 60(4):823–828. https://doi.org/10.1007/BF02787436
47. Swisher RD (1987) Surfactant biodegradation. In: 2nd ed, Surfactant Science Series, Vol 18; Marcel Dekker: New York, 1987
48. Chevalier Y et al (1991) Micellization of betaine surfactants: effect of spacer length. Colloids Surf 58(2):95–104. https://doi.org/10.1016/j.cocis.2017.10.002
49. Wang Y et al (2020) Critical micelle concentration and micellization thermodynamics of zwitterionic surfactants. Colloids Surf B 193:111242. https://doi.org/10.1016/j.colsurfb.2020.111242
50. Zieba M et al (2018) Sulfobetaine surfactants in shampoo formulations: surface activity and foaming properties. J Dispersion Sci Technol 39(10):1447–1455. https://doi.org/10.1080/01932691.2018.1503545
51. Kumar N, Mandal A (2017) Studies on interfacial tension and phase behavior of crude oil–water system using novel zwitterionic surfactants. J Mol Liq 244:446–456. https://doi.org/10.1016/j.molliq.2017.08.032
52. Ahsaei M et al (2019) Evaluation of zwitterionic surfactants for enhanced oil recovery under high-temperature and high-salinity conditions. Energy Sources Part A Recov Util Environ Effects 42(7):846–856. https://doi.org/10.1080/15567036.2019.1651790

53. Shrestha RG et al (2021) Wormlike micelles and gel formation in zwitterionic surfactant systems. J Colloid Interface Sci 601:264–275. https://doi.org/10.1016/j.jcis.2021.07.111
54. Ramírez P et al (2011) Interfacial dilational rheology of amphoteric surfactants. J Colloid Interface Sci 358(1):333–339. https://doi.org/10.1016/j.jcis.2011.01.009
55. Qu X et al (2011) Effect of hydrophobic chain length on interfacial properties of zwitterionic surfactants. J Colloid Interface Sci 361(2):425–432. https://doi.org/10.1016/j.jcis.2011.08.004
56. Hu Y et al (2012) Spacer effects on adsorption and surface properties of zwitterionic surfactants. Colloids Surf A 414:120–126. https://doi.org/10.1016/j.colsurfa.2012.07.056
57. Nguyen T et al (2012) Salt effects on zwitterionic surfactant micellization. Langmuir 28(21):7863–7872. https://doi.org/10.1021/la301150d
58. Varade D, Ghosh P (2017) Electrolyte effects on zwitterionic surfactant solutions. J Colloid Interface Sci 490:736–744. https://doi.org/10.1016/j.jcis.2016.12.068
59. Drinkel EE et al (2013) Chameleon effect in zwitterionic micelles: perchlorate-induced charge inversion. RSC Adv 3(44):21467–21474. https://doi.org/10.1039/C3RA43761D
60. Gerola AP et al (2017) Zwitterionic surfactants: adsorption, self-assembly, and applications. Curr Opin Colloid Interface Sci 32:30–45. https://doi.org/10.1016/j.cocis.2017.10.002
61. Kristen-Hochrein A et al (2011) Foaming behavior of zwitterionic surfactants: role of electrolyte concentration. J Colloid Interface Sci 362(2):470–477. https://doi.org/10.1016/j.jcis.2010.11.064
62. Basheva ES et al (2000) Foam film rupture by oils: mechanisms and stability analysis. Colloids Surf A 177(1):79–91. https://doi.org/10.1016/S0927-7757(00)00637-2
63. Holland PM, Rubingh DN (1983) Nonideal multicomponent mixed micelle model. J Phy Chem 87(11):1984–1990
64. Li Z et al (1998) Synergistic micellization in SDS–sulfobetaine systems. Langmuir 14(1):268–272. https://doi.org/10.1021/la970985m
65. Zhang Q et al (2020) Zwitterionic surfactants in drug delivery: stability and biofouling resistance. Adv Drug Deliv Rev 156:1–20. https://doi.org/10.1016/j.ajps.2023.100883
66. Chen S et al (2022) Phosphorylcholine-based zwitterionic amphiphiles for biomedical applications. Langmuir 38(7):2277–2288. https://doi.org/10.1021/acs.langmuir.2c00001
67. Wang X et al (2022) Environmental safety and biodegradability of zwitterionic surfactants. Environ Int 162:107423. https://doi.org/10.1016/j.envint.2022.107423
68. Liu X, Yu Q, Song A et al (2020) Progress in nuclear magnetic resonance studies of surfactant systems. Curr Opin Colloid Interface Sci 45:14–27
69. Hummel DO (2000) Handbook of surfactant analysis: chemical, physico-chemical, and physical methods
70. Liu L, Lu M, Cheng X, Yu G, Huang J (2022) Suspect screening and nontargeted analysis of per-and polyfluoroalkyl substances in representative fluorocarbon surfactants, aqueous film-forming foams, and impacted water in China. Environ Int 167:107398
71. Cai J, Wu J, Yu X, Wan Z, Yang X (2024) Interfacial assembly and rheology of multi-responsive glycyrrhizic acid at liquid interfaces. Soft Matter 20(6):1173–185
72. Garcia MT, Campos E, Marsal A, Ribosa I (2008) Fate and effects of amphoteric surfactants in the aquatic environment. Environ Int 34:1001–1005. https://doi.org/10.1016/j.envint.2008.03.010
73. Arora U, Khuntia HK, Chanakya HN, Kapley A (2023) Surfactants: combating the fate, impact, and aftermath of their release in the environment. Int J Environ Sci Technol 20:11551–11574. https://doi.org/10.1007/S13762-022-04549-2
74. Palmer M, Hatley H (2018) The role of surfactants in wastewater treatment: impact, removal and future techniques: a critical review. Water Res 147:60–72. https://doi.org/10.1016/j.watres.2018.09.039
75. Jesus F, Passos H, Ferreira AM et al (2021) Zwitterionic compounds are less ecotoxic than their analogous ionic liquids. Green Chem 23:3683–3692. https://doi.org/10.1039/D1GC00477H
76. Zhou J, Srivastava M, Hahn R et al (2020) Evaluation of an amphoteric surfactant for CO_2 foam applications: a comparative study. In: Proceedings—SPE symposium on improved oil recovery. https://doi.org/10.2118/200315-MS

An Integrated Study of Surfactant Applications in Chemical and Petrochemical Industry

Smriti Mishra and Sukalyan Dash

Abstract The structures and types of known categories of surfactants have been discussed in brief, besides some new categories of surfactants, such as gemini, natural and biosurfactants and some specially designed ones like polymeric, viscoelastic, magnetic and fluorinated surfactants. Introductory information on some fundamental properties shown by surfactant systems, such as critical micelle concentration (CMC), Kraft point (KP), cloud point (CP), solubilisation parameters, hydrophile–lipophile balance (HLB), hydrophilic-lipophilic deviation (HLD) as well as microemulsion formation, has been provided. Applications of surfactants in some chemical and petrochemical industries have been discussed. In chemical industries, specifically their applications as corrosion inhibitors, for waste water treatment, in cosmetic industries and in pharmaceutical industries have been described. In petrochemical applications, their utilisation for alternative fuel production, enhanced oil recovery, demulsification of crude oil emulsions, oil sorbent removal and remediation and for water-based drilling fluid production has been discussed. A clear notion regarding the multifaceted applications of surfactants in emerging and established areas has been presented in the present chapter, which can trigger further innovative applications of surfactants.

Keywords Surfactants · Critical micelle concentration · Krafft point · Cloud point · Petrochemicals · Microemulsion

1 Introduction

Dating back to antiquity, surfactants have evolved in numerous applications, not just as cleansers but also applied in various functions as solubilising agents, emulsifying agents, wetting agents, corrosion inhibitors, anti-fogging agents and so on in various fields. The petrochemical field is no longer an exception when it comes to the

S. Mishra · S. Dash (✉)
Department of Chemistry, Veer Surendra Sai University of Technology, Burla, India
e-mail: sukalyan_dash@yahoo.com

M. Abdul Rub et al. (eds.), *Industrial Applications of Individual and Mixed Surfactant Systems—Volume 2*, Sustainable Materials and Technology,
https://doi.org/10.1007/978-981-95-9195-4_16

widespread use of surfactants as multifunctional agents across numerous operational domains.

Surfactant is highly desirable in various chemical and petrochemical industries due to its amphiphilic molecular structure, which comprises both hydrophobic and hydrophilic domains. This structural duality allows surfactant molecules to accumulate at interfaces and reduce the interfacial tension, getting immiscible phases such as water and oil solubilised, resulting in an isotropic and transparent mixture [1]. Consequently, surfactants are extensively employed in a variety of chemical and petrochemical processes such as emulsification and demulsification, drilling fluids, oil remediation, corrosion inhibition and enhanced oil recovery [2]. Their chemical versatility and changeable physicochemical properties make them adaptive to extreme environmental or operational conditions commonly encountered in chemical and petrochemical industries.

The present chapter provides a comprehensive idea of the structure and fundamental properties of surfactant molecules that govern their performance and effectiveness in the various chemical and petrochemical processes. The intent of this chapter is to discuss the detailed overview of surfactant and mixed surfactant related terminologies such as critical micelle concentration (CMC), cloud point, Krafft temperature, hydrophile–lipophile balance (HLB), microemulsion system including the Winsor regions, solubilisation parameters, hydrophilic-lipophilic deviation and their classification based on chemical structure and charge, besides some of the recent discoveries of various surfactants. This chapter will also explore the underlying mechanisms by which the surfactant and mixed surfactant function in chemical and petrochemical processes to improve the outcome of the respective industries.

2 Surfactant Types and their Characteristic Properties

A brief overview of surfactant types and their typical characteristics is discussed in this section to support the subsequent discussion on their applications in chemical and petrochemical processes.

2.1 Types of Surfactants

Surfactant, also known as a surface-active agent, is generally an organic compound that comprises two distinctive parts, a hydrophilic head and a hydrophobic tail. Consequently, it becomes amphiphilic and can get solubilised in both polar and non-polar solvent media [3]. Based on the nature of the uncharged or charged head group, surfactants can commonly be classified into four categories: non-ionic, cationic, anionic and zwitterionic (amphoteric) species (Fig. 1).

Non-ionic surfactants carry a head group without any charge. These head groups normally consist of crown-ether-type structures having ethylene oxide units. They

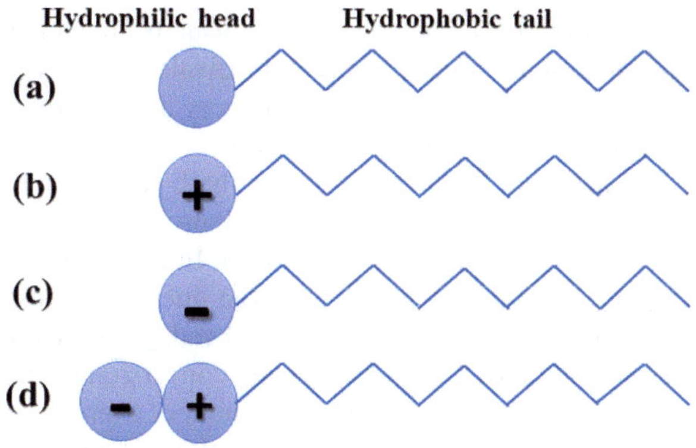

Fig. 1 Surfactant classified based on the charge over the hydrophilic head: **a** non-ionic, **b** cationic, **c** anionic and **d** zwitterionic surfactants [4]

do not ionise in aqueous medium and their molecular interaction is primarily due to hydrogen bonds and van der Waals forces. One of the key physicochemical properties of non-ionic surfactants is their clouding behaviour, which is governed by the number of ethylene oxides present in their molecular chain [5]. Their standalone use is rare because of their instability in high salinity conditions and high temperature environments, in which they may exhibit clouding phenomena and phase separation [6]. Hence, though these surfactants are considered promising due to their cost-effectiveness, high resistance and biodegradability to divalent ions, they are typically employed along with an ionic surfactant to improve the compatibility and stability of the formulation in which they participate. Notable examples include Triton-X-100, polyoxyethylene/polysorbate/Tween 20,40,60,80, sorbitan monolaurate (Span 20), alkyl polyglycoside (APG), etc.

Cationic surfactants possess one or more positive charges over the head groups along with a suitable counter anion. They dissociate in aqueous medium to give positive and negative ions. One of the significant drawbacks of these surfactants is their high cost, making them financially unviable for practical use in field operations [7]. When these surfactants are exposed to high temperature, they lose their chemical stability, leading to their degradation via Hoffman elimination [8]. Long-chain quaternary ammonium compounds, also known as Quats, are typical examples of cationic surfactants. These compounds are synthesised in a multistep process which includes the reaction of fatty acid with ammonia, producing alkyl nitrile, followed by reduction, giving rise to alkyl amines and dialkyl amines, which further undergo quaternisation to give the cationic surfactant. Some common examples of this category of surfactants are cetyltrimethylammonium bromide (commonly referred to as CTAB), dodecyltrimethylammonium bromide (DTAB), tetrabutylammonium bromide (TBAB), etc. [9].

Anionic surfactants carry a negative charge over the head groups along with a suitable counter cation. They dissociate in aqueous medium, releasing negative and positive ions. They are the largest class of surfactants and are recognised for their notable physicochemical properties, such as excellent chemical stability at elevated temperatures and adequate solubility in various media [10]. However, they also have a few limitations, such as they are usually incompatible with hard brine solutions and precipitate upon contact with divalent ions, especially at low concentrations [11]. Notable examples of this category are sodium stearate, alpha-olefin sulfonate, sodium dodecyl sulphate (SDS), etc.

Zwitterionic surfactants possess both positive and negative charges over the head group. pH of the solution containing these surfactants plays a vital role in determining the ionic nature of their head groups. In acidic conditions ($pH < 7$), the head group primarily acquires a positive charge due to protonation, and when the solution is in basic conditions ($pH > 7$), the head group acquires a negative charge due to deprotonation of the acidic part and conversion to salt upon dilution [12]. This class of surfactants exhibits multiple advantages over the conventional surfactants, such as improved aqueous solubility, low biological and environmental solubility, inherent biodegradability and offering more stability in terms of foam, salt and heat [13]. The typical examples of cationic head groups are imidazolium, quaternary ammonium, phosphonium and pyridinium, and those of anionic head groups are carboxylate, sulphonate and sulphate groups.

Current advancements in surfactant chemistry have led to the evolution and exploration of new varieties of surfactants, which include Gemini, natural and biosurfactants, as well as specialised surfactants like polymeric, viscoelastic, magnetic and fluorinated surfactants. ***Gemini surfactants*** are molecules having two hydrophilic heads and two hydrophobic tails connected to each head group, and the head groups are covalently connected through a spacer (Fig. 2). This unique molecular architecture imparts enhanced surface activity and superior physicochemical properties compared to conventional surfactants. The spacer in the Gemini surfactants can vary in length, flexibility and polarity [14]. Its length and nature influence the micelle shape and size formed from the surfactant, which in turn imparts the performance of the surfactant. In cationic Gemini surfactant, the longer spacer eventually reduces the adsorption ability, leading to improved performance [15]. The reverse phenomenon occurs in the case of anionic Gemini surfactant. Gemini surfactants offer lower critical micelle concentrations (CMC), greater surface activity, improved surface tension reduction, stronger biological effects and enhanced wetting, foaming and solubilisation capabilities as compared to conventional surfactants [14]. Gemini surfactants are utilised as foaming agents, detergents, softeners, coaters and as an ingredient in cosmetics and biomedicines [16].

Natural surfactants are developed from extracts of plants [17]. One of the key benefits of this categorisation is its non-toxicity and environmentally benign nature [18]. As a result, it has gained considerable attention and has been explored immensely. Some common sources of such surfactants are mulberry leaves, soapnut, bitter leaf, palm leaf, Jatropha curcas plant, henna leaves, Ziziphus spina christy, olive, etc. [19]. ***Biosurfactants*** are surface-active agents derived from biological origins such as yeast, fungi, bacteria, oils and also from plant and animal parts [20]. Biosurfactants are water-soluble and form micelles above CMC by lowering surface and interfacial tension. These surfactants are amphiphilic molecules where the hydrophilic part consists of amino acids, peptides, polysaccharides, carbohydrates and the hydrophobic tail consists of fatty acids [21]. This genre of surfactants

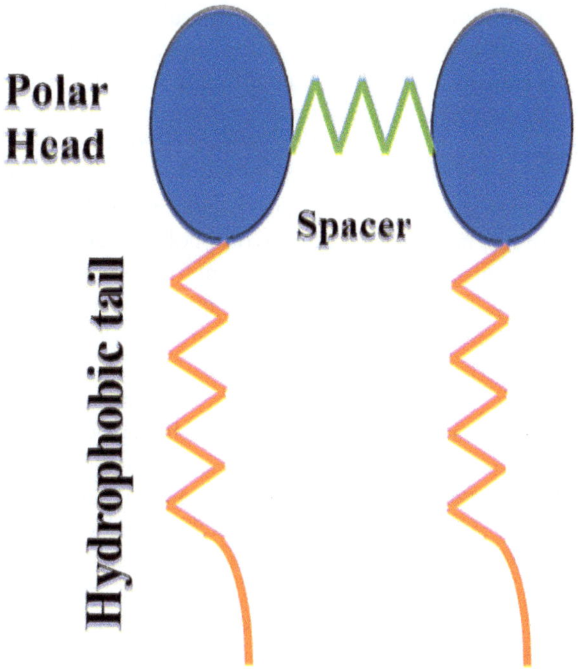

Fig. 2 Basic structure of a Gemini surfactant [15]

is also biodegradable, biocompatible, remains surface active at high temperature, salinity and pH and also has multifunctional properties. Due to such promising characteristics, biosurfactants can be used in the removal of petrochemicals from contaminated soil [22]. Biosurfactants can be sub-categorised into rhamnolipids, glycolipids, lipopeptides, lipoproteins, fatty acids, phospholipids and neutral lipids [23].

Polymeric surfactants have a molecular structure consisting of polymer macromolecules [24]. This category of surfactants can be dissolved in selective solvents where they show unique rheological and interfacial properties, which occur due to their combined polymeric and surface-active nature. They are sub-categorised into polysoap polymers formed from repeating surfactant units and macro surfactants, which are macromolecular analogues of low-molecular-weight surfactants. These surfactants can have linear or complex architectures such as grafts, stars, or dendrimers [25]. ***Viscoelastic surfactants*** exhibit notable viscoelastic properties in aqueous solution. The formation of their self-assembly depends on their concentration. They form spherical micelles at low concentration conditions and rod-shaped or worm-like micelles at high or moderate concentrations. When subjected to shear, the surfactant network deforms for a while but comes back to its original structure and viscosity due to intermolecular interactions, after the external force is removed [26].

Magnetic surfactants are distinguished by their magneto-responsive property. They possess a magnetic segment, typically derived from lanthanide cations or high-spin transition metal ions, resulting in their magnetic responsiveness when exposed to an external magnetic field, enabling control over their orientation, assembly and interfacial activity [27]. ***Fluorinated surfactants*** have a fully or partially fluorinated hydrophobic tail, where one of the hydrogens is replaced by a fluorine atom [28]. The property of the surfactants is determined by the position and number of fluorine atoms replaced within their hydrophobic backbone. This kind of surfactant exhibits sterling properties, such as excellent foamability as well as high thermal and chemical stability. The major drawback of this category of surfactants is their toxicity to the environment, whose remediation strategies have been an area of research in recent years [29].

There are several drawbacks of conventional surfactants, such as the lower efficiency of cationic surfactants, the higher critical micelle concentration (CMC) value of anionic surfactants and the sorption of non-ionic surfactant over the soil [30]. These surfactants can be mixed in judicious proportions to give rise to ***mixed surfactants***, which can tackle the drawbacks of individual members, thereby improving the efficiency and performance of the surfactants considerably. The combination of two or more surfactants in aqueous solution leads to the formation of mixed micelles, with ionic–non-ionic surfactant mixtures being the most prevalent. These combinations exhibit synergistic effects, resulting in enhanced solubilisation capacity, reduced surface tension, Krafft point, CMC, and increased cloud point compared to conventional surfactants. This synergism offers the lowering of the overall surfactant usage, leading to improved efficiency, cost-effectiveness and less environmental impact [31, 32].

2.2 Common Properties/Phenomena of Surfactants

During dealing with surfactants, their solutions and consequent applications, a knowledge of some common phenomena is necessary, some of which are discussed in this section.

1. **Critical Micelle Concentration (CMC).**

Surfactant molecules are dispersed as monomer units in solution at very low concentration. With a gradual increase in the concentration of the solution, these molecules begin to aggregate. At a particular concentration, different for each type of surfactant molecule and each type of medium, the surfactant molecules organise themselves in some specific manner, an orientation generally known as a micelle. At this concentration, their physical properties show a drastic change with respect to the unorganised surfactant solution. Such a concentration is known as the critical micelle concentration or CMC of the surfactant. (Fig. 3) [33].

Micelles are self-assembled structures with their hydrophobic tails oriented inward, forming a non-polar core, and hydrophilic heads projected outward,

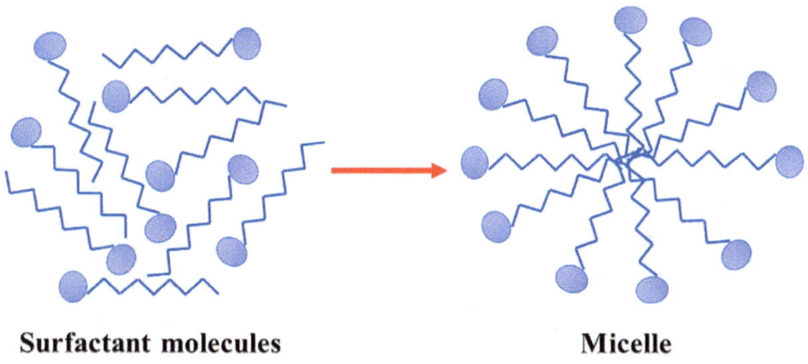

Fig. 3 Micelle formation upon reaching the CMC [34]

remaining in contact with the aqueous phase. The formation of micelles enhances the solubilisation and mobilisation of hydrophobic compounds in polar solvents such as water [35]. However, in non-polar solvents, the orientation may be reversed, wherein the hydrophobic tails face outward, remaining in contact with the non-polar solvent and the hydrophilic heads are sequestered inward, surrounding the aqueous part. This inverted structure is normally termed a reverse micelle; such structures are able to facilitate the blending and transport of polar compounds in non-polar environments (Fig. 4) [36].

At CMC, the surfactant shows the lowest surface tension/interfacial tension, and the solubilisation capacity of oil and aqueous phase is sharply increased due to the creation of micelle structure (Fig. 5). Various aspects affect the CMC of surfactants, such as their molecular structure, type and composition, ionic strength, temperature and the type and concentration of organic ingredients in the solution [37]. It is the formation of a micelle that makes a surfactant exceptional in solubilising the hydrophobic organic compound (HOC) in an aqueous medium.

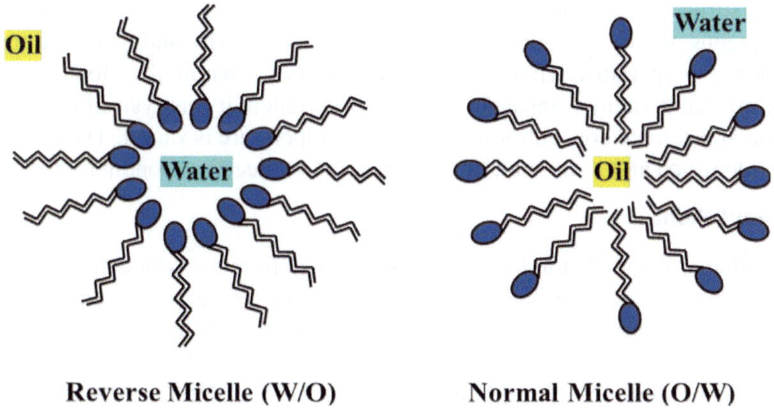

Fig. 4 Schematic structure of a reverse and normal micelle [4]

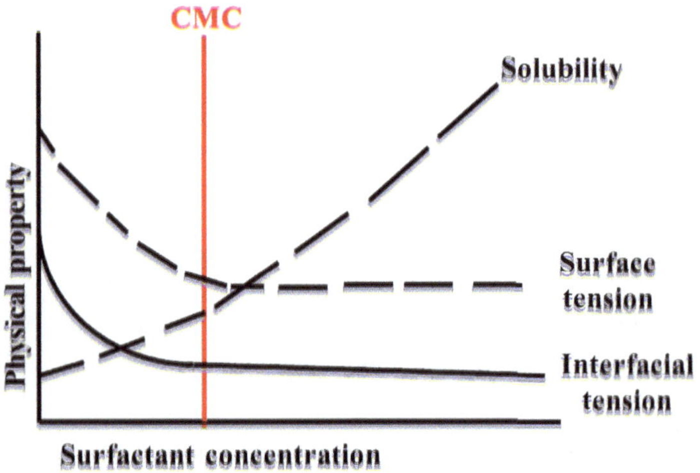

Fig. 5 Correlation between surface tension, interfacial tension and the critical micelle concentration with surfactant concentration [21]

2. **Krafft Point or Krafft Temperature (T_k).**

The specific temperature at which the solubility of micelle-forming surfactant shows a strong increase is termed the Krafft point or Krafft temperature. A single molecule of surfactant has very low solubility compared to its micellar aggregation. At a temperature below the Krafft point, the solubility of a surfactant is too low to allow for micelle formation. Solubility rises as the temperature increases, up to the Krafft point, where the critical micelle concentration is attained (Fig. 6). Beyond this temperature and concentration, a large amount of surfactant molecules can be solubilised, leading to micelle formation, thereby increasing the overall solubility [38]. Non-ionic surfactants do not show a Krafft point; rather, their miscibility is decreased with rising temperature. They show clouding behaviour due to phase separation, where a surfactant-rich phase consisting of swollen micelles separates out from the bulk aqueous phase, which leads to a loss of surface activity and increased turbidity and such a temperature where clouding occurs is known as the cloud point [39]. The Krafft temperature is measurable through conductivity analysis of the surfactant solution at a defined concentration, when the temperature is varied. The Krafft point can be identified at the cut point of a temperature vs electrical conductivity plot [40].

3. **Cloud Point.**

The phenomenon of clouding is common with non-ionic surfactants, due to which a sudden cloudiness or turbidity occurs at a definite temperature, generally termed as cloud point or lower consolute temperature. Clouding behaviour, also referred to as lower consolute behaviour, is a characteristic physical phenomenon observed in homogeneous solutions of amphiphilic substances. The solution gets separated into a surfactant-rich part and a surfactant-poor part at a particular temperature. Non-ionic surfactants organised in micellar aggregates involve hydrophobic interactions

Fig. 6 Schematic representation of Krafft temperature [41]

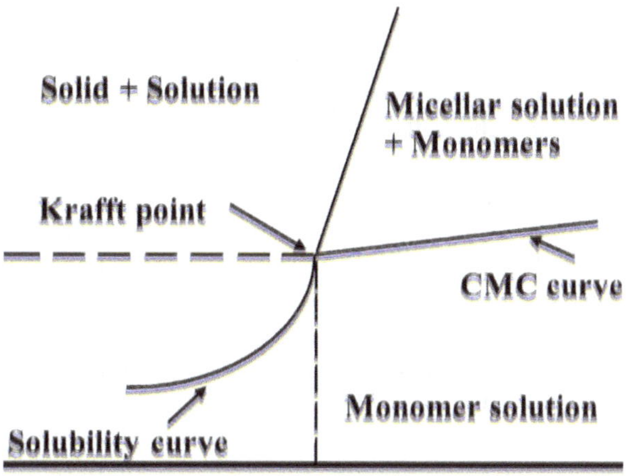

above the critical micelle concentration (CMC). As the temperature is raised, the CMC also increases, resulting in the weakening of these interactions and micelle collapse, followed by phase separation and clouding [42]. Polyoxyethylene non-ionic surfactants exhibit a significant hydrophilic property, having a spiral-shaped meander structure. This type of arrangement is formed when oxygen atoms of the ethylene oxide group wrap the water molecule into the lipophilic part to connect the water molecule more perfectly. As the temperature rises, water molecules which is bonded to the oxygen atoms of the ether of non-ionic surfactants are removed by stretching the coiled structure. This leads to dehydration, resulting in decreased solubility of water and surfactant. This ultimately gives way to liquid–liquid phase separation, which manifests as clouding of the solution (Fig. 7) [43].

4. **Solubilisation Parameter.**

The addition of surfactants to water enables the incorporation of both soluble and insoluble lipids into the aqueous phase through a mechanism called solubilisation. This occurs in a thermodynamically stable system where the non-soluble materials are solubilised into the lipophilic core of micelles. Solubilisation happens only when the surfactant strength surpasses the CMC. In a spherical micelle, the

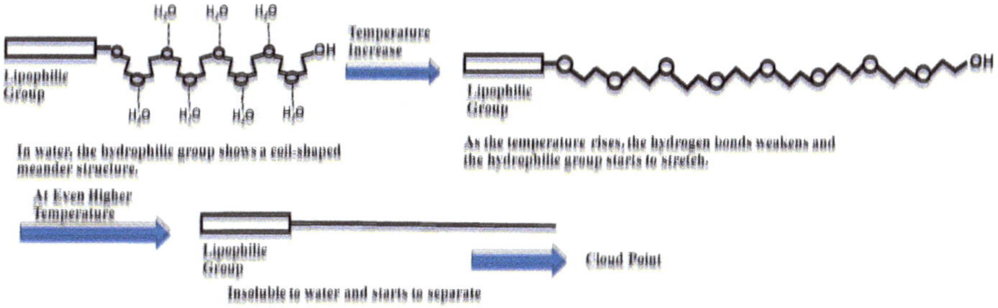

Fig. 7 Clouding phenomenon of polyoxyethylene surfactants [43]

solubilisation site varies depending on the polarity of the solubilised substance. For instance, hydrocarbons get solubilised in the core of the micelle, which is termed core-shell solubilisation. Likewise, substances having higher polarity, such as alcohols and higher fatty acids, get solubilised to the outside surface of the micelle structure, a process known as palisade-type solubilisation. The solubilisation capacity of a surfactant depends upon several aspects such as molecular structure, temperature and various inorganic salts used as additives. Surfactants with longer lipophilic chains generally form larger micelle cores, which improve the solubilisation process. In ionic surfactant solutions, when salts are added, they reduce electrostatic repulsion between hydrophilic head groups, lowering the CMC value and increasing micelle size, further improving solubilisation potential. Similarly, in the case of non-ionic surfactants, the aggregation number of the molecules increases when the temperature of the aqueous solution rises, leading to the formation of larger-sized micelles, which permits the solubilisation of a larger amount of oil [44].

The solubilisation ratio (SR) of oil or water is defined as the volume of oil (or water) solubilised per unit volume of surfactant present in the microemulsion phase [45]. It can be expressed as follows [46]:

Solubilisation ratio of oil = Volume of oil/Volume of surfactant.

Solubilisation ratio of water = Volume of water/Volume of surfactant

5. **Hydrophile–Lipophile Balance (HLB).**

Hydrophile–lipophile balance (HLB) is one of the most vital criteria for the characterisation of surfactants and to create a stable emulsion. HLB represents a scale from 0 to 20, which depicts the propensity of a surfactant to dissolve in oil and water. On the HLB scale, 0 corresponds to a completely hydrophobic or lipophilic molecule, which has no affinity toward water; an HLB value of 20 corresponds to a completely hydrophilic molecule with no affinity toward oil. Table 1 represents the relationship between the HLB value and its applications [47].

Several equations have been proposed to calculate the HLB values of different surfactants. In 1949, Griffin has given Eq. 1.

$$\mathrm{HLB}_{\mathrm{Griffin}} = \frac{1}{5}\left(\frac{M_H}{M_T} * 100\right) \tag{1}$$

Table 1 Relationship between the HLB value and its properties/applications [47]

HLB value	Property/application
0–3	Antifoaming agent
4–6	W/O (water in oil) emulsifier
7–9	Wetting agent
8–18	O/W emulsifier
13–15	Detergent
10–18	Hydrotrope or solubiliser

where M_H = molecular weight of the hydrophilic moiety and M_T = total molecular weight of the surfactant [48].

Surfactant's HLB value can be determined from their chemical formulas using group numbers, as described by Eq. 2.

$$\begin{aligned}\mathrm{HLB}_{\mathrm{Davies}} &= 7 + \sum(\text{hydrophilic group numbers}) \\ &\quad - \sum(\text{lipophilic group number})\end{aligned} \tag{2}$$

If the surfactant contains n number of CH_2 groups, then the HLB value can be calculated using Eq. 3.

$$\begin{aligned}\mathrm{HLB}_{\mathrm{Davies}} &= 7 + \sum(\text{hydrophilic group numbers}) \\ &\quad - n(\text{group number per } \mathrm{CH_2} \text{ group})\end{aligned} \tag{3}$$

In Eq. 3, the group number for each CH_2 unit is replaced with 0.475 and the hydrophilic group numbers are drawn from the data given by Davies in 1957 [34] (Table 2).

6. **Microemulsion systems**.

Microemulsions are thermodynamically stable, optically transparent dispersions of two immiscible liquids, such as oil and water, stabilised by an interfacial

Table 2 HLB values of different surfactants [49]

Name of surfactants	HLB value
Sorbitan laurate (Span 20)	8.6
Sorbitan palmitate (Span 40)	6.7
Sorbitan stearate (Span 60)	4.7
Sorbitan oleate (Span 80)	4.3
Sorbitan trioleate (Span 85)	1.8
Polyoxyethylene sorbitan laurate (Tween 20)	16.7
Polyoxyethylene sorbitan palmitate (Tween 40)	15.6
Polyoxyethylene sorbitan stearate (Tween 60)	14.9
Polyoxyethylene sorbitan oleate (Tween 80)	15
Polyoxyethylene sorbitan trioleate (Tween 85)	11
Brij 30	9.5
Brij 35	16.9
Sodium oleate	18
Potassium oleate	20
Triton-X-100	13.4

film composed of surfactants and co-surfactants. During microemulsion formulation, the surfactant and co-surfactant work synergistically to decline the interfacial tension between the middle of two phases, facilitating the formation of well-organised micellar structures that stabilise the system. In 1948, Winsor classified the microemulsion system into four kinds of phase equilibria, where two (or more) phase portions exist, apart from the O/W and W/O systems [50].

Winsor type I constitutes oil in water (O/W) microemulsion, where a small portion of the organic component, i.e., oil, is preferably dissolved in a surfactant-rich aqueous phase, i.e., water. In contrast, Winsor type II constitutes a water-in-oil (W/O) microemulsion, where a small amount of water is dissolved in surfactant-loaded oil phase. Winsor III has surfactant loaded both water and oil, where all three phases co-exist. Finally, the Winsor type IV is the extension of Winsor type III; it is a homogeneous single-phase system formed entirely of one microemulsion (Fig. 8).

Winsor's *R* ratio is utilised to determine the phase behaviour of the microemulsion and is represented in Eq. 4.

$$R = \frac{A_{CO}}{A_{CW}} \quad (4)$$

where A_{CO} and A_{CW} represent the interaction of the surfactant adsorbed over the boundary of oil and water, respectively [51]. The value of *R* and the type of Winsor region are given in Table 3, where O, S and W represent oil, surfactant and water, respectively.

7. **Hydrophilic and Lipophilic Deviation (HLD).**

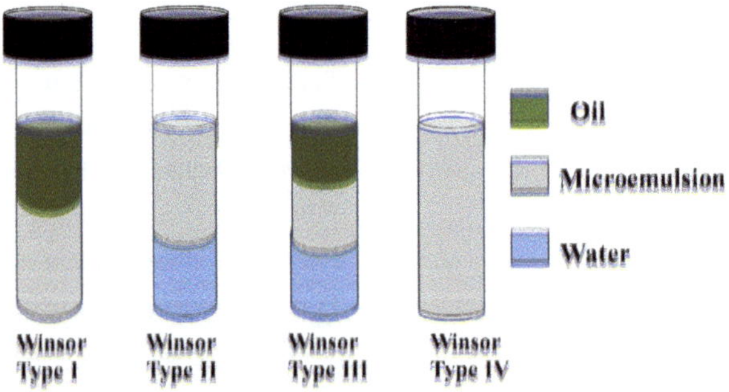

Fig. 8 Winsor's classification of microemulsion [34]

Table 3 Interpretation of *R* value and the type of microemulsion [51]

R ratio	Interpretation of *R*	Microemulsion system
$R < 1$	W–S interaction is stronger than O–S interaction	Type I
$R > 1$	O–S interaction is stronger than W–S interaction	Type II
$R = 1$	W–S and O–S interactions are balanced	Type III

The HLB value and Winsor *R* ratio have drawbacks in predicting the surfactant formulation used for several processes. HLB value does not account for factors such as salinity, temperature, oil type or co-surfactants used, while for the *R* ratio, the parameters are difficult to estimate [52]. Hence, further analyses have been taken up over the years to improve the accuracy of HLB values through the introduction of hydrophilic and lipophilic deviation. Hydrophilic and lipophilic deviation gives numerical expressions for ionic and non-ionic surfactants through Eqs. 5 and 6.

$$\text{HLD} \equiv \ln(\text{S}) - k\text{ACN} + f(A) + \sigma - a_T(T - T_0) \quad (5)$$

$$\text{HLD} \equiv \alpha - \text{EON} + b\text{S} - k\text{ACN} + f(A) + c_T(T - T_0) \quad (6)$$

where S represents the salinity, ACN depicts the alkane carbon number for an alkane oil, b, k, a_T and c_T are empirical constants, $f(A)$ is a function of type and concentration of alcohol, σ and α are parameters of the surfactant structure, T is the temperature in °C and T_0 has a value of 25 °C [53].

When the HLD value is zero, the interfacial tension becomes extremely less. When the HLD value is negative or positive, it represents O/W and W/O, respectively. Since HLD value takes all the relevant parameters into consideration, it becomes an effective tool for predicting the type of microemulsion [54].

2.3 *Surfactant Applications*

Surfactants find tremendous applications in various chemical and petrochemical industries. Herein, some well-established applications are discussed.

3 Applications in Chemical Industries

1. ***Corrosion Inhibition.***

Corrosion is the deterioration of materials due to interactions with the environmental factors like CO_2, H_2S, organic acids and extreme well conditions. These interactions can be electrochemical, chemical and metallurgical in nature [55]. Oxygen can be injected through drilling muds, brine solution, acid operation and CO_2 or water injection. Pipelines, which are crucial for the transportation of oil and gas, are vulnerable to both external and internal corrosion, making effective corrosion control vital to prevent environmental damage and economic loss [56].

Use of corrosion inhibitors is amongst the widely used techniques due to its easy handling, profitability and high efficacy. Corrosion can be controlled by removing oxygen and water components, controlling ambient temperature and pressure for a corrosive environment and by routine inspection. Corrosion inhibitors are the chemical agents that are included to the corrosive environment to decline the rate of corrosion of metal [57]. A protective layer in the middle of the metal and the external corrosive environment can be developed by the use of surfactants, where they mould themselves with the hydrophilic head toward the metal surface and the hydrophobic tail toward the corrosive environment. As a consequence, substances like acid, water and oxygen are prevented from reaching the metal surface. Surfactants can also form micelles, which can trap corrosive ions and molecules, further decreasing their interaction with the metal surface. Additionally, they lower surface tension, reducing the environment's aggressiveness and may promote passive oxide layers that further protect the metal [58].

Several surfactants have been reported as potential corrosion inhibitors, such as anionic sodium methyl ester sulfonate, cationic thiazine, dodecyl thiazin bromide (DTB) and hexyl thiazin bromide (HTB); non-ionic surfactants such as Triton-X-100, Triton-X-165, Triton-X-305; as well as several polymeric and biosurfactants [59, 60]. The effectiveness of the corrosion inhibition depends on the amount of surfactant coated over the metal surface. The performance of the surfactant also depends on the temperature: as the temperature rises, the efficiency is elevated in the case of non-ionic surfactants, which is contradictory to certain polymeric surfactants [61, 62]. The corrosion inhibition tendency may also increase when the surfactant has a larger hydrophobic moiety [63].

2. ***Wastewater Treatment***.

Surfactants find applications in enhancing water treatment efficiency. In the process of waste-activated sludge treatment, the usage of surfactant can conquer the drawbacks of conventional techniques, like low productivity and lack of flexibility. Waste-activated sludge treatment requires three prime methods: dewatering, anaerobic digestion and pollutants elimination (Fig. 9) [64]. During the sludge dewatering, i.e., removal of moisture from sludge and making it a dry semi-solid state, surfactant weakens the holding force between the sludge flocs and facilitates the breakdown of extracellular polymeric substances, thereby releasing bound water and improving dewatering efficiency [65].

Surfactants can improve the anaerobic digestion process by releasing trapped enzymes and boosting the hydrolytic activity. Due to their micelle-forming ability and encapsulating ability, surfactants can capture the polycyclic aromatic hydrocarbon, which eventually improves solubility and biodegradability [67]. Surfactants are also used for capturing the heavy metals that contaminate the environment and pose a hazard to human health. Removal of these heavy metals is essential, which can be done by using a surfactant-based mechanism. Surfactants, due to their self-aggregating ability and binding ability, can facilitate the leaching of heavy metals from sludge, enhancing overall treatment efficiency. The surfactant molecules get adsorbed over the sludge surface at head-head and tail-tail positions, followed by accumulation of hemimicelles and desorption of metal surfactant complexes so that heavy metals are separated (Fig. 10) [68].

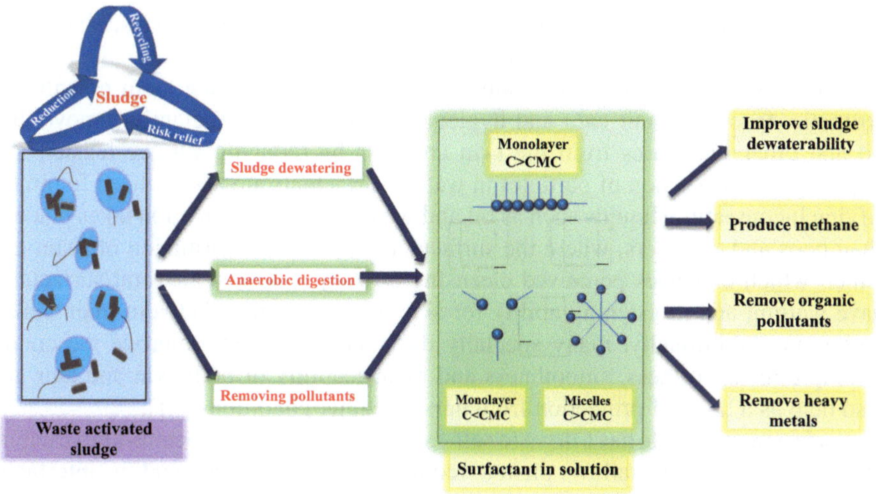

Fig. 9 Process of waste-activated sludge treatment using surfactant [66]

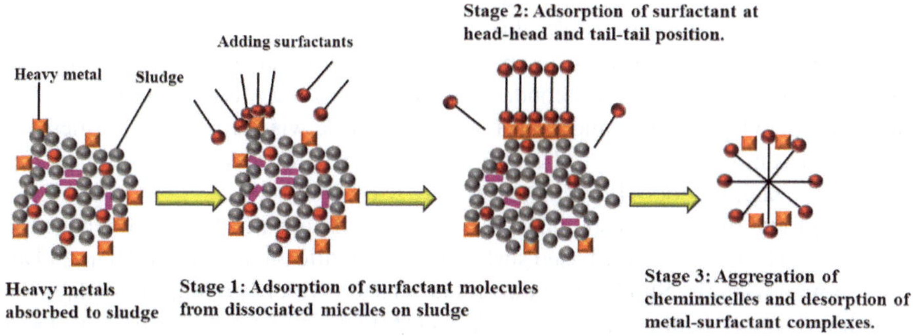

Fig. 10 Mechanism of metals leaching from sludge by surfactants [66]

Cationic surfactants, particularly quaternary ammonium compounds, are commonly utilised as disinfectants due to their ability to disrupt the integrity of microbial cell membranes. In Gram-negative bacteria, the negatively charged outer layer binds electrostatically with the positively charged head part of the surfactant, which increases the membrane permeability and leakage of cell content and eventually cell death and hence is used as a disinfectant for sanitation [69].

3. ***In Cosmetics.***

Human skin is dynamic and well responsive to various environments, supporting overall health and serving as a critical component of the immune system of the body. Surfactants are popularly used in cosmetic products and personal care products for their ability to reduce surface tension, emulsify oils, cleanse the skin and generate foam. They are actively used in face cleansers, body washes, soaps, shampoos, conditioners, lotions, face masks and lipsticks [70].

Chemical surfactants exhibit multiple functions which make them essential in cosmetic, face care, hair care and hygiene products. They efficiently remove dirt, oil and other impurities from the skin and hair by reducing surface tension and enabling the suspension of particles in water. Surfactants are also utilised for emulsifying incompatible liquids such as oil and water. This emulsification takes place in shampoos and cleansers, where the surfactants improve the formation of foam and lather, which enhances perceived cleansing efficacy. Likewise, several surfactants have wetting and moistening ability for skin and hair, which helps in maintaining hydration. Additionally, many specially designed surfactants improve the conditioning effects, softness, smoothness and manageability of both skin and hair and also help as a layer of protection from harsh weather conditions [71].

Surfactant molecule has a drive to self-assemble and create a micelle-like structure above the CMC, which allows them to reduce the surface tension and the interfacial tension between two incompatible components. Every micelle typically contains between 10 and 100 monomers, with the exact number defined by the aggregation number [72]. The growth and spreadability of micelles of various shapes, such as vesicles, rod-like, plate-like, spherical, cylindrical, hemimicelles and many others, can take place in milliseconds [73, 74].

The primary surfactant used in shampoo is sodium lauryl sulphate, along with cocamidopropyl betaine as a secondary surfactant. Face cleansers use mild surfactants like decyl glucoside and lauryl glucoside. Likewise, for body washes and shower gels, ammonium lauryl ether sulphate, sodium lauryl sulphate, sodium lauroyl sarcosinate, sodium cocoyl glycinate, etc., are used [70].

Biodegradability and sustainability of surfactants are the two major concerns related to surfactant use. Bio-based surfactants are gentle on the face and hair and are environment friendly. Sulphate-free formulations are widely popular these days due to their reduced potential for irritation. Advancements in green chemistry have led to several developments of enzymatically produced surfactants that provide improved performance with minimal environmental impact.

While surfactants are vital in personal care and cosmetic products, harsh types of chemicals like sodium lauryl sulphate can strip natural lipids, which can cause dryness and irritation of skin and hair. To reduce such effects, formulators use milder surfactants, conditioning agents, humectants and pH adjustments to improve skin and hair compatibility [70].

4. ***In Pharmaceuticals.***

Surfactants are essential in pharmaceutical formulations by enhancing their stability, solubility, biodegradability and dissolution rate of active pharmaceutical ingredients. Many of the active pharmaceutical ingredients are poorly water-soluble, which diminishes their efficacy and absorption. Surfactant reduces the surface and interfacial tension, facilitate the dispersion of active pharmaceutical ingredients in an aqueous system and enhance drug absorption and overall therapeutic efficacy of the drug [75].

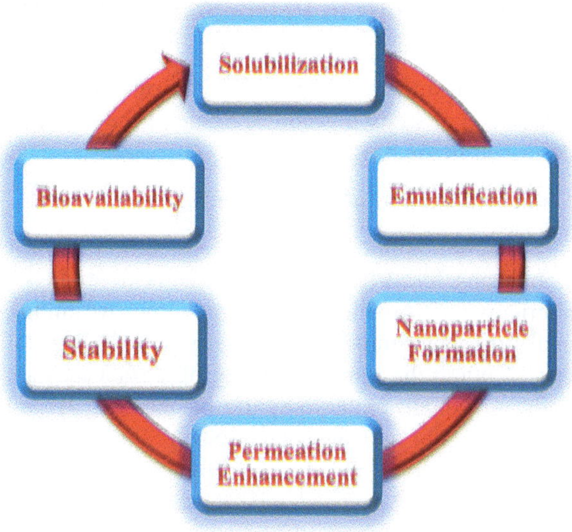

Fig. 11 Systematic representation of the mechanism of surfactant in drug delivery [70]

Drugs are chemicals that alter the body composition and function and provide relief from symptoms, treat and diagnose diseases by interacting with biological targets like cells, enzymes, or receptors [76] (Fig. 11). Surfactants play a vital role in drug delivery by enhancing solubility, stability and bioavailability [77]. They reduce surface and interfacial strain between two immiscible components, enabling the formation of stable emulsions, micelles and liposomes that can hold the hydrophobic drugs and make them effective. Surfactants protect active ingredients from environmental degradation, such as oxidation and hydrolysis and improve stability, biological membrane permeability, aiding drug absorption at targeted cells or tissues. Surfactants are employed to deliver injectable, oral and topical medications, where they support controlled release and targeted therapies such as directing drugs to tumour cells and ultimately improving treatment outcomes and patient compliance [78].

Surfactant-aided drug delivery involves micellar encapsulation of hydrophobic drug molecules, thus improving their dispersion and apparent solubilising ability in an aqueous setting [79]. Surfactants induce the formation of stable emulsions and facilitate the emulsification in drug delivery [80]. In nanoparticle drug formulations, surfactants are vital since they stabilise particles and maintain their size by reducing surface tension and preventing them from aggregating in the surrounding liquid. Surfactant also improves the permeation of medicines by breaking down biological barriers [81]. Due to their amphiphilic nature, they can interact with lipid membranes and fluidise them, decreasing their stiffness and increasing permeability [82]. The degradation, precipitation and aggregation of the drugs are controlled by surfactants. They also shield drugs from oxidation, pH shifts, enzymatic breakdown and other environmental conditions [83]. Surfactant also enhances the bioavailability of the drug to targeted cells and tissues and facilitates their uniform distribution over the affected area [84].

4 Applications in Petrochemical Industries

The use of surfactants in petrochemical processes of various natures has been a thrust area of study nowadays. Some of the common applications in this regard are described in the present chapter.

1. ***Preparation of alternative fuels.***

Surfactant application in the formulation of various emulsions and microemulsions using fossil fuel and water has been studied by various workers to establish alternative fuels, which may not only improve engine efficiency but also render safer and more environmentally friendly fuels for sustainable human growth. Rastogi et al. studied the kinetic stability of surfactant-stabilised water-in-diesel emulsion fuels. The hydrophile–lipophile balance (HLB) value is changed by varying the composition of two surfactants, Span 80 and Tween 80. The study was carried out using microscopy and dynamic light scattering (DLS) [85]. Charoensaeng et al. explored the influence of pure diesel fuel, palm-oil-blended diesel and biodiesel-blended diesel, along with three co-surfactants, on fuel utilisation and exhaust gas emissions in a direct-injection (DI) diesel engine. In their formulation, palm oil and diesel were served as the oil phase, ethanol as the aqueous phase, methyl oleate as the surfactant and various alkanols as the co-surfactants. The resulting microemulsified biofuel was found to have slightly more fuel consumption due to low heat of combustion and demonstrated significantly reduced NO_x and CO_2 emissions and exhaust gas temperatures in contrast to conventional diesel fuel [86].

Dash et al. investigated the phase behaviour and fuel applications of some microemulsions using diesel and kerosene as the oil phase, Tween 80 and n-butanol/i-butanol as the surfactant and co-surfactant mixture, respectively, and water/2% brine as the aqueous phase. The findings highlight the improved fuel compatibility of microemulsion with branched chain co-surfactants, i.e. iso-butanol, relative to straight-chain co-surfactant, i.e., n-butanol [87]. When Triton-X-100 was used as the surfactant and several alcohols such as n-butanol, iso-butanol, n-amyl alcohol and n-octanol were used as co-surfactants, the straight-chain n-butanol outperformed the branched iso-butanol and all other alcohols [88]. A recent work on study on fuel properties like kinematic viscosity, density, flash point and calorific value using castor-oil-blended diesel microemulsified using Tween 80 and butanol as surfactants and co-surfactants has been reported. Various formulations from this work suggested remarkable alternative fuel characteristics of the microemulsion systems [89].

2. ***Enhanced Oil Recovery.***

Enhanced oil recovery (EOR) process is carried out since, after the pressure depletion and water flooding, some residual oil is entrapped inside the rock due to capillary forces. During the recovery process, surface flooding is carried out because it decreases the oil portion in the pore space of the matrix rock. Several varieties of surfactants, such as cationic, anionic, amphoteric and non-ionic, have been introduced to move the residual oil by decreasing the interfacial tension between oil and water [90, 91]. The introduction of surfactant during such processes alters the wettability of the oil-wet rock system.

In reservoir rocks, oil and water exist together, which generates an interfacial tension over the interface. This tension hinders their mobility and the oil gets trapped in the reservoir system. When the surfactant is injected inside the trap, it rearranges itself at the boundary and binds its hydrophilic moiety with the polar part and hydrophobic moiety with the non-polar part. As a result, it creates a homogeneous and thermodynamically stable microemulsion and reduces the interfacial tension and increases the mobility toward the production well [56]. The microemulsion formulation is severely affected by the pH, salinity, temperature, pressure, properties and concentration of oil and surfactants and also the co-surfactant used [92].

Surfactants help in wettability alteration, a process of intentional modification of the surface of the reservoir rock to alter its affinity from oil-wet to water-wet or vice-versa and oil recovery can be performed effectively [93–97]. Moreover, surfactants are also utilised for foam generation, which can reduce gas mobility, gravity segregation, viscous fingering and also diverts fluid from high permeability regions to low permeability regions, thereby improving the oil recovery [98–100]. However, achieving optimal performance at low concentrations and cost remains a key challenge in advancing surfactant-based EOR technologies [56].

3. ***Demulsification of Crude Oil Emulsions.***

During hydrocarbon exploration and production, crude oil emulsions are formed as a by-product due to the interaction between oil, water and solid particles such as asphaltenes and resins, which may induce corrosion, erosion, catalytic fouling, as well as deactivation of the process [101, 102]. Surfactants help in the removal of crude oil emulsions through the demulsification technique.

A popular technique of chemical demulsification, a three-step process, is generally adopted, wherein the first step involves the disruption and removal of the asphaltenic film from the water/oil interface using a surfactant [103] and the second step involves flocculation, which promotes aggregation of dispersed droplets forming larger flocs leading to coalescence wherein reduced interfacial viscosity and film rupture takes place causing the droplets to merge. Finally, phase separation occurs through sedimentation in water-in-oil (W/O) emulsions, where water droplets settle downward and through creaming in oil-in-water (O/W) emulsions, where oil droplets rise.

Hydrophobic chain length and the surfactant content are important for demulsification [104, 105]. The performance of the surfactant also relies upon the water/oil ratio, asphaltenes content, temperature and salinity of the medium. At high temperature conditions, the efficiency of the surfactant becomes more due to increased kinetic energy of the system [106]. Saline environment improves the adsorption of surfactant over the interface and accelerates the process [107].

4. ***Oil Sorbent Removal and Remediation.***

During storage and transportation, crude oil may leak, leading to pollution of soil and water bodies. Contamination of soil and water bodies can be resolved using surfactants, which are one of the most remarkable agents for the remediation process.

Remediation of soil can be done by in-situ and ex-situ methods [108]. The in-situ method is performed by injecting a surfactant solution into the contaminated area through injected wells, where it solubilises the oil present in the contaminated soil, allowing it to be pumped out to the surface via production wells. In the ex-situ approach, contaminated soil is excavated and mixed with a suitable surfactant solution. After it gets settled, the liquid phase containing the contaminants is separated, allowing the surfactant to be recovered and the waste to be safely disposed of.

When water bodies are contaminated with crude oil, the removal happens through a solubilisation process, where the suitable surfactant is incorporated at CMC, the hydrophilic head gets attached to the aqueous phase and the lipophilic tail to the petroleum contaminant, thus generating a Winsor type III microemulsion. Similarly, when contaminants are adsorbed onto soil particles, removal occurs through mobilisation or "soil roll-up" procedure (Fig. 12). The surfactant lowers interfacial tension and increases the contact angle between the contaminant and soil, displacing the oil from soil surfaces. This process results in a Winsor Type I microemulsion, where surfactant is primarily surrounded by oil [109].

The efficiency of removal of contamination from soil is controlled by the type of soil and also by the type and chain length of the surfactant utilised. For example, clay has a larger surface area due to extremely small particle size, which provides a larger space for adsorption of the hydrocarbon contaminant and making it tough for removal [111]. Since a negatively charged surface is available on clay, cationic surfactants tend to strongly adsorb, contrary to anionic and non-ionic surfactants, which exhibit significantly lower adsorption. Hence, anionic surfactants generally exhibit better performance than the cationic ones.

Out of all classes of surfactants, the non-ionic and biosurfactants are more commonly employed in remediation work. This may be attributed to the difference in their HLB values. Non-ionic surfactants typically have HLB values ranging from 0 to 20, while ionic surfactants can reach up to 50. As a result, micelle volume produced from ionic surfactants tends to decrease, resulting in reduced solubilisation and mobilisation of oil contaminants compared to the non-ionic surfactants [112]. Moreover, increased temperature condition also helps in better mobility and interaction with the surfactants, thus enhancing the remediation process [113].

5. ***Water-based Drilling Fluid***.

For the lubrication of the drill bit, oil-based or water-based drilling fluids may be used. However, in the case of the former, there is a greater chance of toxic contamination and it needs more environmental regulations for its use. On the contrary, water-based drilling fluids have gained more attention for drilling operations despite some intricate problems associated with them [114, 115]. The problems can, however, be reduced by using surfactants, which are widely added to reduce the reaction between the clay-water system and thus prevent dispersion and swelling. The surfactant molecules coat the surface of the clay minerals, with their hydrophilic head groups aligned with the clay through electrostatic forces and the hydrophobic tail of the surfactant forms a protective layer around the clay, thus preventing water

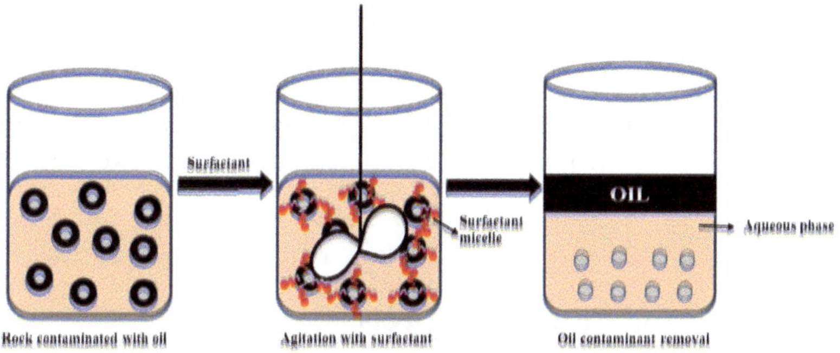

Fig. 12 Surfactant-assisted extraction of oil contaminants from aqueous and solid phases [110]

molecules from interacting with the surface and thereby reducing hydration and swelling trouble (Fig. 13) [116].

A large number of investigations on the problem have been reported regarding the types of surfactants for swelling inhibition. Use of cationic surfactant lowers the negative charges of clay, thereby preventing the adsorption of water molecules over it. The alkyl chain length of the surfactant also impacts the efficiency of swelling inhibition [117]. Gemini surfactants with a biphenyl spacer and bromide and chloride as counter ions were also developed as swelling inhibitors. Gemini surfactants with unsaturated hydrophobic chains are more effective than their saturated alkyl chain analogues [118]. Cationic Gemini surfactant with different spacers was investigated, and it was observed that the more the spacer length, the greater the degree of hydrophobicity and the higher the degree of swelling inhibition property of drilling

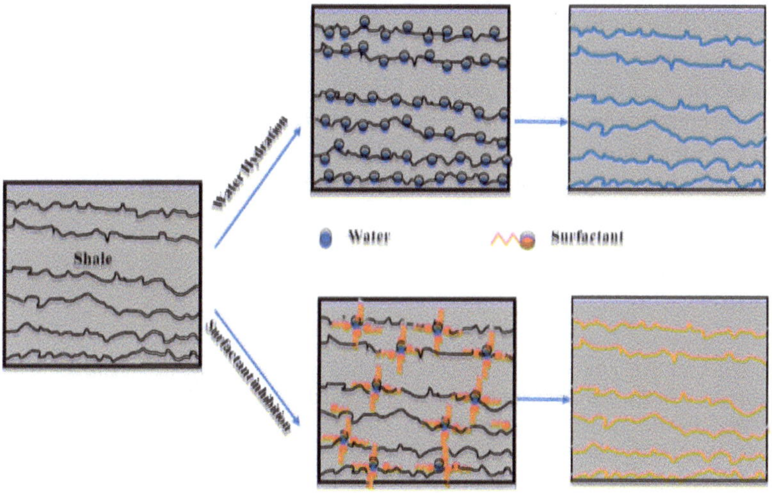

Fig. 13 Mechanism of shale swelling inhibition by surfactant [56]

fluids [119]. Non-ionic surfactants derived from plants have excellent swelling inhibition properties, and moreover, they are non-toxic and eco-friendly. The inhibition of swelling has been recognised to be due to the interactivity of the hydrogen bonds of the hydroxyl part of the plant-based surfactants and the silica content of the clay [120]. Magnetic and zwitterionic surfactants also possess clay-swelling inhibitor properties [121].

5 Conclusion and Future Prospects

The present chapter describes a comprehensive understanding of the fundamentals of surfactants, including their classification, key terminologies and underlying mechanisms of action. Emphasis was placed on their remarkable ability to change surface and interfacial properties, which underpins their wide-ranging chemical and petrochemical applications. Furthermore, this chapter throws light on some precise and multifaceted usage of surfactants across some urgent chemical and petrochemical applications. These applications underscore the remarkable functional adaptability of surfactants and their contribution to improve the process efficiency, environmental compliance and operational performance of the respective industries. Nonetheless, several extreme external factors, such as salinity, temperature, pressure, pH and surfactant concentration, which are very common in these industries, must be carefully taken into consideration for optimal performance. Additionally, the molecular structure and configuration of surfactants play a vital role in determining their effectiveness across various applications. Although conventional surfactants may offer superior efficiency, alternative surfactants are increasingly favoured due to their lower toxicity and greater environmental compatibility. Continued research is imperative to create advanced application and remediation strategies for better degradation and removal of residual surfactants from both the water and soil ecosystems, thereby improving environmental sustainability. Moreover, there is a growing need to design and engineer next-generation surfactants with enhanced stability, low ecotoxicity and superior performance under extreme conditions such as high salinity, temperature and pressure commonly encountered in industrial and oilfield operations.

References

1. Olajire AA (2014) Review of ASP EOR (alkaline surfactant polymer enhanced oil recovery) technology in the petroleum industry: prospects and challenges. Energy 77:963–982. https://doi.org/10.1016/j.energy.2014.09.005
2. Murtaza M, Gbadamosi A, Ahmad HM, Hussain SMS, Kamal MS, Mahmoud M, Patil S (2023) A magnetic surfactant having one degree of unsaturation in the hydrophobic tail as a shale swelling inhibitor. Molecules 28(4):1878. https://doi.org/10.3390/molecules28041878

3. Nagtode VS, Cardoza C, Yasin HKA, Mali SN, Tambe SM, Roy P, Singh K, Goel A, Amin PD, Thorat BR, Cruz JN, Pratap AP (2023) Green surfactants (biosurfactants): a petroleum-free substitute for sustainability–comparison, applications, market, and future prospects. ACS Omega 8(13):11674–11699. https://doi.org/10.1021/acsomega.3c00591
4. Shiri MSZ, Henderson W, Mucalo MR (2019) A Review of the lesser-studied microemulsion-based synthesis methodologies used for preparing nanoparticle systems of the noble metals, OS, RE, IR and RH. Materials 12(12):1896. https://doi.org/10.3390/ma12121896
5. Bajpai D, Tyagi VK (2010) Nonionic surfactants: an overview. Tenside Surfactants Deterg 47(3):190–196. https://doi.org/10.3139/113.110062
6. Morvan M, Koetitz R, Moreau P, Pavageau B, Rivoal P, Roux B (2008) A combinatorial approach for identification of performance EOR surfactants. In: SPE symposium on improved oil recovery. https://doi.org/10.2118/113705-MS
7. Cui L, Khramov DM, Bielawski CW, Hunter D, Yoon P, Paul. (2008) Effect of organoclay purity and degradation on nanocomposite performance, part 1: surfactant degradation. Polymer 49(17):3751–3761. https://doi.org/10.1016/j.polymer.2008.06.029
8. Galimberti M, Martino M, Guenzi M, Leonardi G, Citterio A (2009) Thermal stability of ammonium salts as compatibilizers in polymer/layered silicate nanocomposites. e-Polymers *9*(1). https://doi.org/10.1515/epoly.2009.9.1.686
9. Gonçalves RA, Holmberg K, Lindman B (2023) Cationic surfactants: a review. J Mol Liq *375*:121335. https://doi.org/10.1016/j.molliq.2023.121335
10. Yusuff AS, Gbadamosi AO, Bode-Olajide FB, Igbafe AI (2024) Adsorption of methyl ester sulfonate surfactant on Berea sandstone: parametric optimization, kinetics, isotherm and thermodynamics studies. Colloids SurfS Phys Chemical Eng Asp 686:133363. https://doi.org/10.1016/j.colsurfa.2024.133363
11. Souayeh M, Al-Maamari RS, Karimi M, Aoudia M (2020) Wettability alteration and oil recovery by surfactant assisted low salinity water in carbonate rock: the impact of nonionic/anionic surfactants. J Petrol Sci Eng 197:108108. https://doi.org/10.1016/j.petrol.2020.108108
12. Li P, Yang C, Cui Z, Song B, Jiang J, Wang Z (2016) A new type of sulfobetaine surfactant with double alkyl polyoxyethylene ether chains for enhanced oil recovery. J Surfact Deterg 19(5):967–977. https://doi.org/10.1007/s11743-016-1839-2
13. Cheng Y, Yang Y, Niu C, Feng Z, Zhao W, Lu S (2019) Progress in synthesis and application of zwitterionic Gemini surfactants. Front Mater Sci 13(3):242–257. https://doi.org/10.1007/s11706-019-0473-0
14. Akbaş H, Boz M, Elemenli A (2014) Interaction between cationic Gemini surfactant and related single-chain surfactant in aqueous solutions. Fluid Phase Equilibria 370:95–100. https://doi.org/10.1016/j.fluid.2014.02.024
15. Befkadu AA, Chen Q (2018) Surfactant-enhanced soil washing for removal of petroleum hydrocarbons from contaminated soils: a review. Pedosphere 28(3):383–410. https://doi.org/10.1016/s1002-0160(18)60027-x
16. Chen L, Xie H, Li Y, Yu W (2008) Applications of cationic Gemini surfactant in preparing multi-walled carbon nanotube contained nanofluids. Colloids SurfS Phys Chemical Eng Asp *330*(2–3):176–179. https://doi.org/10.1016/j.colsurfa.2008.07.047
17. Atta DY, Negash BM, Yekeen N, Habte AD (2020) A state-of-the-art review on the application of natural surfactants in enhanced oil recovery. J Mol Liq 321:114888. https://doi.org/10.1016/j.molliq.2020.114888
18. Gbadamosi A, Yusuff A, Agi A, Oseh J (2022) An overview of natural surfactant application for enhanced oil recovery. In: Enhanced oil recovery - selected topics. IntechOpen. https://doi.org/10.5772/intechopen.104935
19. Hama SM, Manshad AK, Ali JA (2023) Review of the application of natural surfactants in enhanced oil recovery: state-of-the-art and perspectives. Energy Fuels 37(14):10061–10086. https://doi.org/10.1021/acs.energyfuels.3c01347
20. Lee M, Woo S, Ten LN (2012) Characterization of novel diesel-degrading strains Acinetobacter haemolyticus MJ01 and Acinetobacter johnsonii MJ4 isolated from oil-contaminated

soil. World J Microbiol Biotechnol 28(5):2057–2067. https://doi.org/10.1007/s11274-012-1008-3
21. Mulligan CN (2004) Environmental applications for biosurfactants. Environ Poll 133(2):183–198. https://doi.org/10.1016/j.envpol.2004.06.009
22. Zhou W, Wang X, Chen C, Zhu L (2013) Enhanced soil washing of phenanthrene by a plant-derived natural biosurfactant, Sapindus saponin. Colloids SurfS Phys Chemical Eng Asp 425:122–128. https://doi.org/10.1016/j.colsurfa.2013.02.055
23. Ambaye TG, Vaccari M, Prasad S, Rtimi S (2021) Preparation, characterization and application of biosurfactant in various industries: a critical review on progress, challenges and perspectives. Environ Technol Innov 24:102090. https://doi.org/10.1016/j.eti.2021.102090
24. Raffa P, Wever D, a. Z., Picchioni, F., & Broekhuis, A. A. (2015) Polymeric surfactants: synthesis, properties, and links to applications. Chem Rev 115(16):8504–8563. https://doi.org/10.1021/cr500129h
25. Afolabi F, Mahmood SM, Yekeen N, Akbari S, Sharifigaliuk H (2021) Polymeric surfactants for enhanced oil recovery: a review of recent progress. J Petrol Sci Eng 208:109358. https://doi.org/10.1016/j.petrol.2021.109358
26. Hu R, Tang S, Mpelwa M, Jiang Z, Feng S (2021) Research progress of viscoelastic surfactants for enhanced oil recovery. Energy Explor Exploit 39(4):1324–1348. https://doi.org/10.1177/0144598720980209
27. Shehzad F, Hussain SMS, Adewunmi AA., Mahboob A, Murtaza M, & Kamal MS (2021) Magnetic surfactants: a review of recent progress in synthesis and applications. Adv Colloid Interface Sci 293:102441. https://doi.org/10.1016/j.cis.2021.102441
28. Hussain SMS, Adewunmi AA, Mahboob A, Murtaza M, Zhou X, Kamal MS (2022) Fluorinated surfactants: a review on recent progress on synthesis and oilfield applications. Adv Colloid Interface Sci 303:102634. https://doi.org/10.1016/j.cis.2022.102634
29. Buck RC, Murphy PM, Pabon M (2011) Chemistry, properties, and uses of commercial fluorinated surfactants. In: The handbook of environmental chemistry, pp 1–24 https://doi.org/10.1007/978-3-642-21872-9_1
30. Parekh P, Varade D, Parikh J, Bahadur P (2011) Anionic–cationic mixed surfactant systems: micellar interaction of sodium dodecyl trioxyethylene sulfate with cationic Gemini surfactants. Colloids SurfS Phys Chem Eng Asp 385(1–3):111–120. https://doi.org/10.1016/j.colsurfa.2011.05.057
31. Zhou W (2004) Solubilization of pyrene by anionic? nonionic mixed surfactants. J Hazard Mater 109(1–3):213–220. https://doi.org/10.1016/j.jhazmat.2004.03.018
32. Zhou W, Zhu L (2005) Solubilization of polycyclic aromatic hydrocarbons by anionic–nonionic mixed surfactant. Colloids SurfS Phys Chem Eng Asp 255(1–3):145–152. https://doi.org/10.1016/j.colsurfa.2004.12.039
33. Vishnyakov A, Lee M, Neimark AV (2013) Prediction of the critical micelle concentration of nonionic surfactants by dissipative particle dynamics simulations. J Phys Chem Lett 4(5):797–802. https://doi.org/10.1021/jz400066k
34. Massarweh O, Abushaikha AS (2020) The use of surfactants in enhanced oil recovery: a review of recent advances. Energy Rep 6:3150–3178. https://doi.org/10.1016/j.egyr.2020.11.009
35. Elgh-Dalgren K, Arwidsson Z, Camdzija A, Sjöberg R, Ribé V, Waara S, Allard B, Von Kronhelm T, Van Hees PA (2009) Laboratory and pilot scale soil washing of PAH and arsenic from a wood preservation site: changes in concentration and toxicity. J Hazard Mater 172(2–3):1033–1040. https://doi.org/10.1016/j.jhazmat.2009.07.092
36. Malik MA, Wani MY, Hashim MA (2010) Microemulsion method: a novel route to synthesize organic and inorganic nanomaterials. Arab J Chem 5(4):397–417. https://doi.org/10.1016/j.arabjc.2010.09.027
37. Perger T, Bešter-Rogač M (2007) Thermodynamics of micelle formation of alkyltrimethylammonium chlorides from high performance electric conductivity measurements. J Colloid Interface Sci 313(1):288–295. https://doi.org/10.1016/j.jcis.2007.04.043

38. Malik NA, Ali A (2015) Krafft temperature and thermodynamic study of interaction of glycine, diglycine, and triglycine with hexadecylpyridinium chloride and hexadecylpyridinium bromide: a conductometric approach. J Mol Liq 213:213–220. https://doi.org/10.1016/j.molliq.2015.11.017
39. Schramm LL, Marangoni DG (2000) Surfactants and their solutions: basic principles. In: Surfactants: fundamentals and applications in the petroleum industry, pp 3–50. https://doi.org/10.1017/cbo9780511524844.002
40. Bales BL, Benrraou M, Zana R (2002) Krafft Temperature and micelle ionization of aqueous solutions of cesium dodecyl sulfate. J Phys Chem B 106(35):9033–9035. https://doi.org/10.1021/jp021297l
41. Nabi ME (2005) Investigations on the behaviour and properties of different types of unilamellar vesicles
42. Mukherjee P, Padhan SK, Dash S, Patel S, Mishra BK (2011) Clouding behaviour in surfactant systems. Adv Colloid Interface Sci 162(1–2):59–79. https://doi.org/10.1016/j.cis.2010.12.005
43. Nakama Y (2017) Surfactants. Cosmet Sci Technol:231–244. https://doi.org/10.1016/b978-0-12-802005-0.00015-x
44. Shinoda K (1978) Principles of solution and solubility
45. Das A, Nguyen N, Nguyen QP (2019) Low tension gas flooding for secondary oil recovery in low-permeability, high-salinity reservoirs. Fuel 264:116601. https://doi.org/10.1016/j.fuel.2019.116601
46. Abalkhail N, Liyanage PJ, Upamali KA, Pope GA, Mohanty KK (2020) Alkaline-surfactant-polymer formulation development for a HTHS carbonate reservoir. J Petrol Sci Eng 191:107236. https://doi.org/10.1016/j.petrol.2020.107236
47. Sheng J (2010) Modern chemical enhanced oil recovery: theory and practice. Gulf Professional Publishing
48. Royer M, Nollet M, Catté M, Collinet M, Pierlot C (2017) Towards a new universal way to describe the required hydrophilic lipophilic balance of oils using the phase inversion temperature of C 10 E 4/n-octane/water emulsions. Colloids Surfaces Physicochem Eng Aspects 536:165–171. https://doi.org/10.1016/j.colsurfa.2017.07.024
49. Jigarvyas, Pandey RK (2021) Determination of HLB value by saponification method: a brief review. Nat J Pharm Sci 1(2):23–24
50. Gadhave AD (2014) A short review on microemulsion and its application in extraction of vegetable oil. Int J Res Eng Technol 03(09):147–158. https://doi.org/10.15623/ijret.2014.0309022
51. Nguyen TT, Edelen A, Neighbors B, Sabatini DA (2010) Biocompatible lecithin-based microemulsions with rhamnolipid and sophorolipid biosurfactants: formulation and potential applications. J Colloid Interface Sci 348(2):498–504. https://doi.org/10.1016/j.jcis.2010.04.053
52. Bouton F, Durand M, Nardello-Rataj V, Serry M, Aubry J (2008) Classification of terpene oils using the fish diagrams and the Equivalent Alkane Carbon (EACN) scale. Colloids Surfaces Physicochem Eng Aspects 338(1–3):142–147. https://doi.org/10.1016/j.colsurfa.2008.05.027
53. Salager J, Antón RE, Sabatini DA, Harwell JH, Acosta EJ, Tolosa LI (2005) Enhancing solubilization in microemulsions—state of the art and current trends. J Surfact Deterg 8(1):3–21. https://doi.org/10.1007/s11743-005-0328-4
54. Salager JL, Antón RE, Briceño MI, Choplin L, Márquez L, Pizzino A, Rodriguez MP (2003) The emergence of formulation engineering in emulsion making—transferring know-how from research laboratory to plant. Polym Int 52(4):471–478. https://doi.org/10.1002/pi.1112
55. Obot IB, Haruna K, Saleh TA (2018) Atomistic simulation: a unique and powerful computational tool for corrosion inhibition research. Arab J Sci Eng 44(1):1–32. https://doi.org/10.1007/s13369-018-3605-4
56. Gbadamosi A, Badmus S, Haruna K, Patil S, Kamal MS, Mahmoud M, Saleh T (2025) Surfactant applications in oil and gas industry: efficiency, toxicity, and remediation techniques. J Mol Liq 427:127440. https://doi.org/10.1016/j.molliq.2025.127440

57. Dwivedi D, Lepková K, Becker T (2017) Carbon steel corrosion: a review of key surface properties and characterization methods. RSC Adv 7(8):4580–4610. https://doi.org/10.1039/c6ra25094g
58. Verma C, Hussain CM, Quraishi M, Alfantazi A (2022) Green surfactants for corrosion control: design, performance and applications. Adv Colloid Interface Sci 311:102822. https://doi.org/10.1016/j.cis.2022.102822
59. Moselhy MA, Zaki EG, El-Maksoud S, a. E. H. A., & Migahed, M. A. (2022) The role of some cationic surfactants based on thiazine as corrosion inhibitors in petroleum applications: experimental and theoretical approach. ACS Omega 7(36):32014–32025. https://doi.org/10.1021/acsomega.2c02961
60. Amel A, M'Yassa AC, Tazerouti A (2024) Application of sodium methyl ester sulfonate anionic surfactants as corrosion inhibitors for carbon steel in oilfield injection water. Russ J Appl Chem 97(2):264–277. https://doi.org/10.1134/s1070427224020046
61. Amin MA, Ahmed M, Arida H, Arslan T, Saracoglu M, Kandemirli F (2010) Monitoring corrosion and corrosion control of iron in HCl by non-ionic surfactants of the Triton-X series—Part II. Temperature effect, activation energies and thermodynamics of adsorption. Corr Sci 53(2):540–548. https://doi.org/10.1016/j.corsci.2010.09.019
62. Abdel-Hameed R, Aleid GMS, Ragab H, Alshafey H, Alosaimi AM (2023) Synthesis, characterization, and evaluation of polymeric surfactants derived from PET plastic waste as green corrosion inhibitor of steel surfaces in marine environment for heavy industry. RSC Adv 13(45):31969–31988. https://doi.org/10.1039/d3ra04518h
63. El-Lateef HMA, Abo-Riya MA, Tantawy AH (2016) Empirical and quantum chemical studies on the corrosion inhibition performance of some novel synthesized cationic Gemini surfactants on carbon steel pipelines in acid pickling processes. Corr Sci 108:94–110. https://doi.org/10.1016/j.corsci.2016.03.004
64. Jiang L, Liang J, Yuan X, Li H, Li C, Xiao Z, Huang H, Wang H, Zeng G (2014) Co-pelletization of sewage sludge and biomass: the density and hardness of pellet. Bioresour Technol 166:435–443. https://doi.org/10.1016/j.biortech.2014.05.077
65. Hong C, Wang Z, Si Y, Yang Q, Xing Y (2016) Improving sludge dewaterability by combined conditioning with Fenton's reagent and surfactant. Appl Microbiol Biotechnol 101(2):809–816. https://doi.org/10.1007/s00253-016-7939-0
66. Guan R, Yuan X, Wu Z, Wang H, Jiang L, Li Y, Zeng G (2017) Functionality of surfactants in waste-activated sludge treatment: a review. Sci Total Environ 609:1433–1442. https://doi.org/10.1016/j.scitotenv.2017.07.189
67. Song GJ, Feng XY (2011) Review of enzymatic sludge hydrolysis. J Bioremed Biodegr 02(05). https://doi.org/10.4172/2155-6199.1000130
68. Kuczajowska-Zadrożna M, Filipkowska U, Jóźwiak T (2015) Application of biosurfactants for heavy metals leaching from immobilized activated sludge. Arch Environ Protect 41(1):43–52. https://doi.org/10.1515/aep-2015-0006
69. Palmer M, Hatley H (2018) The role of surfactants in wastewater treatment: impact, removal and future techniques: a critical review. Water Res 147:60–72. https://doi.org/10.1016/j.watres.2018.09.039
70. Banjare MK (2025) Role of surfactants in drug delivery, pesticides, antimicrobial activity and bioremediation. https://doi.org/10.70593/978-93-7185-025-4
71. Ross S, Morrison ID (1988) Colloidal systems and interfaces. Wiley-Interscience
72. Arai H (1966) Study of detergency. I. Effect of the concentration and the kind of detergent in hard water. J Am Oil Chem Soc 43(5):312–314. https://doi.org/10.1007/bf02609680
73. Rosen MJ (2004) Micelle formation by surfactants. In: Surfactants and interfacial phenomena, pp 105–177. https://doi.org/10.1002/0471670561.ch3
74. Rosen MJ (2004) Emulsification by surfactants. In: Surfactants and interfacial phenomena, pp 303–331. https://doi.org/10.1002/0471670561.ch8
75. Aisha N, Batool I, Iftekhar S, Taj MB, Carabineiro SAC, Ahmad F, Khan MI, Shanableh A, Alshater H (2025) Wetting the surface: a deep dive into chemistry and applications of surfactants. Clean Chem Eng 11:100197. https://doi.org/10.1016/j.clce.2025.100197

76. Banjare MK, Tandon DK (2024) Study the effect of L-arginine and L-tyrosine on the surface adsorption and micellar properties of long-chain imidazolium-based ionic liquid. Results Chem 7:101452. https://doi.org/10.1016/j.rechem.2024.101452
77. Banjare MK, Banjare RK, Ghosh KK (2020) Molecular interaction on imidazolium based ionic liquids and serum albumin: a spectroscopy approach. Int J Adv Chem 8(2):209–216. https://doi.org/10.14419/ijac.v8i2.30994
78. Pandya SJ, Kapitanov IV, Banjare MK, Behera K, Borovkov V, Ghosh KK, Karpichev Y (2022) Mixed Oxime-functionalized IL/16-S-16 Gemini surfactants system: physicochemical study and structural transitions in the presence of promethazine as a potential chiral pollutant. Chemosensors 10(2):46. https://doi.org/10.3390/chemosensors10020046
79. Hoque MA, Rahman MM, Alam MM, Mahbub S, Khan MA, Kumar D, Albaqami MD, Wabaidur SM (2021) Interaction of cephalexin monohydrate with surfactants in aqueous and sodium chloride solution at variable temperatures: conductivity and spectroscopic measurements. J Mol Liq 326:115337. https://doi.org/10.1016/j.molliq.2021.115337
80. Kumar H, Kaur J (2020) Influence of electrolyte and temperature on the aggregation behaviour of mixed system consisting drug and anionic surfactant. J Phys Conf Ser 1531(1):012102. https://doi.org/10.1088/1742-6596/1531/1/012102
81. Singh BK, Walker A (2006) Microbial degradation of organophosphorus compounds. FEMS Microbiol Rev 30(3):428–471. https://doi.org/10.1111/j.1574-6976.2006.00018.x
82. Shamsipur M, Yazdanfar N, Ghambarian M (2016) Combination of solid-phase extraction with dispersive liquid–liquid microextraction followed by GC–MS for determination of pesticide residues from water, milk, honey and fruit juice. Food Chem 204:289–297. https://doi.org/10.1016/j.foodchem.2016.02.090
83. Xue N, Xu X, Jin Z (2005) Screening 31 endocrine-disrupting pesticides in water and surface sediment samples from Beijing Guanting reservoir. Chemosphere 61(11):1594–1606. https://doi.org/10.1016/j.chemosphere.2005.04.091
84. Thiaré D, Khonté A, Diop A, Cissé L, Coly A, Tine A, Delattre F (2015) Determination of ground and excited state dipole moments of amino-benzimidazole by solvatochromic shift methods and theoretical calculations. J Mol Liq 211:640–646. https://doi.org/10.1016/j.molliq.2015.07.071
85. Preetika R, Mehta PS, Kaisare NS, Basavaraj MG (2018) Kinetic stability of surfactant stabilized water-in-diesel emulsion fuels. Fuel 236:1415–1422. https://doi.org/10.1016/j.fuel.2018.09.074
86. Charoensaeng A, Khaodhiar S, Sabatini DA, Arpornpong N (2018) Exhaust emissions of a diesel engine using ethanol-in-palm oil/diesel microemulsion-based biofuels. Environ Eng Res 23(3):242–249. https://doi.org/10.4491/eer.2017.204
87. Acharya B, Guru PS, Dash S (2015) Tween-80–n-butanol/isobutanol–(diesel+kerosene)–water microemulsions—phase behavior and fuel applications. Fuel 171:87–93. https://doi.org/10.1016/j.fuel.2015.12.013
88. Acharya B, Dash S (2018) Tuning commercial diesel to microemulsified and blended form: phase behavior and implications. J Dispers Sci Technol 40(8):1159–1168. https://doi.org/10.1080/01932691.2018.1500479
89. Mishra S, Acharya B, Sharma R, Dash S (2025) Phase behavior and alternative fuel characteristics of some castor oil blended diesel-tween 80-n-butanol/i-butanol-water/brine microemulsions. J Mol Liq 437:128312. https://doi.org/10.1016/j.molliq.2025.128312
90. Kamal MS, Hussein IA, Sultan AS (2017) Review on surfactant flooding: phase behavior, retention, IFT, and field applications. Energy Fuels 31(8):7701–7720. https://doi.org/10.1021/acs.energyfuels.7b00353
91. Gbadamosi A, Patil S, Shehri DA, Kamal MS, Hussain SS, Al-Shalabi EW, Hassan AM (2022) Recent advances on the application of low salinity waterflooding and chemical enhanced oil recovery. Energy Rep 8:9969–9996. https://doi.org/10.1016/j.egyr.2022.08.001
92. Pal S, Mushtaq M, Banat F, Sumaiti AMA (2017) Review of surfactant-assisted chemical enhanced oil recovery for carbonate reservoirs: challenges and future perspectives. Petrol Sci 15(1):77–102. https://doi.org/10.1007/s12182-017-0198-6

93. Deng X, Kamal MS, Patil S, Hussain SMS, Zhou X, Mahmoud M (2021) Wettability alteration of locally synthesized cationic Gemini surfactants on carbonate rock. J Mol Liq 344:117817. https://doi.org/10.1016/j.molliq.2021.117817
94. Adila AS, Al-Shalabi EW, Alameri W (2021) Geochemical investigation of hybrid Surfactant and low salinity/engineered water injections in carbonates: a numerical study. J Petrol Sci Eng 208:109367. https://doi.org/10.1016/j.petrol.2021.109367
95. Standnes DC, Austad T (2003) Wettability alteration in carbonates. Colloids Surfaces Physicochem Eng Aspects 216(1–3):243–259. https://doi.org/10.1016/s0927-7757(02)00580-0
96. Adebayo AR (2021) Foam flow in different pore systems—part 2: the roles of pore attributes on the limiting capillary pressure, trapping coefficient, and relative permeability of foamed gas. SPE J 26(06):3926–3948. https://doi.org/10.2118/205523-pa
97. Hou B, Wang Y, Huang Y (2015) Mechanistic study of wettability alteration of oil-wet sandstone surface using different surfactants. Appl Surface Sci 330:56–64. https://doi.org/10.1016/j.apsusc.2014.12.185
98. Yekeen N, Manan MA, Idris AK, Padmanabhan E, Junin R, Samin AM, Gbadamosi AO, Oguamah I (2018) A comprehensive review of experimental studies of nanoparticles-stabilized foam for enhanced oil recovery. J Petrol Sci Eng 164:43–74. https://doi.org/10.1016/j.petrol.2018.01.035
99. Rafati R, Haddad AS, Hamidi H (2016) Experimental study on stability and rheological properties of aqueous foam in the presence of reservoir natural solid particles. Colloids Surfaces Physicochem Eng Aspects 509:19–31. https://doi.org/10.1016/j.colsurfa.2016.08.087
100. Abdelgawad KZ, Adebayo AR, Isah A, Muhammed NS (2022) A literature review of strength and stability of foam and their relationship with the absolute permeability of porous media. J Petrol Sci Eng 211:110195. https://doi.org/10.1016/j.petrol.2022.110195
101. Husain A, Adewunmi AA, Gbadamosi A, Al-Harthi MA, Patil S, Kamal MS (2023) Demulsification of asphaltene stabilized crude oil emulsions by biodegradable ethylcellulose polymers with varying viscosities. Sci Rep 13(1):1090. https://doi.org/10.1038/s41598-023-27973-x
102. Adewunmi AA, Kamal MS, Gbadamosi A, Patil S (2023) Natural extracted waste materials for breaking crude oil emulsion. In: SPE western regional meeting. https://doi.org/10.2118/213007-ms
103. Raya SA, Saaid IM, Ahmed AA, Umar AA (2020) A critical review of development and demulsification mechanisms of crude oil emulsion in the petroleum industry. J Petrol Explor Prod Technol 10(4):1711–1728. https://doi.org/10.1007/s13202-020-00830-7
104. Javadian S, Bahri M, Sadrpoor SM, Rezaei Z, Kakemam J (2022) Structure effect in the demulsification performance of cationic surfactants. J Petrol Sci Eng 218:110895. https://doi.org/10.1016/j.petrol.2022.110895
105. Mirvakili A, Rahimpour MR, Jahanmiri A (2012) Effect of a cationic surfactant as a chemical destabilization of crude oil based emulsions and asphaltene stabilized. J Chem Eng Data 57(6):1689–1699. https://doi.org/10.1021/je2013268
106. Javadian S, Sadrpoor SM, Khosravian M (2023) Taking a look accurately at the alteration of interfacial asphaltene film exposed to the ionic surfactants as demulsifiers. Sci Rep 13(1):12837. https://doi.org/10.1038/s41598-023-39731-0
107. Ye F, Mi Y, Liu H, Zeng G, Shen L, Feng X, Yang Y, Zhang Z, Yuan H, Yan X (2021) Demulsification of water-in-crude oil emulsion using natural lotus leaf treated via a simple hydrothermal process. Fuel 295:120596. https://doi.org/10.1016/j.fuel.2021.120596
108. Umar HA, Khanan MFA, Shiru MS, Ahmed MS, Abdullahi J, Obute MA, Ahmad A (2024) Application of modified drastic model for oil spills pollution affecting water quality system in part of Niger delta region of Nigeria. J Hydrol 637:131446. https://doi.org/10.1016/j.jhydrol.2024.131446
109. Mao X, Jiang R, Xiao W, Yu J (2014) Use of surfactants for the remediation of contaminated soils: a review. J Hazard Mater 285:419–435. https://doi.org/10.1016/j.jhazmat.2014.12.009
110. Javanbakht G, Goual L (2016) Mobilization and micellar solubilization of NAPL contaminants in aquifer rocks. J Contam Hydrol 185–186:61–73. https://doi.org/10.1016/j.jconhyd.2016.01.003

111. Li G, Guo S, Hu J (2015) The influence of clay minerals and surfactants on hydrocarbon removal during the washing of petroleum-contaminated soil. Chem Eng J 286:191–197. https://doi.org/10.1016/j.cej.2015.10.006
112. Ceschia E, Harjani JR, Liang C, Ghoshouni Z, Andrea T, Brown RS, Jessop PG (2013) Switchable anionic surfactants for the remediation of oil-contaminated sand by soil washing. RSC Adv 4(9):4638–4645. https://doi.org/10.1039/c3ra47158f
113. Urum K, Pekdemir T, Çopur M (2004) Surfactants treatment of crude oil contaminated soils. J Colloid Interface Sci 276(2):456–464. https://doi.org/10.1016/j.jcis.2004.03.057
114. Muhammed NS, Olayiwola T, Elkatatny S (2021) A review on clay chemistry, characterization and shale inhibitors for water-based drilling fluids. J Petrol Sci Eng 206:109043. https://doi.org/10.1016/j.petrol.2021.109043
115. Rana A, Arfaj MK, Saleh TA (2019) Advanced developments in shale inhibitors for oil production with low environmental footprints—a review. Fuel 247:237–249. https://doi.org/10.1016/j.fuel.2019.03.006
116. Murtaza M, Khan RA, Kamal MS, Hussain SMS, Mahmoud M (2020) Poly (Oxyethylene)-amidoamine based Gemini cationic surfactants with hydrophilic spacers as clay stabilizers. Energy Fuels 34(9):10619–10630. https://doi.org/10.1021/acs.energyfuels.0c01010
117. Hosseini SE, Nowrouzi I, Shahbazi K, Kamari M, Mohammadi AH, Manshad AK (2023) An investigation into shale swelling inhibition properties of dodecyltrimethylammonium chloride (DTAC) for water-based drilling fluids. Geoenergy Sci Eng 223:211465. https://doi.org/10.1016/j.geoen.2023.211465
118. Ahmad HM, Murtaza M, Kamal MS, Hussain SS, Mahmoud M (2020) Cationic Gemini surfactants containing biphenyl spacer as shale swelling inhibitor. J Mol Liq 325:115164. https://doi.org/10.1016/j.molliq.2020.115164
119. Murtaza M, Kamal MS, Hussain SMS, Mahmoud M (2020) Clay swelling inhibition using novel cationic Gemini surfactants with different spacers. J Surfact Deterg 23(5):963–972. https://doi.org/10.1002/jsde.12420
120. Moslemizadeh A, Shadizadeh SR, Moomenie M (2015) Experimental investigation of the effect of henna extract on the swelling of sodium bentonite in aqueous solution. Appl Clay Sci 105–106:78–88. https://doi.org/10.1016/j.clay.2014.12.025
121. Ahmad HM, Murtaza M, Gbadamosi A, Kamal MS, Hussain SMS, Mahmoud M, Patil S (2023) Application of novel magnetic surfactant-based drilling fluids for clay swelling inhibition. Energy Fuels 37(12):8212–8223. https://doi.org/10.1021/acs.energyfuels.3c00704

Advantages of Biosurfactants over Chemical Surfactants

Shikha Indoria, Arpita Padhan, Rajat Sinhmar, and Vickramjeet Singh

Abstract Surfactant-based formulations have applications from biological fields to industry. Chemical surfactants are often harsh and have high toxicity, low biocompatibility, and poor biodegradability. As a result, they cause significant environmental problems and ecological damage. Therefore, eco-friendly biosurfactants, mainly derived from microorganisms, are preferred as they help maintain environmental safety. Biosurfactants require lower concentrations to reduce surface and interfacial tension, indicating a lower CMC than commonly used surfactants. Biosurfactants remain stable in various salt concentrations, different pH levels, and temperatures, and they stabilize emulsions more effectively than chemical surfactants. The chapter also covers the production of biosurfactants, including microbial strains, fermentation strategies, and factors affecting yield. Biological activities of biosurfactants, including antimicrobial and antifungal effects, indicate their potential applications in medicine, agriculture, and the pharmaceutical industry. Overall, biosurfactants offer significant industrial and environmental advantages, and future research may focus on large-scale production, economic feasibility, and the development of novel applications.

Keywords Biosurfactants · Chemical surfactants · Biodegradability · Microorganisms · Low toxicity · Antimicrobial

---

S. Indoria
Department of Chemistry, Government College Ropar, Rupnagar, Punjab, India

A. Padhan · R. Sinhmar · V. Singh (✉)
Department of Chemistry, Dr. B. R. Ambedkar National Institute of Technology, Jalandhar, Punjab, India
e-mail: singhvj@nitj.ac.in

M. Abdul Rub et al. (eds.), *Industrial Applications of Individual and Mixed Surfactant Systems—Volume 2*, Sustainable Materials and Technology,
https://doi.org/10.1007/978-981-95-9195-4_17

1 Introduction to Surfactants

The term "surfactant" is derived from the words "surface-active agents"; these molecules position themselves at the boundary of water and a surface or at the interface (solid–liquid, solid–gas, liquid–liquid, or liquid–gas), reducing surface tension and thereby improving its cleansing ability. Since each molecules contain a polar head group and a non-polar tail group, they are often referred to as amphiphilic molecules. Their structural arrangement allows them to interfere with the interaction among solvent (water) molecules, which in turn reduces the surface tension. It can be employed for a wide range of applications, in addition to reducing surface tension, such as emulsifiers, foam stabilizers, and corrosion-preventing additives, among others. Conventionally, surfactants were solely produced from natural sources such as plant oils and animal fats [1–3].

1.1 Classification of Surfactants

1. **Based on Charges on the Head Group.**

(a) **Anionic Surfactant.**

Anionic surfactants are characterized by a negative charge on their head group, which includes phosphate, sulfate, carboxylate, and sulfonate. When they are dissolved in water, they transform into a negative charge surfactant ion. These surfactants were employed in various fields, including laundry and shampoo. Some of the most widely known examples of anionic surfactants are sodium lauryl ether sulfate, dioctyl sodium sulfosuccinate, and linear alkyl benzene sulfates. Anionic surfactants are usually considered to have very low toxicity [4–6].

(b) **Cationic Surfactant.**

Cationic surfactants possess a positively charged head group and are widely employed due to their great emulsifying activity. These surfactants can be employed as a topical antiseptic and possess strong bactericidal activity, as revealed through various research works. They have potential applications in hand sanitizers, and they are also employed as fabric softeners, hydrotropes, and hydrophobic agents, among others. The cationic head part has the ability to disrupt bacterial cell membranes; hence, it can be utilized as an antibacterial agent. Cationic surfactants must be used within the safety limits set by regulatory bodies. A few examples of cationic surfactants are pyridines and imidazoline derivatives, quaternary ammonium compounds, and their ester derivatives [7–9].

(c) **Zwitterionic Surfactant.**

Zwitterionic surfactants carry both positive and negative charge groups within one molecular structure. The anionic part can be modified; however, the cationic part may contain amines. Some examples of zwitterionic surfactants include surfactin, miltefosine, and lauryl dimethyl amine N-oxide [10, 11].

(d) **Nonionic Surfactant.**

Nonionic surfactants remain unionized when introduced to an aqueous solution; hence, they maintain good stability and are minimally influenced by strong electrolytes. These surfactants exhibit excellent compatibility with other surfactants and demonstrate high solubility in both water and organic media. Nonionic surfactants are the most widely employed surfactants for drug delivery. Common examples include pluronics, polyoxyethylene ester, poloxamer, and polyol ester. At a certain temperature, nonionic surfactants begin to separate from the cleaning solution, a phenomenon known as the cloud point, which is considered the optimal temperature for detergency [12, 13].

2. **Based on Solubility.**

Surfactants that are readily dissolved in water are termed hydrophilic, and those that dissolve in oily media are hydrophobic or lipophilic. For instance, ionic surfactants tend to be hydrophilic; however, nonionic surfactants can exhibit both hydrophilic and hydrophobic properties. These properties of nonionic surfactants rely on the relative balance between their hydrophilic and lipophilic portion. This balance is typically represented by the hydrophilic–lipophilic balance (HLB) value, while the cloud point can be utilized for characterizing nonionic surfactants, the Krafft point is used for ionic surfactants [14, 15].

1.2 Biosurfactant

Historically, traditional surfactants in the marketplace have been composed predominantly of synthetic surfactants produced from petroleum, i.e., alkylbenzene sulfonate (ABSs) and ethoxylated alcohols, which are lower cost than naturally derived surfactants but also result in numerous concerns related to their persistence in the environment, toxicity, and dependence on nonrenewable resources. In light of increasing public interest in sustainable practices, growing consumer demand for socially responsible products, and increasing pressure from government and nonprofit organizations to provide clean, sustainable products, biosurfactants are becoming an increasingly appealing, naturally derived alternative to traditional synthetic surfactants [16].

Biological surfactants (biosurfactants), or surfactants naturally produced by, or derived from, living microorganisms (bacteria, yeast, and fungi), possess unique properties when compared to synthetic surfactant products. Because biosurfactants are derived from microorganisms (and therefore represent a diverse range of groups), they possess inherent environmentally sustainable characteristics, making them ideal candidates for green chemistry and circular bio-economy business concepts [17]. These benefits include being biodegradable, less toxic, more robust in function, and having greater potential to be produced from renewable materials. Recent studies have started validating these benefits of biosurfactants [16, 18].

Biosurfactants are highly biodegradable, which is one of their major benefits. While many synthetic surfactants will remain for long periods within the environment, biosurfactants can be more easily broken down by microbes, thereby minimizing their accumulation over time within the environment [17]. In addition, the rapid degradation of biosurfactants within a natural setting helps to reduce pollution and support sustainable development [16]. Generally, compared to synthetic surfactants, biosurfactants have higher biocompatibility and lower toxicity to humans and the environment. Recent reviews suggest that the cytotoxicity of biosurfactants, such as rhamnolipids, sophorolipids, and lipopeptides, is relatively low when evaluated with mammalian cells, and ecotoxicological impacts of these compounds are much lower than those of synthetic surfactants. Therefore, biosurfactants can be safely used in personal care products, agricultural operations, and environmental cleanup processes [16, 18].

The performance of biosurfactants is not diminished in any functional form. They possess a low CMC, wide-ranging surfactant application flexibility (a broad range of pH, temperature, and salt concentrations) [19]. In fact, many studies have cited the ability of certain biosurfactants to lower surface tension to values lower than those of synthetic surfactants and retain their function under high environmental conditions.

Another differentiating factor for biosurfactants is their sustainability in terms of production as microbiological production has the potential for using renewable sources such as agricultural and industrial wastes, previously used oils, and used products, thereby linking biosurfactant production and waste valorization [17], which results in reduced costs and supports the development of circular economy concepts. Research and development work continues in developing strain improvement and optimizing fermentation conditions and in situ processing technologies to increase yield and decrease production cost [17].

Biosurfactants also display numerous other (non-surfactant) types of activity (e.g., antimicrobial activity, inhibition of biofilm formation, prevention of corrosion) that make them ideal candidates for use in a variety of industries, including biocontrol (control of pests and pathogens), corrosion prevention (maintenance of metal integrity), and medical formulations (treatment/management of diseases). Several recent studies have described biosurfactants as environmentally friendly biocides that inhibit microbial biofilms on metallic surfaces, offering a less toxic alternative to traditional chemical biocides [20].

Biosurfactants also support the mobilization and solubilization of hydrophobic contaminants (petroleum hydrocarbons, heavy metals) during environmental remediation, as well as improving biodegradation and cleanup effectiveness of these contaminants [21]. Their low toxic potential and high biodegradability make them ideal for large-scale, in situ bioremediation implementations.

Commercialization of biosurfactants offers many advantages; however, there are still many challenges to overcome. One of the major obstacles is developing an economically viable large-scale production system, especially with respect to downstream purification, process stability, and regulatory approval [16, 17]. The main challenge is to eliminate these bottlenecks so that biosurfactants can be more widely used and could potentially replace synthetically made surfactants in industry. Despite these barriers, biosurfactants will likely play a significant role in future as they offer several benefits, including biodegradability, low toxicity, superior functional performance, and sustainable production. This chapter critically highlights recent advances in these areas, highlights emerging trends, and outlines research needs for scaling up biosurfactant production and application.

1.3 Sources and Production of Biosurfactants

Synthetic surfactants are mainly synthesized from petrochemical sources, which allow for their large-scale production at relatively very low prices. Although they have many economic advantages, this type of production method is consistently being criticized for enhancing environmental unsustainability [22–24]. Also, it is conflicted with Sustainable Development Goals, promotion of green economies, and the preference of consumers for eco-friendly products [1]. In addition, synthetic surfactants are made up of harsh chemicals having higher toxicity posing risks to both human health and the environment [23, 24]. In contrast, biosurfactants are bio-based alternatives produced by various microorganisms such as bacteria, yeasts, and fungi through fermentation processes [25]. These compounds are naturally derived, environment friendly, and comparatively less toxic than that of synthetic chemicals [26]. Due to these advantages, the research and demand of biosurfactants have been increased in scientific areas and global markets over the last fifteen years [27, 28].

Biosurfactants are formed in diverse chemical forms. These can be synthesized by microbes which are grown on various raw materials including insoluble substrates like oils and hydrocarbons or soluble substrates like sugars [29, 30]. It has been evaluated that replacing synthetic chemicals with biosurfactants can lead to an 8% reduction in lifetime CO_2 emissions prohibiting the release of around 1.5 million tons of carbon dioxide into the atmosphere [31, 32]. Currently, the production of biosurfactants accounts for 10% of the global surfactant production, which is estimated at ten million tons per year. Applications for these natural surfactants are numerous and include petroleum, agriculture, food, medicines, personal care, and environmental management (e.g., wastewater treatment) [22, 27, 33].

The research on biosurfactants began in 1960s and has been significantly evolved, resulting in the commercial availability of many biosurfactant-based products [34]. In recent years, growing concern over sustainability has led to intensified research in this impact, selectivity, and ability to function under extreme conditions such as high salinity and low area. Biosurfactants are particularly appealing due to their biodegradability, low temperatures, or varied pH levels [35]. Moreover, their production using renewable feedstocks enhances both their environmental and economic value, positioning them as a promising alternative to traditional surfactants [29].

Biosurfactants and biological surfactants are sometimes assumed as synonyms, but they have distinct fine gradation in their definition and applications. Microorganisms produce biosurfactants, which are essentially a subset of biological surfactants, during their development or metabolism [27]. These molecules reduce surface tension and enhance liquid emulsification like synthetic surfactants, but are less toxic and biodegradable. Biological surfactants besides biosurfactants also include the surfactants produced from plants and animals. But all these encompass one similarity that is their functionality; both these categories reduce surface tension and facilitate emulsification of liquids [36]. However, the major difference lies in their origin and process of production. Biosurfactants are produced by microbial biotechnologies (yeast, bacteria, and fungi), while biological surfactants can be produced from a wide range of natural resources, including plants and animals. However, biosurfactants are gaining great applications due to their instantaneous production, multifunctional properties, widespread applicability, and scalability [36].

Several research studies have shown that microorganisms capable of producing biosurfactants can be found in both clean and polluted soils [37]. In clean (uncontaminated) soils, approximately 2–3% of microbes are known to produce biosurfactants, whereas this number increases to around 25% in polluted (contaminated) soils. Several types of bacteria are well known for biosurfactant production, including *Actinobacteria, Pseudomonas, Sphingomonas, Bacillus, Halomonas, Pseudoalteromonas, Mycobacterium, Alcanivorax, Rhodococcus, and Arthrobacter* [37, 38].

Although bacteria have been widely studied for their ability to produce biosurfactants, less attention has been given to fungi. However, some fungal species, such as *Candida lipolytica, Candida bombicola, Candida batistae, Candida ishiwadae, Aspergillus ustus, and Trichosporon asahii*, have shown potential as biosurfactant producers [39–41].

Biosurfactants usually have molecular weights between 500 and 1500 Daltons and are grouped into low- and high-molecular-weight types. Low-molecular-weight biosurfactants are good at lowering surface tension and helping to form stable emulsions, while high-molecular-weight types, often called bioemulsifiers, are better at stabilizing oil-in-water emulsions [42]. The low-molecular-weight biosurfactants are categorized as glycolipids, lipopeptides, and phospholipids [43]. Among these, glycolipids are the most studied biosurfactants. They are made up of a carbohydrate part (hydrophilic) connected to fatty acid chains (hydrophobic). Common types of glycolipids are rhamnolipids, trehalolipids, mannosylerythritol lipids, and sophorolipids (Table 1).

1.4 Classification and Factors Influencing the Production of Biosurfactants

The classification of biosurfactants is influenced by the wide variety of microorganisms involved in their production, as well as their distinct biological and chemical properties. Major categories of biosurfactants include glycolipids, phospholipids, liposaccharides, lipopeptides, fatty acids, and neutral lipids [43, 79, 80]. Each class exhibits specific features that have been extensively explored in scientific

Table 1 Classification of biosurfactants and producing microorganisms

Biosurfactant type	Subtype	Microbial producers	Citations
Glycolipids	Rhamnolipids	Produced by various strains of *pseudomonas*, including *P. aeruginosa*, *P. cepacia*, and others	[44]
		Lysinibacillus sphaericus and *Serratia rubidaea*	[45, 46]
	Trehalolipids	*Nocardia farcinica* and species from the *Rhodococcus* genus	[47]
	Sophorolipids	*Candida bombicola*, *C. Sphaerica*, and *Starmerella bombicola*	[48–50]
		Cutaneotrichosporon mucoides also contributes to sophorolipid production	[51]
	Mannosylerythritol lipids	*Pseudzyma aphidis*	[52]
	Unclassified	*Meyerozyma guilliermondii*, *Saccharomyces cerevisiae*, and *Candida utilis*	[53–55]
Lipopeptides	Surfactin	*Marinobacter hydrocarbonoclasticus*, *Bacillus subtilis*, and *B. Nealsonii*	[56–58]
	Lichenysin	*Bacillus licheniformis*	[59, 60]
	Unclassified	*Pseudomonas azotoformans*, *Bacillus velezensis*, *Virgibacillus salarius*, *B. pumilus*, and *Halomonas* species	[61, 62]
Phospholipids	–	Typically produced by *Thiobacillus thiooxidans* and *Klebsiella pneumoniae*	[63–65]
Polymeric surfactants	Liposan	*Candida lipolytica*	[66–68]
Complex (carbohydrate/ protein/lipid)	Carbohydrate–protein–lipid emulsifier	*Candida ingens*	[68]
	Carbohydrate–protein–lipid emulsifier	*Candida utilis*	[69]
	Mannoprotein emulsifier	*Saccharomyces cerevisiae*	[70]
Complex (carbohydrate/ protein)	Mannoprotein emulsifier	*Candida tropicalis* ATCC86842, *Saccharomyces cerevisiae*, *Kluyveromyces marxianus*, *Candida utilis*, *Candida glabrata*	[71, 72]
Lipopeptide	*Anionic lipopeptide*	*Candida glabrata*	[73, 74]
Glycolipid	Sophorolipids (free acids and lactones, often di-acetylated)	*Candida bombicola*	[75]

(continued)

Table 1 (continued)

Biosurfactant type	Subtype	Microbial producers	Citations
	Not specified	*Candida boidinii*	[76]
	Presumed sophorolipids	*Candida bogoriensis*	[76]
	–	*Candida sphingicola* UCP0995	[61]
	–	*Candida glabrata*	[77]
	–	*Candida ishiwadae*	[78]

studies. Table 1 summarizes the main categories of biosurfactants and identifies their microbial sources.

The trend of using biosurfactants for commercial is increasing. Microbial metabolites that are amphiphilic are called biosurfactants. Examples of glycolipids are: rhamnolipids, sophorolipids, and mannosylerythritol lipids; examples of lipopeptides are: surfactin and iturin. Each class of biosurfactant has its own specific fermentation characteristics. Rhamnolipids (two sugar units + β-hydroxy fatty acids) are typically produced by *Pseudomonas* spp. (notably *P. aeruginosa*) and can reduce water's surface tension from ~ 72 to ~ 30 mN/m [81]. Surfactin is a cyclic lipopeptide from *Bacillus subtilis* and related species [82], known for very low critical micelle concentration (~ 21 mg/L) and strong foaming. Sophorolipids are yeast-derived glycolipids (sophorose disaccharide linked to fatty acids) produced by *Starmerella bombicola* (formerly *C. bombicola*) and related yeasts [80, 83]. In optimized fermentations, yields of these biosurfactants can be very high: e.g., *C. bombicola* reached ~ 34 g/L sophorolipids in a 50-L bioreactor [84] and engineered *P. aeruginosa* strains have achieved ~ 15–25 g/L rhamnolipids in batch and up to ~ 43–70 g/L in fed-batch [85]. By comparison, surfactin yields are typically a few grams per liter (e.g., ~2.4 g/L surfactin from *B. subtilis* in optimized conditions [62]. Other biosurfactants (e.g., trehalolipids, mannosylerythritol lipids) exist, but rhamnolipids, sophorolipids, and surfactin dominate research on fermentation.

Biosurfactants are microbial surface-active molecules with hydrophilic and hydrophobic regions, including glycolipids (rhamnolipids, sophorolipids, MELs) and lipopeptides (surfactin, iturin). Each of these biosurfactant classes exhibits unique fermentation characteristics. Rhamnolipids, typically synthesized by *Pseudomonas* species (especially *P. aeruginosa*) [86], are composed of one or two rhamnose sugar units linked to β-hydroxy fatty acids. Surfactin, a cyclic lipopeptide primarily produced by *B. subtilis* and its close relatives, exhibits strong surface activity with a low critical micelle concentration (CMC) of approximately 21 mg/L. It has been shown to produce stable foams, which can affect beer fermentation on a commercial scale [87]. Sophorolipids are produced by yeasts; for example, the yeast *S. bombicola* (previously called *C. bombicola*) produces very high levels of sophorolipids of about 34 g/L when grown in 50 L fermenters [88]. The genetically engineered strains of *P. aeruginosa* have been reported to produce rhamnolipid at levels of 15–25 g/L in batch mode and 43–70 g/L in fed-batch mode. The yields of surfactin from *B. subtilis* are generally lower, with optimized conditions producing around 2.4 g/L. Trehalolipids and MELs are also of interest, but rhamnolipids, sophorolipids, and surfactin have received the most attention in research on microbial fermentation [89, 90].

2 Fermentation Traits of Different Biosurfactants

Most rhamnolipids (glycolipids) are produced by species of Pseudomonas, particularly *P. aeruginosa*, although some non-pathogenic strains (for example, *Pseudomonas putida* and *Pseudomonas cepacia*) also produce rhamnolipids. Rhamnolipid production can occur aerobically; however, substantial foam formation is frequently seen during rhamnolipid production. Rhamnolipids are classified as ether-linked extracellular biosurfactant (occasionally called secondary) metabolites, and their level of production is increased usually during the stationary growth phase due to the limitation of nitrogen availability [89, 90]. Rhamnolipids are stable over a large pH range (4–12) and can endure being heated to 100 °C or greater. Typical fermentation media are represented by "Landy" medium or its variants, using sugar or molasses as a source of carbon and protein as a source of nitrogen. Optimal fermentation conditions are generally considered to be a neutral to slightly acidic pH and temperatures in the range of 30–37 °C, although engineered strains have been reported to have higher thermal optima (e.g., one study reported the ideal thermal conditions for an engineered strain to be 42.9 °C). Shake flask cultures typically yield 0.5–2 g/L; for instance, a shake flask fermentation of *B. subtilis* strain YPS-32 produced 1.82 g/L and a 5-L bioreactor w/foam reflux under optimized conditions produced 2.39 g/L [62]. Sophorolipids (glycolipids) are biosurfactants produced by yeasts of the Starmerella (formerly Candida) genus. Sophorolipids are composed of a sophorose disaccharide linked to fatty acids, and they are generally produced in the form of acidic and lactonic forms. The production of sophorolipids requires dual carbon substrates; generally, one substrate is a hydrophilic source (e.g., glucose or molasses), and the other is a hydrophobic source (e.g., oleic acid or vegetable oil) [90]. Fermentation is carried out aerobically at a temperature of about 25–30 °C, with a moderate pH of approximately 3–6. Biphasic media (two-phase) are often used to promote production. Yields can be very high; e.g., *S. bombicola* produced approximately 58.4 g/l total sophorolipids, when grown on inexpensive substrates like cottonseed oil or cottonseed molasses [90]. In fermentation, foaming and phase separation are often observed. The ratio of lactonic to acidic sophorolipids can be altered by the pH and type of substrate used [90]. In addition to MSLs, other microbial surfactants include mannosylerythritol lipids (MELs) from Pseudozyma species, trehalolipids from Rhodococcus, and lipopeptides like iturin and fengycin from Bacillus. Fungal biosurfactants have also been investigated. Solid-state fermentation techniques using fungi such as Aspergillus and Phialemonium, grown on agro-industrial waste, have yielded both biosurfactants i.e. glycolipid and lipopeptide. Although SSF generally gives lower product titers than submerged fermentation, SSF has certain advantages, such as a lower water requirement and simpler downstream processing [91, 92].

2.1 Fermentation Methodology

(a) **Laboratory-Scale Fermentation.**

The first stage of the development of biosurfactants is usually performed in a laboratory setting, using shake flasks or small-scale bioreactors (100 mL to 5 L in size). Screening studies to identify the best conditions for the production of biosurfactants are often carried out in Erlenmeyer flasks (150–500 mL working volume) which contain specifically defined mineral salts media and are agitated via an orbital shaker. The inoculum used in the screening studies may be obtained from cryopreserved culture collections or from agar slants, with 1–2% (v/v) inoculum being added to the seed flasks, in which the culture grows overnight in preparation for the next step. Following a period of 24–48 h. of incubation, the seed bottle contents are transferred into production flasks. Specific operating parameters are kept constant as they pertain to incubation temperature, pH, agitation (usually set at 100–300 rpm), and aeration. As an example, the optimal formulation for the production of surfactin in shake flasks has been identified using Landy medium containing 20 g/L molasses, 15 g/L L-glutamic acid, and 4.5 g/L soy meal, along with the additions of KCl, K_2HPO_4, $MgSO_4$, and $Fe_2(SO_4)_3$. Under conditions of 42.9 °C, pH 5.0, 2% inoculum, and 43 hours of fermentation, significant yields were achieved [62]. Results obtained at this stage are used to inform scale-up strategies.

(b) **Pilot Scale.**

Scaling transitions to stirred-tank bioreactors occur with working volumes ranging from 1 to 50 L at pilot scale and up to 1000 L or more for industrial applications. Aerobic fermentation systems rely on mechanical agitation (via impellers) and gas sparging to regulate dissolved oxygen levels. Parameters such as pH, temperature, agitation speed, and aeration rate are monitored and adjusted through integrated sensors and control systems. Biosurfactant-producing strains tend to cause excessive foaming; therefore, bioreactors are typically equipped with foam traps, condensers, or reflux systems. One study demonstrated that a 5-L stirred tank with a 2.5-L working volume, 2% inoculum, aeration at 1 L/min, and stirring at 300 rpm (adjusted for foam levels), operated at pH 7.0 controlled with phosphoric acid, enabled surfactin production with in situ foam reflux to recover lost medium. In the case of rhamnolipid production, continuous foam fractionation or antifoam agents are essential to manage excessive foam generation and prevent culture loss [86].

Industrial biosurfactant production is predominantly carried out using fed-batch strategies. In this approach, substrates typically carbon sources are added incrementally to avoid substrate inhibition and extend the productive phase of the culture. For example, rhamnolipid yields as high as 48.7 g/L were attained via repeated fed-batch fermentation coupled with periodic harvesting [93]. Although continuous fermentation offers advantages in productivity, its industrial use remains limited due to the challenge of foam management. Emerging technologies such as membrane-assisted continuous systems are being investigated to address these issues. While stirred-tank reactors are most commonly employed, alternative systems such as airlift reactors, bubble columns, and membrane bioreactors are used to mitigate shear stress and manage foam. Solid-state fermentation (SSF) setups including tray fermenters and packed-bed columns operate without free liquid but require controlled aeration and humidity. For sophorolipid production, *S. bombicola* has been cultivated in large-scale batch fermenters ranging from 10 to 100 m^3. Rhamnolipid production has reached pilot-scale implementation. Industrial-scale production favors non-pathogenic strains like recombinant *P. putida* and low-cost substrates to improve safety and economic feasibility [94, 95].

The selection of media has a critical impact on both cost and productivity. Carbon sources constitute the bulk of the production expense, and choices vary from refined sugars (glucose, sucrose) to economical alternatives such as molasses, crude glycerol, and industrial waste oils. Media are typically based on minimal salts, tailored to suit the microbial strain in use. For *Pseudomonas* species, a widely used formulation includes 40 g/L glycerol (or glucose/oil), 4 g/L $NaNO_3$ as the nitrogen source, along with buffering and mineral salts: 4 g/L K_2HPO_4, 3 g/L KH_2PO_4, 0.5 g/L $MgSO_4 \cdot 7H_2O$, 0.5 g/L KCl, and 0.2 g/L $CaCl_2$. Trace elements such as iron, zinc, and manganese are often added for enhanced metabolic activity [96, 97].

For *Bacillus* strains, the Landy medium is common, consisting of molasses as a carbon source, 15 g/L L-glutamic acid, and soy meal for nitrogen, supplemented with KCl, K_2HPO_4 $MgSO_4$, and $FeSO_4$ [62]. In yeast-based systems like *S. bombicola* for sophorolipid production, media typically include glucose or cheaper sugars, along with 5–10% v/v of oils such as soybean, palm, or oleic acid [82]. To enhance biosynthesis of glycolipids, optimal nitrogen levels for the growth of microorganisms are maintained at low levels, e.g., 3 g/L of yeast extract. Statistical techniques such as response surface methodology (RSM) have been used by many investigators to develop optimal growth media to yield higher amounts of glycolipids [98].

Starting from pure cultures or cryopreserved stocks, researchers may grow a seed culture in a nutrient-rich medium (LB broth or nutrient broth) and then establish that culture through multiple sequential stages (pre-seed and seed) in shake flasks or smaller bioreactors. The size of inoculum typically varies from 1 to 10% (v/v) depending on the species of microorganism being cultured and their desired cell density; e.g., in one study of surfactin production, the inoculum size of a seed culture transferred into the 5 L bioreactor was 2% [98]. In addition, for fermentations carried out with yeasts, inoculum is usually prepared as a specific cell density in a suspension. With spore-forming organisms such as Bacillus, spore suspensions can also be used to establish inoculum. Ultimately, the goal of this step in the process is to ensure that the culture being used for fermentation is both actively growing and well adapted to the environmental conditions encountered during fermentation operations to ensure successful fermentation performance.

Bioreactors for production are first sterilized using either autoclaving or steam-in-place (SIP) techniques before sterile media can be added. Various operating parameters such as temperature, pH, agitation, and aeration are adjusted according to the characteristics of the microbial system. For example, Pseudomonas and Bacillus prefer neutral pH (~ 7) and grow best at temperatures of approximately 30–37 °C; however, yeast grow under mildly acidic conditions (pH 4–5), which helps support the production of lactonic sophorolipids. The aeration rate of aerobic fermentation is usually between 0.5 and 2 vvm (volume of air/volume of broth/minute). 100–400 rpm is the typical range for agitation speed of fermentation vessels with volumes of 1–10 L. Usually, foaming occurs when using aerobes to produce biosurfactants, especially in the case of Pseudomonas and Bacillus; condenser, gas trap, and mechanical foam breaker devices are frequently incorporated in the production systems to manage this issue. In addition, some fermentation systems utilize fed-batch nutrient feeding systems that periodically add nutrients to extend productivity, whereas others utilize continuous chemostat nutrient addition. Solid-state fermentation (SSF) is conducted by moistening solid substrates (e.g., soybean meal or bran) to a water content of 60–80% and inoculating them before incubation in either trays or packed beds at controlled temperatures and humidity levels using humidified air [62, 99]. To recover and purify biosurfactants from fermentation processes, it is important to efficiently isolate and recover them from the liquid or powdered state. The collection of biosurfactants usually starts right after verification of fermentation to prevent degradation or loss of the biosurfactant.

To remove insoluble debris and biomass from the fermentation broth, centrifugation (or filtration) is performed as the first step. In the case of liquid formulations, the centrifugation is carried out at 4000–10,000 rpm for about 10–30 min to obtain a clear supernatant containing dissolved biosurfactants. If solid substrate fermentation systems are used, the products are extracted from the solid matrix into an extraction solution via either water or solvent extraction. A good number of biosurfactants can be precipitated as a result of the acidification process. A common example of this is the precipitate formation of the lipopeptide surfactin when its pH is decreased to approximately 2.0 using HCl [100]. Additionally, rhamnolipids can precipitate solubilizing into insoluble salts at low pH, or at the same time when multivalent cations such as Ca^{2+} are present as well. Recovery from this type of precipitate can be performed through centrifugation and decantation methods. In the case of sophorolipids, acidification will create two different phases: The solidified lactonic form will appear as a precipitate, while the lactonic and acidic forms will remain soluble [101].

A crude biosurfactant can be removed from its original medium using organic solvent extracts like ethyl acetate, methanol, and mixtures of chloroform and methanol; for example, Rhamnolipids and Surfactin have commonly been extracted using ethyl acetate [81, 100]. Another method of concentrating these products is by laundering the foam that has settled during foam fractionation. Next, additional purification options are available via solvent precipitation and chromatographic techniques. For example, a classic multistep technique for purifying rhamnolipids usually includes ethanol precipitation, followed by acid precipitation, and finally calcium salt precipitation [81]. To eliminate any residual proteins and salts, adsorptive columns (e.g., silica, alumina) together with membranes (e.g., microfiltration or ultrafiltration) could be employed; however, while HPLC (high-performance liquid chromatography) or TLC (thin-layer chromatography) can be employed analytically for purity analysis, they are not used on the scale required in the industry. Lastly, as another option to the purification processes noted above, foam fractionation could also provide an additional method of enriching products in an ongoing approach. The use of foam fractionation as an enrichment tool may reduce the amount of solvent used and improve concentration of the active agents.

The biosurfactants are extracted and concentrated into powder or plastic-like forms through freeze drying or spray drying. Extraction of biosurfactants from fermentation broth is typically done by evaporation of the water from the broth and concentrating the extract into a liquid extract. Therefore, the biosurfactants can be delivered as liquids, solids, and/or emulsified products based on their intended use. The importance of using environmentally friendly (non-toxic) solvents such as ethyl acetate instead of toxic solvents like chloroform continues to grow. It has also been reported that 50–60% of the total biosurfactants cost is attributed to the recovery from the fermentation broth. Therefore, creating more efficient biosurfactants production processes has become an area of focus for industry [90]. Therefore, the focus of research in the area of biosurfactants will be on how to combine optimization of bioprocess parameters (such as strains, growth media, and fermentation mode) along with optimal recovery methods [101, 102]. The contemporary trend for the formulation and production of biosurfactants will focus on three approaches to help to produce the highest titer of biosurfactants [103]. The first approach involves the use of both sugar and oil as substrates for biosurfactant production. The second approach involves engineered feeding to improve the production of high titer biosurfactants. The third modeling approach will include safety limits for foaming in production fermentors. Properties of production fermentors and other aspects of production are addressed in many recent publications [94, 96, 99, 102, 104]. The downstream processing of biosurfactants is a significant driving factor behind the economic feasibility of biosurfactant production. In fact, it typically accounts for 60–80% of the total costs of biosurfactant production. Researchers have developed several different strategies to recover and purify various types of biosurfactants (e.g., rhamnolipids, surfactin, sophorolipids) through methods such as acid precipitation, solvent extraction (ethyl acetate, methanol), foam fractionation, and membrane-based separation techniques. Numerous studies indicate that the pure recovery of biosurfactants may be maximized through the combined use of acidification with either alcohol or calcium salts. Solvent extraction continues to be one of the most common techniques for concentrating crude biosurfactants. Foam fractionation is an effective and viable method for recovering and purifying surfactants that naturally generate foams when used during fermentation processes. There is increasing interest in using membrane separations such as ultrafiltration and nanofiltration for concentrating biosurfactants using fewer or no solvents and working in more sustainable environmental practices. The method of choice for recovering and quantifying the content of a specific biosurfactant will ultimately depend on the specific biosurfactant, the fermentation system used to produce the surfactant, and the intended application of the biosurfactant. Investigating multistep recovery strategies that can be incorporated into eco-efficient processing workflows while minimizing or eliminating the use of toxic solvents is a central focus of research moving forward. Hence, these approaches will encourage the sustainable development and improvement of surfactant production for industrial use.

3 Physicochemical Properties

3.1 *Interfacial Properties*

Amphiphilic molecules rapidly reduce surface and interfacial tension due to their fast adsorption at the air–liquid interface. Typically, biosurfactants such as surfactin, glycolipids, and trehalolipid are considered capable of reducing surface tension to much lower values, reaching 25–30 mN/m. Chemical surfactants, such as SDS, on the other hand, can decrease the surface tension of water from approximately 72 mN/m to about 30–35 mN/m [104, 105]. Regarding interfacial tension (IFT), especially in oil–water systems, biosurfactants present an outstanding performance, sometimes reducing IFT from 40–50 mN/m to values as low as 0.05–1.0 mN/m. This is particularly valuable in processes such as enhanced oil recovery and environmental remediation [106–108]. While generally quite effective, chemical surfactants typically provide IFT reductions in the range of 1–5 mN/m. The higher reductions in IFT due to biosurfactants translate into the ability of liquids to spread on surfaces more easily and uniformly, resulting in a lower contact angle and therefore much better wetting. This increased effectiveness at spreading makes biosurfactants especially effective on hydrophobic and porous surfaces where surfactants have poor performance. One of the most important properties of surfactants is the self-assembling structure formed by the amphiphilic molecules. Both biosurfactants and chemical surfactants are amphiphilic, enabling them to form nanostructured micellar structures [109–111]. The most important property that explains this increased effectiveness of biosurfactants can be tied to their CMC. Micelle formation happens at that critical concentration of surfactants where the surface tension attains its minimum value. Since biosurfactants require much less amount compared to that of chemical surfactant to form micelle, they have increased efficiency to reduce both ST and IFT. CMC for trehalolipid was found to be 0.025 g/L, while for rhamnolipid, it was found to be 0.09 g/L. On the other hand, the commonly used chemical surfactant SDS has shown a CMC value of 1.8 g/L [112, 113]. This low CMC promotes dissolution and mobilization of hydrophobic substances like oils and hydrocarbons, aside from increasing wettability. In processes such as oil recovery and bioremediation, biosurfactants become especially applicable because they can allow better penetration into surfaces as wettability increases, and there is more effective solubilization. Chemical surfactants with their higher CMC need greater amounts to be applied to produce similar reductions in ST and IFT. Their increased use, particularly in sensitive applications, could result in higher costs and more environmental problems. The biosurfactants, therefore, do not only provide better physicochemical effectiveness but also a more eco-friendly and cost-effective alternative for enhancing wettability, solubilization, and mobilization, especially under harsh conditions.

3.2 *Emulsifying Activities*

In an emulsion, oil and water mix as small droplets, and biosurfactants prevent the separation by lowering the tension between the two. Surfactant-generated self-assembled structures may further stabilize emulsions by providing steric and electrostatic barriers. The emulsification index indicates the ability of surfactants to form a stable emulsion under specific conditions. Biosurfactants are more effective in stabilizing emulsions compared to chemical surfactants such as SDS or alkylbenzene sulfonates that typically present an emulsification index of 50–80%. Biosurfactants usually present EI within the range of 60–90% [114, 115]. In applications such as oil recovery and bioremediation, their amphiphilic nature, low CMC, and efficient interfacial tension reduction contribute to the stabilization of emulsions. Chemical surfactants, though effective, have the potential for environmental problems due to toxicity and non-biodegradability besides often requiring higher concentrations. Environmentally compatible biosurfactants, on the contrary, represent an alternative in applications that require stable emulsions.

Even at high salt concentrations, biosurfactants (marine biosurfactants) exhibit the same function and retain their stability and surface-active qualities [116–118]. On the other hand, charge screening, micelle destabilization, and possible precipitation at salinities exceeding 2–5% NaCl cause many chemical surfactants, including sodium dodecyl sulfate (SDS) or Tween 80, to lose their effectiveness in saline environments. Even at high temperatures, biosurfactants often remain stable and effective, and their CMC typically marginally drops. Chemical surfactants, on the other hand, are less effective at high temperatures because they frequently lose stability and their CMC may rise. Biosurfactants generally have a wide pH stability range, working well from pH 2–10, and can tolerate both acidic and alkaline conditions. In contrast, chemical surfactants have a narrower pH range and may lose effectiveness in very acidic or alkaline environments.

4 Environmental Advantages

Due to the advantages of biosurfactants over traditional or synthetic surfactants, such as being more sustainable and having a greater potential for use in bioremediation, many researchers consider these materials to be the best material to use in the process of bioremediation. With respect to their chemical characteristics (i.e., their molecular structure), biosurfactants offer greater biocompatibility and biodegradability than synthetic surfactants [119]. In addition to these advantages, in general, bioremediation materials provide numerous additional benefits over synthetic surfactants, including biodegradability, biocompatibility, digestibility, low toxicity, broad temperature and pH ranges of applicability, flexibility, naturally abundant sources of supply, and antimicrobial properties.

On the other hand, conventional synthetic surfactants most often are derived from petroleum sources, which makes these surfactants difficult to break down in the environment and thus toxic to the environment. Synthetic surfactants have the potential to accumulate in organisms, and their production methods, along with the resulting by-products, can be harmful to the environment. Due to surging awareness about the importance of conserving this ecosystem, surfactants from natural sources present viable alternatives to those chemically synthesized [2]. Biosurfactants appear to degrade more readily than synthetic alternatives. For instance, rhamnolipids have showcased a 50% lethal concentration (LC) toward Photobacterium phosphoreum, almost 10 times surpassing synthetic surfactants. A synthetic surfactant named Marlon A-305 and a biosurfactant generated by *P. aeruginosa* were comparatively analyzed, which unveiled the high toxicity of the synthetic surfactant in all aspects [120]. Biodegradation tests were also conducted on sophorolipid biosurfactants, which are generated by the non-pathogenic yeast *C. bombicola*, as per the guidelines of OECD 301C Modified MITI Test; it was found that the microbial degradation process initiates soon after the cultivation phase. In this case, microbial degradation was expressed as the ratio of BOD (biological oxygen demand) to TOD (total oxygen demand). Studies have uncovered that biosurfactant presence can augment biodegradability, primarily by expediting the solubilization of pollutants [121]. For example, biosurfactants were deployed to remediate soil fouled by octane. Within 15 days of treatment, biosurfactants derived from *Lactobacillus pentosus* reduced soil octane levels by about 60%, and by day 30, nearly 76% of the octane had undergone biodegradation. Without the presence of biosurfactants, the rate of removal was at least three times more sluggish [122]. Three types of biosurfactants, sophorolipids, rhamnolipids, and trehalose-containing lipids, were tested for their potential to solubilize HCH isomers and improve their bioavailability through biodegradation of the isomers in liquid culture and soil. Sophorolipid was found to be efficient in solubilizing and enhancing the breakdown of HCH isomers in both liquid and soil environments.

4.1 Toxicity

Surfactants can invade the effluent water treatment process and, while being used, may end up back in the natural environment or water sources, where they can negatively affect aquatic ecosystems [123]. The degree of harmful effects caused by surfactants is closely related to the concentration of surfactants. There is a direct relationship between time and higher levels of surfactant in aquatic ecosystems, affecting the developmental stages of algae and other microorganisms. At higher concentrations, surfactants may allow for greater penetration through cellular membranes, subsequently leading to the degradation of the internal components or building blocks of microorganisms [124, 125]. When surfactant's increase in specific hydrophobicity (a characteristic that makes a chemical less soluble), the toxicity of a chemical to aquatic organisms will also increase. The more hydrophobic a surfactant, the higher the toxicity to aquatic organisms. Once the concentration of surfactants becomes higher than the acceptable limits in aquatic habitats or systems, surfactants may travel up the trophic level or food chain through a biological transfer pathway (food web), producing ecological effects, for example, on fish. Fish are particularly susceptible to surfactants due to their ability to absorb surfactants on their skin and through their gills from both aqueous and food sources. When fish consume surfactants, these substances are delivered to the circulatory system [124]. Historically, surfactants were primarily absorbed through the diet.

In the case of products intended for direct human consumption or application, including pharmaceuticals, food, skincare products, and nutraceuticals, the digestibility and low toxicity of biosurfactants are two of the most important advantages they have over their synthetic counterparts. The majority of synthetic surfactants have negative effects on the body, which has led to regulatory limitations on their use in products for human ingestion; therefore, biosurfactants reduce or eliminate these regulatory restrictions due to their low toxicity and are considered to be safe for use in food and food products [126]. Additionally, the antimicrobial characteristics and ability to reduce biofilm formation make biosurfactants advantageous in manufacturing products for both food and pharmaceuticals. Temptation of toxic microbes inside production equipment from these products could also be avoided with biosurfactant products [126]. Most surfactants are released into aquatic systems from the disposal and/or use of manufactured products, or from industrial processes or unintentional discharges. Because of their biodegradability, biosurfactants have an important advantage over traditional surfactants because they can quickly and completely degrade once they are released into the environment. Traditional synthetic surfactants can persist and have the potential to build up to toxic levels in the environment. This makes biosurfactants an excellent product to use in skin care products and detergents [127, 128]. Also, researchers have demonstrated that biosurfactants are effective in the treatment of wastewater. In fact, by using a lipopeptide biosurfactant during water treatment, researchers were able to significantly reduce sulfate concentrations in their treated effluent [129].

4.2 Waste Reduction and Valorization

The selection of the appropriate carbon and nitrogen sources is a critical factor in the production of biosurfactants, as they have a significant impact on the cost of production. The two major challenges faced in the commercialization of surfactants include decreased production efficiency and increased costs of production, which ultimately limit the market for surfactants across the diverse industries. The use of inexpensive substrates during surfactant production has the potential to reduce production costs by 10–30% [26]. Therefore, selecting low-cost raw materials is essential for optimal surfactant production. As for optimum surfactant production, microorganisms require an appropriate ratio of lipids, proteins, and carbohydrates in order to grow and produce bioproduct. Industrial waste products that contain a high concentration of lipids and carbohydrates provide the necessary nutrients for microorganisms to produce biosurfactants. The production of biosurfactants from waste streams is important not only due to its sustainability but also because of the possibility of using renewable resources [130]. Using waste in this way reduces the potential threat to both human health and the environment [131]. The use of residual feedstocks in the production of microbial surfactants reduces processing costs and increases both the environmental sustainability of the product and of the process leading to the production of that product [132]. For this reason, industrial wastes are being increasingly viewed as a major source for the production of biosurfactants, because they possess a high amount of carbon, nitrogen, sulfur, and phosphorus. Industrial waste generated by biological processes has the best chance of providing biosurfactants [133]. Increased industrialization has resulted in large quantities of waste (solid and liquid) being produced. Industries like poultry farming, tanning, sugar cane processing, distilling, and dairy production all produce waste that is rich in nutrients (and organic material) [134]. The waste (bagasse) from these industries is made up of hemicellulose, cellulose, and lignin, which makes it a good source of raw material for producing biosurfactants and other products like microbial products [135, 136]. However, before bagasse can be effectively converted to value-added products, it needs to be pretreated [137]. If you utilize industrial waste as raw materials, it offers a variety of advantages such as improving how to manage waste and conserving the environment through the deletion of organic and inorganic toxins. This approach decreases industrial waste generation, which is caused by industries, while at the same time generating profit [138]. The production of biosurfactants from the use of industrial waste also has many advantages, including better utilization of waste materials, lower manufacturing costs, possible scalability, low levels of toxicity, environmental friendliness, and an enhancement of microbial activity [139]. Fruit waste usually comes in the form of the skin of fruits, like pineapple, orange, apple, and banana, which can be used to create biosurfactants. These fruit wastes are nutritionally rich in such components as minerals, water-soluble carbohydrates, fiber, vitamins, polyphenols, and other bioactive compounds [140]. The polyphenol component of the fruit waste is biodegradable; it can function as a vital carbon source for generating biosurfactants. There are a number of ways of treating fruit waste including composting, incineration, landfilling, or feeding them to animals. Each of these methods produces hazardous gases that would be dangerous to both humans and the environment, and could potentially kill [141]. Rhamnolipids could be produced by *P. aeruginosa* DR1 from mango kernels [142]. Biosurfactant can also be synthesized through solid-state fermentation from orange peel by implementing *Bacillus licheniformis*. Lipopeptide biosurfactant has also been prepared from banana peel by *Halobacteriaceae archaeon* AS65 [143, 144]. The biosynthesis of lipopeptide biosurfactants from beet molasses, apple peel, brewery wastewaters, and carrot peel has also been documented. Using fruit waste has enabled large-scale production and served as a substitute for chemical surfactants [145]. A significant amount of dairy waste is generated from dairy-related activities, which can also be utilized for biosurfactant synthesis. Dairy wastewater primarily consists of residues from raw milk, milk powders, yogurt, cream, cheese, and other dairy products. Most dairy waste results from bacterial activity, poor processing techniques, and the multiple stages involved in dairy product manufacturing. Because dairy waste consists of a large amount of fat, protein, sugar, minerals, and vitamins, it increases the functioning capabilities of microbial cells. Since there is an abundance of fat and oil in dairy waste, it is an excellent candidate for producing biosurfactants [146]. Biosurfactants from dairy waste were created from lactose fatty acid esters [147]. By utilizing whey, an inexpensive source of lactose and its hydrolyzed derivatives (glucose and galactose), biosurfactants may be produced. Whey is typically comprised of 30–50% lactose along with various components such as organic acids/vitamins/proteins, which depend on the milk's origin and composition. Therefore, whey may also be used to produce biosurfactants. For example, cheese whey, a source for producing sophorolipids, is a great source of biosurfactants, as there are many agro-wastes with similar properties (high carbohydrate and lipid content) that may also be utilized to produce biosurfactants. Also, the oil industry produces a large volume of waste oils that include fats; these waste oils are typically by-products of oil refining processes and contain leftover/free fatty acids, oil-soy cakes, soap stocks, and aqueous by-products from oil refining. Oil processing plants produce different types of oil waste such as tallow oil, marine oils, soap stock, and free fatty acids, which because of their high-lipid content are not biodegradable and/or degrade with time and/or pose an environmental risk when released into the environment. Oil wastes released from oil processing plants into soils and waterways have disastrous impacts on aquatic habitat, lead to backup of municipal sewer systems, and contribute to greenhouse gas production. Biosurfactants have been successfully produced from waste oils produced from oil industries [148, 149]. As such, converting waste oils into biosurfactant products will offer environmentally friendly and sustainable methods to recycle and reduce polluting wastes from our planet.

4.3 Antibacterial and Antiadhesive

Biosurfactants are a class of biocompatible, amphiphilic molecules naturally made by certain types of microorganisms. These molecules are able to alter the chemical and physical properties of the lipid-containing cellular membranes through their interaction with these lipid components [150, 151]. Research has indicated that a number of biosurfactants exhibit some type of biological activity, for example, as antimicrobial and antifungal agents [152]. This suggests that biosurfactants may have application in medicine, agriculture, and pharmaceuticals [153]. A large number of studies have demonstrated the antimicrobial and antifungal activity of various biosurfactants and their potential to inhibit and reduce biofilm formation. Specifically, glycolipids have been demonstrated to induce damage to the cytoplasmic membranes of both Gram-negative and Gram-positive microorganisms (e.g., *Klebsiella pneumoniae*, *Staphylococcus aureus*, *B. subtilis*, and *Escherichia coli*), resulting in a bactericidal effect [154]. In contrast, rhamnolipids displayed heightened antibacterial activity against Gram-positive bacteria under acidic conditions [155]. Sophorolipids have demonstrated antibacterial properties as well as the ability to disrupt biofilm. Research indicates that sophorolipids exhibit fungicidal properties and effectively disrupt biofilm at low concentrations (e.g., 50 g/L) [156]. The use of a 5% rhamnolipid solution on microtiter plates resulted in a 50% reduction in biofilm growth. The rhamnolipid solution has been shown to be effective at eliminating previously established biofilm at similar concentrations [157]. The use of biosurfactant from *B. subtilis* to determine the biofilm adhesion resistance demonstrated that biofilm development was reduced by 90% when tested with *S. aureus* and 97% when tested with *E. coli* [158]. The ability of biosurfactants to interact with a variety of surfaces may alter the ability of microorganisms to adhere to and form biofilms based on their hydrophobic nature [159]. Sophorolipids possess a broad range of bactericidal properties, while traditional antimicrobial agents primarily possess bacteriostatic properties [156]. It has been suggested that the hydrophobic and positively charged characteristics of these molecules contribute to their effectiveness in combating biofilm. A possible approach for preventing the formation of biofilm is to pre-coat medical devices with purified sophorolipid, as evidenced by the application of purified sophorolipid on silicone disks for medical use. This demonstrated that the application of a sophorolipid caused a 75% decrease in microbial cell attachment after two hours and effectively inhibited *Staphylococcus* spp. biofilm formation [160]. Rhamnolipids can also disrupt biofilms. Research on the ability of rhamnolipids to inhibit the colonization of biofilm-forming microorganisms on devices used in the medical industry has been conducted similar to that of sophorolipids [161]. Studies have shown that the effectiveness of rhamnolipids as a preventive agent for biofilms is dependent on the concentration used. For example, when mono-RL compounds were used to treat biofilms of *S. cerevisiae* and *S. parapsilosis* at 2X the MIC value (15.6 µg/mL), there was an almost 50% reduction in cell viability (MIC50) compared with untreated controls; whereas, when using 10X the MIC value, the reduction was greater than 75% [160]. Although these surface-active agents have many advantages and applications, the ability to incorporate them into the medical field is currently very limited [162]. There are opportunities to create new strategies for the development of biosurfactant-based methods that could be adapted for many other applications [163].

4.4 Synergistic Effect

An important area of research is on how biosurfactants can be combined with antibiotic treatments to create synergies in combating pathogenic microorganisms

[164]. For example, in a study, the combination of rhamnolipid-based biosurfactants and antibiotic treatments resulted in enhanced antibacterial activity against Escherichia coli, where four of the six tested antibiotics (e.g., kanamycin, ampicillin, and chloramphenicol) showed increased effectiveness when combined with rhamnolipids at below-critical concentrations [164]. The synergistic activity between sophorolipids and antibiotics is believed to occur via the biosurfactant's interactions with hydrophobically and hydrophilically modified surfaces of bacterial cells to enhance the permeability of the outer membrane to antimicrobial compounds through enhanced interactions and penetration pathways [164]. Another study has reported that the simultaneous administration of tetracycline and sophorolipids at subcritical micelle concentrations provides significant inhibition of bacterial growth compared to tetracycline alone, while also inducing morphological changes in cells and facilitating cell lysis [165]. The antimicrobial compound GBB12 created from Shewanella algae has demonstrated that it provides exceptional antimicrobial effects. The GBB12 is synergistic when used with either ciprofloxacin or gentamicin. GBB12 was also able to inhibit biofilm formation by both MRSA and A. baumannii, demonstrating an inhibition rate of approximately 84–93%. GBB12 also disrupted established biofilms formed by MRSA and A. baumannii in combination with gentamicin or ciprofloxacin, demonstrating a disruption ability of approximately 99%. In many instances, GBB12 has been shown to cause cell membrane integrity loss, leading to leaks from the cytoplasm and eventually cell death [166].

4.5 Emulsification of Biosurfactants

Surfactants exhibit froth formation and froth stability as their most distinguishing characteristics. Examples of industrial applications where surfactants create the least amount of foam while providing the most varied surface activity include paper manufacture and dyeing textiles, as well as spray cleaning. Due to this reason, many block-copolymer type surfactants (nonionic), most of which are based on petrochemicals and contain polyethylene oxide (PEO) and polypropylene oxide (PPO), have been incorporated in many of these industries. Unfortunately, however, due to the extreme branching of their molecular structures, their degradability is greatly limited [167]. Certain biosurfactants have the necessary ability to emulsify. The emulsifying action of biosurfactants has a positive effect on the electrical field flow of crude oil within the reservoir, improving its extraction. The emulsifying action of biosurfactants will also play a role in the success of microbial enhanced oil recovery (MEOR) [168]. MEOR is a method of recovering heavy oil (at or below pH 5.0–6.5) from the capillary pores of sandstone rocks and from regions not accessible by either traditional OR modern methods of OER. Biosurfactants, which can be obtained from a mixture of hemicellulosic sugars obtained from a three-lignocellulosic waste product, namely walnut shells, distilled grape marc, and hazelnut, were studied for their emulsifying properties [169]. Emulsification index to measure emulsion formation after 24 h was measured using the Cooper and Goldenberg technique. The emulsification index

(EI%24) is calculated as:

$$\mathrm{EI}\%_{24} = \left(\frac{L_{24}}{L_0}\right) \times 100.$$

L24 refers to the height of the emulsion film, while L0 identifies the entire height of the emulsion at the start of measurement. Measurements used were taken from a millimeter scale [170]. The emulsifying characteristics (emulsification index) of the biosurfactant produced by *Mucor hiemalis* UCP 0039 demonstrated almost constancy around 90% across a wide range of environmental conditions, including high temperatures, higher salt levels, and different pHs. The economic production of biosurfactants was achieved by using whey waste as a substrate for *P. aeruginosa* strain PP2 and Kocuria turfanensis. Biosurfactants from whey waste were found to produce increased surface activity (56–27 mN/m) and approximately 50–54% emulsification of pesticides under adverse environmental conditions [171]. Because molasses contains significant amounts of vitamins, it is often used as a raw material for producing biosurfactants. These biosurfactants are very effective in reducing surface tension and providing enhanced emulsification and are therefore good candidates for use in bioremediation applications [172, 173]. The biosurfactant surfactin has produced impressive results in surface activity; therefore, this compound should allow to more easily release entrapped oil and significantly improve the effectiveness of oil recovery operations [174]. Surfactin also has the ability to emulsify petroleum-type products (kerosene, crude oil, pentadecane, toluene, and hexane), thus helping to suspend the oil in the mobile phase. The stability of surfactin's emulsion may be adversely affected by solidifying surfactin in an acidified medium; however, this is an advantage because of the potential to fracture the stabilized emulsion easily, which allows for increased potential use of surfactant in MEOR [175].

5 Conclusions

Synthetic surface-active agents derived from petrochemical sources have been widely used for decades in industries such as detergents, cosmetics, textiles, pharmaceuticals, and petroleum processing due to their low cost and ease of large-scale production. These chemical surfactants effectively reduce surface and interfacial tension, enabling cleaning, emulsification, foaming, and dispersion. However, many exhibit poor biodegradability and toxicity toward aquatic organisms, raising environmental concerns. Fluorinated surfactants, in particular, are highly persistent and tend to bioaccumulate and biomagnify, posing significant ecological and health risks [173–176].

In contrast, biosurfactants are naturally produced by microorganisms such as bacteria, fungi, and yeasts and are considered more environmentally benign. They are biodegradable, less toxic, and stable across a wide range of pH, temperature, and salinity. Owing to their high efficiency at low concentrations, biosurfactants are increasingly explored for use in environmental remediation, pharmaceuticals, and food applications. However, high production and purification costs currently limit their large-scale commercial adoption.

References

1. Nagtode VS, Cardoza C, Yasin HKA, Mali SN, Tambe SM, Roy P, Singh K, Goel A, Amin PD, Thorat BR, Cruz JN (2023) Green surfactants (biosurfactants): a petroleum-free substitute for sustainability—comparison, applications, market, and future prospects. ACS Omega 8(13):11674–11699
2. De S, Malik S, Ghosh A, Saha R, Saha B (2015) A review on natural surfactants. RSC Adv 5(81):65757–65767
3. Tcholakova S, Denkov ND, Lips A (2008) Comparison of solid particles, globular proteins and surfactants as emulsifiers. Phys Chem Chem Phys 10(12):1608–1627
4. Bazel YR, Antal IP, Lavra VM, Kormosh ZA (2014) Methods for the determination of anionic surfactants. J Anal Chem 69(3):211–236
5. Schwuger MJ, Bartnik FG (1980) Interaction of anionic surfactants with proteins, enzymes and membranes. In: Anionic surfactants—biochemistry, toxicology, dermatology. Marcel Dekker, New York, pp 1–49
6. Lin B, McCormick AV, Davis HT, Strey R (2005) Solubility of sodium soaps in aqueous salt solutions. J Colloid Interface Sci 291(2):543–549
7. Zakharova LY, Pashirova TN, Doktorovova S, Fernandes AR, Sanchez-Lopez E, Silva AM, Souto SB, Souto EB (2019) Cationic surfactants: self-assembly, structure-activity correlation and their biological applications. Int J Mol Sci 20(22):5534
8. Mynercik D, Devinsky F, Lacko I (1996) Influence of counterions on antimicrobial activity of quaternary ammonium salts. Eur J Pharm Sci 4:S191
9. Chen A, Karanastasis A, Casey KR, Necelis M, Carone BR, Caputo GA, Palermo EF (2020) Cationic molecular umbrellas as antibacterial agents with remarkable cell-type selectivity. ACS Appl Mater Interfaces 12(19):21270–21282
10. Gerola AP, Costa PF, Quina FH, Fiedler HD, Nome F (2017) Zwitterionic surfactants in ion binding and catalysis. Curr Opin Colloid Interface Sci 32:39–47
11. Zhou M, Li S, Zhang Z, Wang C, Luo G, Zhao J (2017) Progress in the synthesis of zwitterionic Gemini surfactants. J Surfactant Deterg 20(6):1243–1254
12. Kumar GP, Rajeshwarrao P (2011) Nonionic surfactant vesicular systems for effective drug delivery—an overview. Acta Pharm Sin B 1(4):208–219
13. Cserháti T (1995) Alkyl ethoxylated and alkylphenol ethoxylated nonionic surfactants: interaction with bioactive compounds and biological effects. Environ Health Perspect 103(4):358–364
14. Nishikido N, Kobayashi H, Tanaka M (1982) Pressure effect on the Krafft points of ionic surfactants. J Phys Chem 86(16):3170–3172
15. Bou-Chacra N, Melo KJC, Morales IAC, Stippler ES, Kesisoglou F, Yazdanian M, Löbenberg R (2017) Evolution of choice of solubility and dissolution media after two decades of biopharmaceutical classification system. AAPS J 19(4):989–1001
16. Romero Vega G, Gallo Stampino P (2025) Bio-based surfactants and biosurfactants: an overview and main characteristics. Molecules 30(4):863
17. Garg R, Garg R, Anjum A, Eddy NO, Watandost H (2025) Comprehensive review of the synthesis and sustainable applications of biosurfactants. Discover Appl Sci 7(10):1168
18. Karnwal A, Shrivastava S, Al-Tawaha ARMS, Kumar G, Singh R, Kumar A, Mohan A, Yogita, Malik T (2023) Microbial biosurfactant as an alternate to chemical surfactants for application in cosmetics industries in personal and skin care products: a critical review. Biomed Res Int 2023(1):2375223

19. Simões CR, da Silva MWP, de Souza RFM, Hacha RR, Merma AG, Torem ML, Silvas FPC (2024) Biosurfactants: an overview of their properties, production, and application in mineral flotation. Resources 13(6):81
20. Płaza G, Achal V (2020) Biosurfactants: eco-friendly and innovative biocides against biocorrosion. Int J Mol Sci 21(6):2152
21. Biswas S, Nayak AK, Pal A (2025) Surfactant-influenced biosorption as a sustainable and effective way for the eradication of environmental pollutants: a review. RSC Sustain
22. Johnson P, Trybała A, Starov VM, Pinfield V (2020) Effect of synthetic surfactants on the environment and the potential for substitution by biosurfactants. Adv Colloid Interf Sci 288:102340
23. Badmus et al (2021) Environmental risks and toxicity of surfactants: overview of analysis, assessment, and remediation techniques. Environ Sci Pollut Res 28:62085–62104
24. Fung et al (2023) From wastewater treatment plants to the oceans: a review on synthetic chemical surfactants (SCSs) and perspectives on marine-safe biosurfactants. Sustainability 15(14):11436
25. Karmakar, K., Sarkar, R., Pal, A. et al. Recent Advances and Emerging Trends in Biosurfactants: A Concise Review. J Solution Chem 54, 393–420 (2025)
26. Banat IM, Satpute SK, Cameotra SS, Patil R, Nyayanit NV (2014) Cost-effective technologies and renewable substrates for biosurfactants' production. Front Microbiol 5:697
27. Pacwa-Płociniczak M, Płaza GA, Piotrowska-Seget Z, Cameotra SS (2011) Environmental Applications of Biosurfactants: Recent Advances, Int J Mol Sci 12(1):633
28. Leonie A. Sarubbo, Maria da Gloria C. Silva, Italo José B. Durval, Káren Gercyane O. Bezerra, Beatriz G. Ribeiro, Ivison A. Silva, Matthew S. Twigg, Ibrahim M. Banat (2022) Biosurfactants: Production, properties, applications, trends, and general perspectives, Biochem Eng J 181:108377
29. Mohanty, S.S., Koul, Y., Varjani, S. et al. (2021) A critical review on various feedstocks as sustainable substrates for biosurfactants production: a way towards cleaner production. Microb Cell Fact 20:120
30. Gayathiri E, Prakash P, Karmegam N, Varjani S, Awasthi MK, Ravindran B. (2022) Biosurfactants: Potential and Eco-Friendly Material for Sustainable Agriculture and Environmental Safety-A Review. Agronomy 12(3):662
31. Nikolova C and Gutierrez T (2021) Biosurfactants and Their Applications in the Oil and Gas Industry: Current State of Knowledge and Future Perspectives. Front. Bioeng. Biotechnol. 9:626639
32. Barreto, Óliver S. C., Almeida, E. dos S., & Medeiros, D. L. (2024). Life Cycle Assessment of Biossurfactants: A Critical Analysis. Revista De Gestão Social E Ambiental - RGSA, 18(4), e04708.
33. Venkataraman, S., Rajendran, D.S. & Vaidyanathan, V.K. (2024) An insight into the utilization of microbial biosurfactants pertaining to their industrial applications in the food sector. Food Sci Biotechnol 33:245–273
34. Sharma N, Lavania M and Lal B (2023) Biosurfactant: an emerging tool for the petroleum industries. Front. Microbiol. 14:1254557
35. Mnif I, Ghribi D (2016) Glycolipid biosurfactants: main properties and potential applications in agriculture and food industry, J Sci Food Agric 96(13):4310
36. Abbot V, Paliwal D, Sharma A, Sharma P (2022) A review on the physicochemical and biological applications of biosurfactants in biotechnology and pharmaceuticals. Heliyon 8(8), e10149
37. Bodour AA, Drees KP, Maier RM (2003) Distribution of biosurfactant-producing bacteria in undisturbed and contaminated arid southwestern soils. Appl Environ Microbiol 69(6):3280–3287
38. Yan Jer Ng, Sook Sin Chan, Kuan Shiong Khoo, Heli Siti Halimatul Munawaroh, Hooi Ren Lim, Kit Wayne Chew, Tau Chuan Ling, Anbalagan Saravanan, Zengling Ma, Pau Loke Show (2023) Recent advances and discoveries of microbial-based glycolipids: Prospective alternative for remediation activities, Biotech Adv 68, 108198

39. Rufino RD, Luna JM, Campos-Takaki GM, Sarubbo LA (2014) Characterization and properties of the biosurfactant produced by Candida lipolytica UCP 0988. Electron J Biotechnol 17(1):8–14
40. Roelants SLKW, Saerens KMJ, Derycke T, Li B, Lin YC, Van de Peer Y, De Maeseneire SL, Van Bogaert INA, Soetaert W (2013) Candida bombicola as a platform organism for the production of tailor-made biomolecules. Biotechnol Bioeng 110:2494–2503
41. Accorsini, F R, Mutton, M J R, Lemos, E G M, Benincasa, M (2012) Biosurfactants production by yeasts using soybean oil and glycerol as low cost substrate, Braz J Microbiol 43(1) 116-125.
42. Satpute, S. K., Banpurkar, A. G., Dhakephalkar, P. K., Banat, I. M., & Chopade, B. A. (2010). Methods for investigating biosurfactants and bioemulsifiers: a review. Crit Rev Biotechnol, 30(2), 127–144
43. Eras-Muñoz, E., Farré, A., Sánchez, A., Font, X., & Gea, T. (2022). Microbial biosurfactants: a review of recent environmental applications. Bioengineered, 13(5), 12365–12391
44. Maier RM, Soberon-Chavez G (2000) Pseudomonas aeruginosa rhamnolipids: biosynthesis and potential applications. Appl Microbiol Biotechnol 54(5):625–633
45. Gaur, VK, Bajaj, A, Regar, RK, Kamthan, M, Jha, RR, Srivastava, JK, Manickam, N (2019) Rhamnolipid from a Lysinibacillus sphaericus strain IITR51 and its potential application for dissolution of hydrophobic pesticides, Biores Technol 272:19–25
46. Nalini S, Parthasarathi R (2014) Production and characterization of rhamnolipids produced by Serratia rubidaea SNAU02 under solid-state fermentation and its application as biocontrol agent. Bioresour Technol 173:231–238
47. White DA, Hird LC, Ali ST (2013) Production and characterization of a trehalolipid biosurfactant produced by the novel marine bacterium Rhodococcus sp., strain PML026. J Appl Microbiol 115(3):744–755
48. Elshafie AE, Joshi SJ, Al-Wahaibi YM, et al. (2015) Sophorolipids production by Candida bombicola ATCC 22214 and its potential application in microbial enhanced oil recovery. Front Microbiol 6:1324
49. Wang H, Roelants SLKW, To MH et al (2019) Starmerella bombicola: recent advances on sophorolipid production and prospects of waste stream utilization. J Chem Technol Biotechnol 94(4):999–1007
50. Cho WY, Ng JF, Yap WH, Goh BH (2023) Sophorolipids—bio-based antimicrobial formulating agents for applications in food and health. Molecules 27(17):5556
51. Marcelino PRF, Peres GFD, Teran-Hilares R, Pagnocca FC, Rosa CA, Lacerda TM, dos Santos JC, da silva SS (2019) Biosurfactant production by yeasts using sugarcane bagasse hemicellulosic hydrolysate as a sustainable alternative for lignocellulosic biorefineries. Ind Crop Prod 129:212–223
52. Rau U, Nguyen LA, Schulz, S, Wray, V, Nimtz, M, Roeper, H, Koch H, Lang S (2005) Formation and analysis of mannosylerythritol lipids secreted by Pseudozyma aphidis. Appl Microbiol Biotechnol 66(5):551–559
53. Songdech P, Jayasekara LACB, Watchaputi K, Butkinaree C, Yingchutrakul Y, Soontorngun N (2025) Elucidating a novel metabolic pathway for enhanced antimicrobial glycolipid biosurfactant production in the yeast Meyerozyma guilliermondii MX. Sci Rep 15(1):18233
54. Ribeiro BG et al. (2020) Study of the biosurfactant production by Saccharomyces cerevisiae URM 6670 using agroindustrial waste. Chem Eng Trans 79:61-66.
55. Stamford TLM, Campos JM, Sarubbo LA, Azevedo DCS (2019) Characterization and application of a biosurfactant isolated from Candida utilis in salad dressings. Biodegradation 30:313–324
56. Dikit P, Maneerat S, Saimmai A (2019) The effective emulsifying property of biosurfactant–producing Marinobacter hydrocarbonoclasticus ST1 obtained from palm oil contaminated sites. Appl Biochem Microbiol 55(6):615–625
57. Ali Phulpoto I, Yu Z, Hu B et al (2020) Production and characterization of surfactin-like biosurfactant produced by novel strain bacillus nealsonii S2MT and it's potential for oil-contaminated soil remediation. Microb Cell Factories 19:145

58. Nitschke M, Araújo LV, Costa SGVAO, Pires RC, Zeraik AE, Fernandes ACLB, Freire DMG, Contiero J (2009) Surfactin reduces the adhesion of food-borne pathogenic bacteria to solid surfaces. Lett Appl Microbiol 49(2):241–247
59. Madslien EH, Rønning HT, Lindbäck T, Hassel B, Andersson MA, Granum PE (2013) Lichenysin is produced by most Bacillus licheniformis strains. J Appl Microbiol 115(4):1068-1080.
60. Yakimov MM, Timmis KN, Wray V, Fredrickson HL (1995) Characterization of a new lipopeptide surfactant produced by thermotolerant and halotolerant subsurface Bacillus licheniformis BAS50 Appl Environ Microbiol 61(5):1706-1713.
61. Barale SS et al (2022) Purification and characterization of antibacterial surfactin isoforms produced by Bacillus velezensis SK. AMB Express
62. Zhou Y, Yang X, Li Q, Peng Z, Li J, Zhang J, Zhang J (2023) Optimization of fermentation conditions for surfactin production by *Bacillus subtilis* YPS-32. BMC Microbiol 23:117
63. Beebe JL, Umbreit WW (1971) Extracellular lipid of *Thiobacillus thiooxidans*. J Bacteriol 108(1):612–614
64. Shively JM, Benson AA (1967) Phospholipids of Thiobacillus thiooxidans. J Bacteriol 94(5):1679–1683
65. Nwaguma IV, Chikere CB, Okpokwasili GC (2016) Isolation, characterization, and application of biosurfactant by Klebsiella pneumoniae strain IVN51 from hydrocarbon-polluted soil in Ogoniland, Nigeria. Bioresour Bioprocess 3(1):40
66. Cirigliano MC, Carman GM (1984) Isolation of a bioemulsifier from Candida lipolytica. Appl Environ Microbiol 48(4):747–750
67. Cirigliano MC, Carman GM (1985) Purification and characterization of liposan, a bioemulsifier from Candida lipolytica. Appl Environ Microbiol 50(4):846–850
68. Amezcua-Vega C, Poggi-Varaldo HM, Esparza-García F, Ríos-Leal E, Rodríguez-Vázquez R (2007) Effect of culture conditions on fatty acid composition of a biosurfactant produced by Candida ingens and changes of surface tension of culture media. Bioresour Technol 98(1):237–242
69. Campos JM, Stamford TLM, Sarubbo LA (2019) Characterization and application of a biosurfactant isolated from Candida utilis in salad dressings. Biodegradation 30(4):313–324
70. Qiao Y, Xia C, Liu L, Tang L, Wang J, Xu C et al (2022) Structural characterization and emulsifier property of yeast mannoprotein enzymatically prepared with a β-1,6-glucanase. LWT Food Sci Technol 168:113898
71. Cameron DR, Cooper DG, Neufeld RJ (1988) The mannoprotein of Saccharomyces cerevisiae is an effective bioemulsifier. Appl Environ Microbiol 54(6):1420–1425
72. Hajhosseini A et al (2021) Stress response and characterization of oil-in-water emulsions stabilized with Kluyveromyces marxianus mannoprotein. J Food Sci 86(2):454–62
73. Lima RA, Andrade RF, RodrÃguez DM, Araújo HW, Santos VP, Campos-Takaki GM (2017) Production and characterization of biosurfactant isolated from Candida glabrata using renewable substrates. Afr J Microbiol Res 11(6):237-44.
74. Rufino RD, de Luna JM, de Campos Takaki GM, Sarubbo LA (2014) Characterization and properties of the biosurfactant produced by Candida lipolytica UCP 0988 Electron. J Biotechnol 17(1):34-8.
75. Kurtzman CP, Price NPJ, Ray KJ, Kuo TM (2010) Production of sophorolipid biosurfactants by multiple species of the Starmerella (Candida) bombicola yeast clade. FEMS Microbiol Lett 311(2):140–146
76. Kitamoto D, Ikegami T, Suzuki GT, Sasaki A, Takeyama YI, Idemoto Y, Koura N, Yanagishita H (2001) Microbial conversion of n-alkanes into glycolipid biosurfactants, mannosylerythritol lipids, by Pseudozyma (Candida antarctica). Biotechnology Letters 23(20):1709-14.
77. Sarubbo LA (2008) Utilization of two agroindustrial by-products for the production of a surfactant by Candida sphaerica UCP0995. Process Biochem 43(9):912–917
78. Morita T, Konishi M, Fukuoka T, Imura T, Yamamoto S, Kitagawa M, Sogabe A, Kitamoto D (2007) Identification of Pseudozyma graminicola CBS 10092 as a producer of glycolipid biosurfactants, mannosylerythritol lipids. J Oleo Sci 57(2):123–131

79. Fenibo, Emmanuel Oliver, Salome Ibietela Douglas, and Herbert Okechukwu Stanley. (2019) A Review on Microbial Surfactants: Production, Classifications, Properties and Characterization. J Adv Microbiol 18 (3):1-22.
80. Santos DKF, Rufino RD, Luna JM, Santos VA, Sarubbo LA (2016) Biosurfactants: multifunctional biomolecules of the 21st century. Int J Mol Sci 17(3):401
81. Eslami P, Hajfarajollah H, Bazsefidpar S (2020) Recent advancements in the production of rhamnolipid biosurfactants by Pseudomonas aeruginosa. RSC Adv 10(56):34014–34032
82. Wang X, Yu Y, Ma J, Wang Y, Zhang H, Li W, Xu Q (2023) Optimization of fermentation conditions for surfactin production by Bacillus subtilis YPS-32 using modified Landy medium and response surface methodology. BMC Microbiol 23:218
83. Vega GR, Stampino PG (2025) Bio-based surfactants and biosurfactants: an overview and Main characteristics. Molecules 30(4):863
84. Miao Y, To MH, Siddiqui MA, Wang H, Lodens S, Chopra SS, Kaur G, Roelants SL, Lin CS (2024) Sustainable biosurfactant production from secondary feedstock—recent advances, process optimization and perspectives. Frontiers in chemistry 12:1327113.
85. Chakraborty P et al (2023) A comprehensive review on bio-surfactants: types, production methods and industrial applications. RSC Adv 13:12345–12367
86. Yu H, Chen W, Chen Y (2020) Enhanced rhamnolipids production using ex situ foam-control in fed-batch fermentation by Pseudomonas aeruginosa. Biotechnol Biofuels 13:203
87. Bezza FA, Chirwa EMN, Sibanda V (2015) Optimization of surfactin production in Bacillus subtilis isolated from diesel-contaminated soil. Biochem Eng J 104:243–250
88. Shah V, Jurjevic M, Badia D (2007) Utilization of restaurant waste oil as precursor for sophorolipid production by Candida bombicola. Biotechnol Prog 23(2):512–517
89. Rosenberg E, Ron EZ (1999) High- and low-molecular-mass microbial surfactants. Appl Microbiol Biotechnol 52(2):154–162
90. Qin Z, Guo W, Liu J, Zhao G, Liu M, Song X (2023) Reduced-cost production of sophorolipids by Starmerella bombicola CGMCC1576 grown on cottonseed molasses and cottonseed oil-based medium. Int J Mol Sci 24(6):5759
91. Morita T, Fukuoka T, Imura T, Kitamoto D (2015) Mannosylerythritol lipids: production and applications. J Oleo Sci 64(2):133–141
92. Rau U, Nguyen LA, Schulz S, Wray V, Nimtz M, Roper H, Koch H, Lang S (2007) Formation and fed-batch production of mannosylerythritol lipids by Pseudozyma aphidis. Appl Microbiol Biotechnol 68(5):607–613
93. Xu N, Liu S, Xu L, Zhou J, Xin F, Zhang W, Qian X, Li M, Dong W, Jiang M (2020) Enhanced rhamnolipids production using a novel bioreactor system based on integrated foam-control and repeated fed-batch fermentation strategy. Biotechnol Biofuels 13:80
94. Chong H, Li Q. Microbial production of rhamnolipids: opportunities, challenges and strategies. Microb Cell Fact. 2017 Aug 5;16(1):137
95. Wang H et al (2019) Starmerella bombicola: recent advances on sophorolipid production and prospects of waste stream utilization. J Chem Technol Biotechnol 94(4):999–1007
96. Guo-liang Z, Li-zhi H, Yue J (2005) Biodegradation of crude oil by Pseudomonas aeruginosa in the presence of rhamnolipids. Microbiology 151(10):3457–3465
97. Parekh VJ, Pandit AB (2011) Optimization of fermentative production of sophorolipid biosurfactant by Starmerella bombicola NRRL Y-17069 using response surface methodology. Int J Pharm Biolo Sci 1(3):103–116
98. Barretto C, de Luna JM, Sarubbo LA (2016) Production of a biosurfactant by Bacillus subtilis ICA56 using cassava wastewater as substrate: characterization and stability studies. BMC Microbiol 16(1):1–10
99. Cooper DG, MacDonald CR, Duff SJB, Kosaric N (1981) Enhanced production of surfactin from Bacillus subtilis by continuous product removal and metal cation additions. Appl Environ Microbiol 42(3):408–412
100. Najmi Z, Ebrahimipour G, Franzetti A, Banat IM (2018) In situ downstream strategies for cost-effective bio/surfactant recovery. Biotechnol Appl Biochem 65(4):523–532

101. Firdose M et al (2021) Extraction and antimicrobial activity of rhamnolipid biosurfactant produced by Pseudomonas aeruginosa UKMP14T. Malaysian J Microbiol 17(2):103–112
102. Papa P, Marchant R, Cameotra SS (2017) Microbial production of rhamnolipids: opportunities, challenges and strategies. Microb Cell Factories 16:36
103. Oraby A, Weickardt I, Zibek S (2022) Foam fractionation methods in aerobic fermentation processes. Biotechnol Bioeng 119(7):1697–1711
104. Zhou J, Xue R, Liu S, Xu N, Xin F, Zhang W et al (2019) High di-rhamnolipid production using Pseudomonas aeruginosa KT1115, separation of mono/di-rhamnolipids, and evaluation of their properties. Front Bioeng Biotechnol 7:245
105. Mnif I, Ellouz-Chaabouni S, Ghribi D (2018) Glycolipid biosurfactants, main classes, functional properties and related potential applications in environmental biotechnology. J Polym Environ 26:2192–2206
106. Salehi M (2006) Wettability alteration of carbonate rock mediated by biosurfactant produced from high-starch agricultural effluents. 9th International Wettability. https://digital.library.unt.edu/ark:/67531/metadc887730/
107. Iyare PU, Ouki SK, Bond T (2020) Microplastics removal in wastewater treatment plants: a critical review. Environ Sci Water Res Technol 6(10):2664–2675
108. Cai J, Chen X, Chen B, Wu D, Yang S (2024) A review on approach for biosurfactants enhance oil recovery technology. Geomicrobiol J 41(9):910–920
109. Kuczajowska-Zadrożna M, Filipkowska U, Jóźwiak T (2015) Application of biosurfactants for heavy netals leaching from immobilized activated sludge. Arch Environ Protect 41(1):43–52
110. Mabrouk MM, Hamed NA, Mansour FR (2022) Physicochemical and electrochemical methods for determination of critical micelle concentrations of surfactants: a comprehensive review. Monatshefte für Chemie-Chem Monthly 153(2):125–138
111. Raza ZA, Khalid ZM, Khan MS, Banat IM, Rehman A, Naeem A, Saddique MT (2010) Surface properties and sub-surface aggregate assimilation of rhamnolipid surfactants in different aqueous systems. Biotechnol Lett 32:811–816
112. Tomar SK, Singh R, Gupta SC, Arora DK, Joshi BK, Kumar D (2009) Phenotypic and genotypic characterization of lactobacilli from Churpi cheese. Dairy Sci Technol 89(6):531–540
113. Cheffi M, Maalej A, Mahmoudi A, Hentati D, Marques AM, Sayadi S, Chamkha M (2021) Lipopeptides production by a newly Halomonas venusta strain: characterization and biotechnological properties. Bioorg Chem 109:104724
114. Sarkar A, Zhang S, Holmes M, Ettelaie R (2019) Colloidal aspects of digestion of Pickering emulsions: experiments and theoretical models of lipid digestion kinetics. Adv Colloid Interf Sci 263:195–211
115. de Oliveira DWF, França ÍWL, Félix AKN, Martins JJL, Giro MEA, Melo VMM, Gonçalves LRB (2013) Kinetic study of biosurfactant production by Bacillus subtilis LAMI005 grown in clarified cashew apple juice. Colloids Surf B Biointerfaces 101:34–43
116. Vaz DA, Gudina EJ, Alameda EJ, Teixeira JA, Rodrigues LR (2012) Performance of a biosurfactant produced by a Bacillus subtilis strain isolated from crude oil samples as compared to commercial chemical surfactants. Colloids Surf B: Biointerfaces 89:167–174
117. Penfold J, Chen M, Thomas RK, Dong C, Smyth TJ, Perfumo A et al (2011) Solution self-assembly of the sophorolipid biosurfactant and its mixture with anionic surfactant sodium dodecyl benzene sulfonate. Langmuir 27(14):8867–8877
118. Onaizi SA, Nasser MS, Twaiq FA (2012) Micellization and interfacial behavior of a synthetic surfactant–biosurfactant mixture. Colloids Surf A Physicochem Eng Asp 415:388–393
119. Drakontis CE, Amin S (2020) Biosurfactants: formulations, properties, and applications. Curr Opin Colloid Interface Sci 48:77–90
120. Gayathiri E, Prakash P, Karmegam N, Varjani S, Awasthi MK, Ravindran B (2022) Biosurfactants: potential and eco-friendly material for sustainable agriculture and environmental safety—a review. Agronomy 12(3):662
121. Akintunde TA, Abioye OP, Oyeleke SB, Boboye BE, Ijah UJJ (2015) Remediation of iron using rhamnolipid-surfactant produced by Pseudomonas aeruginosa. Res J Environ Sci 9(4):169

122. Moldes AB, Paradelo R, Rubinos D, Devesa-Rey R, Cruz JM, Barral MT (2011) Ex situ treatment of hydrocarbon-contaminated soil using biosurfactants from lactobacillus pentosus. J Agric Food Chem 59(17):9443–9447
123. Ivanković T, Hrenović J (2010) Surfactants in the environment. Arh Hig Rada Toksikol 61(1):95–109
124. Yuan CL, Xu ZZ, Fan MX, Liu HY, Xie YH, Zhu T (2014) Study on characteristics and harm of surfactants. J Chem Pharm Res 6(7):2233–2237
125. Klosowska-Chomiczewska I, Medrzycka K, Karpenko E (2011) Biosurfactants–biodegradability, toxicity, efficiency in comparison with synthetic surfactants. Adv Chem Mech Eng 2:1–9
126. Calvo C, Silva-Castro GA, Uad I, García Fandiño C, Laguna J, González-López J (2008) Efficiency of the EPS emulsifier produced by Ochrobactrum anthropi in different hydrocarbon bioremediation assays. J Ind Microbiol Biotechnol 35(11):1493–1501
127. Zhu Z, Zhang F, Wei Z, Ran W, Shen Q (2013) The usage of rice straw as a major substrate for the production of surfactin by bacillus amyloliquefaciens XZ-173 in solid-state fermentation. J Environ Manag 127:96–102
128. Jain RM, Mody K, Joshi N, Mishra A, Jha B (2013) Production and structural characterization of biosurfactant produced by an alkaliphilic bacterium, Klebsiella sp.: evaluation of different carbon sources. Colloids Surf B: Biointerfaces 108:199–204
129. Janek T, Rodrigues LR, Czyżnikowska Ż (2018) Study of metal-lipopeptide complexes and their self-assembly behavior, micelle formation, interaction with bovine serum albumin and biological properties. J Mol Liq 268:743–753
130. Pandey AK, Gaur VK, Udayan A, Varjani S, Kim SH, Wong JW (2021) Biocatalytic remediation of industrial pollutants for environmental sustainability: research needs and opportunities. Chemosphere 272:129936
131. Feng S, Ngo HH, Guo W, Chang SW, Nguyen DD, Cheng D, Varjani S, Lei Z, Liu Y (2021) Roles and applications of enzymes for resistant pollutants removal in wastewater treatment. Bioresour Technol 335:125278
132. Mishra B, Varjani S, Iragavarapu GP, Ngo HH, Guo W, Vishal B (2019) Microbial fingerprinting of potential biodegrading organisms. Curr Pollut Rep 5:181–197
133. Gudiña EJ, Rodrigues AI, Alves E, Domingues MR, Teixeira JA, Rodrigues LR (2015) Bioconversion of agro-industrial by-products in rhamnolipids toward applications in enhanced oil recovery and bioremediation. Bioresour Technol 177:87–93
134. Dong J, Tang Y, Nzihou A, Chi Y, Weiss-Hortala E, Ni M (2018) Life cycle assessment of pyrolysis, gasification and incineration waste-to-energy technologies: theoretical analysis and case study of commercial plants. Sci Total Environ 626:744–753
135. Sindhu R, Gnansounou E, Binod P, Pandey A (2016) Bioconversion of sugarcane crop residue for value added products–an overview. Renew Energy 98:203–215
136. Mohanakrishna G, Mohan SV, Sarma PN (2010) Bio-electrochemical treatment of distillery wastewater in microbial fuel cell facilitating decolorization and desalination along with power generation. J Hazard Mater 177(1–3):487–494
137. Soares LA, Braga JK, Motteran F, Sakamoto IK, Monteiro PAS, Seleghim P, Varesche MBA (2019) Bioconversion of sugarcane bagasse into value-added products by bioaugmentation of endogenous cellulolytic and fermentative communities. Waste Biomass Valorization 10:1899–1912
138. Rene ER, Ge J, Kumar G, Singh RP, Varjani S (2020) Resource recovery from wastewater, solid waste, and waste gas: engineering and management aspects. Environ Sci Pollut Res 27:17435–17437
139. Jahan R, Bodratti AM, Tsianou M, Alexandridis P (2020) Biosurfactants, natural alternatives to synthetic surfactants: physicochemical properties and applications. Adv Colloid Interf Sci 275:102061
140. Schieber A (2017) Side streams of plant food processing as a source of valuable compounds: selected examples. Annu Rev Food Sci Technol 8(1):97–112

141. Dessie W, Zhang W, Xin F, Dong W, Zhang M, Ma J, Jiang M (2018) Succinic acid production from fruit and vegetable wastes hydrolyzed by on-site enzyme mixtures through solid state fermentation. Bioresour Technol 247:1177–1180
142. Reddy KS, Khan MY, Archana K, Reddy MG, Hameeda B (2016) Utilization of mango kernel oil for the rhamnolipid production by Pseudomonas aeruginosa DR1 towards its application as biocontrol agent. Bioresour Technol 221:291–299
143. Saisa-Ard K, Maneerat S, Saimmai A (2013) Isolation and characterization of biosurfactants-producing bacteria isolated from palm oil industry and evaluation for biosurfactants production using low-cost substrates. Biotechnologia 94(3):275–284
144. Chooklin CS, Maneerat S, Saimmai A (2014) Utilization of banana peel as a novel substrate for biosurfactant production by Halobacteriaceae archaeon AS65. Appl Biochem Biotechnol 173:624–645
145. Kumar PS, Mohanakrishna G, Hemavathy RV, Rangasamy G, Aminabhavi TM (2023) Sustainable production of biosurfactants via valorisation of industrial wastes as alternate feedstocks. Chemosphere 312:137326
146. Jatoi AS, Nguyen HM, Jeyapaul AS (2026) From waste to resource: microbial pathways for sustainable food production. Sustain Food Technol
147. Enayati M, Gong Y, Goddard JM, Abbaspourrad A (2018) Synthesis and characterization of lactose fatty acid ester biosurfactants using free and immobilized lipases in organic solvents. Food Chem 266:508–513
148. Hentati D, Chebbi A, Hadrich F, Frikha I, Rabanal F, Sayadi S, Manresa A, Chamkha M (2019) Production, characterization and biotechnological potential of lipopeptide biosurfactants from a novel marine bacillus stratosphericus strain FLU5. Ecotoxicol Environ Saf 167:441–449
149. Chen C, Sun N, Li D, Long S, Tang X, Xiao G, Wang L (2018) Optimization and characterization of biosurfactant production from kitchen waste oil using Pseudomonas aeruginosa. Environ Sci Pollut Res 25:14934–14943
150. Puyol McKenna P, Naughton PJ, Dooley JS, Ternan NG, Lemoine P, Banat IM (2024) Microbial biosurfactants: antimicrobial activity and potential biomedical and therapeutic exploits. Pharmaceuticals 17(1):138
151. Bjerk TR, Severino P, Jain S, Marques C, Silva AM, Pashirova T, Souto EB (2021) Biosurfactants: properties and applications in drug delivery, biotechnology and ecotoxicology. Bioengineering 8(8):115
152. Essghaier B, Mallat N, Khwaldia K, Mottola F, Rocco L, Hannachi H (2023) Production and characterization of new biosurfactants/bioemulsifiers from Pantoea alhagi and their antioxidant, antimicrobial and anti-biofilm potentiality evaluations. Molecules 28(4):1912
153. Kashif A, Rehman R, Fuwad A, Shahid MK, Dayarathne N, Jamal A, Aftab MN, Mainali B, Choi Y (2022) Current advances in the classification, production, properties and applications of microbial biosurfactants–a critical review. Adv Colloid Interf Sci 306:102718
154. Shu Q, Lou H, Wei T, Liu X, Chen Q (2021) Contributions of glycolipid biosurfactants and glycolipid-modified materials to antimicrobial strategy: a review. Pharmaceutics 13(2):227
155. de Freitas Ferreira J, Vieira EA, Nitschke M (2019) The antibacterial activity of rhamnolipid biosurfactant is pH dependent. Food Res Int 116:737–744
156. De Rienzo MD, Banat IM, Cuesta ML, Williams M, Fracchia L (2018) Evaluation of the activity of sophorolipids against microbial biofilms on medical-grade silicone. New Biotechnol 44:S13
157. Turbhekar R, Malik N, Dey D, Thakare D (2015) Disruption of Candida albicans biofilms by rhamnolipid obtained from Pseudomonas aeruginosa. RT IJRSB 3(3):73–78
158. Rivardo F, Turner RJ, Allegrone G, Ceri H, Martinotti MG (2009) Anti-adhesion activity of two biosurfactants produced by bacillus spp. prevents biofilm formation of human bacterial pathogens. Appl Microbiol Biotechnol 83:541–553
159. Kaczorek E, Pacholak A, Zdarta A, Smułek W (2018) The impact of biosurfactants on microbial cell properties leading to hydrocarbon bioavailability increase. Colloids Interfaces 2(3):35

160. Da Silva AR (2019) Nuevos tensioactivos catiónicos basados en ramnolípidos y aminoácidos: propiedades fisicoquímicas y biológicas
161. Nickzad A, Déziel E (2014) The involvement of rhamnolipids in microbial cell adhesion and biofilm development–an approach for control? Lett Appl Microbiol 58(5):447–453
162. Rodríguez-López L, Rincón-Fontán M, Vecino X, Moldes AB, Cruz JM (2020) Biodegradability study of the biosurfactant contained in a crude extract from corn steep water. J Surfactant Deterg 23(1):79–90
163. Adu SA, Naughton PJ, Marchant R, Banat IM (2020) Microbial biosurfactants in cosmetic and personal skincare pharmaceutical formulations. Pharmaceutics 12(11):1099
164. Shusterman E, Mottahedeh A, McCarthy M (2021) The synergistic effects of Rhamnolipids and antibiotics against bacteria. J Stud Res 10:1495
165. Naughton PJ, Marchant R, Naughton V, Banat IM (2019) Microbial biosurfactants: current trends and applications in agricultural and biomedical industries. J Appl Microbiol 127(1):12–28
166. Amirinejad N, Shahriary P, Hassanshahian M (2023) Investigation of the synergistic effect of glycolipid biosurfactant produced by Shewanella algae with some antibiotics against planktonic and biofilm forms of MRSA and antibiotic resistant Acinetobacter baumannii. World J Microbiol Biotechnol 39(2):45
167. Hirata Y, Ryu M, Oda Y, Igarashi K, Nagatsuka A, Furuta T, Sugiura M (2009) Novel characteristics of sophorolipids, yeast glycolipid biosurfactants, as biodegradable low-foaming surfactants. J Biosci Bioeng 108(2):142–146
168. Wu B, Xiu J, Yu L, Huang L, Yi L, Ma Y (2022) Biosurfactant production by Bacillus subtilis SL and its potential for enhanced oil recovery in low permeability reservoirs. Sci Rep 12(1):7785
169. Portilla-Rivera O, Torrado A, Domínguez JM, Moldes AB (2008) Stability and emulsifying capacity of biosurfactants obtained from lignocellulosic sources using lactobacillus pentosus. J Agric Food Chem 56(17):8074–8080
170. Lara VM, Vallejo M, Parada R, Henao Ossa JS, Gliemmo MF, Campos CA (2022) Characterization of the emulsifying activity of biosurfactants produced by lactic acid bacteria isolated from the Argentinian Patagonia. J Dispers Sci Technol 43(6):902–909
171. Dubey KV, Charde PN, Meshram SU, Shendre LP, Dubey VS, Juwarkar AA (2012) Surface-active potential of biosurfactants produced in curd whey by Pseudomonas aeruginosa strain-PP2 and Kocuria turfanesis strain-J at extreme environmental conditions. Bioresour Technol 126:368–374
172. Martins PC, Martins VG (2018) Biosurfactant production from industrial wastes with potential remove of insoluble paint. Int Biodeterior Biodegradation 127:10–16
173. Domínguez Rivera Á, Martínez Urbina MÁ, López y López VE (2019) Advances on research in the use of agro-industrial waste in biosurfactant production. World J Microbiol Biotechnol 35(10):155
174. Gudina EJ, Pereira JF, Costa R, Coutinho JA, Teixeira JA, Rodrigues LR (2013) Biosurfactant-producing and oil-degrading Bacillus subtilis strains enhance oil recovery in laboratory sand-pack columns. J Hazard Mater 261:106–113
175. Long X, He N, He Y, Jiang J, Wu T (2017) Biosurfactant surfactin with pH-regulated emulsification activity for efficient oil separation when used as emulsifier. Bioresour Technol 241:200–206
176. Hussain SMS, Adewunmi AA, Mahboob A, Murtaza M, Zhou X, Kamal MS (2022) Fluorinated surfactants: a review on recent Progress on synthesis and oilfield applications. Adv Colloid Interf Sci 303:102634
177. Coronel-León J, Marqués AM, Bastida J, Manresa A (2016) Optimizing the production of the biosurfactant lichenysin and its application in biofilm control. J Appl Microbiol 120(1):99–111
178. Juma A, Lemoine P, Simpson AB, Murray J, O'Hagan BM, Naughton PJ et al (2020) Microscopic investigation of the combined use of antibiotics and biosurfactants on methicillin resistant Staphylococcus aureus. Front Microbiol 11:1477

MIX
Papier aus verantwortungsvollen Quellen
Paper from responsible sources
FSC® C105338

If you have any concerns about our products,
you can contact us on
ProductSafety@springernature.com

In case Publisher is established outside the EU,
the EU authorized representative is:
Springer Nature Customer Service Center GmbH
Europaplatz 3, 69115 Heidelberg, Germany

Printed by Libri Plureos GmbH
in Hamburg, Germany